Interfaces Under Laser Irradiation

NATO ASI Series

Advanced Science Institutes Series

A Series presenting the results of activities sponsored by the NATO Science Committee, which aims at the dissemination of advanced scientific and technological knowledge, with a view to strengthening links between scientific communities.

The Series is published by an international board of publishers in conjunction with the NATO Scientific Affairs Division

A	Life Sciences	Plenum Publishing Corporation
B	Physics	London/New York
C	Mathematical and Physical Sciences	D. Reidel Publishing Company Dordrecht/Boston/Lancaster/Tokyo
D	Behavioural and Social Sciences	Martinus Nijhoff Publishers Dordrecht/Boston/Lancaster
E	Applied Sciences	
F	Computer and Systems Sciences	Springer-Verlag Berlin/Heidelberg/New York/ London/Paris/Tokyo
G	Ecological Sciences	
H	Cell Biology	

Series E: Applied Sciences – No. 134

Interfaces Under Laser Irradiation

edited by

L.D. Laude

Faculté des Sciences
Université de l'Etat
Mons, Belgium

D. Bäuerle

Angewandte Physik
Johannes Kepler Universität
Linz, Austria

M. Wautelet

Faculté des Sciences
Université de l'Etat
Mons, Belgium

1987 **Martinus Nijhoff Publishers**
Dordrecht / Boston / Lancaster
Published in cooperation with NATO Scientific Affairs Division

Proceedings of the NATO Advanced Study Institute on "Interfaces Under Laser Irradiation", Acquafredda di Maratea, Italy, July 14-25, 1986

Library of Congress Cataloging-in-Publication Data

Interfaces under laser irradiation.

(NATO advanced science institutes series. Series E, Applied sciences ; no. 134)
1987 proceedings of the NATO Advances Study Institute on Interfaces Under Laser Irradiation, Acquafredda di Maratea, Italy, July 14-25, 1986; co-sponsored by the European Physical Society ... et al.
Published in cooperation with NATO Scientific Affairs Division.
Includes index.
1. Surfaces (Physics)--Effect of radiation on--Congresses. 2. Surface chemistry--Congresses. 3. Laser beams--Congresses. 4. Semiconductors--Surfaces--Congresses. I. Laude, Lucien D. II. Baüerle, D. (Dieter), 1940- . III. Wautelet, M. IV. NATO Advanced Study Institute on Interfaces Under Laser Irradiation (1986 : Acquafredda di Maratea, Italy) V. European Physical Society. VI. North Atlantic Treaty Organization. Scientific Affairs Division. VII. Series: NATO ASI series. Series E, Applied sciences; no. 134.
QC173.4.S94I575 1987 530.4 87-14119

ISBN 90-247-3569-6 (this volume)
ISBN 90-247-2689-1 (series)

Distributors for the United States and Canada: Kluwer Academic Publishers, P.O. Box 358, Accord-Station, Hingham, MA 02018-0358, USA

Distributors for the UK and Ireland: Kluwer Academic Publishers, MTP Press Ltd, Falcon House, Queen Square, Lancaster LA1 1RN, UK

Distributors for all other countries: Kluwer Academic Publishers Group, Distribution Center, P.O. Box 322, 3300 AH Dordrecht, The Netherlands

Printed in The Netherlands

CONTENTS

PREFACE

Known and developed over the past twenty five years, lasers have been experimented in a variety of processes with an uneven success. Apart from fundamental physics experiments in which the various aspects of coherence are systematically exploited, applications in the field of Materials Science have been scattered recently over so many situations that it is apparently difficult today to conceive a comprehensive interpretation of all physical processes encountered. In some domains of research like photochemistry, development has been fast and rather self-supporting. In others, like solid-state processing, progress has been either very specific or deviated towards marginal applications, or else emerged as a joint-venture between physicists and chemists. This yielded a number of professional meetings, where day-to-day research activities are presented. In 1982, the Cargèse ASI on "Cohesive properties of semiconductors under laser irradiation" was one of such meetings at which a prospective of the field was discussed at length in ebullient round-table sessions. Quoted from the proceedings, "the Institute helped to discern clearly the limits of existing theoretical approaches and the directions along which work is urgently needed within the next few years". Four years have passed and the field has literally exploded. It must be mentioned that some of the most striking developments over the past two years were accurately predicted at the Institute in Cargèse. In particular, all those effects described in hydrodynamic terms (convection, dissipative structures in systems undergoing chemical reactions, bifurcation schemes) emerged there as issues for experimentalists to corroborate with models and, effectively, evidence was later brought up in most convincing fashion in favor of such speculative ideas at the time. Attention was also raised there on systems driven far from equilibrium and on interface behaviours as probably major domains to explore with regard to phase transformation. The interface issue rapidly expanded and current results indicate that i) some of the laser-induced surface reactions do present clear athermal connotation calling for a brand new set of concepts; and ii) reaction rates obtained under laser exceed by several orders of magnitude current rates derived from conventional settings, like electrolysis. In view of the tremendous current evolution of the field, it was apparent that the time had come for a thorough survey of problems encountered in laser-matter interaction at interfaces. This was the subject of the Maratea ASI which consisted of an integrated set of lectures on liquid-solid, gas-solid and solid-solid interface behaviours.

Among the successful assessment of lasers is the processing of near-perfect homogeneous atom systems like gases. In there, plain thermodynamic approaches are fully justified and help to understand the chemical reaction rates experimentally obtained in laser-gas interactions. This situation is not enjoyed in heterogeneous systems, like boundaries. There, thermodynamic concepts need to be revised and either adapted or replaced by others. Attemps in that direction have been very few in the past. The local equilibrium approximation is invalidated in the presence of large gradients or inhomogeneities in time or space and phase instability needs proper evaluation. Laser-solid interactions investigations have always been limited to macroscopic approaches regarding, for instance, energy balance, optical properties, phonon dispersion, electron-phonon interactions, ... However, there are situations in which a local description has been successfully developed, like phenomena taking place at surfaces or near dislocations. What we learn form Surface Physics is that properties of matter undergo large fluctuations at and near solid surfaces as compared to bulk. This is also the case in the gas and liquid phases in which boundary conditions of the atom system are extremely important in setting the actual behaviour of atom species away from boundaries (atom velocity, convective currents, instabilities in hydrodynamic regime). In fact, what we are dealing with in catalysis , electrolysis and other reactions, is the physics of largely heterogeneous systems and not only in spatial terms but also versus time, i.e. typically problems of low-dimensionality in which macroscopic models just do not make sense. From that point-of-view, any source interacting with matter at any phase discontinuity should not be treated as a crude way of simply injecting energy into a three-dimensional system but rather as a tool for disturbing boundary conditions over a scale depending on the source intensity. This is precisely the case in laser-matter interactions and the object of this ASI is specifically to address that issue : can we predict and understand interface behaviours when submitted to intense photon irradiation ?

On the other hand, Materials Science tells us that control of macroscopic (bulk) phase changes in matter stongly depends on solving boundary problems first, as for instance wetting phenomena, atom segregation or surface tension. Along with the process, the hydrodynamics of the melt plays an important role in setting up convective fluxes the parameters of which (Reynolds number, etc.) are systematically affected by the dimensions of the atom system and energy gradients present in it. Again we are facing problems of inhomogeneity which must be solved before any reliable applications is thought of with the grown material. If we have to respond positively to the increasingly large demand for new materials in fields as various as heterogeneous catalysis and micro-optoelectronics, then any process which offers efficiency and versatility in producing either stable or metastable single phase structures should be explored and fully understood.

Since 1982, and quite surprisingly in view of the preceeding arguments, none of the summer schools of the ASI category properly stressed or even mentioned the importance of the interfaces in laser-matter interaction. However, distinct progress has been made in understanding the physical implications of phase changes on a <u>microscopic</u> rather than macroscopic scale. One appreciates here that such classical thermodynamic concepts like entropy, enthalpy, chemical potential and even temperature, all being established for macroscopic and homogeneous systems, have first to be thoroughly questioned before proposing another energy partition in highly inhomogeneous and/or non-equilibrium atom systems. To some extent, this has been done but more advanced investigations, both theoretical and experimental, are needed. For instance, time-resolved phase evolution with CW or pulsed laser sources are currently studied. Also, convection fluxes in laser-processed materials have also been detected to participate to laser-assisted metal-semiconductor phase changes. Most of all, laser technology has made distinct progresses as well which allowed to put in the market a new generation of reliable and very stable pulsed lasers, as for instance excimer lasers, opening new ranges of products as a most direct consequence of novel physical situations never encountered before at solid-gas and solid-liquid interfaces. Some of such items have been presented recently, scattered among other research activities, at general or more specialized conferences. There again is a need for re-centering these activities and shedding proper light on their common background, i.e. interfaces behaviour under intense photon irradiation. The time has come to devise new metastable materials, new device production processes, new low-scale integrated systems all being initiated and/or assisted by the laser light. All such possibilities would not exist only three years ago and progress is still going very fast in these fields. Therefore, it was time also to generalize theoretical concepts which describe comprehensively, all at once, these laser-initiated interfacial reactions. This was the ultimate justification for setting up this Advanced Study Institute which was particularly appropriate both in time and content to correspond to a large demand both from people involved in several fields of applications and from theoreticians who perceive the uniqueness of the physics involved.

The school was opened to a wide variety of attendants from theoreticians invested in non-equilibrium physics to chemists in the fields of catalysis and electrolysis. Although in the lectures the accent was definitely on fundamental approaches, specific applications would find room in the program and were the matter of a number of informal and lively "unscheduled" discussions.

The following program goes crescendo as from the first part in which the known intrinsic physical properties of solid surfaces are presented as a prerequisite to a description of their interaction processes with liquids and gases under laser irradiation. These processes are extensively studied in the following parts, with emphasis on the solid-gas interface for which the present state of development justifies the number and variety of contributions.

In the early stage of the preparation of this Institute, the advice and support of Dr. G. Dewel have been extremely useful in setting up the programme. The collaboration of Dr. Andrew during various stages of this Institute has been essential to the success of the meeting. The efficient and warm assistance of Mr. A. Guzzardi and his staff at the Hotel Villa del Mare, Maratea, Italy is gratefully acknowledged. Particular thanks are due to Mrs. Françoise Masy and Mrs. Françoise Moreau-Hanus for their aid in the preparation of the manuscripts.

L.D. Laude D. Bäuerle M. Wautelet

January, 1987

[illegible] cases. These [illegible] used for concentrating the [illegible] in their research work [illegible] [illegible] [illegible] the fire has [illegible] [illegible] [illegible] [illegible] [illegible] [illegible] [illegible] [illegible] [illegible] [illegible] [illegible] [illegible] [illegible] [illegible] [illegible] [illegible] [illegible]

[illegible]

[illegible]

[illegible]

In [illegible] of [illegible] [illegible] [illegible] [illegible] [illegible] [illegible] [illegible]

[illegible] [illegible] [illegible]

[illegible]

Acknowledgements

This book is dedicated to the memory of Dr. Mario di Lullo, who was instrumental in the early stages of organization of this Institute.

The organizers benefitted from the co-sponsorship of the following:

The European Physical Society
The European Economic Community, Brussels
The Council of Europe, Strasbourg
The National Science Foundation, Washington
IBM-Belgium
Université de Mons- Hainaut, Mons

The organizers of this NATO Advanced Research Institute are grateful to all these co-sponsors for their support.

Particular thanks go to Dr. Craig Sinclair of NATO Scientific Affairs Division for his encouragement and assistance.

ELECTRONIC STRUCTURE AT SEMICONDUCTOR SURFACES AND INTERFACES

G. ALLAN
Laboratoire de Physique des solides (LA CNRS)
Institut Supérieur d'Electronique du Nord
41, Bd Vauban, F-59046 LILLE CEDEX (France)

1. INTRODUCTION

The electronic structure of bulk infinite solids with a perfect periodic structure now is generally well-known. Simple bulk defects like substitutional or interstitial impurities can also be studied using for example the Green function formalism. In such cases, the perturbation of the perfect infinite crystal is localized to the impurity site and a few of its nearest neighbours leading to a small finite size of the perturbative potential. This is no more the case near a surface which breaks the periodicity of the crystal in one direction. The perturbation size is now infinite and one must still take advantage of the remaining two-dimensional periodicity along the surface or use new special methods. In any case, the problem is more difficult. The amount of necessary calculation greatly increases. Then one may use approximations to simulate the surface like slabs keeping a good sophisticated band model. An other possibility is to treat a perfect semi-infinite crystal whereas the approximations are made in the band model.

It is not possible to here study in details the surface electronic structure. This could be the single subject of a summer school like this one. So I shall limit myself to a very fast description of exact models and I shall prefer to emphasize ideas rather than calculations. I shall say a few words about the tight binding approximation as it is one of the most popular method to study defects notably surfaces and it allows to simply demonstrate some general surface electronic structure properties. Self-consistency also is a difficult point when one calculates the surface electronic structure. We shall show that a complete screening of the charge (also known as a zero charge approximation) is a rather good approximation to determine potentials near the surface. Using these models, one must also use new methods like the recursion method or a more recent like the decimation one which seems still more powerful at least to get very accurate electronic charges needed to determine potential near semiconductor interfaces. In the next part, we shall apply these methods to semiconductor surface electronic structure starting from the very simple molecular model and then introducing improvements to the model. If the atomic structure is in principle a result of the electronic structure calculation and is given by the energy minimization, one generally uses the atom experimental positions to get the electronic structure. We shall briefly describe how the electronic structure perbutation near the surface can induce surface reconstruction or superstructures for Silicon or III-V componds. Finally, in the last part of this lecture, we shall look at semiconductor heterostructures and quantum wells. If the preceeding methods can be used to get properties localized near the interface like interface states or charge transfer, the perfect crystal energy levels and the defects (impurity or exciton) ones are obtained by the effective mass approximation. We shall briefly describe different heterostructures and some of their properties. Finally a brief desciption of an heterostructure application is given to semiconductor laser.

2.SOME GENERAL METHODS

1.1 The tight binding approximation

Among all the band models, the tight binding approximation remains the most popular to study bulk defects and surfaces. It can be applied to transition metals or semiconductors. Its simplicity allows to study complex defects.

The one-electron wave function is written as a linear combination of atomic orbitals centered on each lattice site $\vec{R}_i$

$$\Phi(\vec{r}) = \sum_{i,j} a_{ij} \; |\varphi_j(\vec{r}-\vec{R}_i)\rangle \tag{1}$$

where the summation over j is extended to all the atomic functions of the basis. For a transition metal, these are the 5 d orbitals as the s orbital is generally neglected due to its density of states smaller than the d one. In the semiconductor case, one takes the s and p orbitals. Nowadays one often adds to these four orbitals an s^* orbital to simulate higher energy states. This procedure considerably improves the bulk conduction band structure. The overlap between two orbitals on different sites is neglected:

$$\langle\varphi_j(\vec{r}-\vec{R}_i)|\varphi_k(\vec{r}-\vec{R}_l)\rangle = 0 \text{ for } j\neq k \text{ and } i\neq l \tag{2}$$

The lattice potential is equal to the sum of atomic potentials:

$$U(\vec{r}) = \sum_i V(\vec{r}-\vec{R}_i) \tag{3}$$

To solve the Shrödinger equation, one gets a linear equation system satisfied by the cofficients a_{ij}:

$$(E_0+\alpha_{im}-E)a_{im} + \sum_{j\neq i,m'} \beta_{im}^{jm'} a_{jm'} = 0 \tag{4}$$

where E_m^0 is the atomic orbital level, α_{im} the shift integral (also called crystal field integral) defined by:

$$\alpha_{im} = \langle\varphi_m(\vec{r}-\vec{R}_i)|\sum_{j\neq i} V(\vec{r}-\vec{R}_j)|\varphi_m(\vec{r}-\vec{R}_i)\rangle \tag{5}$$

and $\beta_{im}^{jm'}$ the hopping integral between sites i and j:

$$\beta_{im}^{jm'} = \langle\varphi_m(\vec{r}-\vec{R}_i)|\sum_{k\neq j} V(\vec{r}-\vec{R}_k)|\varphi_{m'}(\vec{r}-\vec{R}_j)\rangle \tag{6}$$

Among all the β integrals, only those with $\vec{R}_i$ and $\vec{R}_j$ nearest neighbours or next nearest neighbours are retained as the atomic orbitals decrease exponentially with the distance. Generally one also keeps only two-center integrals (only k equal to i is kept in the summation in (6)). Then remain only a few independent integrals ($ss\sigma, sp\sigma, pp\sigma$ and $pp\pi$ between s and p orbitals or $dd\sigma, dd\pi$ and $dd\delta$ for the d ones). These integrals are then fitted to the energy dispersion curves obtained by more sophisticated calculations (pseudopotential and local density approximations for example in the semiconductor case and Augmented Plane Waves for transition metal). In a bulk perfect infinite crystal, the infinite matrix (4) can be block diagonalized using the Bloch theorem and writing a_{im} as c_m exp(i $\vec{k}.\vec{R}_i$). For each $\vec{k}$ value in the bulk Brillouin zone, one has to diagonalize a n*n matrix (n being the number of orbitals on the bulk unit cell atoms) . In the surface case, we use the periodicity along the surface plane to introduce a wave vector $\vec{k}_{/\!/}$ parallel to the surface and defined in a surface Brillouin zone. Then one writes a_{im} as $c_{mR_{iz}}.\exp(i\vec{k}_{/\!/}\vec{R}_{i/\!/})$ where R_{iz} and $\vec{R}_{i/\!/}$ are the components of $\vec{R}_i$ respectively perpendicular and parallel to the surface plane. Then one has to diagonalize for each $\vec{k}_{/\!/}$ a $(N_\perp * n_{/\!/})$ by $(N_\perp * n_{/\!/})$ matrix where $N_\perp$ is the number of planes parallel to the surface and $n_{/\!/}$ the number of orbitals on the surface unit cell atoms. This is completly similar to the problem of a $N_\perp$ atom linear chain with effective interactions between atoms depending of $\vec{k}_{/\!/}$. We typically get:

$$\begin{pmatrix} H_{00}(\vec{k}_{/\!/}) & H_{01}(\vec{k}_{/\!/}) & 0 & 0 & \cdots \\ H_{10}(\vec{k}_{/\!/}) & H_{11}(\vec{k}_{/\!/}) & H_{12}(\vec{k}_{/\!/}) & 0 & \cdots \\ 0 & H_{21}(\vec{k}_{/\!/}) & H_{22}(\vec{k}_{/\!/}) & H_{23}(\vec{k}_{/\!/}) & 0 \\ \cdots & 0 & H_{32}(\vec{k}_{/\!/}) & H_{33}(\vec{k}_{/\!/}) & H_{34}(\vec{k}_{/\!/}) \end{pmatrix} \qquad (7)$$

if we have interactions only between atoms in adjacent planes. For longer range interactions β, $H_{02}H,_{03},..$ are no more equal to zero. Except if we use a few plane slab, this matrix cannot be diagonalized. One must take at least a 6 plane slab in order to get a small coupling between the surfaces of the slab. If we want to get independent surface density of states, one must use other methods.

1.2 Surface density of states.

To calculate the surface density of states, one can start from a perfect infinite crystal and then cuts this crystal in two pieces. This can be done by the Green function method. The matrix elements of the Green function can be calculated for a perfect crystal by integration along the k_z (normal to the surface) direction. Let us recall that the Green function method is an exact treatment of a perturbation which here cuts the infinite chain into two semi-infinite ones. Using notably the Dyson equation, one gets the Green function of the perturbed crystal and then local densities of states. We shall prefer a more direct method which is called "decimation" and well suited to this kind of problems[1]. First let us recall the Green function definition for a system whose hamiltonian is H:

$$(EI-H)G=I \tag{8}$$

where E is the energy and I the idendity matrix. For each $\vec{k}_{/\!/}$, one can project (8) on the layers m and 0:

$$EG_{m,0}-\sum_{m'} H_{m,m'}G_{m',0} = \delta_{m,0} \tag{9}$$

If we use (7) , we get:

$$(E-H(0)).G(0) - H(\bar{1}).G(1) = I \tag{10.a}$$

$$-H(1).G(0) + (E-H(0)).G(1) - H(\bar{1}).G(2) = 0 \tag{10.b}$$

$$-H(1).G(m-1)+(E-H(0)).G(m) -H(\bar{1}).G(m+1) =0 \tag{10.c}$$

where $H_{m,m'}=H(m-m')$ and $G(m) = G_{m,0}$.

These equations can be written like:

$$W.G_0+\tau_1.G_1=I \tag{11.a}$$

$$\tau_2.G_{m-1}+W.G_m+\tau_1.G_{m+1}= 0 \tag{11.b}$$

with $\tau_1=-H(\bar{1})$, $\tau_2=-H(1)$ and $W=(E-H(0))$.

If we consider the following equations:

$$\tau_2.G_0+W.G_1+\tau_1.G_2=0 \tag{12.a}$$

$$\tau_2.G_2+W.G_3+\tau_1.G_4=0 \tag{12.b}$$

$$\tau_2.G_4+W.G_5+\tau_1.G_6=0 \tag{12.c}$$

we can also write:

$$G_1=-W^{-1}.(\tau_2.G_0+\tau_1.G_2) \tag{13.a}$$

$$G_3=-W^{-1}.(\tau_2.G_2+\tau_1.G_4) \tag{13.b}$$

$$G_5=-W^{-1}.(\tau_2.G_4+\tau_1.G_6) \tag{13.c}$$

These equations are used to eliminate $G_1, G_3, G_5\ldots$ from equations (11):

$$(W-\tau_1.W^{-1}.\tau_2).G_0-(\tau_1.W^{-1}.\tau_1).G_2=I \tag{14.a}$$

$$(-\tau_2.W^{-1}.\tau_2).G_0+(W-\tau_2.W^{-1}.\tau_1-\tau_1.W^{-1}.\tau_2).G_0-(\tau_1.W^{-1}.\tau_1).G_4=I \tag{14.b}$$

If we define new matrices:

$$W_s' = W-\tau_1.W^{-1}.\tau_2 \tag{15.a}$$

$$W_b' = W-\tau_1.W^{-1}.\tau_2-\tau_2.W^{-1}.\tau_1 \tag{15.b}$$

$$\tau_1'=-\tau_1.W^{-1}.\tau_1 \tag{15.c}$$

$$\tau_2'=-\tau_2.W^{-1}.\tau_2 \tag{15.d}$$

we formally recover equations (11) with new renormalized interactions. The procedure can be iterated giving after p steps effective interactions between layers 2^p. If these interactions are negligible, W_s and W_b give the effective matrices of (E-H) for the surface and bulk layers. If we inverse W_s for example, we get $G_{00}(\vec{k}_{/\!/})$. As the density of states in the surface plane is just the imaginary part of the trace of $\mathbf{G}_{00}$, we can get the total density of states by integration over the surface Brillouin zone. This integration is generally done using a simple sum over a few special points in the reduced surface Brillouin zone [2].Moreover, if we have calculated two matrices $\mathbf{W}_s$ for respectively the right surface of a solid and the left surface of an other solid, we can switch on the interaction between these crystals to create an interface.Taking a larger surface unit cell, one can "adsorbed" on the surface a reconstructed layer.So starting from a perfect surface, this method is very versatile to study the real surface electronic structure. If we add a small imaginary part to the energy E, we can accuratly get the electronic charge by integrating $\mathbf{G}$ in the complex plane along a contour surrounding the required energy interval.For example, near a semiconductor interface the charge is obtained with an error less than 0.001 electron. When we have calculated the charges, we can in principle recalculate the potential and by iteration we can reach the self-consistency.

1.2 Self-consistency and core level shifts.

Near a defect like a surface, the atoms have not the same electronic charge as in the bulk. This is simply an extension of Friedel charge oscillations around impurities in the bulk. These oscillations decrease very rapidly as one goes inside the bulk of a transition metal or of a semiconductor. This is the effect of an efficient screening of the defect due to a high density of electrons and a short Thomas-Fermi screening length. Even in the plane just below the surface, the effect of the perturbation on the density of states is already very small. However this plane must take the charge transferred from the surface plane atoms in order to keep the crystal neutral. This creates a dipole layer and a potential shift on the atomic orbital levels of the surface atoms (Fig.1).

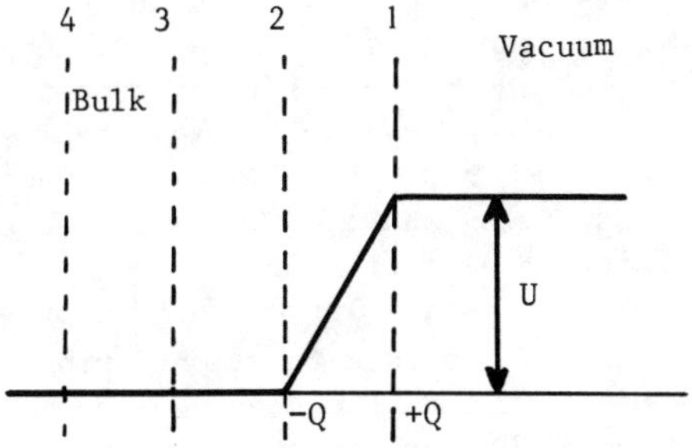

Figure 1. Surface dipole layer.

If the charge oscillations are limited to both these planes, we have:

$$Q = CU \tag{16}$$

where Q is the net surface atom charge. If this potential is small, one can linearize the charge:

$$Q = Q_0 + \chi U \tag{17}$$

where χ is the interplane susceptibility and Q_0 the non-self consistent charge (calculated for U=0). This suceptibility can be calculated as the derivative of the charge transfer to the potential. It can be evalated using the decimation technique and the Green function matrix elements $\mathbf{G}_{ij}$. It will be calculated below for a semiconductor simple model. Using (16) and (17), we get:

$$U = \frac{Q_0}{C} \frac{1}{1-\dfrac{\chi}{C}} \tag{18}$$

If $\chi/C \gg 1$, one gets:

$$U \simeq -Q_0/\chi \tag{20.a}$$

$$Q \simeq 0 \tag{20.b}$$

To first order, one in fact gets:

$$Q \simeq \frac{Q_0}{(\chi/C)} \tag{21}$$

which shows that the self-consistency considerably reduces the charge one gets neglecting the charge rearrangements (U=0). In the bulk for a point defect, the quantity $(1-\chi/C)$ is equal to the dielectric permittivity ϵ_r (close to 11). The capacity C relating the charge transfer to the dipole layer U is more difficult to evaluate. One can relate the potential U_i on a site R_i to the charge on site R_j :

$$U_i = U_c \, \delta\rho_i + \sum_{j\neq i} \frac{\delta\rho_j}{|\vec{R}_i - \vec{R}_j|} \tag{22}$$

where U_c is the intratomic Coulomb integral. One must take also into account the correlations which reduces the free atom value of U_c. The value one gets for the dipole layer is quite large and close to 30 eV per electron transferred from a surface plane atom to an underlayer one. Even if χ is of the order of .1 electron per eV for a semiconductor surface (and 1 electron per eV for a transition metal one), the approximation $\chi/C \gg 1$ is not too bad. Near a semiconductor surface, if we use classical electrostatics and the image charge model at an insulator interface, the screening of a point charge is reduced to $\epsilon_r/2$ which is still larger than unity. Moreover if we looks at the variation of the work function during adsortion, we expect a more efficient screening (and then a smaller charge transfer) at large coverages (monolayer) than at low coverage (point defect). The equation (20.b) is often used to calculate the "self-consistent" potential.Its accuracy seems not too bad. Equation (20.a) is less accurate due to a small non-linearity of the charge as a function of the potential. This is known as a complete charge screening model or a zero charge model. As we shall see below,this approximation seems rather good even to calculate rather small quantities as the dipole layer at a semiconductor interface. An almost direct observation of this potential can be done by photoemission spectroscopy of the bulk and core levels. The core electrons also feel the valence electron potential. They are still more localized than the valence ones so one can easily imagine that the electrostatic potential they feel is slightly different of the valence one U_i. However it is not difficult to show that this difference is small. So if one can in the same photoemission experiment separate the signal due to bulk atom core levels and the surface ones, he will measure directly the self-consistent potential. This has been at the beginning observed for transition metal surfaces and rapidly extended to other material free surfaces and adsorbates. The measurements are in good agreement with theoretical results obtained by the zero charge model. Moreover a recent reexamination of the core level shifts of III-V compound (110) surface shows that the anion and cathion surface atoms have the same charges as the bulk ones

1.3 Surface states.

The perturbation due to the surface induces changes in the surface density of states and even states whose energy lie outside the bulk bands. These states are localized near the surface. When we change the potential,these states can go into the bulk bands giving rise to resonances. Let us examine some of these properties in a very simple case. Let us consider the (100) surface of a simple cubic lattice with only one s orbital per atom. The different terms H_{ij} of the hamiltonian (7) are:

$$H_{00}(\vec{k}_{/\!/}) = U - 2\beta\ (\cos(k_x a)+\cos(k_y a)) \tag{23.a}$$

$$H_{ii}(\vec{k}_{/\!/}) = -2\beta\ \ (\cos(k_x a)+\cos(k_y a)) \qquad \text{for } i=1,2,3.... \tag{23.b}$$

$$H_{ij}\ (\vec{k}_{/\!/}) = -\ \beta \quad \text{for } j=i\pm 1 \tag{23.c}$$

Let us first neglect the quantity $-2\beta\ (\cos(k_x a)+\cos(k_y a))$ which simply shifts all diagonal terms. We shall consider this shift latter. We write the hamiltonian wave function as

$$\Psi = \sum_{l=0}^{\infty} a_l\ \varphi_l \tag{24}$$

To solve the problem, we write the equations:

$$\langle\varphi_k|E-H|\Psi\rangle = 0 \tag{25}$$

for any value of k. This equivalent to the following set of equations:

$$\begin{aligned} (E-U)\ a_0 &= -\beta\ a_1 \\ E\ a_1 &= -\beta\ (a_0+a_2) \\ E\ a_k &= -\beta\ (a_{k-1}+a_{k+1}) \end{aligned} \tag{26}$$

If we look for the possibility of a localized state with a wave function having an exponential decay, we try a solution with a_n equal to

$$a_n = a_0\ K^n \tag{27}$$

the set of equations (26) reduces to

$$\begin{aligned} E &= U - \beta\ K \\ E &= -\ \beta\ (K+\frac{1}{K}) \end{aligned} \tag{28}$$

whose solutions are:

$$\begin{aligned} K &= -\beta/U \\ E_L &= U+\beta^2/U \end{aligned} \tag{29}$$

Such solutions are only valid if $|K|$ is smaller than unity in order to have an exponential decrease at infinity. This means that $|U/\beta| > 1$ which leads to $|E_L| > 2\ |\beta|$. When $|U/\beta|$ decreases, these states penetrate into the band giving rise to a resonance and a peak in surface density of states. Let us now consider the diagonal shift $-2\beta(\cos....)$. As k_x and k_y vary in the surface Brillouin zone, we get a band of surface states. Let us remark that this band can overlap the bulk band as these overlapping states arise for different $\vec{k}_{/\!/}$. In general for a semiconductor surface, we shall see that the dispersion of these states versus the wave vector parallel to the surface is small ($\simeq$1 eV) so it gives rise to a large density of states. Some surprising surface properties can arise from this large surface density of states: magnetic surface whereas the bulk is paramagnetic or surface superstructures.

3. SEMICONDUCTOR SURFACE ELECTRONIC STRUCTURE.

3.1 <u>The molecular model</u>

In a diamond lattice, the atoms occupy the center and the vertices of tetrahedrons. This structure is made of two face centered lattices shifted by one fourth of the cube diagonal. Experimental results or pseudopotential calculations show that the valence electrons are mainly localized between the atoms along the four (111) directions. So with the s and p orbitals we make sp^3 hybrid orbitals. For the (111) direction, we take for example the combination $(s+p_x+p_y+p_z)/2$. These hybrid orbitals have a strong lobe pointing in the (111) directions. From the definitions (5) and (6), it is obvious that only pairs of such hybrid whose positive lobes are overlapping will lead to large β values. This new basis does not reduce the size of the hamiltonian matrix. It simply allows to determine the most important hopping integrals and to neglect the others. Figure 2 shows the largest integrals which occur between nearest and next nearest neighbours. In the molecular model, we neglect all these terms except β which is much larger than the others.

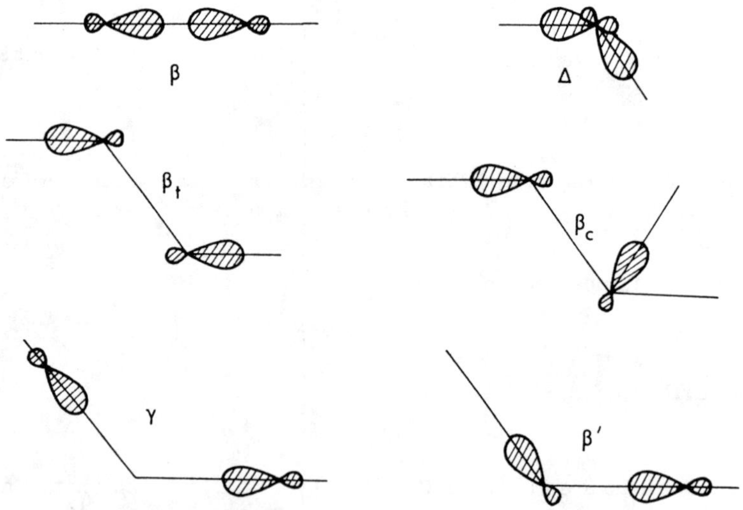

FIGURE 2. Two-center integrals β, β', $β_t$ and $β_c$ between nearest neighbour atoms. We also recall the definition of Δ $(=(E_s-E_p)/4)$ which is an intraatomic term and of γ which is a next nearest neighbour integral.

The crystal hamiltonian is block diagonalized as we only have coupling between hybrid orbitals which point toward an other. This pair of orbital is completly independent of the others. If the crystal contains 2M atoms, we have to solve 4M identical hamiltonians. Each one is for a "diatomic molecule":

$$H=\begin{pmatrix}\alpha_A & -\beta \\ -\beta & \alpha_B\end{pmatrix} \tag{30}$$

where α_A and α_B are the atom hybrid orbital level equal to

$$\alpha_{A,B} = \frac{E_s^{A,B} + 3\, E_p^{A,B}}{4} \tag{31}$$

This leads to a bonding E_B and an antibonding E_B levels

$$E_{B,A} = \frac{\alpha_A+\alpha_B}{2} \mp \sqrt{\frac{(\alpha_A-\alpha_B)^2}{2} + \beta^2} \tag{32}$$

which are 4M times degenerated (in fact 8M times if we consider the spin degeneracy). The bonding level is filled whereas the antibonding is empty. This a rough description of the semiconductor valence and conduction band. The bonding eigenstate Ψ_B is equal to :

$$\Psi_B = \frac{\varphi_A + \lambda\, \varphi_B}{\sqrt{1 + \lambda^2}} \qquad \text{with } \lambda = \frac{\beta}{(\alpha_A - E_B)} \tag{33}$$

A fraction $1/(1+\lambda^2)$ of an electron described by Ψ_B belongs to atom A and a fraction $\lambda^2/(1+\lambda^2)$ to atom B. As there two electrons per bond and four bonds per atom, the net charge $Q_{A,B}$ on atoms A and B are

$$Q_A = \frac{8}{1+\lambda^2} - Z_A$$

$$Q_B = \frac{8\lambda^2}{1+\lambda^2} - Z_B = -Q_A \tag{34}$$

where Z_A and Z_B are the number of valence electrons (3 for the cathion atom and 5 for the anion). In a covalent, the perfect bulk atoms are obviously neutral. In a III-V compound, these charges modify the atomic levels through a Magdelung term. Let us remark these charges may positive or negative according to the ionicity of the compound. From core levels[3], it seems that there are more than 5 electrons on the anions. The net charge is of the order of 0.2 electron for usual III-V compounds (GaAs, GaP, GaSb, InSb). Further interactions will broaden the bonding and antibonding levels into valance and conduction bands.

3.2 The dangling bond

3.2.1 Semiconductor surface localized states

For a surface atom, the number of nearest neighbours is decreased with respect to the bulk value. Figure 3 represents the local situation for a (111) surface of a semiconductor. In the case of an unrelaxed and unreconstructed surface, (the surface atoms occupy the perfect bulk crystal positions) one can still use hybrid sp^3 orbitals. Three orbitals participate in bonds (called "back bonds") with the the underlayer atoms. We are left with an unpaired hybrid wich is called "dangling bond". In the molecular model, its energy is equal to the sp^3 levels α_A or α_B in the region of the energy gap. Further interactions broaden these levels in a surface state band. For a covalent semiconductor, to keep the crystal neutral one must put one electron per atom in the surface band. As this band would be filled with two electrons, the surface is metallic (non zero density of states at the Fermi level). The surface states as the surface Fermi level are well defined in the gap. Their positions are given by short-range conditions. The bulk Fermi level which may vary according to doping must then be aligned with the surface one and "pinned" by the surface band states. An other effect of these surface states is the surface reconstruction. Even if at that time it is not obvious that the semiconductor surface reconstruction is always due to surface states, let us look to their possible effect in the case of the (2*1) Silicon (111) surface reconstruction and in the case of a polar III-V compound surface.

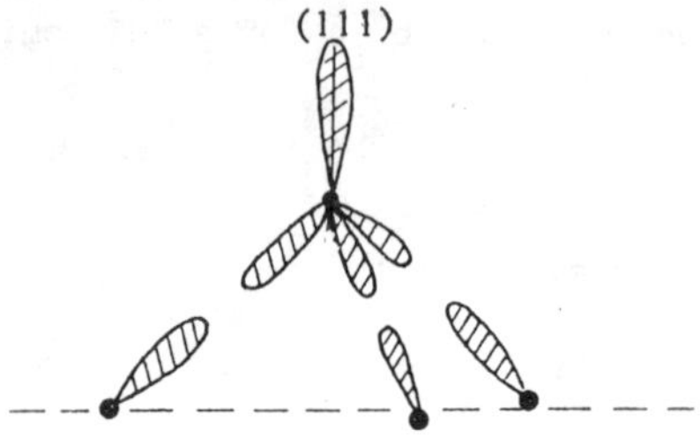

Figure 3. Dangling bond orbital at a (111) surface.

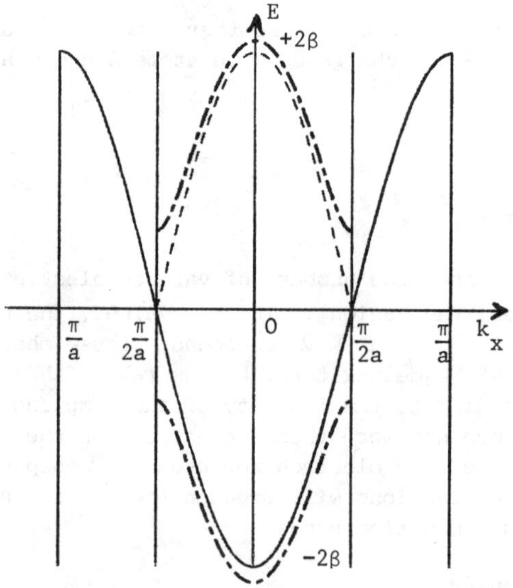

FIGURE 4. Bulk band structure of the linear chain in the tight binding approximation with one s orbital per atom: a) lattice unit cell equal to a (full line). b) lattice unit cell equal to 2a (dotted line). c) with a perturbation of period 2a (dash dotted line).

3.2.2 Silicon (111) surface (2*1) reconstruction Many papers have been published dealing with the origin and the nature of this surface reconstruction. Let us explain in a simple model the possible effect of surface states. Let us consider a linear chain with one atom per unit cell and one s orbital per atom. Figure 4 shows in the Brilloiun zone the energy dispersion curve one obtains in the tight-binding approximation. This band is symmetrical and if we have one electron per atom, the Fermi level is just at the middle of the band (E=0 if the origin of the energy is taken at the atomic level). Let us now consider the same linear chain but with two atoms per unit cell. The Brillouin zone now is between $-\pi/2a$ and $+\pi/2a$ instead of $-\pi/a$ and $+\pi/a$ as the lattice parameter is 2a. But the chain is not changed and we must have the same band structure. However as we have two atoms per unit cell, we have two bands one gets by a band folding. The Fermi level is just at the degenerate energy wich happens at $\pm\pi/2a$. If we make a small perturbation on the perfect chain with a period equal to 2a, we remove the energy degeneracy at the Brilluoin zone edge. The band remains symmetrical and the linear chain is now an insulator as the Fermi level is in the band gap. All the states below the Fermi level have lowered their energy. So if the energy needed to create the perturbation is smaller than the energy gain by the band splitting, the linear chain is unstable and reconstructs. This is exactly the case of the Silicon (111) surface state band. The surface state band can be treated as uncoupled with the bulk states as it is mainly localized on the dangling bonds. One only has to take into account of the indirect interaction between the dangling bonds through an effective interaction parameter. Moreover we have shown that this band is just half-filled. So the required conditions are just

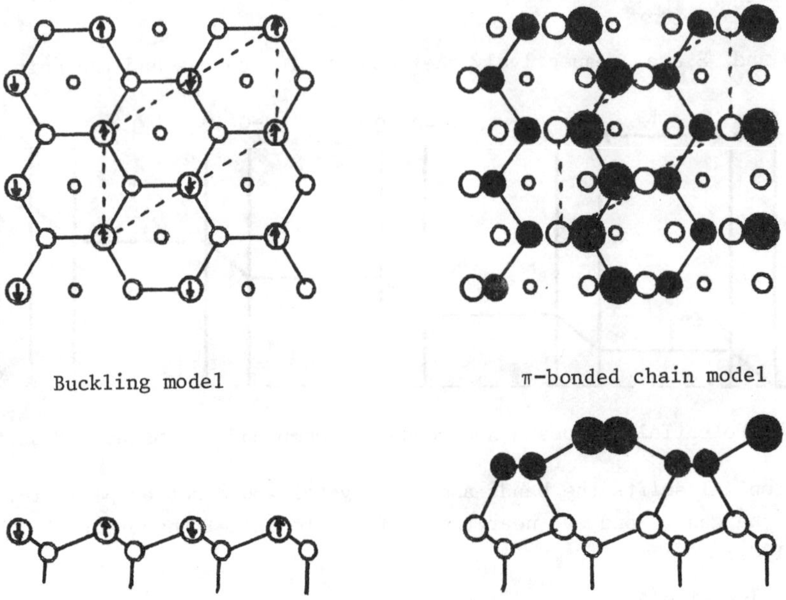

FIGURE 4. Silicon (111) surface reconstruction.

fullfilled as the Fermi level density of states is small and the band is narrow. Different models have been suggested: the buckling model where all the atoms alternatively lowered and raised, an antiferromagnetic structure could explained the observed reconstruction and surface band splitting. However in all these models, the band splitting was too small. Everyone actually seems to agree on a quite different structure: the Pandey's π-bonded chain model[4]. The atomic displacements are in this model quite large. The atoms are along chains like in the (110) plane. Even if the relaxation of the bond lengths is equal to 0.1 Å, the energy gain due to the π bonding of these chains is sufficient to stabilize this structure. The surface state energy dispersion is also consistent with the photoemission measurements. A (7*7) reconstruction also exists for the Si(111) surface. But due to the large number of atoms in the unit cell, the determination of the atomic positions is still more difficult. Steps, terraces, adatoms or staking faults are used to explain this structure. Scanning tunneling microscopy, ion channeling, X-ray diffraction are used to determine the atomic positions. (For recent references, see reference [5]).

3.2.3 <u>III-V compound polar surface reconstruction</u> Equation (34) shows that the cathion and the anion in a compound have opposite charges. If we consider the polar planes (100) or (111), all the atoms of such a plane have the same charge. In this sense, the non polar plane (110) is neutral. If we spread the charges in plane parallel to the surface, it is easy to calculate the potential along a direction normal to the surface. If the crystal has M positive and M negative layers (Fig.5), the potential betwwen the surface planes is:

$$V = \frac{Q}{S} \frac{M\, R/3}{\epsilon_0} \tag{35}$$

where Q and S are respectively the net atomic charge and the unit cell area.

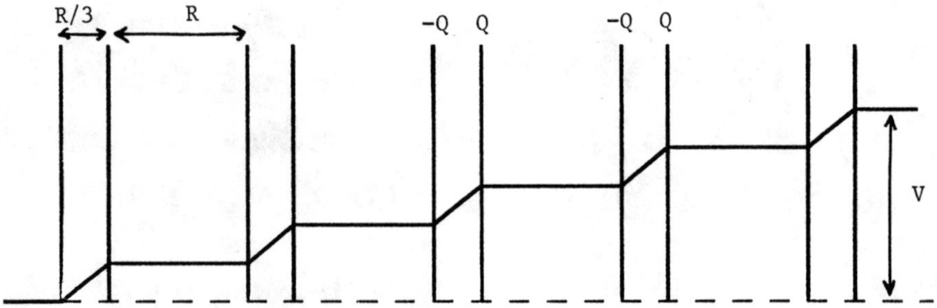

Figure 5. Potential across a semiconductor compound in absence of surface charges.

This potential shifts the bands and the crystal would not be periodic. One must add charges Q' and -Q' near the surfaces to cancel thepotential V by a potential V':

$$V' = \frac{Q'}{S} \frac{M\, 4R/3 + R/3}{\epsilon_0} \tag{36}$$

as the distance between the (111) surface planes is $(4M+1)R/3$. For large values of M, we get $Q'=-Q/4$. It is easy to show that we must then put exactly $Z_{A,B}/4$ electrons in the dangling bonds. This is not always possible (notably as we shall see below near a semiconductor interface) and the surface reconstructs to accomodate the right electron number in the dangling bond.

3. HETEROSTRUCTURES AND QUANTUM WELLS

3.1 Quantum well band structure

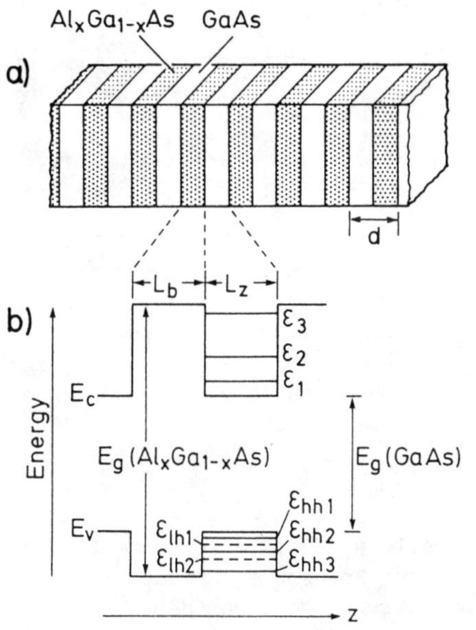

Figure 6. Periodic layer structure of a GaAs-AlGaAs superlattice (a) and its subband energy for the conduction and valence band states.

The recent developments in the thin-film growth techniques of Molecular Beam Epitaxy (MBE) and Metalorganic Vapour Phase Epitaxy (MOVPE) or Chemical Vapour Deposition (MOCVD) make possible the synthesis of ultrathin multilayer structures of crystalline semiconductor. Such structures have been first proposed by Esaki and Tsu[6]. In these artificial superlattices (Fig. 6.a), novel physical phenomena occur due to quantum size effects as the layer thickness is reduced to the order of carrier de Broglie wavelength ($\lambda \simeq L_z$ <500 Å). The energy spectrum is determined in the effective mass approximation assuming that the (single particle) electron or hole Hamiltonian can be separated into a component normal to the growth axis and into the usual Bloch functions in the layer plane. Tunneling between layers can be often neglected at least for the deeper levels. In the wells, the energy E_n is quantized along the superlattice growth direction. In the direction parallel to the interfaces, we have a continuum of a free electron gas with an effective electron mass m_n^* and heavy (h) or light (l) hole m_p^*:

$$E(n,k_x,k_y) = E_n + \frac{\hbar^2}{2m_{n,p}^*}(k_x^2+k_y^2) \tag{37}$$

The hole and electron levels can be tailored to the desired energy by simply preselecting the well thickness L_z and the materials. The materials determine the well depth ΔE_v or ΔE_c. To get perfect superlattices and very thin layers, one must use semiconductors with very close bulk lattice parameters. The lattice mismatch for an unstrained superlattice must be less than 0.2 per cent. This can be realized with different materials. Most commonly the used semiconductors are GaAs and $Al_xGa_{1-x}As$. The alloy Al

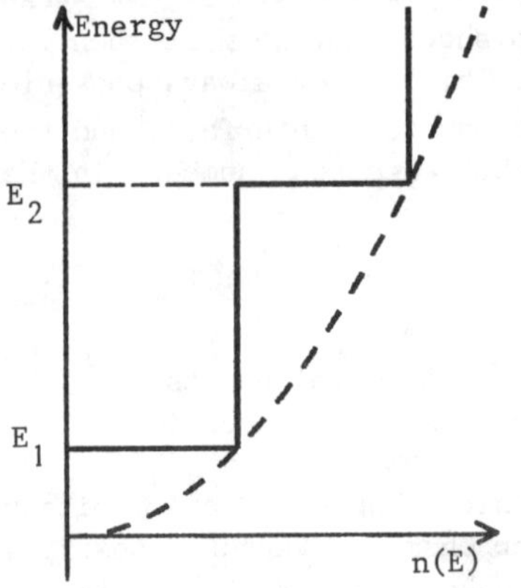

Figure 7. Stair-case density of states

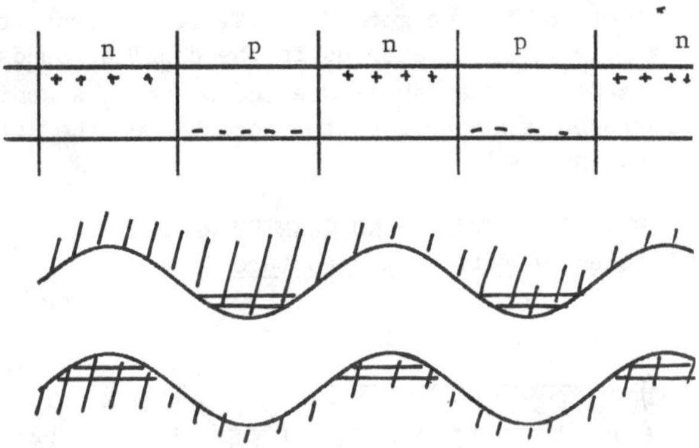

Figure 8. nipi superlattice (top) and the potential with the electron and holes levels.

concentration x can also be choosen according to the properties the device designer wants. Due to the two-dimensional behaviour of the carrier, the density of states is a "stair-case" characteristic of a quantum well structure (Fig.7). Such a structure is approximate for very thin layers as we use bulk material effective masses. Moreover the effective mass approximation is only valid for slowly varying potentials whereas we shall get an interface width of the order of one atomic layer.

An other interesting structure proposed by Esaki and Tsu was the doping superlattice (also called nipi). It is a layer structure with thin alternating p and n doped layers. At low temperature (Fig. 8), all the dopant impurities are ionized giving rise to an oscillating potential one can calculate simply by integration of the Poisson equation. We get an indirect bandgap in the real space. The distance between the valence band maximum and the conduction band minimum is large (200-500 Å) giving a small overlap of the electron and hole wave function. Then one expects a very large electron-hole lifetime (several days seem feasible). Moreover carrier injection in the layers reduce the charge and then the potential: the bandgap is then tunable. GaAs nipi have been realized. Silicon nipi would also present these properties and an indirect gap in reciprocal space.

3.2 Band offsets

One important parameter is the valence band offset ΔE_v. Knowing the band gaps, one can easily obtain the conduction band offset ΔE_c . Three situations can arise according to the carrier localization (Fig. 9). Type I superlattices are semiconductors. However if the electron and holes are confined in the same layer for a GaAs layer sandwiched between two $Al_xGa_{1-x}As$ layers, electrons and holes in $In_xGa_{1-x}Sb$-$Sb_yGa_{1-y}As$ are confined in different layers. Moreover carriers and dopant impurities can be easily separated giving rise to a high mobility in the layers as the scattering from ionized dopant impurities is the dominant mechanism at low temperatures. High Electron Mobility Transistor has been developed from these doped heterostructures for high-speed logic application. Type II

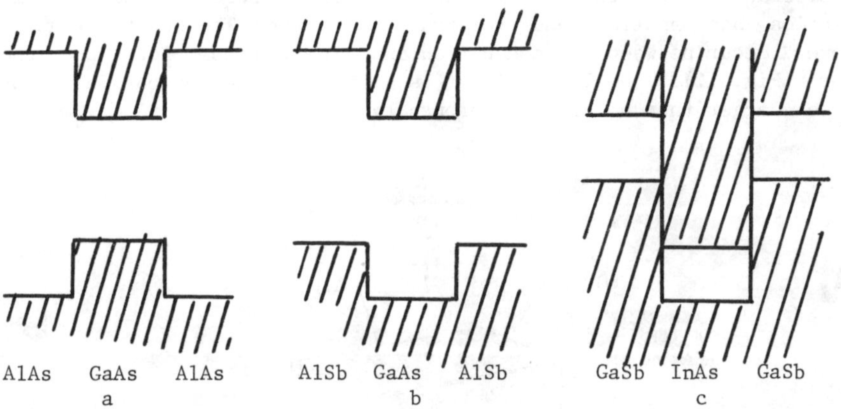

Figure 9. Schematic energy gap of superlattices a,b) Type I c) Type II

superlattice (Fig. 9.c) is a semi-metal like Bismuth. The valence band offset is also an important technological parameter as it determines the electric field near the heterojunction and then the electron ionization rate. Accuracy better than 0.1 eV is required. Simple theories based on the work function difference between both materials gives 0.2 or 0.3 eV[7-8]. In fact the surface is a much stronger crystal perturbation than the heterojunction where the crystal continuity is ensured. One can hardly neglect the surface and interface dipole layer even if the charge transfers are small. *A priori*[9] calculations reaches the same accuracy. Here also a tight-binding calculation of the valence band offset can be done using the zero charge approximation. For non polar interfaces, the agreement with the experimental values is quite good (of the order of 0.1 eV) (Fig. 10). For a polar interface, it is easy to show that one must localize a quarter of the semiconductor bulk charges near the interface. Using the molecular model, this seems impossible to realize if there is no interface state. In the case of the Ge-GaAs interfaces, as the valence band offset does not depend on the growth axis, a drastic atomic re-arrangement is assumed[10].

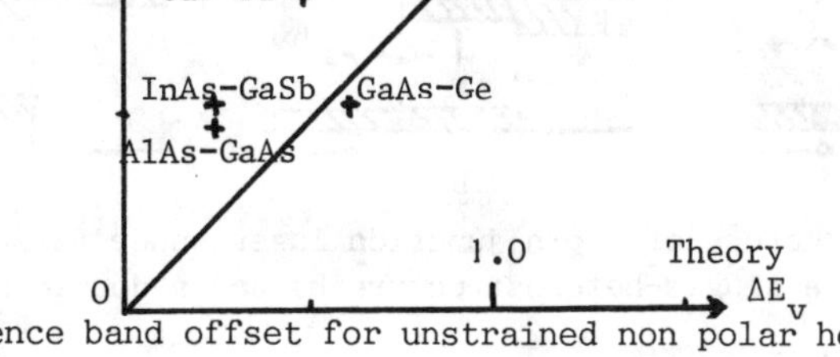

Figure 10. Valence band offset for unstrained non polar heterojunctions.

3.3 Semiconductor lasers[11-12]

Semiconductor lasers are similar to others lasers in that the emitted radiation has spatial and temporal coherence. The quantum transitions however occur between the conduction and valence bands. For a p-n junction laser (Fig. 11), the laser light is simply obtained by passing a forward current larger than a thresold through the diode itself.

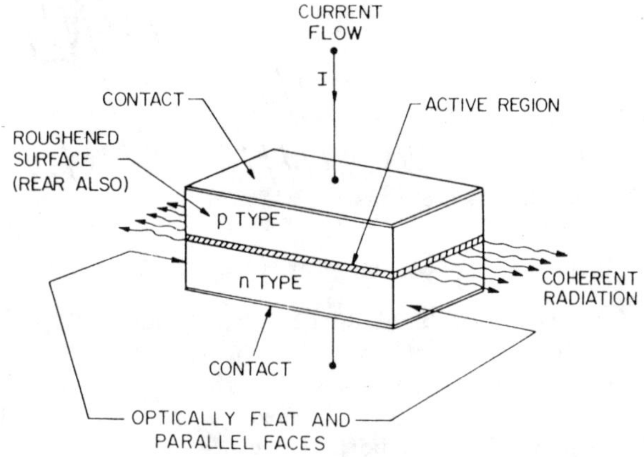

Figure 11. Basic structure of a p-n junction laser[11].

Then modulated light is even at high frequencies easy to realize by modulating the current. This property and the device compact size has made the semiconductor laser the most important light source for optical fiber communication. More over such a laser can be integrated in a device with the light receiver and amplifiers. In a semiconductor laser the light emission wavelength is determined by the material band gap. This can be modify by alloying a semiconductor to make ternary or quaternary compounds. All used semiconductors have a direct band gap to get a high transition probality. The band structure of a p-n junction laser is shown on figure 12. When a forward bias voltage is applied to the diode, stimulated

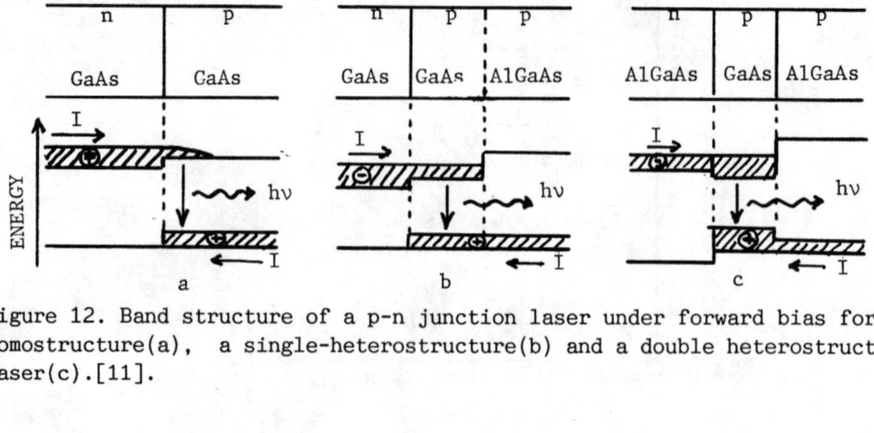

Figure 12. Band structure of a p-n junction laser under forward bias for an homostructure(a), a single-heterostructure(b) and a double heterostructure laser(c).[11].

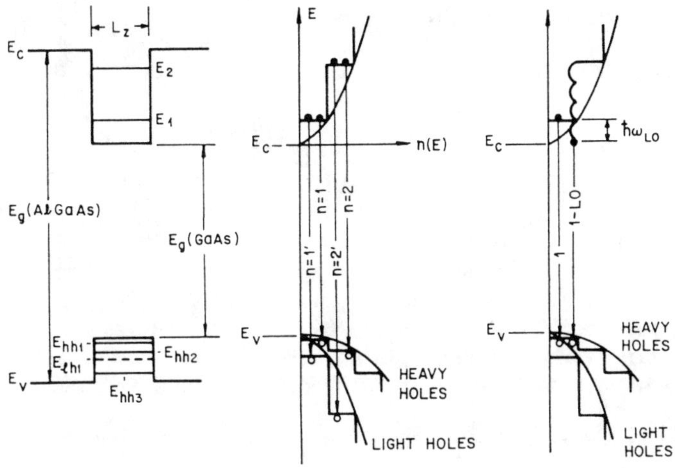

Figure 13. Band structure of a quantum well heterostructure laser.[11]

emission occurs and a monochromatic light beam is emitted from the junction. Several structures have been used (Fig. 12) mainly to reduce the threshold current. In the double-heterojunction laser, the carriers are confined in the active region where the optical field is also confined by reduction of the refractive index outside the junction. Figure 13 shows the band structure for a quantum well semiconductor laser. Transitions occur between the conduction electron eneygy levels and the heavy or light hole levels. The electrons are thermalized more efficiently in such a structure (the density of states is constant in a step) than in a classical semiconductor p-n junction laser (whose conduction band density of states is parabolic). The thermalized electron density of states is large (it does not vanishes as for a parabolic density). The carriers are also strongly confined in the wells and the electron-hole lifetime reduced. So the thresold current is stongly reduced and less sensitive to the temperature. The difference between the electron and hole levels is easier to adjust to the required value but is larger than the bulk material band gap. However from the lowest conduction band step in the deensity of states a virtual transition is possible which makes recombination and laser operation a LO phonon energy $\hbar\omega_{LO}$ below the confined particle energy gap.

REFERENCES

For these lecture notes, no attempt is being made to present a complete listing of citations. We rather refer the reader interested by surfaces to books like "Introduction to the theory of solid surfaces",by F. Garcia-Moliner and F. Flores, Cambridge University Press,1979 or"Handbook of surfaces and interfaces",L. Dobrzynski Ed,Garland STPM Press 1978. Readers interested by heterostructures will find more details in "Two-Dimensional Systems, Heterostructures and Superlattices", Springer series in Solid State Science 53, Edts G. Bauer, F. Kuchar and H. Heinrich or in the proceedings of a winter school in Les Houches "Heterojunctions and Semiconductor Superlattices" Springer Verlag 1986, N. Boccara, G. Allan, G. Bastard, M.Lannoo, M. Voos Edts.

1. F. Guinea,J. Sanchez-Dehesa, F. Flores, J.Phys. C:Solid State Phys. **16**,6499(1983).
2. S.L. Cunningham, Phys. Rev. **B10**,4988(1974).
3. W. Mönch, Solid State Commun. **58**,215(1986).
4. K.C. Pandey, Phys. Rev. Lett. **47**,1913(1981).
5. I.K. Robinson, W.K. Waskiewicz, P.H. Fuoss, J.B. Stark, P.A. Bennett, Phys. Rev. **B33**,7013(1986).
6. L. Esaki,R. Tsu, IBM J.Res.Dev. **14**,61(1970)
7. W.A. Harrison, J.Vac.Sci.Technol. **14**,1016(1977)
8. G. Margaritondo, Phys.Rev. **B31**,2526(1985)
9. R.M. Martin (To be published in Phys.Rev.)
10. W.A. Harrison,E.A. Kraut,J.R. Waldrop,R.W. Grant, Phys.Rev. **B18**,4402(1978)
11. S.M. Sze,Physics of Semiconductor Devices,John Wiley & Sons Edts, New York 1981.
12. N. Holonyak Jr.,Sov. Phys. Semicond. **19**,943(1985)

MOLECULE-SURFACE INTERACTION: VIBRATIONAL EXCITATIONS

G. BENEDEK

Dipartimento di Fisica dell'Università, Via Celoria 16, 20133 Milano, Italy

1. INTRODUCTION

Among the various surface and interface phenomena which are triggered or influenced by laser light irradiation, gas-surface interaction occupies an important position. In the last decades much progress has been made in the subject of gas-surface interaction, not only in the traditional realm of aerospace research (1), but also, and prominently, in the fields of heterogeneous catalysis, surface chemistry, molecular beam epitaxy (MBE), ion implantation, etc. In all cases process rates can be deeply modified by laser irradiation, like in surface annealing, photocatalysis, photoadsorption and desorption phenomena.

More recently molecular beams have revealed themselves as a powerful tool for surface characterization, i.e., for the determination of the surface static and dynamical structures. The analysis of the elastic and diffractive scattering of atom monochromatic beams provides information on surface geometrical parameters such as periodicity and corrugation (2). Resonant scattering involving molecule-surface bound states tells us about the gas-surface attractive potential (3) - a basic piece of information for understanding energy acommodation and sticking (4,5), and inferring the precursor states for chemisorption. Even the signature of individual, isolated chemisorbed atoms has been recently identified in atom scattering (6) so that a scanning surface microscopy using atoms as a probe should be possible in principle.

On the other hand the inelastic molecular beam scattering bears a spectral information on surface low-energy excitations, such as surface vibrations (7-12), and possibly hole-electron pair excitations (13). Surface phonon dispersion curves have been determined in the last five years using He scattering approximately in the same way as neutrons are used for bulk phonons. Insulator (9-12) and metal surfaces (13-15) as well as covalent (16) and layered (17) semiconductors, and charge-density-wave (CDW) systems (18) have been investigated in this way.

The combination of elastic (diffractive) and inelastic time-of-flight (TOF) molecular beam spectroscopy - a technique strictly surface-sensitive and nondestructive - offers in principle an excellent method for monitoring the surface in the same ultra-high vacuum cell during various processes of technical interest.

An example is offered by surface melting, a phenomenon which has been recently investigated theoretically by Pietronero and Tosatti (19) and experimentally observed in lead by F. van der Veen and coworkers by means of blocking-ion spectroscopy (20). A short qualitative discussion on how He atom inelastic scattering can monitor surface melting is given in the last Section.

In the intermediate Sections I shall present the basic concepts of inelastic He scattering experiments, with illustration of a selected set of experimental data. I will restrict myself to the surfaces of those materials which are likely to be more involved in laser-induced processes, such as noble metals, silicon and layered chalcogenides of both semiconducting (GaSe) and metallic ($TaSe_2$) nature.

A common denominator appearing in the surface-phonon dispersion of all these small-gap or free-electron materials is the important role of electron-phonon interaction. Coupling mechanisms that are forbidden or hindered in the bulk for symmetry reasons are activated or enhanced at the surface because of symmetry and atomic coordination reduction. For instance, the electron-phonon interaction is responsible for the appearance of anomalies in the surface-phonon dispersion even in crystals which do not show any anomaly in the bulk phonon dispersion curves.

The existence and nature of surface electronic states are seen to produce by themselves effects on the surface vibrational spectrum. The few examples discussed in the next Section want to elucidate these important aspects of the surface vibrational spectroscopy.

2. He INELASTIC SCATTERING EXPERIMENTS

The spectroscopy of surface phonons by atom scattering recalls immediately the neutron spectroscopy of bulk phonons. There is however an important difference. A neutron TOF spectrum, as a consequence of wavevector conservation in three dimensions, displays a few sharp peaks in correspondence of single bulk phonons.

In atom scattering the wavevector is conserved only in two dimensions parallel to the surface and therefore TOF spectra present, in addition to sharp peaks for surface phonons, the features of the surface-projected bulk phonon density, possibly including surface resonant modes. There is clearly a larger amount of information in the atom scattering TOF spectra, as not just the surface phonon dispersion curves but also the surface-phonon densities have here a spectroscopic relevance.

This has posed two difficult problems. On the experimental side, a particularly good resolution is required, say of the order of few wavenumbers. On the theoretical side, on has to know much about the inelastic scattering cross section as function of energy and parallel momentum transfers, in order to translate the observed scattering intensities into the corresponding surface phonon densities. This step depends in turn upon our knowledge of the atom-surface potentials.

Although the theoretical foundations date back to the work of Cabrera, Celli, Goodman and Manson (7), only recently the two above requirements have been accomplished and put together in a systematic analysis of TOF spectra (8,9). A schematic view of one of the He-scattering TOF spectrometers used by J.P. Toennies and coworkers in Göttingen (11) is shown in fig. 1. A nozzle beam source yields an intense, highly monochromatic He beam with energy in the range 10 to 100 meV and a velocity distribution as narrow as 0.4%. Next to the skimmer the beam is chopped for lock-in amplification and then scattered by the crystal surface in a horizontal plane. A manipulator in the scattering chamber allows all possible orientations of the crystal and for collimation with the various surface analysis and surface preparation equipments. Surface cleavage in situ is also possible. The scattered beam after a path of about 1 m enters the detector formed by a ionization chamber and a mass spectrometer. The detector branch can be rotated in the horizontal plane around the crystal surface thanks to a flexible connection to the scattering chamber. Most of measurements, however, have been taken in the 90° scattering geometry, where the final angle θ_f is complementary to the incidence angle θ_i, and the latter is varied by rotating the crystal. Nine pumping stages (P_n) keep the large pressure gradient required in the experiment, ranging from hundreds atmospheres for the stagnation pressure inside the source to 10^{-11} torr in the scattering chamber and detection branch.

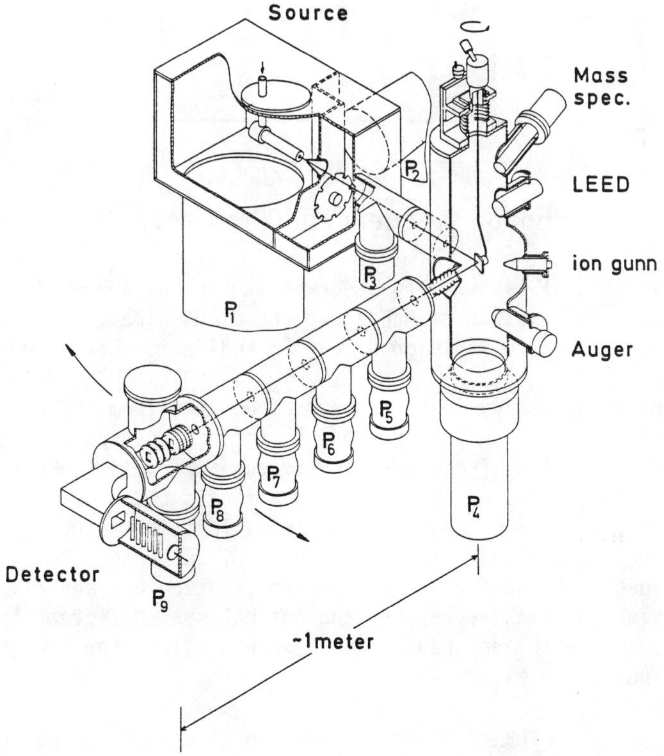

FIGURE 1. Schematic view of one of the He-scattering TOF spectrometers used by J.P. Toennies and coworkers at the Max-Planck-Institut SF in Göttingen, FRG (from ref. 11). This apparatus is presently used for metal and semiconductor surface investigation.

2.1. Ionic surfaces

He scattering from well known surfaces such as those of alkali halides has been studied first in order to assess the technique. An example for the (001) surface of NaF along the <100> direction is shown in fig. 2.

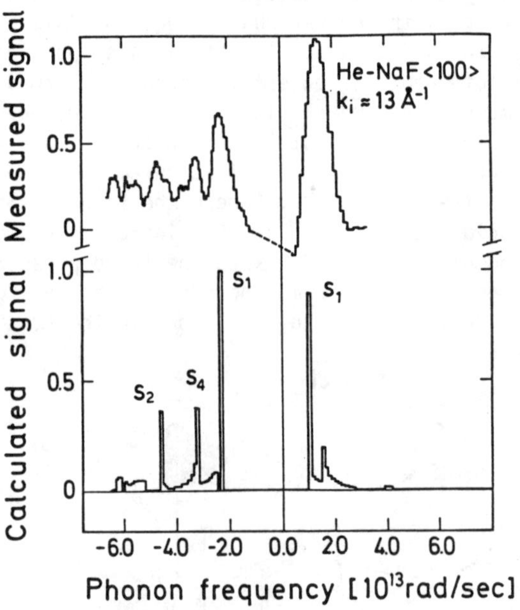

FIGURE 2. Experimental (above) and theoretical (below) energy loss spectrum of He from the (001) surface of NaF along the <100> direction. The peaks labelled S_1 correspond to Rayleigh waves; S_2 and S_4 to surface optical modes.

An experimental energy loss spectrum as derived from TOF data for an incident momentum k_i = 13 A^{-1} is reproduced in the upper part of the figure as function of the phonon frequency, and is compared with the calculated one-phonon scattering intensity. The calculation is based on the breathing shell model for NaF bulk dynamics, the Green's function method (21) and a hard corrugated surface model for atom-surface interaction (10). The main experimental peaks are seen to correspond to creation and annihilation processes involving Rayleigh waves (S_1) and optical surface modes (S_2, S_4).

The comparison with the theoretical spectrum allows for the interpretation of the minor features in terms of bulk phonons (e.g., the LO region at -6. x 10^{13} rad/sec).

In fig. 3 we show the dispersion curves for the sagittal polarization in NaF(001), calculated along the symmetry directions by the Green's function method (21). The heavy lines represent the surface-localized modes and surface resonances. The experimental points corresponding to sharp peaks in TOF spectra are seen to follow quite well the theoretical dispersion curves along the two directions.

The investigation of alkali halide surfaces, where good agreement is found between experiment and theoretical predictions, has shown the capability and limits of the He scattering technique. Materials that are much more interesting from both the practical and conceptual point of view, such as transition metals and semiconductors, are now systematically studied. An example is offered by noble metal surfaces (14-16), where surface electronic states and surface electron-phonon interaction seem to play a dramatic role (22).

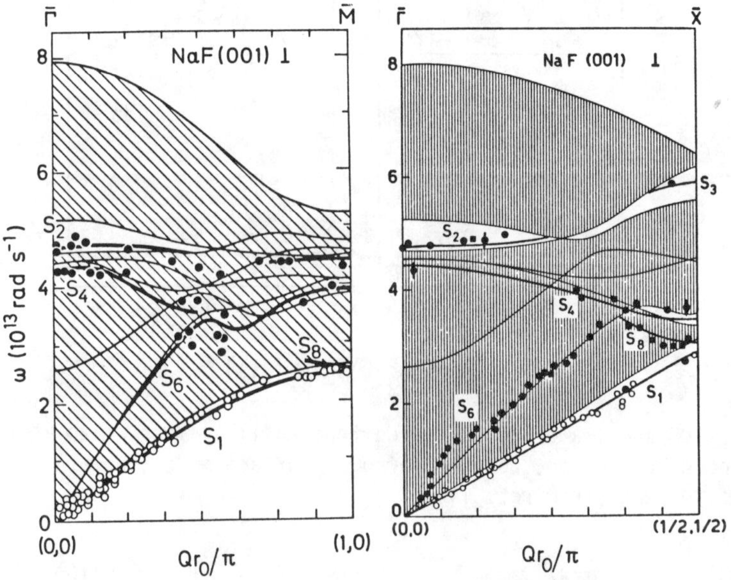

FIGURE 3. Surface phonon dispersion curves of NaF(001) along the symmetry directions for sagittal polarization. Heavy lines represent surface phonons; hatched regions are the surface-projected bulk phonon bands. The experimental points are obtained by He inelastic scattering.

2.2. Noble metal surfaces

High-resolution He TOF spectra from Ag(111) (fig. 4) show a double peak, the stronger one corresponding to the Rayleigh wave, the weaker one to a resonance below the LA bulk edge (14). The intensity and the downward shift of this resonance with respect to the LA edge are anomalously large, in evident disagreement with ideal surface calculations (23) (fig. 5: the LA edge is X along $\langle 11\bar{2}\rangle$ and Y along $\langle 110\rangle$).

Both anomalies have been reproduced by the Modena group (24) in a slab calculation with a suitable parametrization of bulk dynamics and a softening of in-plane surface force constants. The microscopic origin of such softening may well be the phonon-induced d-s hybridization of the electronic states, i.e. the large quadrupolar deformability of silver ions (22).

The case of Pt(111) seems to be even more interesting (15,25) because the TA bulk phonon branch presents a small, sharp Kohn anomaly due to the electron-phonon interaction at the Fermi surface. The dispersion of the

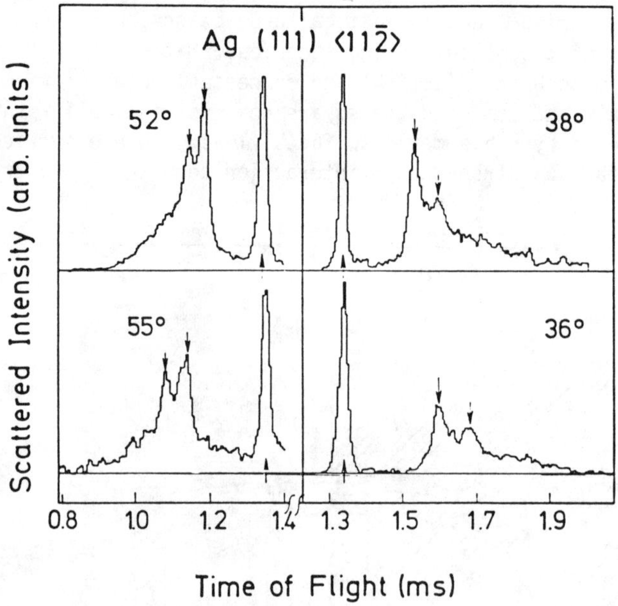

FIGURE 4. TOF spectra of He scattered from Ag(111) along ⟨11$\bar{2}$⟩ at four different incidence angles. Downward arrows: surface modes; upward arrows: elastic diffuse peak (from ref. 14).

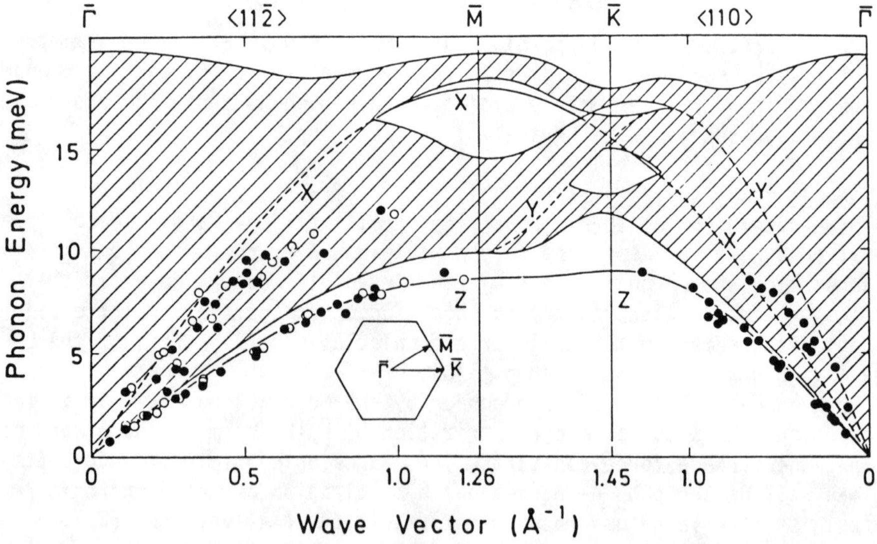

FIGURE 5. Dispersion curves of surface phonons in Ag(111) along the symmetry directions (from ref. 14) Experimental points: He scattering data. Theoretical: Green's function generating coefficients method (from ref. 23).

Rayleigh mode as determined by He scattering does show an anomaly as well, but at a slightly different wavevector with respect to that of the bulk anomaly. This fact seems to be a signature of some surface-localized electronic state at the Fermi surface, whose Fermi wavevector is different from that of the bulk electronic bands.

2.3 Silicon (111) surface

The dispersion of surface phonons in the (111) surface of silicon, as experimentally determined by Toennies and coworkers (16), discloses interesting features of the surface reconstruction. Figure 6 represents the measured surface phonons along <$11\bar{2}$> in the clean Si(111) 2x1 reconstructed surface (b) as compared to those in Si(111) unreconstructed surface stabilized by a 1x1 atomic hydrogen monolayer (a).

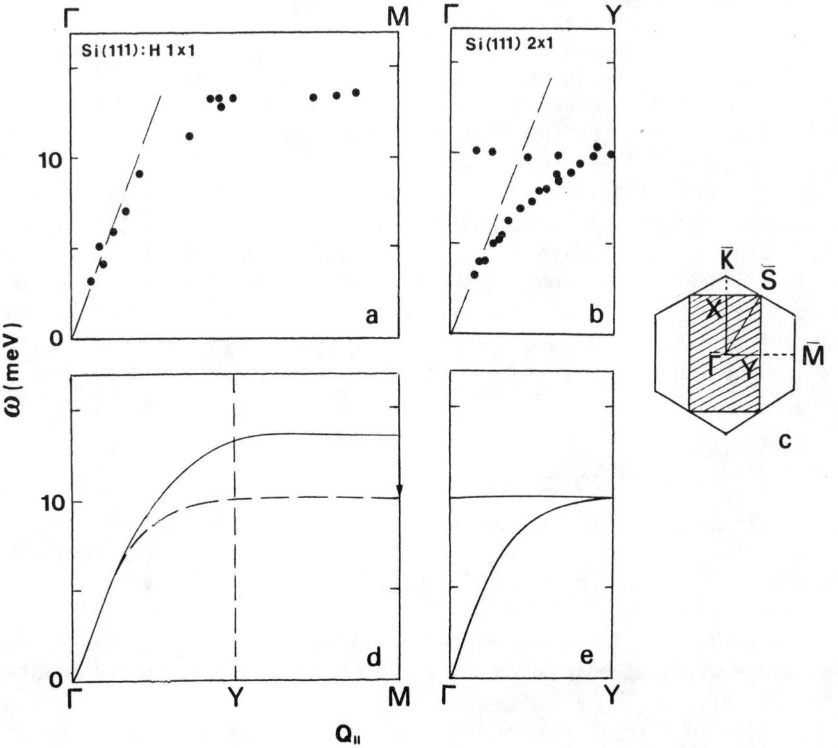

FIGURE 6. (a) Experimental surface-phonon dispersion curve of Si(111):H 1x1 (ref. 16) and (b) of the clean Si(111) 2x1 reconstructed surface (ref. 16); (c) folding of the surface Brillouin zone for the 2x1 reconstruction.
(d) Schematic representation of the Rayleigh wave dispersion curve for the hydrogen-covered Si(111) (full line) and ideal unreconstructed Si(111) surface (broken line), and (e) corresponding 2x1 folding.

The Rayleigh wave branches have the same slope in both cases, due to the large penetration which makes the Rayleigh wave insensitive to the surface structure in the long wave limit. Note that the extension of the surface Brillouin zone of the reconstructed surface is halved with respect to that of the unreconstructed H-covered surface (fig. 6.c), and this implies a folding of the surface phonon branches in both $\langle 11\bar{2} \rangle$ and $\langle 1\bar{1}0 \rangle$ directions. This provides a qualitative explanation of the additional optical branch observed in the Si(111) 2x1 case at about 10 meV in both directions. The location of this branch, corresponding to the rocking of the saturated dangling bond, can be explained as follows. In the ideal clean Si(111) surface there is an attraction between each dangling bond and the six surrounding Si atoms at the surface. Since surface nearest neighbor atoms are next nearest neighbors in the bulk, such attraction yields a dip in the Rayleigh branch at the zone boundary (broken line in fig. 6.d). If we assume that the bond saturation with hydrogen cancels the attraction, the size of the softening with respect to the H-covered branch is a measure of such attraction. The 2x1 reconstruction gives a folding of this branch, as schematically shown in fig 6.e. In Toennies at al data (16) a gap is found between the Rayleigh mode and the 10 meV flat branch at the symmetry points X and S. This gap width is comparable to the stiffening produced by the dangling bond saturation from hydrogen 1x1 coverage.

The nature of the 10 meV optical branch (which is perfectly flat if we ignore the anticrossing with the Rayleigh wave) has been further clarified by the recent optical absorption experiment of Selci, Chiarotti et al (26). These authors have studied the temperature dependence of the linewidth for the transition between bonding and antibonding states of the saturated dangling bond in Si(111) 2x1. The effective phonon frequency involved in that temperature dependence is as low as 10 meV, and can be identified with the rocking mode just from symmetry considerations on the electron-phonon coupling.

2.4. $TaSe_2$(001) and GaSe(001)

The specific role of surface electronic states can be even more conveniently put in evidence in layered crystals. In principle in strongly anisotropic crystals little difference would be expected between surface and bulk properties. Thus, if strongly localized surface modes are observed, they should amount exclusively to intrinsic, local deviations from the ideal (geometrical) surface, that is to the appearance of surface electronic states, to the change of electronic susceptibility at the surface, to the elastic relaxation, etc.

A striking example that we observed recently in Göttingen (18) is the charge-density-wave system $TaSe_2$ (in the form 2H). $TaSe_2$ is a layered compound showing a CDW phase transition at 118 K. Below that temperature a CDW with a period three times larger than the crystallographic cell along the layers appears and is easily detected also at the surface layer by means of He diffraction. The electron-phonon mechanism driving the CDW instability has been associated with the deep anomaly found by neutron scattering data at about 2/3 of the reduced Brillouin zone in the bulk LA branch (broken line in fig. 7) (27). Due to the weak metallic character of $TaSe_2$, the anomaly has been classified of the Kohn type (27). The dispersion curves of the Rayleigh wave measured at different temperatures above and close to the CDW transition temperature (fig. 7) show a surprising shift of the anomaly back to 1/2 of the Brillouin zone and a negative temperature effect (frequency decrease with temperature decrease)..

Despite of the fact that little difference would be expected between surface and bulk properties in such a strongly anisotropic crystal, the anomaly shift is indicative of some modification of the Fermi surface in the surface region. Another possible mechanism is a large change in the average range of the electron-phonon interaction at the surface as compared to that in the bulk, as illustrated by a theoretical calculation made for the (001) surface of the refractory superconductor TiN (28). The position of the surface anomaly is not in contradiction with the periodicity of the bulk CDW as observed at the surface layer by He scattering; it simply tells us that a competitive electron-phonon interaction exists at the surface, which may in principle yield a surface CDW transition at a lower temperature, with different periodicity.

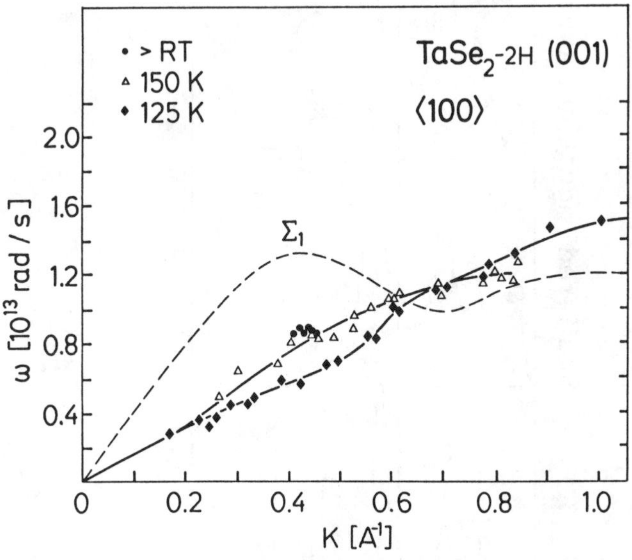

FIGURE 7. Experimental dispersion curve of the Rayleigh wave on the (001) surface of 2H-$TaSe_2$ along the (100) direction, measured by means of He scattering at different surface temperatures. The Rayleigh wave anomaly is shifted with respect to the position found in the bulk LA branch (broken line). (From ref.18)

Recently we have accomplished a He-scattering study of the layered semiconductor GaSe (17). TOF spectra of He scattering from the (001) surface show fairly large intensities from surface optical modes S_2 and S_3 (fig.8) and from the surface longitudinal acoustic mode S_8. To such a large intensities corresponds a strong localization: the corresponding dispersion curves fall much below their bulk bands. This is seen in fig. 9, where the surface projected bulk bands, as derived from Jandl et al calculation (25), are also plotted for reference. In a layer compound such a big softening of the surface branch would not be expected. Again a change in the electronic structure occurring at the surface layer has to be considered in order to explain such anomalous behavior.

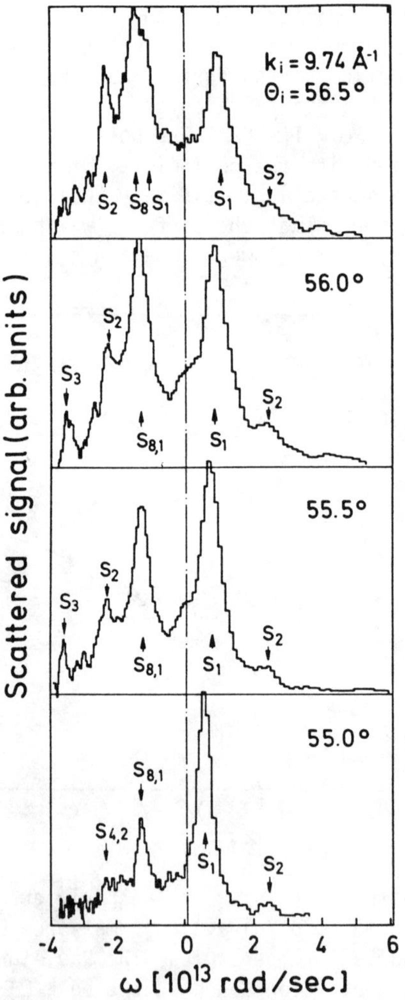

FIGURE 8. TOF spectra of He scattering from the (001) surface of GaSe along <100> direction for different kinematical parameters of the incident beam. S_1: Rayleigh wave; S_2 and S_3: optical modes; S_8: surface LA mode (ref. 17).

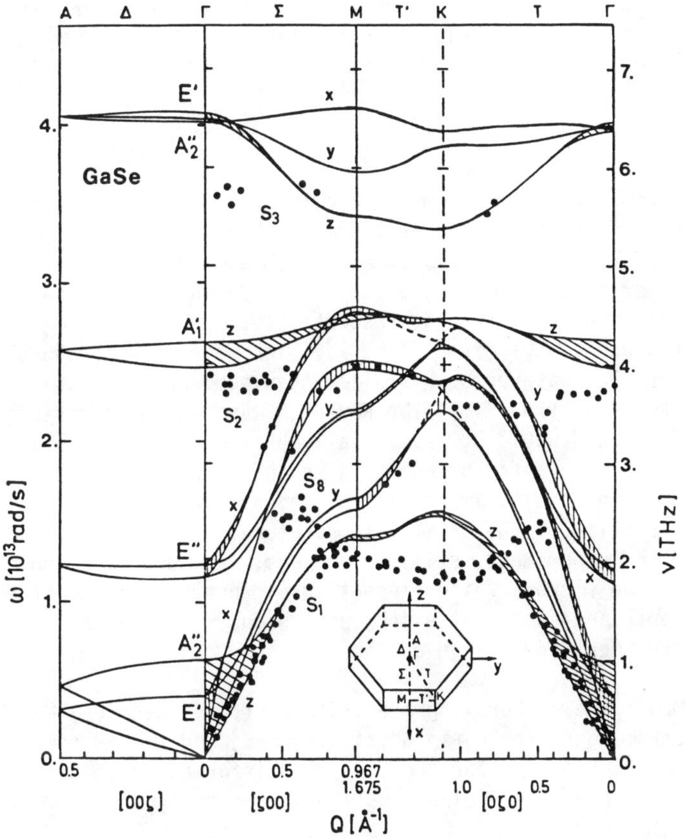

FIGURE 9. Dispersion curves of GaSe (from Jandl et al, ref. 25). Hatched regions represent the surface projected bulk bands for sagittal polarization. Black points: He scattering data.

We note that there is some similarity between the surface LA mode S_8 in GaSe and the bulk LA anomalous branch in $TaSe_2$ in the sense that both exhibit a strong softening towards the zone edge and cut the Rayleigh branch. One could speculate that the topmost layers of GaSe have a metallic behavior due to some half-filled surface states at the Fermi level.

3. MONITORING SURFACE MELTING

The question whether melting of solids does start, under certain conditions, from the surface has been theoretically considered by Tosatti and Pietronero (19), and adfirmatively answered. Various authors (20) confirmed that conclusion by means of a molecular dynamics simulation for a Lennard-Jones slab. Experimental evidence of surface melting has been recently collected in Pb(110) by van der Veen and coworkers (20), who have exploited the shadowing and blocking effects in the ion Rutherford backscattering. Surface melting appears to be an intrinsic phenomenon, no sizeable effect being attributed to impurities segregated at the surface. The depth d of the surface melt appears to be related to the temperature T by the logarithmic relationship (20)

$$d = \zeta \ln[T_o/(T-T_m)] ,$$

where T_m is the bulk melting temperature, ζ and T_o are constants (for Pb(110) $T_o = 55^\circ K$ and $\zeta = 6.2$ A (20)). For T some 50 K below T_m, the melt depth is of the order of a few atomic distances. In this case the RW penetration can be either deeper or smaller than d, depending on whether its wavelength is large or small with respect to d, respectively.

As a consequence at a given T the Rayleigh wave peak is sharply defined only for a parallel wavevector K smaller than some critical value $K_c(T)$, whereas for K larger than $K_c(T)$ the Rayleigh wave would be restricted within the liquid layer and would be dramatically broadened. Since $K_c(T)$ is proportional to d^{-1} through a factor which is known in principle, the logarithmic relationship could be experimentally checked by measuring the critical wavevector as a function of the crystal temperature.

This kind of experiment, still in project, was worth mentioning here in order to give the flavor of the present, great potentialities of molecular beam experiments and atom-scattering spectroscopy in the challenging field of surface treatment and design on the microscopic scale.

ACKNOWLEDGEMENT: The author is grateful to Prof.J.P.Toennies, Max-Planck-Institut für Strömungsforschung in Göttingen, FRG, and to his coworkers, Drs. C. Wöll and G. Brusdeylins for the continued information on He scattering experiments and several useful discussions.

REFERENCES

1. Logan RM: Solid State Surface Science, Green M ed., vol. 3. New York: Dekker, 1973; p. 1.
2. Benedek G and Valbusa U(eds), Dynamics of Gas-Surface Interaction. Berlin Heidelberg: Springer, 1982.
3. Hoinkes H: Rev. Modern Physics 52, 933 (1980).
4. Goodman FO: J. Phys. Chem. 84, 1431 (1980).
5. Benedek G: in Nonequilibrium Phonon Dynamics, Bron WE (ed). New York and London: Plenum, 1985; p. 601.
6. Toennies JP et al: to be published
7. Cabrera N, Celli V and Manson R: Phys. Rev. Letters 22, 346 (1969); also Manson R and Celli V: Surface Sci. 24, 495 (1971).
8. Cantini P and Tatarek R: Phys. Rev. B23, 3030 (1981); Boato G, Cantini P, Guidi C, Tatarek R and Felcher GP: Phys. Rev. B20, 3975 (1979).
9. Brusdeylins G, Doak RB and Toennies JP: Phys. Rev. Letters 44, 1417 (1980) and 46, 437 (1981); Phys. Rev. B27, 3662 (1983).
10. Benedek G, Toennies JP and Doak RB: Phys. Rev. B28, 7277 (1983). Brusdeylins G, Rechsteiner R, Skofronick JG, Toennies JP, Benedek G and Miglio L: Phys. Rev. Letters 54, 466 (1985).
11. Harten U, Toennies JP and Wöll Ch: Faraday Disc. Chem. Soc. 80 (1985).

12. Benedek G: Physica 127B, 59 (1984).
13. Feuerbacher B and Willis RF: Phys. Rev. Letters 47, 526 (1981).
14. Doak RB, Harten U and Toennies JP: Phys. Rev. Letters 51, 578 (1983).
15. Harten U, Toennies JP, Wöll Ch and Zhang G: Phys. Rev. Letters 55, 2308 (1985). See also Kern K, David R, Palmer RL, Comsa G and Rahman TS: Phys. Rev. B33, 4334 (1986).
16. Toennies JP: in Proc. Int. Conference Phonon Physics, Kollàr J, Kroò N, Menyhard N and Siklòs T (eds). Singapore: World Scientific, 1985; p. 575 and to be published.
17. Brusdeylins G, Rechsteiner R, Skofronick JG, Toennies JP, Benedek G and Miglio L: Phys. Rev. B34, July 15 (1986).
18. Brusdeylins G, Heimlich C, Skofronick JG, Toennies JP, Benedek G and Miglio L: Proc. 6th Int. Conf. Solid Surfaces, Baltimore, Oct. 1986 (to be published).
19. Pietronero L and Tosatti E: Solid State Comm. 32, 255 (1979).
20. Frenken JWM and van der Veen JF: Phys. Rev. Letters 54, 134 (1985). A review containing also references to Monte Carlo simulations of surface melting has been given by van der Veen JF at ECOSS 8, Jülich 1986, Europhysics Abstracts 10B, 7 (1986) and to be published in Surface Sci.
21. Benedek G, Brivio GP, Miglio L and Velasco VR: Phys. Rev. B26, 487 (1982).
22. Jayanthi CS, Bilz H and Kress W: in Phonon Physics, Kollàr J, Kroò N, Menyhàrd N and Siklòs T (eds). Singapore: World Scientific, 1985; p. 630 and to be published.
23. Armand G: Solid State Comm. 48, 261 (1983).
24. Bortolani V, Franchini A, Nizzoli F and Santoro G: Phys. Rev. Letters 52, 429 (1984).
25. Neuhaus D, Joo F and Feuerbacher B: in Phonon Physics, Kollàr J, Kroò N, Menyhàrd N and Siklòs T (eds). Singapore: World Scientific, 1985; p.636.
26. Selci S, Ciccacci F, Chiarotti G, Chiaradia P, Habib Z, Felici AC and Goletti C: Vuoto 16, 17 (1986) and to appear in Phys. Rev. Letters.
27. Moncton DE, Axe JD and Di Salvo FJ: Phys. Rev B16, 801 (1977).
28. Benedek G, Miura M, Kress W and Bilz H: Phys. Rev. Letters 52, 1907 (1984).
29. Jandl S, Brebner JL and Powell BM, Phys. Rev. B13, 686 (1976).

MELTING AND SURFACES

G. FRITSCH
Institut für Physik, Universität der Bundeswehr München, D-8014 Neubiberg FRG

1. INTRODUCTION

Condensed matter can exist in various states, characterized by the difference in geometrical structure by the mobility of the constituents and by the binding forces. Some examples are the glassy, the liquid, the crystalline or the liquid crystal states. Normally, these states are separated by lines in the relevant phase diagram, i.e. by phase transitions. I would like to consider the melting transition, defined as the line between the crystalline and the liquid states. This phase transition is of first order i.e. there are jumps in certain thermodynamic properties such as volume and entropy. Figure 1 illustrates the situation in case of a pure substance (a) and of an binary alloy (b). In the latter case a coexistence region of solid and liquid exists due to a different solubility of the two components in the solid and liquid state, respectively.

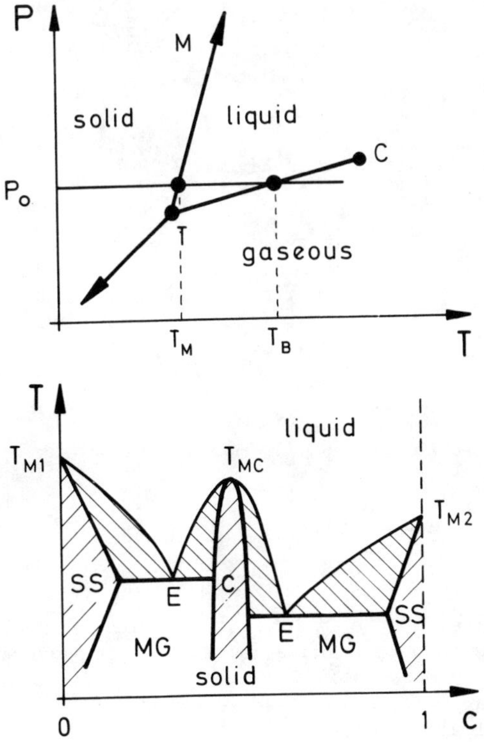

FIGURE 1. a) The melting transition is indicated by M. T: Triple point, C: Critical point, P_o: Ambient Pressure, T_M: Melting- and T_B: Boiling-Temperature. b) Melting Transition in an alloy. T_{M1}, T_{M2}: Melting points of the pure substances, T_{MC}: Melting point of a compound, E: eutectic points, C: compound region, MG: miscibility gaps and SS: region of solid solutions.

In the following I would like to clarify the thermodynamics of the melting transition and to discuss some models in order to calculate the line termed M in Figure 1 a. Then, the role of surfaces in the melting transition is analysed.

2. THERMODYNAMIC MELTING

The thermodynamic state of condensed matter can be described by the Gibbs free energy G per mole substance, depending on the variables pressure P and temperature T: $G(P,T) = H(P,T) - T \cdot S(P,T)$, where $H(P,T)$ denotes the enthalpy and $S(P,T)$ the entropy. Considering a two phase situation the entropy maximum principle yields the equilibrium conditions:

$$P_s = P_l = P_M \quad T_s = T_l = T_M \quad G_s(P_s,T_s) = G_l(P_l,T_l). \tag{1}$$

The indices refer to s: solid, l: liquid and M: melting. These three conditions define the line denoted by M in Figure 1 a:

$$P_M = P_M\,(T_M). \tag{2}$$

The derivatives of the Gibbs free energies are

$$S(P,T) = -\frac{\partial G(P,T)}{\partial T} > 0 \qquad V(P,T) = \frac{\partial G(P,T)}{\partial P} > 0 \tag{3}$$

as well as

$$\varkappa_T(P,T) = -V(P,T)^{-1} \cdot \frac{\partial^2 G(P,T)}{\partial P^2} > 0$$

$$c_p(P,T) = -T \cdot \frac{\partial^2 G(P,T)}{\partial T^2} > 0 \tag{4}$$

$$\alpha_p(P,T) = V(P,T)^{-1} \cdot \frac{\partial^2 G(P,T)}{\partial P \partial T} > 0.$$

The first order melting transition is pictured in Figure 2.

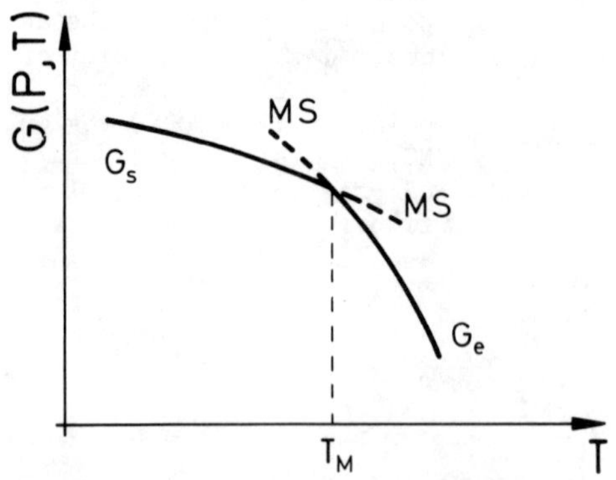

FIGURE 2. Gibbs free energy diagram of the melting transition. G_s, G_l: Gibbs free energies of the solid and liquid states, MS: regions of metastability.

Hence, we see that the derivatives of the Gibbs free energies of the solid and liquid states are discontinuous, we have jumps in volume and entropy as well as in the isothermal compressibility $\varkappa_T$, the isobaric specific heat capacity c_p and the isobaric expansion coefficient α_p.

In a strictly thermodynamic calculation the regions termed MS (Metastability) should not occur, since a indefinite time average is taken. However in reality they can be observed since the proper time average cannot be made. On the contrary, time enters, and states with finite lifetimes will be seen. If the metastable states are terminated by an instability point (spinoidal point) is not known. Before saying more on this subject, let us complete the simple thermodynamic picture.

In differentiating equs. (1), with respect to T_M, we arrive at the Clausius Clapeyron equ.:

$$\frac{dP_M(T_M)}{dT_M} = \frac{S_l(T_M) - S_s(T_M)}{V_l(T_M) - V_s(T_M)} = \frac{\Delta S(T_M)}{\Delta V(T_M)}. \tag{5}$$

Since the entropy of the liquid state is always higher than that of the solid one, the sign of dP_M/dT_M is determined by the difference of the molare volumes. It can be positive or negative (H_2O, Bi etc.) depending on the crystalline phase. From equ. 1 and the definition of G(P,T), we have:

$$\Delta H_M(T_M) = T_M \cdot \Delta S(T_M) = L_M, \tag{6}$$

where L_M is the latent heat of melting and this yields:

$$\frac{dP_M(T_M)}{dT_M} = \frac{L_M}{T_M \Delta V} . \tag{7}$$

From the above discussion, it is clear, that one must calculate the Gibbs free energies G(P,T) of the solid and of the liquid state in order to get the melting parameters $P_M = P_M(T_M)$, ΔV, L_M etc.. This task may be performed by trying to find a single closed expression for both $G_l(P,T)$ and $G_s(P,T)$ or by starting from approximate models for each phase separately.

Such a calculation for sodium has first been performed successfully by Stroud and Ashcroft [1]. These authors calculated the Gibbs free energies of the solid and liquid phases and used equs. 1. In the solid the internal energy of the static lattice, (electron gas, Madelung energy, band structure terms) and the free energy of the phonons (self consistent phonon theory) were added, using the pseudopotential formalism. For the liquid, the static structure factor describing the ionic positions must be known. This quantity was taken from the hard sphere liquid, derived from the Percus-Yevick approximation. The only parameter involved is the hard sphere packing fraction, which was calculated by minimizing the free energy. The result, up to 40 kbar, is shown in Figure 3.

This work as discussed later on by Pelissier [2] and Angelie and Pelissier [3], concerning the various approximation used. Their result is in better agreement with experiment (Figure 3).

Similar results were reported using molecular dynamics calculations for the Gibbs free energies of the solid and liquid states by Holican et al [4]. The points these authors obtained are also shown in Figure 3. Hence, one may say that the melting of simple substance is described reasonably well within this approach.

3. DYNAMICS OF MELTING

However, there remains the problem of metastable states (MS). These can be calculated from the theories mentioned above and from computer simulations, because those models are not exact in a strict thermodynamic sense. The Gibbs free energies as well as that portion of the equations of state

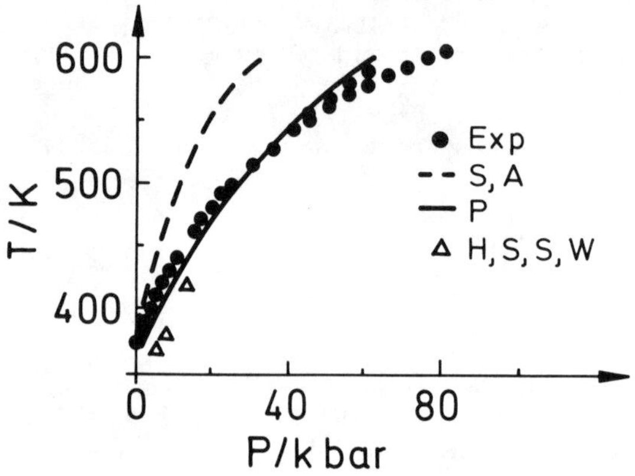

FIGURE 3. The function $T_M = T_M(P_M)$ as calculated from first principles, once the electron-ion pseudopotential is assumed to be known. S,A: Ref. 1, P: Ref. 2 and 3, H,S,S,W: Ref. 4 and Exp: Experimental values.

refering to the MS are simple continuations without any singularities. The main problem lies in the time dependence of these states. Undercooling can be performed most easily in many substances [5], particularly in Ga, however, overheating a crystalline solid is extremely difficult. Normally, theories of nucleation and growth are used to describe the situation. Homogeneous and heterogeneous nucleation provide seeds for the new phase to grow. The probability of forming a nucleus depends on the driving force, i.e. on the magnitude of undercooling or overheating and on the mechanism of seed formation. But there exists a second point of view, which may be able to explain the difficulty of overheating a solid. In this case it is assumed that the crystalline solid shows some instability which destroys the crystalline order and, hence, the stability against low frequency shear. In the picture of Figure 2, we may propose that the regions lablled MS are asymmetric. Whereas G_l extends far below T_M, G_s ends at an instability point at T_M or close above T_M. At that point the solid decays automatically into the liquid. This assumption has two consequences: First precursors of this instability mechanism should possibly occur for $T < T_M$ and the liquid structure must be formed and described using the mechanism which destroys the solid. I would like to discuss some of these models shortly in the next chapter.

IV. INSTABILITY THEORIES

The solid can become unstable by the introduction of defects such as vacancies or dislocations or by the softening of certain vibrations (phonons). The important point in the defect models is the defect-defect interaction, reducing the free energy of the whole system. At a certain temperature this free energy is zero and a catastrophic increase of the defect concentration will take place. The best known model is the one of Kuhlmann-Wilsdorf [6]. This author considers the free energy of dislocations, being the sum of a core energy and of a long range strain field energy. It is shown, that at a certain temperature the free energy of the dislocation cores becomes negative. Thus, the crystal is filled with dislocation cores to its capacity. In that case the strain field energy preventing dislocations to occur in thermal equilibrium can be neglected. Hence, the liquid is viewed as a solid filled up with dislocation cores. It was shown later on, that the static structure factor of the liquid can successfully be described using these assumptions [7]. Such a model may explain various melting parameters and correlations (Lindemann-, Richards-Laws etc.) correctly. It is, however, difficult to derive the correct dislocation core parameters. Cotterill and collaborators [8] have shown in a molecular dynamics computer simulation how the dynamics of such a transition could operate. Close to the melting point they find dislocation pairs (partial dislocation-dipols). These pairs have a strain field energy which is reduced strongly in comparison to the one of a dislocation line. Hence, they can occur in thermal equilibrium. The unbinding of such pairs may then produce the dislocation cores, envisaged by Kuhlmann-Wilsdorf.

Instabilities in crystal vibrations were discussed first by Born [9]. He thought that the shear stability of the crystal must be broken down by a transverse mode with a propagation velocity tending towards zero when approaching T_M. Other models use the fact that the coupling constant for transverse modes is reduced by the increasing amplitude of such vibrations [10]. One example is the self consistent phonon theory [see Ref. [1]]. However, all these proposals discussed above are not supported by experiment [11].

Instability theories should show precursors in some physical properties already below T_M, i.e. in the bulk. However, up to now clear cut experimental evidence for a bulk instability is scarce. Wenzl et al [12] have shown by anomalous X-Ray transmission in Gallium that no precursors are present below T_M. Hence, it is believed generally that if instabilities of the crystal occur in the bulk, they are decoupled from the thermodynamic melting transition. They would appear far above the thermodynamic melting point. In this sense the success of the Kuhlmann-Wilsdorf model is an indication that the assumption describing the difference between the solid and the liquid state are not to far away from reality.

However, there is another source of precursors that must be present. These are the heterophase-fluctuations. There exists a certain probability that a liquid nucleus will show up in the bulk already below T_M, since the difference in the Gibbs free energies will tend towards zero when approaching T_M. In order to discuss this probability one has to add the interface energy G_{IF} and the strain energy G_{ST} to the difference in the Gibbs free energies:

$$\Delta G_{total}(T) = G_l(T) - G_s(T) + G_{IF} + G_{ST} \approx \approx L_M \cdot (T_M - T)/T_M + G_{IF} + G_{ST}. \tag{8}$$

Ivlev et al [13] have analysed the situation, introducing explicit expressions for G_{IF} as well as G_{ST}. These authors were able to show that due to G_{IF} and G_{ST} the thermodynamic probability for heterophase fluctuations to take place is extremely small.

In this situation the only remaining possibility to explain the dynamic asymmetry in melting is the effect of the crystal surface. This will be discussed next.

5. THE ROLE OF THE CRYSTAL SURFACE

Each real crystal is bounded by a surface. Surfaces are characterized by surface energies, which we denote by $\gamma_{\alpha\beta}$, where α,β may take the three values l: liquid, s: solid and v: vapour. The surface energy γ_{SV} depends on the crystal orientation, it is lowest for low index surface planes [14]. A plot of the magnitude of γ versus orientation is called the γ-plot(Figure 4). The terrace-ledge-kink model for a crystal surface which is reproduced in Figure 5, describes a surface misoriented by an angle θ with respect

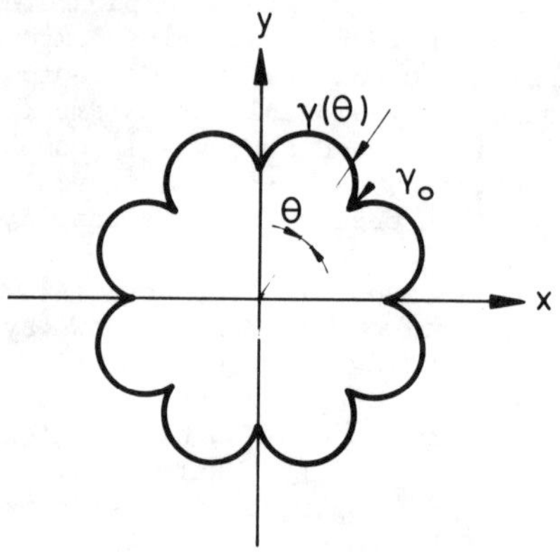

FIGURE 4. The γ-plot for the x, y-plane of a certain crystal.

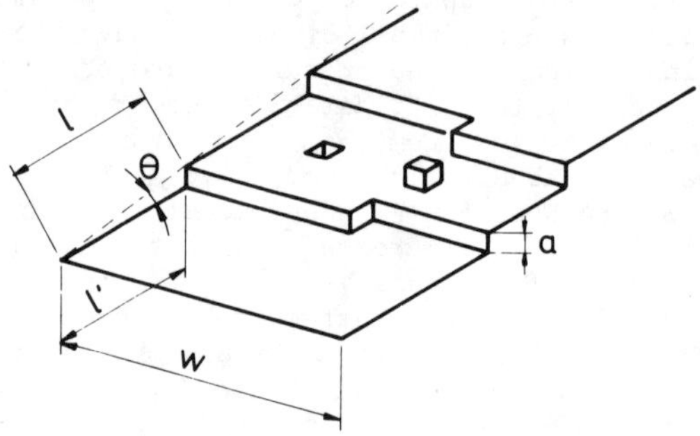

FIGURE 5. The terrace-ledge-kink model of a surface misoriented by an angle θ with respect to a low index plane. The shorter are the plane parts, the larger is the angle θ.

to a low index plane. It contains ledges at a distance l', where the plane parts are the low index plane and the additional kinks may allow for the second misorientation angle. The surface energy $\gamma(\theta)$ can be derived from the surface energy γ_o of the low index plane and from the ledge energy β per unit length as follows (Figure 5):

$$\gamma(\theta) \cdot l \cdot w = \gamma_o \cdot l' \cdot w + \beta \cdot w, \qquad (9)$$

which yields

$$\gamma(\theta) = \gamma_o \cos /\theta/ + (\beta/a) \cdot \sin /\theta/.$$

The last equation explains the cusp like structure in the γ-plot (Figure 4).

It can be shown, that at low temperatures, the flat parts of such a surface are really flat at an atomic scale in thermal equilibrium. We observe a singular surface [14]. At higher temperatures vacancy-adatom pairs may occur. However, the most important fact for our purpose is the possibility that at a certain temperature the surface may change from a singular one to a rough one. This transition is called the roughening transition [15]. Sometimes this effect it is also termed surface melting. The difference between both seems to lie in the mobility of the surface atoms. If it is low one may speak of roughness in the other case the surface appears to be more liquid like in structure [16,17].

The situation is actually much more complicated because surfaces are normally relaxed and reconstructed (Figure 6). In other metals, in addition a coverage by a layer of other substances (oxydes, etc.) can be present.

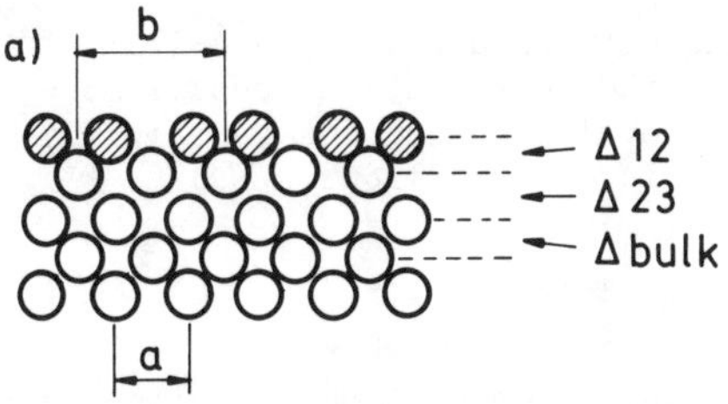

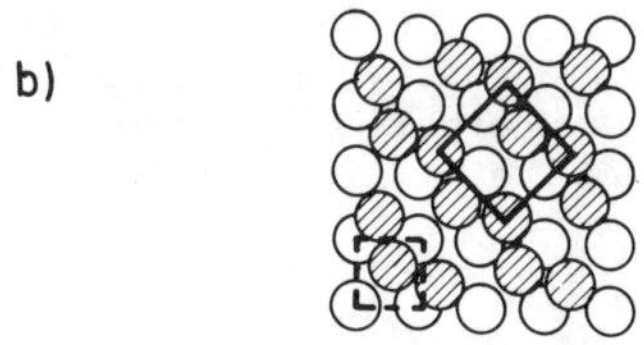

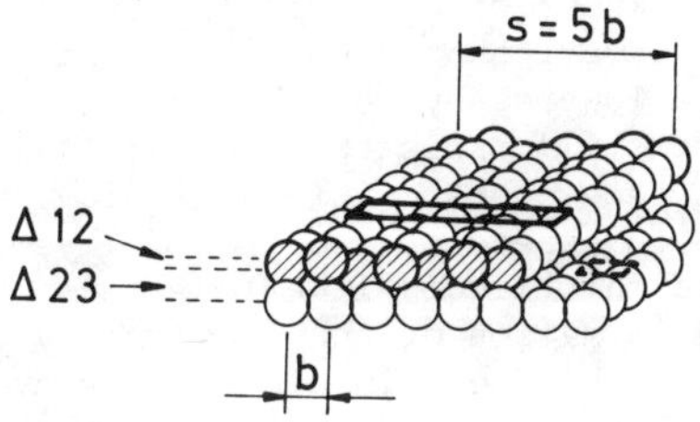

FIGURE 6. a) Relaxed and reconstructed surface: a, Δ bulk: bulk lattice constant and plane distance, respectively; b, Δ12, Δ23: relaxed and reconstructed surface: lattice constant and plane distances, respectively. For example Al(111): Δ12 = 0.9%, Δ23 = 0, Al(110): Δ12 = 8.6%, Δ23 = 5%. b) upper part: reconstructed Tungsten (100)-surface (open circles indicate bulk) and lower part: reconstructed and relaxed Iridium (100)-surface (b: bulk lattice constant).

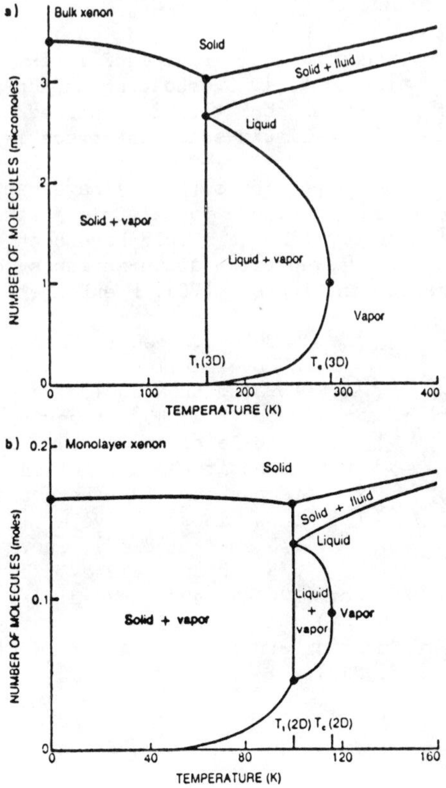

FIGURE 7. Phase diagram of Xenon. a) Bulk Xenon (3D), T_t: Triplepoint-, T_c: critical point-temperature. b) Xenon on grafoil (2D): The quasi-two dimensional Xenon behaves similarly to the bulk, however, the temperature scale is shifted downwards. (After Scientific American).

These layers may show phase transitions by their own (see Figure 7).

Finally, there is the problem of wetting between a crystal and its own melt. Stransky [18] has argued, that if the melt wetts a crystal surface,it's energetically more farourable to develop a liquid layer already below T_M. The wetting condition can be formulated as follows:

$$\gamma_{sv} = \gamma_{sl} + \gamma_{lv} + \Delta\gamma. \tag{10}$$

with $\Delta\gamma > 0$.

If one assumes, that γ_{sl} doesn't depend on the crystal orientation, then $\Delta\gamma$ depends solely on orientation by γ_{sv}. However, it is easy to prove that the thickness of a liquid layer relying on the reduction of the Gibbs free energy by $\Delta\gamma$, should be infinitesimal small, since the free energy can always be lowered by reducing the thickness. Hence, additional terms must be included [17]. Kristensen et al [11] have made the proposal, that the transition from a solid into a liquid-like surface is continuous and depends on the thickness of the liquid layer. This idea is physically appealing since a thin liquid-like layer must be different from the bulk liquid. Then the minimization of the free energy yields a liquid layer already below T_M which thickness is dictated by $\Delta\gamma$ and the temperature. This model was modified [19] and extended also to cover alloys [20] by Fritsch et al.. Hence, the assumption is, that the surface shows a lower melting point as the bulk, depending on the surface orientation. Let us discuss now the experimental evidence for such a behaviour.

6. EXPERIMENTS AND SURFACE MELTING

If these proposals are correct, overheating in thermal equilibrium should only be possible in the bulk material [21]. Overheating in non-equilibrium situations may occur if i) nucleus formation is hindered at the surface or if ii) the propagation of the liquid-solid interface is slowed down by viscosity effects.

Early observations of surfaces by LEED-experiments [22,23] have partly failed to show surface melting effects. However, due to multiple scattering effects it is by no means clear if these experiments should have been able to detect the effect looked for [16]. Lead evaporated at Cu-surfaces does exhibit surface melting as is documented in Figure 8 [22]. Frenken et al

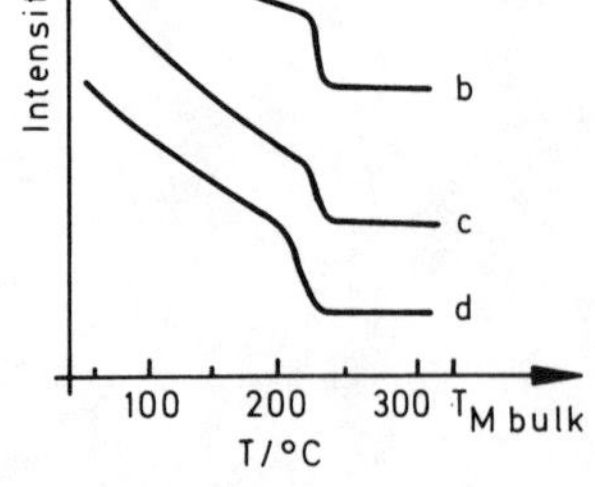

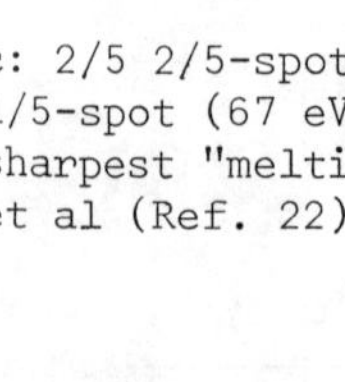

FIGURE 8. Results from a LEED-experiment on Pb, evaporated on a Cu(100)-surface: a) c (5x1) 45° structure of a Pb-layer on Cu. b) LEED-intensity as a function of temperature for the structure shown in a): curve a: 1/5 1/5-spot (38 eV), curve b: 3/5 3/5-spot (65 eV), curve c: 2/5 2/5-spot (124 eV) and curve d: 1/5 1/5-spot (67 eV). This structure shows the sharpest "melting"-transition [After Henrion et al (Ref. 22)].

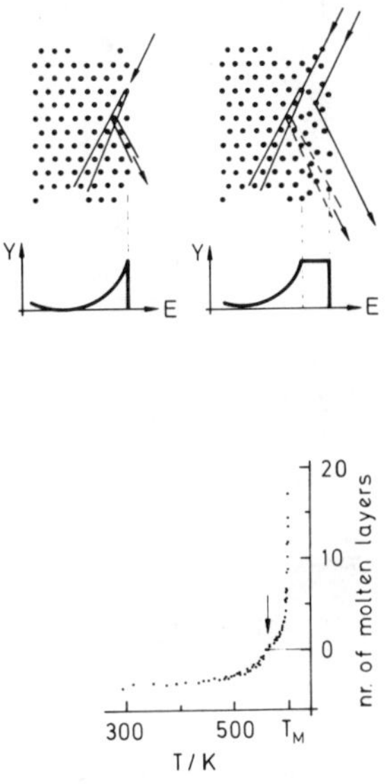

FIGURE 9. Surface melting of Pb after Frenken and van der Veen. a) Schematic drawing of ion shadowing in a crystalline lattice (left) and when a quasi-liquid layer is present (right). The ion beam enters from above. The $\gamma(E)$-plots indicate the backscattering yield in both cases as a function of the beam energy. b) Number of "molten" layers as a function of temperature. The bulk melting point is indicated by T_M.

FIGURE 10. Relative isobaric specific heat capacity of crystalline Gallium when approaching the bulk melting point (29,78°C).

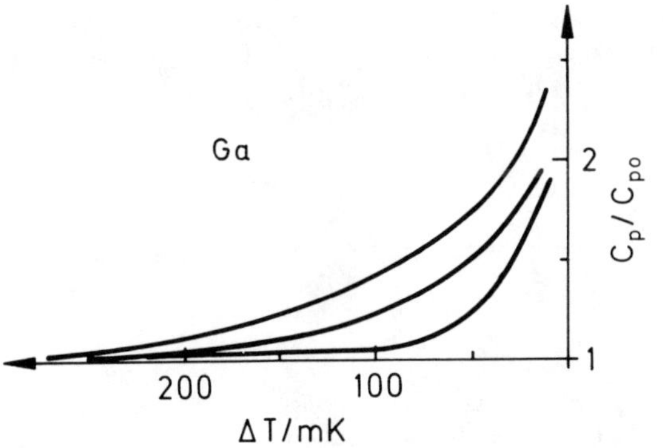

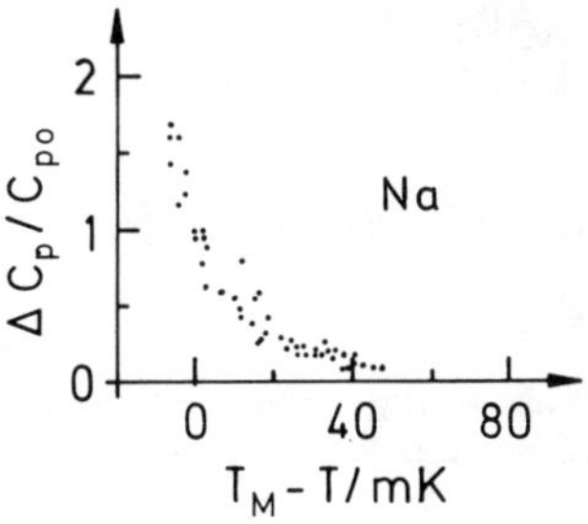

FIGURE 11. Relative isobaric specific heat capacity of crystalline Sodium when approaching the bulk melting point (97, 82°C).

[24] have published results, derived from ion-shadowing and blocking measurements on Pb-crystals. They see effects which they interpret as melting of the < 110 > - surface of lead about 40 K below the bulk melting point of Pb (T_M = 600,7 K). According to their paper the layer thickness amounts to about 17 planes at $\Delta T \approx 0.2$ K below T_M. The results are shown in Figure 9. Fritsch et al [25,26] have determined the specific heat of thin Ga- and Na-samples close to the melting point. The results are shown in Figs. 10 and 11. It is clearly seen, that the specific heat capacity already increases below the bulk melting point. We interpret these findings as surface melting. The liquid-like layer thickness, calculated from this assumption is shown in Figure 12.

Finally, the surface diffusion coefficients of many metals tend to be

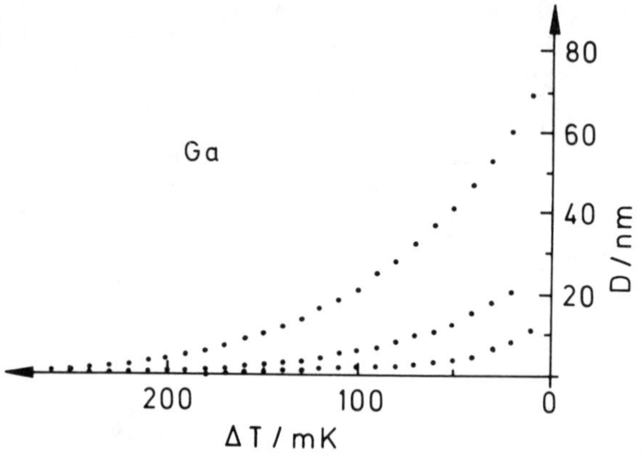

FIGURE 12. Quasi-liquid layer thickness calculated from experimental results for Gallium.

higher or as high as the liquid diffusion coefficient, when approaching T_M [27]. This behaviour may be also a hint to a more liquid-like structure of surfaces. Some results are summarized in Figure 13.

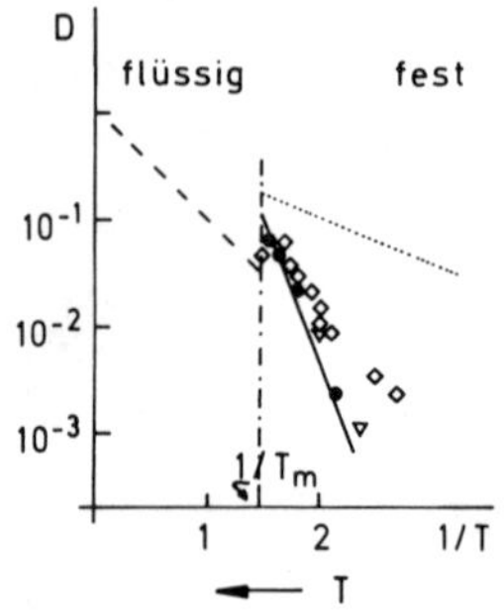

FIGURE 13. Surface Diffusion Coefficients D as a function of temperature. flüssig: liquid, fest: solid. T_M: bulk melting point. --- : diffusion in the liquid, ...: diffusion along dislocations and —— : fit to some data (After [27]).

These experimental results are supported by computer simulations and theoretical work. The molecular dynamics study of Broughton et al [16] was based on 1024 atoms arranged in low (001)-fcc-planes on a crystalline substrate. The atoms interact via a Lennard-Jones potential. These authors find extensive premelting of the upper four layers together with high diffusion coefficients. Some results are reproduced in Figure 14. A similar paper was published by Bushnell-Wye et al [28]. This time a six layer model was considered using the Monte Carlo technique. Finally, a study on surface self diffusion by Mruzik et al [29] applying molecular dynamics indicates also surface melting below T_M. The particles interact via a Lennard-Jones-potential. Each surface was initially constructed by stacking between 10 and 20 defect free rectangular atomic layers. The bottom layers were fixed in their equilibrium atomic positions. The results show extensive mixing of the surface layers with high fluidity. Some results are shown in Figure 15.

The theoretical aspects for surface melting rely on the 2-dim. theory of Kosterlitz and Thouless [30]. In such a system an instability against the unbinding of dislocation dipol-pairs exists. The interaction between the dipol pairs is described by a kind of "dielectric" constant, normalizing this interaction due to the presence of other dislocation dipols.

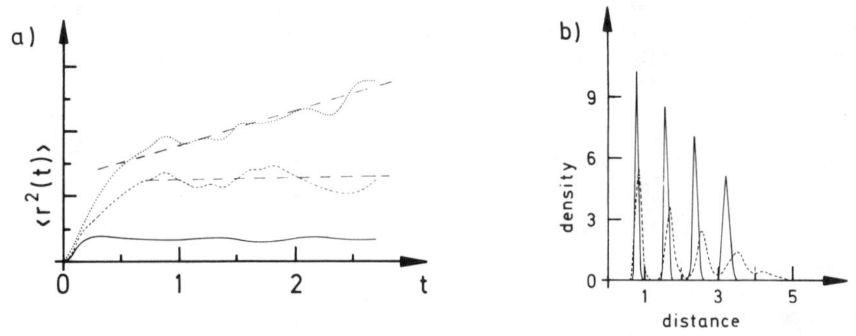

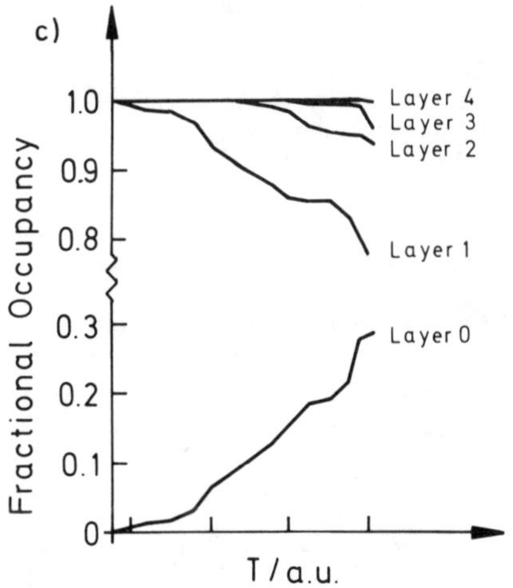

FIGURE 14. Results from a molecular dynamics computer simulation by Broughton et al (Ref. 16), a) Mean square displacement of surface atoms as a function of time. The parameter is the temperature. Horizontal lines are typical for a solid, positive slopes indicate the disordered state. b) Density of particles as a function of distance from the surface. —— : solid surface and ---: molten surface. Relaxation and an extra layer can be seen. c) Fractional occupancy of the surface layer. The extra layer is denoted by "0". The original layers of the solid surface are termed 1 to 4. Compare Fig.14b.

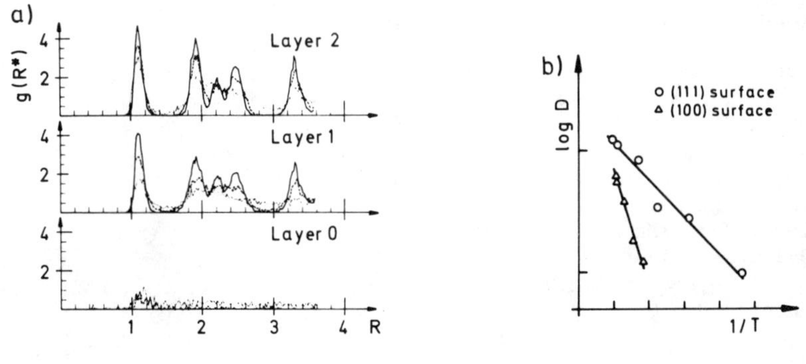

FIGURE 15. Results from a computer simulation of surface diffusion by Mruzik et al (Ref. 29). a) Radial distribution function g(R) for three surface layers. The one denoted by "0" is the upper layer. Three temperatures are shown: ——: $T^x = 0.1$, ---: $T^x = 0.4$ and ...: $T^x = 0.5$. The bulk melting point is at $T^x = 0.68$. b) Surface diffusion coefficients in an Arrhenius-plot for two surface planes.

This in turn facilitates the unbinding of other pairs, modifying the "dielectric function" and so on. This mechanism may be identified with a 2-dimensional melting transition. Whether this concept can also be used for the surface of three dimensional crystals is not known yet [11].

Other work in this direction was published by Pietronero et al [31]. Using an instability model for the thermal vibrations at the surface and by Tognato [32], trying to relate heterophase fluctuations at dislocations with the surface-melting effects. (Here, internal and external surfaces are considered).

7. CONCLUSION

There is now experimental, theoretical and computer-simulation generated evidence that surface melting will occur at certain surface planes. However, a clear theoretical model describing these effects and also their correlation with surface roughening is still missing. This latter problem is strongly related to the difficulty to characterize a thin quasi-liquid layer in structure. Whereas the viscosity criterion relies on a large variation of the atomic mobility between the solid and the liquid states a more stringent criterion is the lack of long range order in a liquid. However, at a surface we may have a liquid like structure showing no long range order parallel to the surface, but still structure perpendicular to it. This is seen particularly in the computer simulations. Also, a high mobility may be simulated by surface diffusion of clusters, partly involving jump distances larger than the atomic spacing. Hence the most wanted evidence for a liquid-like structure would be the measurement of the static structure factor of the quasi-liquid layer parallel as well as perpendicular to the surface.

REFERENCES.

1. Stroud D and Ashcroft N W: Phys. Rev. B 5/2, 371 (1972).
2. Pelissier J L: Physica 121A, 217 (1983).
3. Angelie A and Pelissier J L: Physica 121A, 207 (1983).
4. Holian B L, Straub G K, Swanson R E and Wallace D C: Phys. Rev. B 27/5, 2873 (1983).
5. Turnbull D: J. Appl. Phys. 21, 1022 (1950).
6. Kuhlmann-Wilsdorf D: Phys. Rev. 140/5A, 1599 (1965).
7. Kotzé J A and Kuhlmann-Wilsdorf D: Appl. Phys. Letters 9, 96 (1966).
8. Cotterill R M J: Phil. Mag. 32, 1283 (1975).
9. Born M: J. Chem. Phys. 7, 591 (1939).
10. Ida Y: Phys. B1, 2488 (1970).
11. Kristensen J K and Cotterill R M J: Phil. Mag. 36/2, 437 (1977).
12. Wenzl H and Mair G: Z. Phys. B21, 95 (1975).
13. Ivlev V J and Mal'tseva G K: Soviet Physics - Solid State 12/8, 1810 (1971).
14. Woodruff D P: The Solid-Liquid Interface, Cambridge University Press London 1973.
15. Burton W K, Cabrera N and Frank F C: Phil. Trans. Roy. Soc. 243,299 (1951).
16. Broughton J Q and Woodcock L V: J. Phys. C 11, 2743 (1978)
    Broughton J Q and Gilmer G H: J. Chem. Phys. 79, 5095,5105,5119(1983)
17. Trayanov A and Nenow D: J. Cryst. Growth 74, 370 (1986).
    Selke W and Szpilka A M: Z. Phys. B 62, 381 (1986).
18. Stranski J N: Z. Phys. Chem. 136, 259 (1928),
    Stranski J N: Naturwissenschaften 31, 144 (1943).
19. Fritsch G and Lüscher E: Phil. Mag. A 48/1, 21 (1983).

20. Fritsch G and Lüscher E: Z. Phys. B 53, 301 (1983).
21. Martynyuk M M and Lyakhovets V D: Sov. Phys. Sol. State 16/6, 1180 (1974). Käss M and Magun S: Z. f. Kristall. Miner 116, 354 (1961).
22. Henrion J and Rhead G E: Surface Science 29, 20 (1972).
23. Goodman R A and Somorjai G A: J. Chem. Phys. 52/12, 6325 (1970)
24. Frenken J W M and van der Veen J F: Phys. Rev. Letters 54/2, 134 (1985).
25. Fritsch G, Lachner R, Diletti H and Lüscher E: Phil. Mag. A 46/5, 829 (1982).
26. Fritsch G, Diletti H and Lüscher E: Phil. Mag. A 50/4, 545 (1984).
27. Neumann G and Neumann G M: Surface Self-Diffusion of Metals, Trans Tech Publications, Aedermannsdorf 1972.
28. Bushnell-Wye G and Finney J L: Phil. Mag. A 44/5, 1053 (1981).
29. Mruzik M R and Pound G M: J. Phys. F 11, 403 (1981).
30. Kosterlitz J M and Thouless D J: J. Phys. C 6, 1181 (1973).
31. Pietronero L and Tosatti E: Surface Theory of Melting, Preprint
32. Tognato R: Phys. Stat. Solidi a 80, 147 (1983).
Phys. Stat. Solidi a 73, 237 (1982).

SHORT-PULSE SURFACE INTERACTIONS

D. Hulin*
Laboratoire d'Optique Appliquée
ENSTA - Ecole Polytechnique
91220 Palaiseau, FRANCE

I INTRODUCTION

The picosecond era in laser pulse generation and measurement began in 1966 with the appearance of mode-locked lasers, soon followed by the development of non-linear techniques of time measurement. It is now possible to produce reliably pulses as short as a few tens femtoseconds. In such brief interval of time, practically the only things that move in the material universe are the electrons in atoms, molecules or solids. These ultrashort pulses permit for the first time direct time-resolved studies of extremely rapid phenomena previously indirectly inferred from spectral measurements.

Since the first mode-locked lasers, optical pulse generation techniques have continued to improve. In figure 1, the progress in optical pulse width is plotted against year. As it stands now, the shortest reported pulse is 8 femtoseconds ($8\ 10^{-15}$ second) duration.

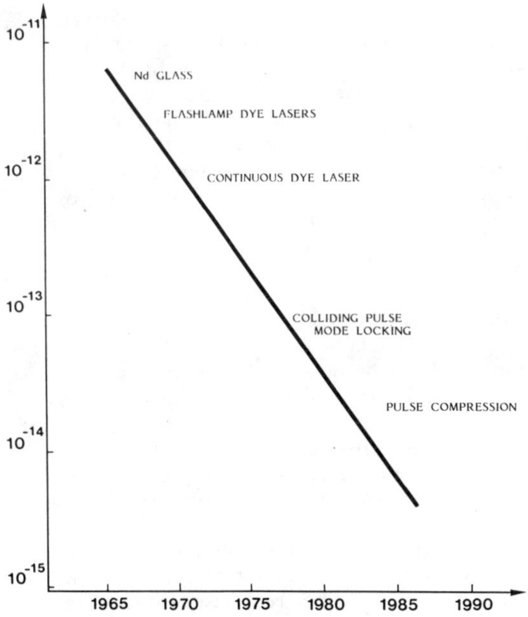

FIGURE 1. Shortest reported pulse width in function of years

Surface studies, like other domains, will benefit from this improved time resolution. In the following sections, some brief descriptions are developed on the short-pulse lasers and the corresponding experimental techniques of measurement. Then three surface related studies involving time-resolved measures will be presented as examples.

II PICOSECOND AND FEMTOSECOND LASER PULSES

Short pulses can be obtained through what is called "mode locking" [1]. In the laser cavity, a certain number of modes are present, determined by laser characteristics (optical length of the resonator, spectral dependence of gain and loss). If the modes are somehow forced to maintain a fixed phase and amplitude relationship so that to add coherently, the output will be a well defined fonction of time and the laser is said to be " mode-locked ". The result is a strongly amplitude modulated output consisting of a regularly spaced train of pulses with a width approximately equal to the reciprocal of the total mode-locked bandwidth. What are the ways in which laser mode-locking may be achieved experimentally ? Often it is necessary to introduce an external driving force or an additional non-linear element to achieve good mode-locking. A great variety of lasers can be obtained using different gain media and mode-locking systems. In the following section, some examples of ultrashort pulse lasers are presented.

The most straightforward of the various way of mode-locking is obtained by the use of a modulator inside the laser resonator. The pulses have to pass through the modulator at a time of less loss. The chief disadvantage of this technique is that it requires extreme stability of the driving frequency with respect to the resonator length. Pulses in the range of few tenths of picoseconds have been produced through this internal modulation.

The mode-locking can be also achieved by a gain modulation, a situation particularly well suited to dye lasers. Since dye lasers are optically pumped and have very broad absorption bands, it is possible to drive the dye laser gain with the output of another short-pulse pumping laser which is at a fixed wavelength. It is then of course necessary to adjust the length of the dye laser resonator to be integrally related to the repetition rate of the pumping laser. The resulting dye laser pulse is often shorter than the pump pulse, with the great advantage that theses picosecond pulses have been obtained over a wide variety of wavelengths. For example, a commercial set-up including an Ar^+ laser (Spectra Physics 171) which is acousto-optically mode-locked at a repetition rate of 82 MHz (pulse duration 100 psec), provides pulses as short as 3 psec (or less) when pumping an adjusted Rhodamine 6G dye laser. This system is known as the **synchronously-pumped dye laser.**

The most widely employed technique for inducing short-pulse mode-locking involves the use of a saturable absorber inside a laser resonator. A saturable absorber is a material whose absorption decreases as light intensity is increased. Thus a short, high-peak-power pulse experiences less loss in the absorber than a longer pulse of the same energy. Through the adjustement of the saturable absorber thickness and concentration one allows in the cavity only the pulses corresponding to modes locked in phase (added coherently). It is also important that the absorber recovery time after bleaching be shorter than the round-trip time of a pulse in the resonator. Otherwise the induced depth of modulation is greatly reduced. The time response (bleaching and recovery times) of a saturable absorber can be much faster than that of any electrically driven modulator. This type of resonator is called **passively mode-locked laser.** This technique has been successfully used for Nd: YAG lasers or Nd:glass lasers leading to pulses of the order of a few picoseconds. However the shortest pulses have been produced by passively mode-locked **dye** lasers and what follows will be concentrated on this model.

Among the continuouswave dye lasers, the best results have been obtained up to now with the colliding-pulse mode-locked (CPM) dye lasers first developped by R.L. Fork et al. [2].

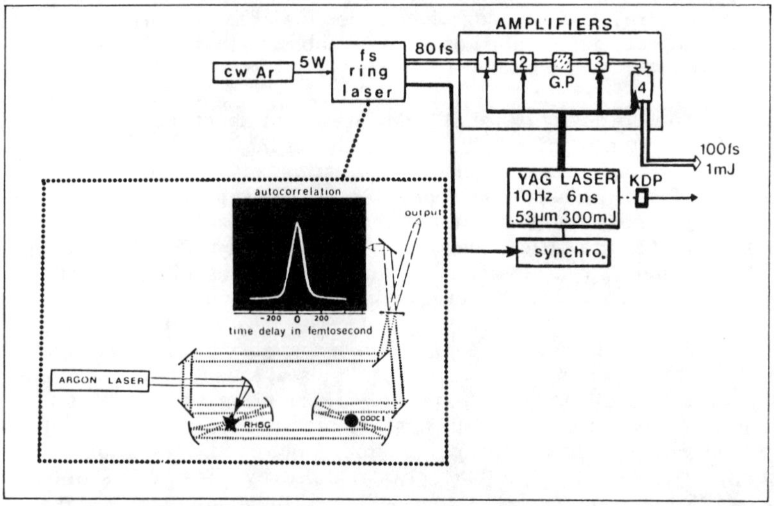

FIGURE 2. Generation and amplification of femtosecond pulses. The bottom insert represents teh CPM laser and the corresponding pulse autocorrelation figure. From ref.7.

This configuration is represented in the inset of figure 2, consisting in a series of mirrors forming a ring cavity. The oscillator simply contains two flowing jets at the two focal points in the cavity : one conventional jet of Rhodamine 6G and one extremely thin jet of saturable absorber (DODCI and malachite green) whose recovery time is about 1.2 ns [3]. The gain medium is optically pumped by 5 W of C.W. argon laser power at 514.5 nm. Since this is a completely passive system, there is no need to accurately stabilize any drive frequency or cavity length. The output pulse emerges through a partially reflecting mirror (around 96 %) and has a duration of typically 100 fsec (1 fsec = 10^{-15} sec) with a spectrum centered at 620 nm.

The mechanism for pulse shortening has been described simply [4]. If we consider an optical pulse traveling around the ring, the pulse is shaped each time it is amplified by the gain dye and absorbed by the saturable absorber. The leading edge of the pulse is preferentially clipped by the saturable absorber. Preferential amplification fo the leading edge of the pulse, combined with cavity loss, effectively clips the rear edge of the pulse with each pass around the cavity until a limiting pulse width is achieved. An additionnal mechanism operative in the colliding pulse ring configuration results from the fact that there are two equally stable but oppositely directed pulses which "collide" as they meet each other travelling around the ring. The energetically most favorable place for the two pulses to meet is in the saturable absorber. Since the pulses are coherent, they can interfere and set up a standing wave pattern in the absorber. The standing wave pattern minimizes the energy loss, because the field is more intense where the absorption is saturated and is weaker in the fiels minima where the absorption is not saturated. The shortest optical pulses are produced with a thin absorbing region that confines the standing wave field. Although the performance of the CPM dye laser depends strongly on the mirrors in the resonator, the introduction of a four-prism device into the cavity (resulting in negative dispersion) has recently been shown to permit the direct generation of pulses shorter than 30 fsec through a solitonlike shaping process.

The output pulse of a CPM laser is of short duration but not very energetic (a few tenths of picojoules). As shown in figure 2, this pulse can be amplified in a four stage amplifier consisting of dye cells pumped by a Q-switched Nd : YAG laser and separated by saturable-absorber jets. Output pulses of energy above 1 mJ and typically 100 fsec duration are obtained at 10 Hz repetition rate (limitation due to the Nd:YAG laser). Recently copper-vapor laser [5] has been used instead of the Nd:YAG laser leading to 10 Khz repetition rate pulses with however less energy per pulse (1 to 10 μJ).

Finally the pulse compression technique has been extended to the femtosecond time regime [6], leading to pulses as short as 8 femtoseconds (about 4 optical cycles). These extremely short pulses are obtained by using nonlinear frequency broadening in a short piece of optical fiber followed by compression with a dispersive delay line (figure 3). The fiber impress a frequency sweep or "chirp" on the optical pulse by the process of self-phase modulation. Pulse compression is then achieved by compensating the frequency sweep in the pulse frequency through the use of a pair of gratings (dispersive delay line)

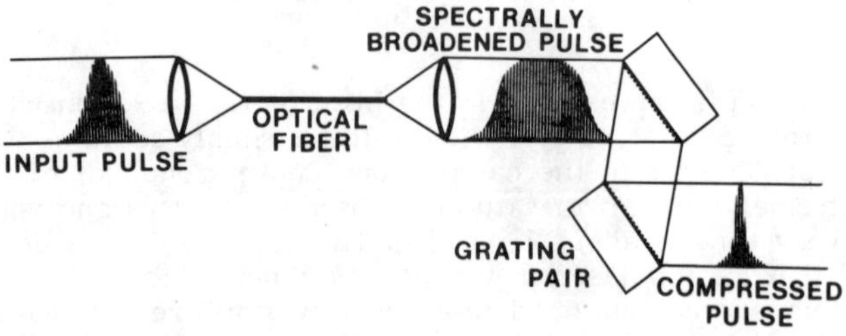

FIGURE 3. Experimental arrangement for compressing femtosecond optical pulses.

Before any picosecond-pulse system can be used with confidence, the pulses must be characterized accurately. A pulse measurement arrangement that is particularly suitable for use with a continuously operated laser is shown in figure 4. The pulse train from the laser is split and the two beams traverse

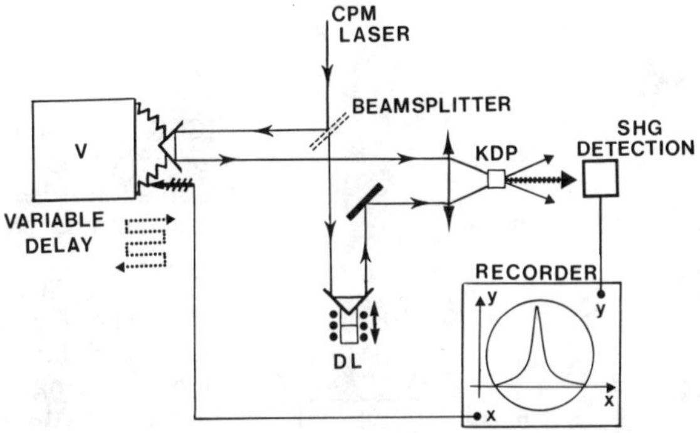

FIGURE 4. Modified interferometer arrangement for the measure of short pulse autocorrelation by using non-collinear second-harmonic generation

different paths in a modified interferometer arrangement. The time delay between the two beams is controlled by measuring the difference of the optical paths. This is obtained by following with a magnetic sensor the displacements of the mirrors mounted on a fast-moving support (indicated as "variable delay" on the figure). The two beams, parallel but not collinear, are focused by a simple lens through the same point in a thin crystal of KDP (potassium-dihydrogen phosphate). The crystal is oriented for phase-matched second-harmonic generation (SHG) at an angle bisecting that made by the two fundamental beams. The second harmonic beam is generated only when the two pulses overlap in the crystal. On the display, the recorded SHG signal in function of the path difference represents the autocorrelation curve. The pulse duration can be directly extracted from the width of the autocorrelation curve.

Experiments performed with ultrashort laser pulses are often related to phenomena too rapid to be followed by classical means of detection. Time-resolved studies are made possible by using the pump and probe configuration. Almost any kind of classic optical measurements can be transposed in the short time domain by this method: absorption, reflectivity, second harmonic generation, electron diffraction, ellipsometry, etc... . As an example, figure 5 exhibits a pump and probe absorption experiment [7], but it could be easily transformed into an other type of measurement. The pulses originate from the ouput of an amplified CPM dye laser (typically 1 mJ per pulse, 100fsec duration, rep. rate 10 Hz, = 620 nm). The beam is divided into two parts by a beam splitter, one being used as the pump beam (upper beam in the figure), the other as the probe beam. High intensity femtosecond pulses are fundamental for spectroscopic purposes in that they can generate white-light pulses of comparable duration. This can be done by focusing each beam into different transparent liquids such as water, alcohol, or ethylene glycol. Phase modulation is the dominant generation process [8]. In the pump beam, a suited wavelength is selected through the use of an interferential filter and then further amplified to the desired level of energy. This method of tunable pump wavelength is very convenient for resonant excitation but can be removed if the need of high power pulses is more important. The probe pulse arrives on the sample S at a delay with the pump pulse which can be controlled by the stepping motor. After the sample, the probe beam is recorded with no need of special time resolution. Here again, like for the determination of the pulse duration, the measure of time difference is transformed into a measure of optical path difference.

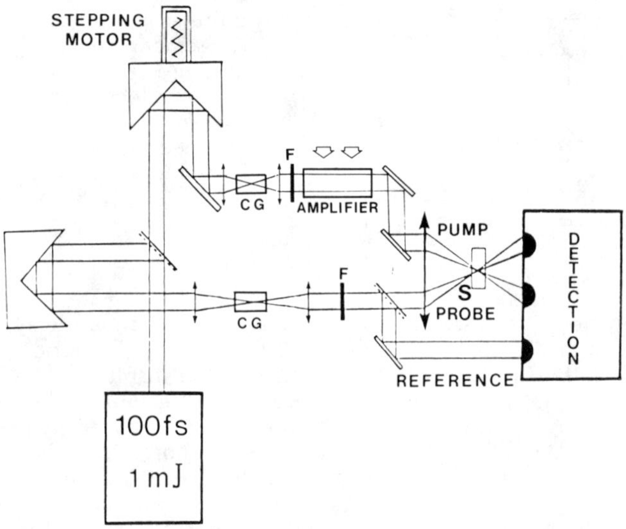

FIGURE 5. General scheme of time-resolved absorption experiments. The input beam at 620 nm is split in two parts. The pump beam is delayed in time by using a translation stage mounted on a stepping motor with an accuracy of 0.1 µm. The continuum generation (CG) occurs in water cells. The detection apparatus consists of a combination of either a spectrometer and an optical multichannel analyzer or monochromators and photodiodes.

III SUBPICOSECOND TIME-RESOLVED REFLECTIVITY

Changes in reflectivity can be induced by modification of the surface state. If these modifications can be triggered by a light pulse, it is easy to monitor the reflectivity changes by an adaptation of the classical pump and probe experiment with a time resolution limited only by the laser pulse duration. As an example, an experiment performed in our laboratory on the kinetics of silicon melting is described in this section. A crystalline wafer of silicon is irradiated by a pump pulse of 100 fs duration at 620 nm with an energy density of the order of 0.4 mJ/ cm^2 for the excitation. The first interaction of this light pulse with the semiconductor is to create [9] an electron-hole plasma of very large density (up to 10^{22} cm^{-3}) and of ultra- high temperature (T around 10000 K). Then the plasma relaxes it excess energy towards the lattice whose temperature increases, leading to the occurence of melting of the silicon crystal. It is even possible that the presence of this large density of electrons excited from the valence band into the conduction band reduces the covalent bonding between the silicon atoms, therefore inducing a possible new kind of melting (cold melting) without need of heat transfer [10,11]. In fact both processes may happen and the use of ultrashort pulses, even shorter than 100 fsec, seems to be the only possibility to discriminate between the two mechanisms, the heat transfer from the hot plasma towards the lattice being not instantaneous. It is therefore interesting to study this transfer in presence of such electron-hole density.

The experimental scheme is indicated in figure 6. The surface area excited by the pump pulse (620 nm) is lightened by the probe pulse (at a wavelength chosen in the continuum) at a fixed delay after irradiation with a repetition rate imposed by the laser (10 Hz). The enlarged image of the spot is then projected on a hole which collect only its center [12] or on a plate or on a photodiode array which gives the reflectivity values along the spot diameter. By slowly varying the delay between the two pulses one can follow the evolution of the excited spot on an extremely expanded time scale. This technique may be called "**femtosecond stroboscopy**".

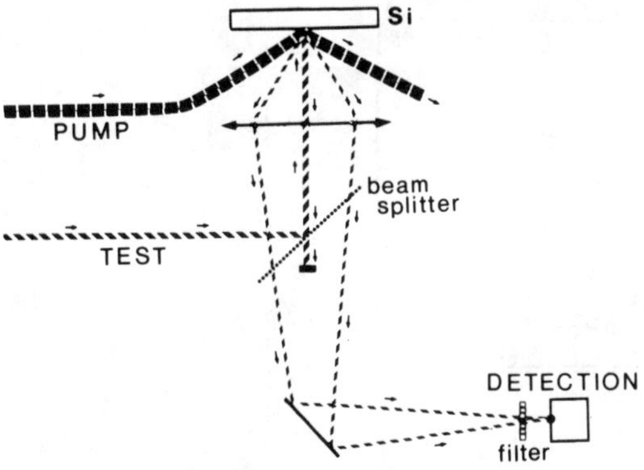

FIGURE 6. Time-resolved imaging of an excited silicon surface.The beam enlightens the surface at a chosen delay after excitation. The enlarged image of the excited spot is then formed on the detector.

Figure 7 exhibits the profile of the induced changes in reflectivity for the excited area in function of time. The exciting pulse at 620 nm has almost a gaussian spatial shape and the probe pulse wavelength is 550 nm. In the first picosecond the reflectivity jumps rapidly and then increase slowly with values always below the liquid value. Although the role of the electron-hole plasma has to be taken into account for the first increase in reflectivity, this is no more the case after 50 psec when the e-h density has decreased beyond 10^{20} cm^{-3} due to Auger recombinaison. The highly reflecting state of the silicon observed at the long delays is then attributed to the occurence of melting. The discrepancy between R / R_o and the well known liquid value [13,14] implies the existence of only a partial melting. Although the homogeneity of the laser spot was checked with great care (double spatial filtering) this is not in contradiction with the heterogeneous nature of the nucleation of melted regions [15]. The fact that the reflectivity increases with time even after 30 psec is an indication that there is still some heat transfer from the electrons and the holes towards the lattice, or that the lattice has been overheated [16] in the early times and releases its excess energy while melted regions are growing up.

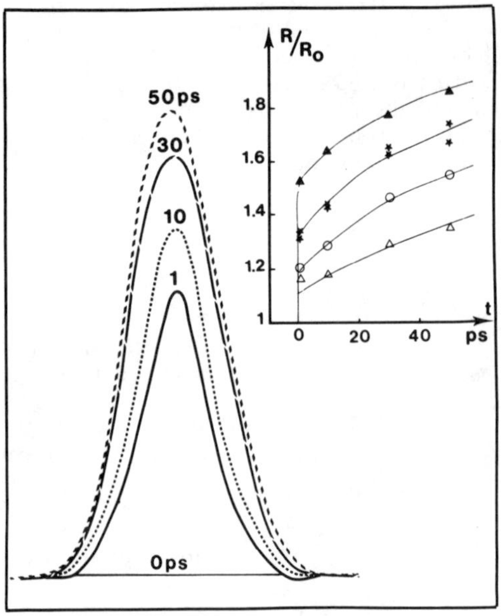

FIGURE 7. Profile of the induced changes in reflectivity (R/R_o) at different delays after excitation. The baseline corresponds to the reflectivity of the unperturbed silicon.The half height width of the perturbed area at 50psec is 35 μm. The insert shows the quantitative value of R/R_o recorded at different positions in the profile, corresponding to different initial excitation intensities.

IV SUBPICOSECOND TIME-RESOLVED SURFACE STRUCTURAL DYNAMICS

Reflectivity of course does not provide all the necessary informations on the surface changes. For example, in the study of silicon melting, no structural information can be obtain from the reflectivity during the phase transition from a crystal to a liquid. C.V. Shank, R. Yen and C. Hirliman (hereafter referred as SYH) have used second harmonic generation from the silicon surface to provide a means of determining the structural dynamics following excitation with an intense optical pulse [17]. Unlike reflectivity measurements, the tensor nature of the second harmonic polarizability makes this quantity sensitive to crystalline order. Therefore the transition from crystalline structure to disorder can be monitored in function of time.

Second harmonic generation from the surface of a medium with inversion symetry, such as silicon, has not a simple origin. Since silicon is a centrosymmetric crystal, it is expected that the dipole volume contribution to the second harmonic would be zero. At the surface higher order effects contribute to the second harmonic. The dominant contribution comes from the surface quadrupole term [18], implying that the second harmonic is primarily generated by the first few atomic layers. It may also be important to include a dipole contribution from the assymetrically distorded electrons in the first atomic layer [19]. In addition volume quadrupole effects could also contribute but their importance is limited by the absorption depth in the silicon of the generated second harmonic. At 4 eV this length is 70 A corresponding to 13 unit cells of the silicon crystal.

In S.Y.H. experiment, a pulse at 620 nm with a duration of 90 fsec excite the [111] surface of a polished silicon wafer and the second harmonic generated by an other much weaker pulse is measured at delayed time. The sample is mounted on a combination rotation and translation stage controlled by a computer so that each pulse of the laser is incident on a new spot on the silicon wafer. This is absolutely necessary because multiple shots from the laser produce cumulative damage to the surface. The probing pulse is incident on the surface at an angle of 45° with the polarization parallel to the plane of incidence.

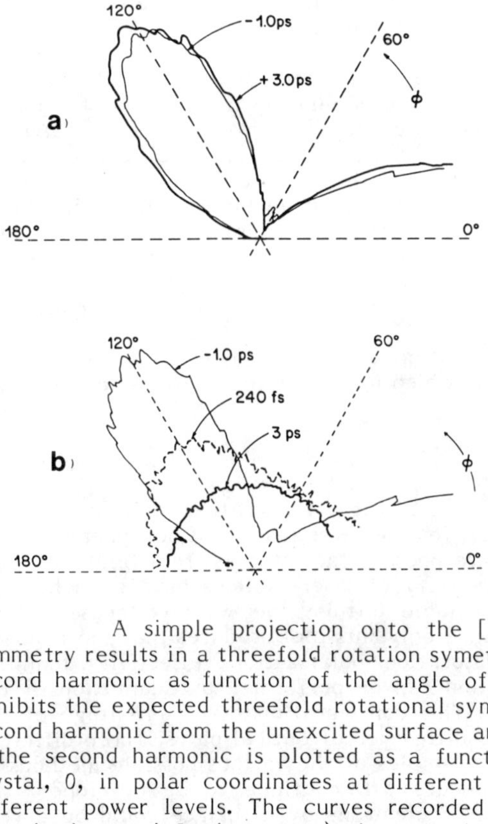

FIGURE 8. Polar plot of the second harmonic intensity as a function of time and angle 0 for two different pump energies: a) the pump intensity is below the threshold for amorphous formation , b) the pump intensity is two times this threshold. (C.V. Schank, R.T. Yen, C. Hirliman, Phys. Rev. Lett. 51, 900 (1983).

A simple projection onto the [111] surface assuming the bulk symmetry results in a threefold rotation symetry. The observed variation of the second harmonic as function of the angle of rotation 0 about the [111] axis exhibits the expected threefold rotational symetry. Maxima and minima in the second harmonic from the unexcited surface are observed every 120°. In figure 8 the second harmonic is plotted as a function of angle of rotation of the crystal, 0, in polar coordinates at different delays after excitation for two different power levels. The curves recorded at -1.0 psec correspond to the unexcited crystal. In the case a), the pump intensity is below the threshold for amorphous formation and the created electron density is approximatively 10^{21} cm^{-3}. Little changes are observed after excitation. In the case b,) the polar plot reveals a large change in the symetry of the second harmonic following excitation (four times more intense than in the first case). The width of the lobe at 0 = 120° brodens and the minima becomes less distinct. Apparently the surface is begining to lose order. Within the first picosecond the second harmonic coming from the surface becomes nearly isotropic. This is precisely what would be expected if the surface order makes a transition from a crystalline structure to a disordered liquid.

V LASER PROBING OF MOLECULAR-SURFACE INTERACTION

Chemical and physical processes occuring at solid-gas or solid-liquid interfaces may be significantly influenced by the rates and pathways of vibrational energy transfer within adsorbate and substrate surface vibrational modes. Knowledge of transfer rates may enhance understanding of reactions at surfaces (e.g. heterogeneous catalysis, etching, corrosion). Interpretation of the vibrational spectroscopy of molecules on surfaces (e.g. spectral linewidths and shifts) requires similar knowledge. Energy accomodation, sticking, mobility and desorption of molecules are all processes related to vibrational energy transfer for molecules on surfaces.

E.J. Heilweil et al. have performed picosecond IR transient bleaching experiments [20] to measure the population lifetime T_1 of vibrationally excited bonds of molecules chemisorbed on high-surface-area materials. These materials include fumed silica (SiO_2 particles) and zeolite catalysts. A Nd:YAG- pumped optical parametric amplifier generates tunables IR pulses (2600- 7200 cm^{-1}) of 15 psec duration. Tuned to an infrared transition of an adsorbate vibrational mode and focused into the sample, the pulse saturates this transition, causing the sample to becomes more transparent (bleached) at this frequency. A probe pulse at the same frequency and delayed with respect to the pump pulse monitors the return of sample transmission to its equilibrium value. The determinations of T_1 for high-frequency OH, OD, NH and CH stretching vibrations have been obtained both at the solid-vacuum and solid-liquid interfaces, whith results which may have implications for catalysis.

VI CONCLUSION

Lasers with pulses of ultrashort duration (picoseconds or femtoseconds) are now reliable tools for the investigation of new phenomena in different research domains. In this respect, surface studies are a field still open to future developments. A great variety of laser systems have been built in laboratories or are commercialy available. In this paper we have focused on the passively mode-locked dye laser since only this tool produces that short pulses. Time-resolved measurements of absorption, polarization, reflectivity, luminescence or second-harmonic generation can be performed by adaptation of the classical pump and probe scheme. This configuration takes advantage of the transformation of time delays into difference of optical paths. Three different experiments are reported here as examples of the various possible methods of investigation. Subpicosecond techniques will continue to be developed. For instance, time-resolved diffraction of electrons by a surface could be a very usefull method when all the experimental parameters will be controlled. Therefore we anticipate a large increase of such studies in the near future.

REFERENCES

* also at Groupe de Physique des Solides de l'E.N.S. , Paris.

1- C.V. Shank and E.P. Ippen, 121, Topics in Appl. Phys. 1 edited by F.P. Schäfer, Springer-Verlag (1973),

2- R.L. Fork, B.I. Greene and C.V. Shank, Appl. Phys. Lett. 41, 671 (1981)

3- C.V. Shank and E.P. Ippen Appl. Phys. Lett. 26, 62 (1975)

4- C.V. Shank, Science, 219 , 1027 (1983)

5- W.H. Knox, M.C. Downer, R.L. Fork C.V. Shank, Optics Lett. 9 552 (1984)

6- C.V. Shank, R.L. Fork, R. Yen, R.H. Stolen and W.J. Tomlinson, Appl. Phys. Lett.

7- A. Migus, A. Antonetti, J. Etchepare, D. Hulin and J. Orszag, J. Opt. Soc. Am. B 2 584 (1985)

8- R.L. Fork, C.V. Shank, R. Yen, C. Hirliman, IEEE J. Quantum Electron. QE19, 500 (1983)

9- D. Hulin, M. Combescot, J. Bok, A. Migus, J.Y. Vinet, A. Antonetti Phys. Rev. Lett. 52, 1998 (1984)

10- M. Combescot and J. Bok, Phys. Rev. Lett. 48, 1413 (1982)

11- D. Hulin, A. Migus, C. Tanguy and A. Antonetti, J. Lum. 30, 262 (1985)

12- C.V. Shank, R.T. Yen and C. Hirliman, Phys. Rev. Lett. 50, 454 (1983)

13- K.M. Shvarev, B.A. Baum,P.V. Gel'd, Sov. Phys. Sol. State,16, 2111 (1975)

14- D. von der Linde and N. Fabricius, Appl. Phys. Lett. 41, 991 (1982)

15- D.Y. Sheng, R.M. Walser, M.F. Becher, J.G. Ambrose, Appl. Phys. Lett. 39, 99 (1981)

16- S. Williamson, G. Mourou, J.C.M. Li, Phys. Rev. Lett. 53, 1837 (1984)

17- C.V. Shank, R.T. Yen, and C. Hirliman, Phys. Rev. Lett. 51, 900 (1983)

18- N. Bloembergen, R.K. Chang, S.S.Hha and C.H. Lee, Phys. Rev. Lett. 174, 813 (1968)

19- N. Bloembergen and Y.R. Shen, Phys. Rev. 141, 298 (1966)

20-E.J. Heilweil, M.P. Casassa, R.R. Cavanagh and J.C. Stephenson, J. Chem. Phys. 82, 5216 (1985) and 84, 2361 (1986)

NONEQUILIBRIUM PHASE TRANSITIONS

G.DEWEL[+]
Service de Chimie Physique II.
Université Libre de Bruxelles
CP 231, Boulevard du Triomphe
B- 1050 Bruxelles

I. INTRODUCTION.

With the advent of moderate and high-power laser sources energy can be deposited into near surface layers of different materials in large amounts and in short times. The laser radiation thereby strongly excites the solid and brings it into a nonequilibrium state. It is now well known that many systems can undergo an instability when they are driven sufficiently far from thermal equilibrium.(1) In many nonequilibrium transitions the solutions which emerge at instability have lower symmetry than appropriate equations and boundary conditions. As there is one state of maximum symmetry and many less symmetrical ones, the most difficult question is the problem of pattern selection: how certain structures out of a multitude are selected by the physical system? Spontaneous breakdown of symmetry also occurs in systems in thermodynamic equilibrium. But there, the selection problem is reduced to a variational principle based on the thermodynamic potentials. The beauty and the basic simplicity of the ideas developed to describe these equilibrium phase transitions have led to the belief that the same or very similar ideas could also be useful to understand the onset of cooperative behavior in nonequilibrium systems.

Progressively an increasing number of scientists became aware of the far reaching similitaries both in physical mechanisms and formal description between instabilities occuring in very different dissipative systems ranging from fluid mechanics to biology. The physical origin of these analogies lies in the fact that in the proximity of an instability, the relaxation of the marginal mode is slowed down. This scale separation allows one to reduce the dynamics of physically dissimilar systems to a few behavioral types. (2)

+ Chercheur Qualifié au Fonds National de la Recherche Scientifique de Belgique.

2. LINEAR STABILITY ANALYSIS.

Linear stability analysis provides the classical tool for detecting the first instability ("Hunt for the soft mode"). In this note I only consider examples of instability which can be described by the phenomenological approach based on the balance equations for the macroscopic variables. A general dissipative system can then be described by a state vector $q(r;t)$, its evolution is governed by nonlinear equations which may be written as

$$\dot{q} = N(q, b, \underline{r}, t) \tag{1}$$

The control parameter b describes the constraints which drive the system out of thermal equilibrium. Since the basic equations (1) are generally partial differential equations, one must also specify the boundary conditions. The onset of any dissipative structure requires a supply of matter and/or energy, the boundaries wich are the gates of this supply are much more than simple geometrical constraints. That is one of the reason why the size and the shape of the system play a more important role in these dynamic instabilities than in equilibrium phase transitions.

In general, Equ.(1) admits a homogeneous steady state q_0 for small values of the constraints (Thermodynamic branch.) The routine procedure then starts with the linearization of the basic equations and boundary conditions in the vicinity of this steady state :

$$q = q_0 + \underline{w}(\underline{r}, t)$$

$$\dot{\underline{w}} = \underline{\underline{L}}\, \underline{w} \tag{2}$$

where L is a linear operator (Fréchet derivative of N). If this operator is time independent the solutions of Equ.(2) can be written as a superposition of eigenmodes of L which are here supposed to be the plane waves.

$$\underline{w} = \sum_k a_k(t) \exp\left[w_k t + i\underline{k}\cdot\underline{r}\right] \tag{3}$$

If all the eigenvalues $\{w_k\}$ of L have a strictly negative real part, the thermodynamic branch q_0 is stable.
On the other hand if an eigenvalue of L has a strictly positive real part the homogeneous solution q_0 is unstable.

The linear stability analysis is then reduced to the evaluation of the principal eigenvalue (that with the largest real part) of the linearized matrix L . Among the various instabilities that can develop when b is increased on can distinguish the following cases :

- L has an isolated zero eigenvalue

$$Re\{w_k\} = 0 \; ; \; Im\{w_k\} = 0 \text{ for } k = 0,\ b = b_c$$

corresponding to a transition to another homogeneous state (Multistability)

- L has a set of eigenvalues such that

$$Re\{\omega_k\} = 0, \; Im\{\omega_k\} = 0 \qquad |\underline{k}| = q_c ; b = b_c$$

this instability generally induces the formation of a stationary periodic structure.

- One complex eigenvalue crosses the imaginary axis

$$Re\{\omega_k\} = 0, \; Im\{\omega_k\} = \Omega_0 \qquad |\underline{k}| = 0 \; ; \; b = b_c$$

This instability can lead to homogeneous time-periodic solutions of the limit-cycle type.
When these conditions are satisfied for $|\underline{k}| = q_c$; spatiotemporal structures (waves) bifurcate at $b = b_c$

In summary, the linear stability analysis determines the critical value b_c of the constraint, the frequency Ω_0 of the oscillations at the threshold of a Hopf instability and the critical wavelength $\lambda_c = 2\pi q_c^{-1}$ at the onset of a transition to a periodic pattern. The loss of stability to infinitesimal perturbations implies a qualitative change in the dynamic behavior. The study and the classification of these qualitative changes in the structure of the solutions is the subject of bifurcation theory.

3. BIFURCATIONS.

3.1 Elementary bifurcations and slow dynamics.

Linear stability indicates the existence of stable modes, once excited they would rapidly decay and unstable modes which present an exponential growth. The nonlinear contributions must now be added in order to prevent the unlimited growth of these modes. In slightly supercritical situations, these nonlinear terms can be evaluated by applying the principle of quasi-stationarity. In order to illustrate the mathematical skeleton of this method we consider now a system involving one stable (s) and one unstable mode (a) which obey the following equations.

$$\dot{s} = -\lambda_s s + g(a, s)$$
$$\dot{a} = \varepsilon a + f(a, s) \qquad (4)$$

with $\lambda_s > 0$ and $\varepsilon = \frac{b - b_c}{b_c}$
On the rapid time scale $\tau_R \simeq \lambda_s^{-1}$, the stable mode relaxes to a level $S_s(a)$ satisfying

$$s_s = \lambda_s^{-1} g(a, s_s(a)) \qquad (5)$$

Following Haken's terminology: the rapid mode is slaved by the slow mode. The long time behavior $t > \lambda_s^{-1}$ is then described by the following amplitude equation obtained by simple substitution.

$$\dot{a} = \varepsilon a + f(u, s_s(a)) = F(a) \qquad (6)$$

Near threshold F (a) can be expanded in a power series in $\sqrt{\varepsilon}$ keeping only the lowest nonlinearities. Of course in real systems the number of slaved modes is generally large and the explicit computation of the nonlinear coupling terms can be rather involved. However the form of this amplitude equation depends only on few fundamental features such as the type of bifurcation, and the symmetry.

For example, in the case of symmetry to reflection of the state variable ($a \to -a$), the simplest generic form is (pitchfork bifurcation)

$$\dot{a} = \varepsilon a - \mu a^3 \tag{7}$$

- if $\mu > 0$: two stable stationary states $a_{\pm} = \pm\left(\frac{\varepsilon}{\mu}\right)^{1/2}$

appear at $\varepsilon > 0$ where the zero state becomes unstable. This is a supercritical bifurcation corresponding to a second order transition (critical point)

- if $\mu < 0$ the nontrivial solutions $a_{\pm} = \pm\left(-\frac{\varepsilon}{\mu}\right)^{1/2}$

are unstable and exist in the stability region of the trivial solution $a = 0$
When one takes into account the next order term in Equ.(6), two new symmetrical branches ($\pm a_s$) appear for $\varepsilon > 0$
This corresponds to a subcritical bifurcation equivalent to a first order transition.
Note that Equ. (6) can be written as

$$\dot{a} = -\frac{\partial V}{\partial a} \tag{8}$$

where

$$V(a) = -\frac{\varepsilon a^2}{2} + \mu \frac{a^4}{4} \tag{9}$$

V is a Lyapounov functional that decreases in any dynamics, and which plays here the role of a thermodynamic potential. This analogy, however, generally loses ground when there are two or more variables. Actually this is the source of different qualitative behaviors in non-equilibrium systems.

3.2 Bifurcation of stationary patterns

In isotropic systems the manifold of eigensolutions of the form (3) exhibits a three-fold degeneracy. First one encounters the problem of wavelength selection. For b slightly above b_c , the homogeneous solution φ_o is linearly unstable for a continuous family of nonequivalent wavelengths, yet a unique wavelength which does not correspond to the critical wavelength $\lambda_c = 2\pi q_c^{-1}$ is selected in some experiments (3,4).

Secondly in unbounded media there is an orientational degeneracy associated to the unrestrained choice for the directions of the critical wavevector. Finally for dimensions higher than one, to a given wavelength there corresponds a large variety of possible planforms. For example, in a two-dimensional system one can have rolls, suqares, hexagons. It has been shown that the nonlinear terms induce a selection among these various structures already in the weakly nonlinear regime (5,6). We now illustrate this property on a simple variational model.
By applying the method outlined in previous section, it is possible to derive an equation for the amplitude a_k of the critical mode. In many cases, it takes a form similar to the Ginzburg-Landau equation:

$$\dot{a}_k = \omega_k a_k - {\sum_{k_1}}' v(k_1,k)\, a_{k_1} a_{-k_1+k} - {\sum_{k_1}}' {\sum_{k_2}}' u(k_1,k_2,k)\, a_{k_1} a_{k_2} a_{-k_1-k_2+k} \tag{10}$$

with

$$\omega_k = \varepsilon - D(k^2 - q_c^2)^2 \tag{11}$$

In Equ.(10) the summation is restricted to wavenumbers close to q_c. As in Equ.(8), one can often define Lyapounov functional. When the nonlinear coupling terms do not depend on the angles between the interacting wavevectors, this functional takes the simple form (Brazowskii's model):

$$V = -{\sum_k}' \omega_k |a_k|^2 + \frac{v}{3!} {\sum_{k_1}}' {\sum_{k_2}}' {\sum_{k_3}}' a_{k_1} a_{k_2} a_{-k_1-k_2} + \frac{u}{4!} {\sum_{k_1}}' {\sum_{k_2}}' {\sum_{k_3}}' a_{k_1} a_{k_2} a_{k_3} a_{-k_1-k_2-k_3} \tag{12}$$

Similar functionals have been derived in the case of the Rayleigh-Bénard instability (7), the Turing instability (6) and hydrodynamic instabilities in nematic liquid crystals(8). Each pattern can be characterized by pairs of wavevectors $(\underline{k}_i, -\underline{k}_i)$. We consider explicitly in the following only the strutures which minimize the potential (12) $|\underline{k}_i| = q_c$. In space variables the corresponding order parameter becomes:

$$a(\underline{r}) = 2 \sum_{i=1}^{m} a_i \cos(\underline{k}_i \cdot \underline{r}) \tag{13}$$

Two classes must be considered :

A) The structures described by m independent pairs. In that case the quadratic terms in the equations of motion do not contribute. They appear supercritically though a second order like phase transition: we get indeed from the stationary condition of equations (10):

$$a_m = 0 \qquad b < b_c$$

$$a_m = \left[\frac{2(b - b_c)}{(2m-1)\mu}\right]^{1/2} \qquad b > b_c \tag{14}$$

B) The structures the wave vectors of which satisfy the triangular condition

$$a_3(\underline{r}) = 2\, a_3\Big[\cos q_c x + \cos \frac{q_c}{2}(x + \sqrt{3}\, y) + \cos \frac{q_c}{2}(x - \sqrt{3}\, y)\Big] \tag{15}$$

Depending on the sign of the cubic term v in (12), the maxima respectively define a triangular $(v<0)$ or a honeycomb lattice $(v>0)$. These structures appear subcritically through a first order like transition. The amplitude jumps indeed discontinuously to a finite value.

The physically possible patterns correspond to the fraction of all the stationary solutions which is stable with respect to arbitrary disturbances of infinitesimal amplitude. In the case of the Brazowskii's model (12), all the solutions of the first class are unstable with the exception of the case $m = 1$, corresponding to a stationary wave periodic in one direction (rolls). In this case the stability analysis strongly restrict the manifold of possible solutions; only two structures remain : rolls and hexagons.

There is a range of parameters where rolls and hexagons coexist. In this region, the potential (12) can be used to calculate the relative stability of the coexisting patterns.

The bifurcation diagram of the Brazowskii's model (μ, v constant) is unfortunately not universal. For instance, a square pattern will be selected whenever in the corresponding stationary amplitude equations :

$$\varepsilon a_i - g_D |a_i|^2 a_i - g_{ND} |a_j|^2 a_i = 0 \tag{16}$$

(i and j correspond to the two orthogonal directions of the square), one has the inequality $g_D > g_{ND}$ as a result of the dependence of the nonlinear terms $\mu(k_i, k_j)$ on the angles between the interacting wavevectors.

This situation occurs in hydrodynamic instabilities in nematics (9).

In anisotropic systems there is an intrinsic mechanism which raises the orientational degeneracy by inducing preferred directions for the critical wavevectors). In many examples the amplification rate then takes the form

$$w_k = \varepsilon - D(k^2 - q_c^2)^2 - A \sin^2 \phi \tag{17}$$

When $A > 0$, this expression corresponds to an easy axis parallel the direction $\phi = 0$..

4. PATTERN FORMATION IN CRYSTAL GROWTH

As an example of symmetry breaking instability, I discuss in this chapter the problem of cell formation during solidification. Many high power laser processing phenomena have been described in terms of a simple melting model. One assumes that a thin molten surface layer is formed above a particular threshold; at the end of the irradiation the liquid can crystallize starting at the substrate. In this model it is important to study the morphological stability of the planar interface during this rapid liquid epitaxial regrowth. Two types of such cells have been experimentally observed after laser irradiation : small mosaïc cells ($\simeq 0.1 \mu m$) attributed to constitutional supercooling and larger cells ($\simeq 1 \mu m$)with a size comparable with the liquid layer thickness and which might be due to convective motions in the melt (frozen Bénard cells). (10) In this lecture, I consider only the formation of morphological cells in absence of convective motions in the melt.

4.1 Constitutional supercooling

Constitutional supercooling is one of the most important problem in crystal growth. It is related to the compositional changes that occur at liquid-solid transitions in multicomponent systems. For a binary alloy, one defines at a given temperature, the equilibrium segregation coefficient $k_0 = C_S / C_L$ where C_S is the equilibrium composition of the solid phase and C_L that of the corresponding liquid. In practice crystallisation is rarely conducted so slowly that the system can readjust its composition homogeneously throughout the bulk of its phases according the equilibrium phase diagram. Hence one generally encounters in the melt dynamic concentration gradients near the interface. Similarly temperature gradients are induced both on the liquid (G_L) and on the solid (G_S) side. Following the original argument of Tiller and al. (11), one can now derive the condition for constitutional supercooling. As shown in Fig.(1), we consider two experiments (I and II) having the same solute profile ($k_0 < 1$) but different temperature gradients. We have plotted also the permitted concentrations.

The latter is the smallest amount of solute that will just keep the liquid out of the two phase region.

Clearly in experiment I, there is a region near the solid-liquid interface where the permitted concentration is above the actual concentration and the melt is thus supercooled in this case.

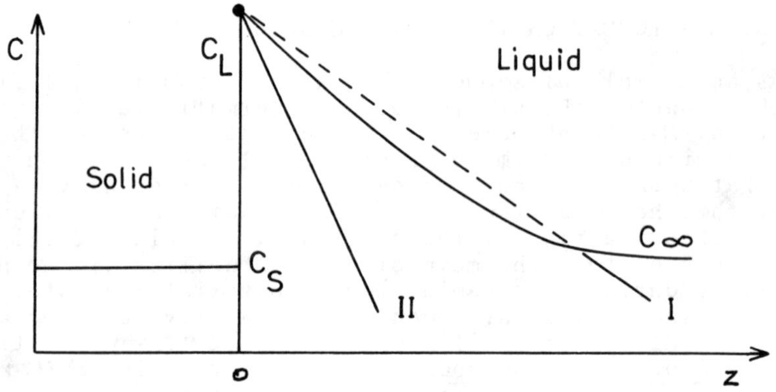

Fig. 1. Actual concentration and permitted concentration in the steady state corresponding to experiments I and II.

On the interface ($z=0$) the permitted concentration is equal to c_L and it decreases linearly with z ($z>0$) according to

$$c_p(z) = c_L + z\frac{G_L}{m} \tag{18}$$

where m is the slope of the liquidus in the phase diagram (dilute alloy). From Fig. 1 the condition for constitutional supercooling is

$$\frac{G_L}{m} > \left.\frac{\partial c}{\partial z}\right|_{z=0} \tag{19}$$

and $\left.\frac{\partial c}{\partial z}\right|_{z=0}$ is determined by the conservation of the solute which is rejected at the interface

$$-D\left(\frac{\partial c}{\partial z}\right)_{z=0} = (c_L - c_S)V = -D\,G_{Lc} \tag{20}$$

where V is the interface velocity, D the diffusion coef- in the melt. The condition Equ. (19) then becomes :

$$\frac{G_L}{V} < \frac{|m| C_\infty (1-k_0)}{k_0 D} \tag{21}$$

where $C \to C_\infty$ as $z \to \infty$ in the melt.

For given values of G_L and V , there is a critical concentration above which cells develop; this critical concentration is an important parameter for the laser annealing of doped semiconductors by ion implantation. (12) We must next discuss what will happen when the liquid is constitutionally supercoold. It is unlikely that the solid will be precipitated in the liquid out of contact with the interface because a large undercooling is required for homogeneous nucleation. In most experiments, the interface rearranges its shape to reduce the supercooling.

4.2 The Mullins-Sekerka instability (13)

The criterion developped in previous section was based on a static analysis. In this section we review briefly the linear stability analysis of a flat solidification front advancing at constant velocity into a supercooled melt (the concentration in the bulk of the liquid is necessarily in the two phase region and for simplicity we take $C_\infty = C_S$. In a frame of reference attached to this interface the basic equations are:

- the diffusion equation governing the solute concentration in the melt

$$\frac{\partial C}{\partial t} = D \nabla^2 C + V \frac{\partial C}{\partial z} \tag{22}$$

the diffusion of solute in the solid is so slow that it can be neglected.

- the heat diffusion equations

$$\frac{\partial T}{\partial t} = K_i \nabla^2 T_i + V \frac{\partial T_i}{\partial z} \quad ; \quad i = s, L \tag{23}$$

K_i is the thermal diffusivity.

These equations must be supplemented by the following boundary conditions at the interface:

- the concentration balance

$$v_n (C_L - C_S) = - D (\underline{\nabla} C) \cdot \underline{\hat{n}} \tag{24}$$

v_n is the normal velocity of the front and $\hat{\underline{n}}$ the unit normal at the interface.
- the condition of heat conservation

$$L v_n = \left[c_p^s K_s (\underline{\nabla} T)_s - c_p^L K_L (\underline{\nabla} T)_L \right] \cdot \hat{\underline{n}} \tag{25}$$

L is the latent heat; c_p^L and c_p^S the specific heats of the liquid and solid.
- The Gibbs-Thompson relation

$$T_L = T_S = T_M + mC - T_M \Gamma K \tag{26}$$

T_M represents the melting temperature of the plane interface in absence of solute, K is the curvature (here understood to be positive if the center of curvature lies on the solid side) and $\Gamma = \gamma / L$ is the capillary constant in which γ is the specific interfacial free energy. The local equilibrium assumption Equ. (26) determines the interface temperature including the curvature induced shift. We have also neglected fluid flow. To further simplify the problem we note that D/K_L and D/K_S are very small ($\simeq 10^{-4}$ for metals), it is thus convenient to take the limit in which D/K_L and D/K_S approach zero.

The reference state corresponds to the linear temperature gradients G_L and G_S and the actual solute profile shown in Fig. (1)

$$c^0(z) = c_S + [c_L - c_S] \exp - \left[\frac{Vz}{D} \right] \tag{27}$$

We then consider a small deformation $\zeta(\underline{x}, t)$ of the plane interface :

$$\zeta(\underline{x}, t) = \sum_k \zeta_k \exp . [i \underline{k} \cdot \underline{x} + w_k t] \tag{28}$$

$\underline{k}$ is a two-dimensional wave vector perpendicular to the z axis and w_k is the amplification rate. By linearizing the kinetic equations and boundary conditions one find after some algebra :

$$w_k = kV \left\{ \left[-\Gamma T_m k^2 - \mathcal{G} \right] \left[\alpha - \frac{(1 - k_0)}{D} V \right] + m G_{LC} \left(\alpha - \frac{V}{D} \right) \right\}$$

$$\times \left\{ m k G_{LC} + \frac{LV}{(K_S + K_L)} \left[\alpha - \frac{1 - k_0}{D} V \right] \right\}^{-1}$$

(29)

where

$$\mathcal{G} = \frac{K_L G_L + K_S G_S}{K_L + K_S}$$

$$\alpha = \frac{V}{2D} + \left[\left(\frac{V}{2D} \right)^2 + k^2 + \frac{\omega}{D} \right]^{1/2}$$

and G_{LC} has been defined in Equ.(20).
The denominator in (29) is always positive.
In the numerator the first term which contains the surface tension dominates in the short wavelength limit; the second term involving a weighted mean of the thermal gradients G_L and G_S is negative if $G_L > 0$ and $G_S > 0$ as it is generally the case in laser treatments. These two contributions tend to restore the flatness of the interface. Whereas the third term is positive and destabilizing, it plays here the role of the supercooling bifurcation parameter.

The marginal stability point leading to the onset of stationary cells occurs when both $Re\{\omega_k\} = 0$ and $Im\{\omega_k\} = 0$ (principle of exchange of stability). From Equ.(29) this condition reduces to

$$\tau - \beta k^2 - \frac{k_0}{p + k_0 - 1} = 0$$

(30)

where

$$\tau = 1 - \frac{\mathcal{G}}{m G_{LC}} \quad ; \quad \beta = \frac{\Gamma T_m}{m G_{LC}}$$

and

$$p = \frac{1}{2} + \left[1/4 + \frac{k^2 D^2}{V^2} \right]^{1/2} \qquad (31)$$

Equ.(30) defines the neutral stability curve and the critical wavenumber q_c can be determined by the parametric equation (30) and the condition of minimum of τ :

$$\frac{\partial \tau}{\partial k^2} = \beta - \frac{k_0 (D^2/V^2)}{(2p-1)(p + k_0 - 1)^2}$$

(32)

A disturbance of wavelength $\lambda_c = 2\pi/q_c$ grows more rapidly into the suprecooled liquid than the rest of the interface; these protrusions develop and give rise to a structure perpendicular to the interface. The rejected solute accumulates in the cell walls which can only freeze at a lower temperature . The formation of such cells has been observed in many materials after pulsed laser irradiation.
J. Narayan (12) has performed systematic studies on average cell sizes and critical solute concentration in In^+, Sb^+, Ga^+ and Bi^+ , implanted silicon after laser annealing. A good agreement is obtained between the experimental results and the calculations when the nonequilibrium segregation coefficient is included in Equ (29) (see Sect.4.3). The nonlinear treatments of this instability may divided into two broad categories (14). One category is based on the corresponding amplitude equations (See Sect. 3.1), the second category is composed of numerical solutions to the full nonlinear equations. In next section, we discuss how the

4.3 Nonequilibrium segregation

In the derivation of the Mullins-Sekerka formula we have made a number of approximations. In particular we have assumed that the conditions of local equilibrium prevail. However pulsed laser melting of doped semiconductors produce such rapid interface motion that the assumption of local compositional equilibrium can hardly be valid. Solute concentrations exceeding the equilibrium solubility limits have been attained (15). The segregation coefficient also shows a marked dependence on the solidification velocity. It can be shown that the impurity has actuality undergone an increase in chemical potential upon solidification. The term solute trapping has been coined to characterize this phenomenon. I want to discuss briefly a model introduced by Aziz that fits the data quite well (16).

A) Stepwise growth

The interface is assumed to be planar and to be executing jumps of length λ (lattice spacing) separated by a time interval $\tau = \lambda/v$. At $t=0$, the advance of the interface incorporates a liquid monolayer including the solute atoms. The impurity atom has then a time τ to diffuse back into the liquid before the next monolayer is added at time τ whereupon the whole process repeats itself.

B) Continuous growth model

In a more realistic model one considers that the interface is advancing at a constant velocity; the potential surrounding the impurity is then constantly changing during the time τ . Assuming steady state conditions in the frame of reference moving with velocity

$$J_{s\to l} - J_{l\to s} = v\,[c_L - c_s] \qquad (33)$$

Chemical rate theory gives the following expressions for the impurity diffusion fluxes.

$$J_{\ell\to s} = k_f\, c_\ell \quad ; \quad J_{s\to \ell} = k_r\, c_s \tag{34}$$

and the equilibrium segregation coefficient is given by $k_o = \frac{k_f}{k_r}$

By substitution into (33) we get

$$c_s \left[1 + \frac{V}{V_D}\right] = c_\ell \left[k_o + \frac{V}{V_D}\right] \tag{35}$$

where $V_D = D/V$ is the diffusive speed. The nonequilibrium partition coefficient is then given by

$$\bar{k} = \frac{c_s}{c_L} = \frac{k_o + V/V_D}{1 + V/V_D} \tag{36}$$

Clearly this simple formula predicts equilibrium segregation when $V \ll V_D$ and complete solute trapping for $V \gg V_D$. By combining transient-conductance and glancing Rutherford backscattering spectroscopy, the dependence of the nonequilibrium partition coefficient of Bi in Si upon solidification velocity has been measured with sufficient accuracy (17) to distinguish between proposed solute trapping mechanisms. These results agree well with Aziz continuous-growth model (16,17).

REFERENCES

1. Nicolis G. and Prigogine I. "Self-Organization in Nonequilibrium Systems" (Wiley-Interscience, New-York 1977)
2. Haken H. " Advanced Synergetics" (Springer-Verlag,Berlin 1983).
3. Koschmieder L., Adv. Chem. Phys. 32 (1975) 109.
4. Langer J.S., Rev. Mod. Phys. 52 (1980), 1
5. Schlüter A., Lortz D., and Busse F, J. Fluid Mech. 23 (1965) 129
6. Walgraef D, Dewel G, Borckmans P, Adv. Chem. Phys. 49 (1982) 311
7. Swift J, and Hohenberg P.C., Phys. Rev. A15 (1977) 315
8. Manneville P, J. de Physique 39 (1978) 911
9. Pieranski P, Dubois-Violette E, Guyon E, Phys. Rev. Lett. 30 (1973) 736
10. See for instance van Gurp G.J. and al. Appl. Phys. Lett. 35 (1979) 273
11. Tiller W.A. and al, Acta Met. 1 (1953) 428
12. Narayan J, J. Cryst. Growth 59 (1982) 583
13. Mullins W.W. and Sekerka R.F. J. Appl. Phys. 35 (1964) 444
14. Coriell S.R. Mc Fadden G.B. Sekerka R.F., Ann. Rev., Mater. Sci. 15 (1985) 119
15. White C.W., Wilson S.R., Appleton B.R., Young F.W., J. Appl. Phys. 51 (1980) 738

16. Aziz J.M. J. Appl. Phys. 53 (1982) 1158;
Appl. Phys. Lett. 43 (1983) 552
17. Aziz J.M., Tsao J.Y., Thompson M.O., Peercy P.S., and
White C.W., Phys. Rev. Lett. 56 (1986) 2489

DISLOCATION MICROSTRUCTURES IN NONEQUILIBRIUM MATERIALS.

D. Walgraef(+)
Service Chimie-Physique II
Université Libre de Bruxelles, Brussels, Belgium.

I. INTRODUCTION.

One of the most fascinating and intriguing aspects of natural phenomena is that complex systems may undergo symmetry-breaking instabilities leading to pattern formation or coherent behavior over macroscopic space and time scales (1).

Therefore the understanding of why order may appear spontaneously in open systems far from equilibrium and which planforms are selected among a large manifold of possibilities has become a major theme of research both theoretically and experimentally (2).

These studies, first related to fundamental questions, appear now to be of technological importance. Effectively, during the last years, the whole field of materials science experienced a complete renewal. Indeed, by using techniques able to operate in strong nonequilibrium conditions, and hence to escape from the constraints of equilibrium thermodynamics, totally new materials structures have been processed. Such techniques include ion implantation, laser beam surface melting as well as electron beam heating. For example, ion implantation processing is able to create surfaces with compositions markedly different from the bulk, leading to materials having new electric, magnetic or chemical properties. In laser annealing, after the tremendously rapid melting and recrystallization of the sample surfaces, microstructures with superior resistance to friction, corrosion,... are frozen into place. Rapid solidification of alloys trigger the formation of quasi-crystalline structures. Ion beam mixing can modify the electrical properties of polymers or improve the adhesion of metallic films to ceramics. On the other hand, dislocation microstructures in stressed materials are associated with strain localization and crack initiation but also controls the advance of crack tips. Hence the control of such microstructures may affect the fatigue life of the specimen (3).

Such techniques which have already been used in commercial manufacturing are developed in a rather empirical way and despite their technological interest, their theoretical understanding is relatively poor. This is mainly due to the

---

(+) Research Associate at the National Fund for Scientific Research (Belgium).

strong nonequilibrium conditions in which they operate and which are required to create and stabilize unstable or at least highly metastable materials or microstructures. Hence equilibrium considerations fail to describe satisfactorily such phenomena. One may however expect significant breakthroughs in this field by the use of typical nonequilibrium concepts related to irreversible thermodynamics, instability, bifurcation analysis, nonlinear reaction-diffusion dynamics which were already successful in other fields where spatio-temporal ordering is the dominant feature (hydrodynamics, biophysics, nonlinear chemistry, ...) (1). The effort in this domain is also intended to answer the need for a larger transfer of knowledge between fundamental science and technology and is aimed to furnish a unifying framework for the treatment of the basic questions suscited by the development of modern materials science.

The general properties of instabilities and self-organisation phenomena in nonequilibrium systems have been reviewed by G.Dewel in these proceedings and I would like to emphasize here some aspects of reaction-diffusion dynamics relevant to the self-organisation of defect populations in driven materials. A concrete example I will discuss in this paper is the nucleation of dislocation microstructures in driven materials under stress. It is effectively well known that, under cyclic or monotonous loading, dislocations form structures (veins, persistent slip bands, labyrinth structures, ...) which induce strain localization and act as initiating centers for crack nucleation and propagation. The origin of such phenomena may be found in the competition between the motion of dislocations in the plastic regime and their cooperative interactions (annihilation, multiplication, locking, ...). They are of considerable practical interest since they may control the crack evolution , induce stochastic aspects in the development of small cracks, ... It is also a further example where instabilities occuring at the microscale may dictate the macroscopic behavior of the material. To this respect, persistent slip bands (PSB) are particularly interesting for the following reasons:

- they develop a ladder-like structure with walls of very high dislocation density perpendicular to the primary slip direction; the wavelength of this structure seems to be a material property independent of the underlying crystalline parameters,
- they induce the formation of extrusions and intrusions at the sample surfaces which are at the origin of micro-crack formation,
- they usually propagate from the surface into the bulk of the material and their nucleation is consequently very sensitive to the boundary properties (dislocation density, residual stresses, ...) and is affected by surface coating, annealing or irradiation.

The aim of this paper is to show that the main properties of dislocation patterning in far from equilibrium materials may be interpreted as the result of dynamical instabilities of the symmetry-breaking type in reaction-diffusion systems. Hence the general aspects of instability and bifurcation in reaction-diffusion dynamics will be reviewed in section 2 while section 3 will be devoted to the detailed study of the various instabilities which may occur in the dislocation populations of driven materials with special emphasis on a dynamical model for the formation of persistent slip bands. Finally the results of the numerical analysis of this model will be presented in section 4 and their relevance to the interpretation of experimental observations will be discussed.

2. INSTABILITIES AND BIFURCATIONS IN REACTION-DIFFUSION DYNAMICS.

In many nonequilibrium systems spatio-temporal organisation results from cooperative effects within populations of individuals like electrons, molecules, cells, insects, dislocations, vacancies, voids, ... The instability of their homogeneous equilibrium steady states is usually due to the interplay between their mobility and local and nonlocal interactions which is modified by the external constraints driving the system away from equilibrium. The macroscopic dynamics of such systems are deduced from phenomenological rate equations or from master equations built on elementary processes between the members of the populations under consideration. Both approaches lead to dynamical systems of the type:

$$\partial_t X_i = F_i(\lambda, X_j) + D_i \nabla^2 X_i \tag{1}$$

where X_i are the local concentrations of the active species, D_i are the diffusion coefficients (or tensors in anisotropic media; in the case of diffusional instabilities, higher order diffusion coefficients may be needed). F_i are the rates of change of X_i due to mutual interactions (usually nonlinear functions of the concentrations), and λ represents a set of parameters describing the external constraints. The systems is also provided with adequate boundary conditions.

When the external constraints are increased beyond the linear regime around thermal equilibrium various instabilities may arise which are located by analyzing the linear stability of the homogeneous steady state X_i^0 . Linearising the system (1) around this reference state one obtains:

$$x^i_{\underline{q}}(t) = L_{ij} x^j_{\underline{q}}(t)$$

where

$$x^i_{\underline{q}}(t) = X^i_{\underline{q}}(t) - X_i^0 \delta_{q,0} = L^{-d/2} \int d^d r \, (X_i(\underline{r}) - X^0_i) \exp i\underline{q}.\underline{r}$$

with

$$L_{ij}(q^2, \lambda) = (\partial F_i / \partial X_j)^0 - q^2 D_i \qquad (2)$$

The stationnary state X_i^0 is asymptotically stable if all the eigenvalues of the matrix L_{ij} have negative real parts. Different types of instabilities may occur when λ is varied, for example:

- transitions between homogeneous steady states when L_{ij} has one eigenvalue $w_q(\lambda)=0$ for $q=0$ at $\lambda = \lambda_c$;

  when L_{ij} has one eigenvalue $w_q(\lambda)=0$ for $|q|=q_c= 0$ at $\lambda = \lambda_c$ one has a soft mode instability inducing the formation of spatial patterns (Turing-like instability);

- when L_{ij} has two complex conjugate eigenvalues with zero real part at $\lambda = \lambda_c$ and $q=0$, one has a hard mode instability leading to time-periodic solutions of the limit cycle type (Hopf bifurcation).

In the region close to the instability a separation of time scales occurs which is associated to the divergence of the relaxation time of the unstable modes. This effect which is known in equilibrium phase transitions as critical slowing down allows a reduction of the complete dynamics of the system to the dynamics of the unstable modes which are the eigenmodes associated to the vanishing eigenvalues of the linear evolution matrix L_{ij}. The corresponding kinetic equations may be obtained via the adiabatic elimination of the stable modes (4) or by the derivation of the associated normal forms within the framework of the center manifold theory (5). Nevertheless, like in equilibrium phase transitions, the sensitivity of the systems to local fluctuations increases tremendously near instabilities and many authors have stressed the need to include them in the description of systems where a coherent behaviour takes place at the macroscopic level. To do so we add appropriate Langevin noise terms $_i(r,t)$ to the balance equations (1) in the spirit of the conventional theory of fluctuating hydrodynamics(6). Because they take care of rapid small scale effects, we assume that:

$$\langle \eta_i(\underline{r},t)\, \eta_j(\underline{r}',t) \rangle = 2\left[\Gamma_{ij}^{c}(\underline{r}) + \nabla_i^2\, \Gamma_{ij}^{d}(\underline{r}) \right] \delta(r-r')\, \delta(t-t') \qquad (3)$$

and also that the system remains in a state of local equilibrium throughout the transition region. The coefficients Γ^c and Γ^d may be determined by a local fluctuation-dissipation theorem or by the underlying master equation description of the phenomenon. This type of approach then tranforms the system into a time-dependent Ginzburg-Landau type of dynamics in the vicinity of the instability and I will discuss in this framework the characteristic aspects of patterning and Hopf bifurcations.

An essential aspect of these bifurcations lies in the fact that they are symmetry-breaking instabilities (translational and/or rotational symmetry in the case of patterning, phase symmetry in the case of oscillations). This leads to formal similarities in the description of the transitions which occur in such systems which may however be of very different physical nature. It is nevertheless possible to go beyond formal similarities to find the origin of symmetry-breaking instabilities in a common interplay between different physical factors. Basically two different mechanisms operating on different time and space scales are needed. One of them would tend to destabilize the symmetrical state while the other one would control the growth of perturbations. An asymmetric state would then emerge as a compromise between these two trends.

For example, pattern formation occurs, in the simplest cases of reaction-diffusion dynamics, as the result of the interaction between a slowly diffusing activator and a rapidly diffusing inhibitor (1). In hydrodynamics, the role of the activator is played by a convective mechanism enhancing its own source while the role of the inhibitor is played by a dissipative process like diffusion, heat conduction, viscosity,...

In the case of dislocation patterning in the plastic regime, the forest of immobile dislocations play the role of the activator since these dislocations are able to lock and immobilize the mobile ones via an autocatalytic type of interaction. On the other hand, the dislocations which break free and become mobile play the role of the rapidly diffusing inhibitor.

Systems which undergo symmetry-breaking transitions are particularly sensitive to any form of perturbations, especially those related to boundary effects. Boundary effects may indeed induce inhomogeneities in the medium or in the bifurcation parameter associated to the external constraints. This may lead to the propagation of patterns from the boundaries within the bulk and this effect has very practical implications in the case of dislocation patterns we are interested in here. While PSBs may in principle nucleate everywhere in the crystal, they usually initiate at the surface as the result of the high dislocation content associated to extrusions or intrusions and propagate into the bulk, this process being much more rapid than the first one. However, if the surface has been treated by laser annealing for example, the residual compressive surface stresses inhibit the initiation of PSBs at the boundaries. As a result, microcrack formation at the surface is more difficult and the fatigue life of the specimen is increased.

2.1 Hopf Bifurcation and the Transition to Temporal Oscillations:

The Hopf bifurcation is a typical example of symmetry breaking instability in nonlinear dynamical systems. As mentioned earliar, such a bifurcation occurs when two eigenvalues of the linear evolution matrix associated with the dynamical system under consideration become purely imaginary while the other ones have negative real parts. The corresponding eigenmodes are, hence, the slow modes of the problem which behave like a two-component order parameter.
The remaining fast modes may be adiabatically eliminated and the evolution of the system on its longest time scale is then given by the following normal form including diffusion:

$$\dot{\sigma}(\underline{r},t) = [r + c\nabla^2]\sigma(\underline{r},t) - u|\sigma|^2\sigma(\underline{r},t) + \eta(\underline{r},t) \tag{4}$$

The structure of the equation is generic but of course the coefficients r, c, u are model dependent and may be computed explicitely for concrete dynamical systems. Quadratic terms in σ do not appear near the instability but are significant for time scales shorter than ω_0^{-1} or at higher values of the bifurcation parameter.

As the parameters of this normal form are complex, eq. (4) exhibits no evident gradient structure. However, as we will see below some aspects of the dynamics may be described by a potential-like functional. Let me first discuss the homogeneous problem which is relevant to small or well-stirred systems. In this case , eq.(4) may be written as:

$$\dot{\sigma}_0 = -\Gamma L^{-d}\frac{\delta\mathcal{F}_0}{\delta\sigma_0^*} - i[r_2\sigma_0 + u_2|\sigma_0|^2\sigma_0] + \eta_0 \tag{5}$$

where

$$\mathcal{F}_0 = \frac{L^d}{\Gamma}\left[r_1|\sigma_0|^2 + \frac{u_1}{2}|\sigma_0|^4\right] \quad (r=r_1+r_2, u=u_1+u_2) \tag{6}$$

is the standard homogeneous potential or Lyapounov functional for systems with a two-component order parameter such as the XY model of ferromagnetism (7) or the single mode laser (8). The asymptotic distribution of the fluctuations is then:

$$P \propto N \exp - \mathcal{F}_0 \tag{7}$$

Hence the shape of this potential corresponds to a Gaussian well in the disordered phase ($r_1 > 0$) and becomes craterlike beyond the bifurcation point defined by $r_1 = 0$. Therefore the maximum of the steady state probability corresponding to the most probable state of the system correponds to the homogeneous steady state for $\lambda < \lambda_c$ and to a limit cycle the orbit of which corresponds to the minimum of the crater-like potential for $\lambda > \lambda_c$. The time evolution of the most probable value of the order parameter is then given by :

$$\overline{\sigma}_0(t) = R(t)\ \exp i\phi(t)$$

with

$$R^2(t) = \begin{cases} -r_1R^2(0)/(u_1R^2(0)-(r_1+u_1R^2(0))\exp 2r_1t) & r_1 \neq 0 \\ R^2(0)/(1+2u_1R^2(0)t) & r_1 = 0 \end{cases} \qquad (8)$$

leading to the stationnary values:

$$R^2(t \to \infty) = \overline{R}^2 = \begin{cases} 0 & r_1 \geqslant 0 \\ |r_1|/u_1 & r_1 < 0 \end{cases} \qquad (9)$$

characteristic of a second orde-like transition. The phase dynamics is then given by:

$$\phi(t) = \begin{cases} \phi(0) - \frac{u_2}{u_1} \ln\left(1 + \frac{u_1R^2(0)}{r_1}\right) - r_2 t & (r_1 > 0) \\ \phi(0) - \frac{u_2}{u_1} \ln \frac{R(0)}{\overline{R}} - (r_2 + u_2\overline{R}^2)t & (r_1 < 0) \end{cases} \qquad (10)$$

As a consequence, this mean field analysis predicts a second order-like transition at $r_1=0$ from a homogeneous steady state to a limit cycle,-its amplitude growing with the classical index 1/2-with the spontaneous breakdown of the phase symmetry (fig.1). The study of the stability of the limit cycle shows that there is no symmetry restoring mechanism for the phase : any perturbation of the system induces a phase shift of the oscillations. Hence homogeneous fluctuations may neither be neglected in the vicinity of the transition nor in the ordered regime and furthermore one may also expect that long range inhomogeneous fluctuations will play an important role beyond threshold.

Effectively, the latter are of prime interest in many oscillating nonlinear media described by general reaction-diffusion equations where no uniformisation process of the medium is available.

In the ordered regime, the Langevin equations (4) may be written, in phase-amplitude variables, as:

$$\dot{R} = u_1 R(R_0^2 - R^2) + c_1(\nabla^2 R - R(\vec{\nabla}\phi)^2) - c_2(2\vec{\nabla}R\vec{\nabla}\phi + R\nabla^2\phi) + \eta_R$$

$$R\dot{\phi} = u_2 R(R_0^2 - R^2) + c_2(\nabla^2 R - R(\vec{\nabla}\phi)^2) + c_1(2\vec{\nabla}R\vec{\nabla}\phi + R\nabla^2\phi) + \eta_\phi \qquad (11)$$

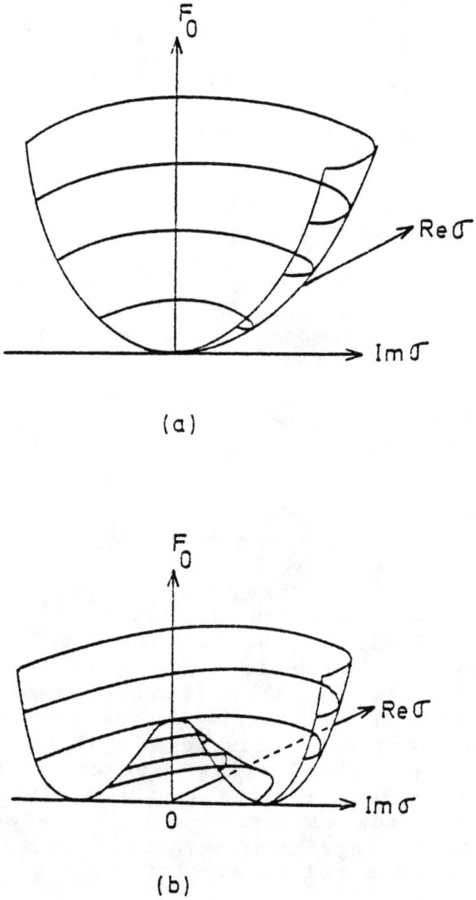

Figure 1. Homogeneous potential near the Hopf bifurcation:
a) below threshold in the disordered phase,
b) above threshold in the ordered phase.

Linear stability analysis performed on this system around the limit cycle defined by $R=R_0, \phi = \phi_0$, shows that when $u_1c_1+u_2c_2 > 0$, it is stable against small radius or phase fluctuations while it is unstable in the opposite case where phase turbulence occurs. Even in the first case the asymptotic dynamics may be reduced to the following phase dynamics:

$$\dot{\phi} = \mu \nabla^2 \phi + \nu (\vec{\nabla}\phi)^2 + \eta_\phi \qquad (12)$$

where μ and ν are constants depending on the parameters of the system ($\mu = c_1+u_2c_2/u_1$, $\nu = (u_1c_2-u_2c_1)/u_1$).

Hence, two types of phase fluctuations may influence the oscillations on the macroscopic level:

1) As phase and amplitudes are strongly coupled on short length scales, even small localized disturbances may induce local frequency shifts of the oscillations. The small regions where these shifts take place oscillate with a period different from that of the bulk and radiate concentric waves of activity or target patterns. One finds that such a fluctuation induces the following phase behaviour far from the center:

$$\phi(\underline{x},t) = \mathrm{Max}\left(0, \nu k^2 (t-t_0) - k|\underline{x}-\underline{x}_i|\right) \qquad (13)$$

where k, the wavenumber of the target pattern increases with the amplitude of the perturbation. Such patterns are observed in several two-dimensional oscillating media.

2) Vortex-like fluctuations associated to phase singularities analogous to the vortices of XY or superfluids may also be nucleated in two-dimensional oscillating systems. Such vortices have an integer topological charge, or winding number N. Their long-time evolution leads to rotating N-armed Archimedean spiral waves of isoactivity:

$$\phi(\underline{x},t) = N \tan^{-1}\frac{(y-y_i)}{(x-x_i)} + \mathrm{Max}\left(0, \nu k_N^2 (t-t_0) - k_N|\underline{x}-\underline{x}_i|\right) \qquad (14)$$

In one-dimensional or highly anisotropic systems the corresponding fluctuations lead to one-dimensional propagative wavefronts.

To conclude this section, let us emphazise the essential aspects of the transition to oscillatory behavior in reaction-diffusion systems. In the case of a Hopf bifurcation, such a transition is strongly analogous to a second order phase transition. However the dynamics is original and specific to each nonequilibrium kinetics. Nevertheless as the instability is a continuous symmetry-breaking instability, the phase plays the role of a Goldstone mode which is diffusive in dissipative systems. The associated phase fluctuations may destroy the long range order in low-dimensional systems via the spontaneous emission of defects which manifest themselves by the emission of waves of activity. We see in this first example that in the case of symmetry-breaking instability the deterministic analysis alone is unable to render the complete macroscopic properties of the system and that phase fluctuations are able to modify this analysis.

2.2 Transitions to Spatial Patterns:

Many nonlinear systems where interaction terms are coupled with transport processes may also undergo symmetry-breaking transitions leading to macroscopic spatial structures. The instability occurs in this case when one root of the characteristic equation associated with the linearized kinetic equation around the reference state has a positive real part for nonuniform perturbations. Linear stability analysis hence fixes the critical value of the control parameter and the critical wavelength of the patterns. Contrary to most hydrodynamical instabilities, the critical wavelength in reaction-diffusion systems is determined by intrinsic properties and is given in terms of the two competing mechanisms (reaction rates and diffusion constants) the dynamic equilibrium of which leads to the patterning of the medium. Systems with spatial dimensionnalities higher than one exhibit a great degeneracy. In isotropic media this degeneracy is two-fold (9):
1) for a given wavelength different planforms (rolls, squares, hexagons,...) may be simultaneously stable;
2) the localisation and the orientation of the patterns remain indetermined in unbounded systems.

In anisotropic media, however, an intrinsic mechanism raises the orientational degeneracy by inducing preferred directions in the medium for the wavevectors cheracterising the structure.

As pointed out in the introduction, close to the instability the complete nonlinear dynamics may be reduced to the dynamics of the unstable modes and takes the form:

$$\dot{\sigma}_q = \omega_q \sigma_q - \sum_{q'} v(\{q\}) \sigma_{q-q'} \sigma_{q'} - \sum_{q',q''} u(\{q\}) \sigma_{q-q'} \sigma_{q'-q''} \sigma_{q''} \tag{15}$$

The linear growth rate, ω_q, tends to zero when approaching the instability. In isotropic systems it depends only on the length of the wavevector while in anisotropic media it also depends on its orientation. For example in uniaxial materials we have:

$$\omega_q = r_0 - D(q^2 - q_c^2)^2 - A q^2 \sin^2 \varphi \tag{16}$$

where $r_0=(\lambda-\lambda_c)/\lambda_c$, while φ is the angle between the wavevector and the pricipal axis of the system. A is the anisotropy constant; D,u and v depend on the parameters of the dynamical system and may be obtained explicitely via the adiabatic elimination methods or multiple-scale analysis. Such kinetic equations have been derived near Rayleigh-Benard instabilities, near convective instabilities in nematic liquid crystals or in nonlinear chemical networks. Numerical and experimental analysis also show that such models correctly describe the behavior of the system even well beyond threshold.

In the isotropic case and for u and v constants, the kinetic equation (15) may be shown to derive from the following generalized potential:

$$\mathcal{F} = -\frac{1}{2}\sum_{q} \omega_q |\sigma_q|^2 + \frac{1}{3!}\sum_{\{q\}} v(\{q\})\,\sigma_q \sigma_{q'} \sigma_{-q-q'} + \frac{1}{4!}\sum_{\{q\}} u\,\sigma_q \sigma_{q'} \sigma_{q''} \sigma_{-q-q'-q''} \quad (17)$$

The preferred steady state which minimizes F can be characterized by m pairs of wavevectors of critical length. The spatial variations of the order parameter are then given by:

$$\sigma(\underline{r}) = 2\sum_{i=1}^{m} |\sigma_i| \cos\left(q_c(\underline{1}_i \cdot \underline{r}) + \varphi_i\right) \quad (18)$$

where the phases φ_i are free constants defining the position and orientation of the pattern while the amplitudes σ_i satisfy the stationarity condition:

$$\delta \mathcal{F} / \delta \sigma_{-q_i} = 0 \quad (19)$$

Two types of structures have to be considered:

a) structures appearing supercritically through a second order transition. They are characterized by independent pairs of wavevectors, i.e. with no definite angular relation between them. In this case there is no contribution from the cubic invariant of the potential and the amplitudes are given by:

$$\sigma_i \ (i=1,\dots m) = \sigma^{(m)} = \begin{cases} 0 & \lambda < \lambda_c \\ \left[\dfrac{2r_0}{(2m-1)u}\right]^{1/2} & \lambda > \lambda_c \end{cases} \quad (20)$$

The linear stability analysis shows that the only stable structures of this type corresponds to modulations in one direction only or roll patterns constructed on one pair of wavevectors.

b) structures appearing subcritically: in this case the transition is of the first order type because of the contribution of the cubic terms of the potential. Such patterns are constructed on pairs of wavevectors arranged to form equilateral triangles. In two-dimensional systems they correspond to honeycomb or triangular lattices. In three-dimensional systems they correspond to rodlike structures with triangular symmetry. The corresponding bifurcation diagram is displayed in fig.2.

Let us now discuss anisotropy effects on the following slow mode dynamics:

$$\dot{\sigma}_q(t) = [r_0 - D(q^2 - q_c^2)^2 - Aq_y^2]\sigma_q(t) - \sum_{\{q\}} u(q+q')\sigma_{q'}\sigma_{q''}\sigma_{q-q'-q''} \quad (21)$$

where the anisotropy coefficient A is positive and u(q+q') is minimum when q and q' are orthogonal. This model is well adapted to the description of convective instabilities in liquid crystals especially in the case of nematics submitted to elliptical shears (10). A typical realisation of a space-dependent nonlinear coupling may be found in the Sivashinsky model for Rayleigh-Benard convection between poorly conducting plates (11). In the absence of anisotropy the nonlinear terms of this model favor square patterns of amplitude given by:

$$|\sigma_S|^2 = r_0/(u_D + 2u_{ND}) \quad (22)$$

where u_d is the coupling between parallel wavevectors and u_{nd} between perpendicular ones. The stability analysis shows that the square pattern is stable when $u_d - 2u_{nd} > 0$ whereas roll patterns are stable otherwise. Hence the angular dependence of the nonlinear couplings furnishes in this case a selection mechanism. What happens now in the presence of anisotropy?

When A = 0 the instability corresponds to r_o=0, q_x=q_c and q_y=0 leading to a roll pattern with rolls orthogonal to the y direction. On the other hand, the nonlinear terms still favor square patterns defined by:

$$r_0 - [u_D|\sigma_1|^2 + 2u_{ND}|\sigma_2|^2] = 0$$

$$[r_0 - Aq_c^2] - [u_D|\sigma_2|^2 + 2u_{ND}|\sigma_1|^2] = 0$$

$$\sigma_S = 2[|\sigma_1|\cos(q_c x + \varphi_1) + |\sigma_2|\cos(q_c y + \varphi_2)] \quad (23)$$

The stability analysis combined with eq.(21) shows that the roll pattern induced by the anisotropy becomes unstable for:

$$u_D > 2u_{ND}, \quad r_0 > u_D A q_c^2/(u_D - 2u_{ND}) \quad (24)$$

where one has a direct transition to a square pattern. We see explicitely in this model the competition between anisotropy and nonlinear effects. First, rolls induced by the anisotropy are selected but as the bifurcation parameter is increased, the nonlinearities become more important and favor the existence od a square structure. This model presents strong

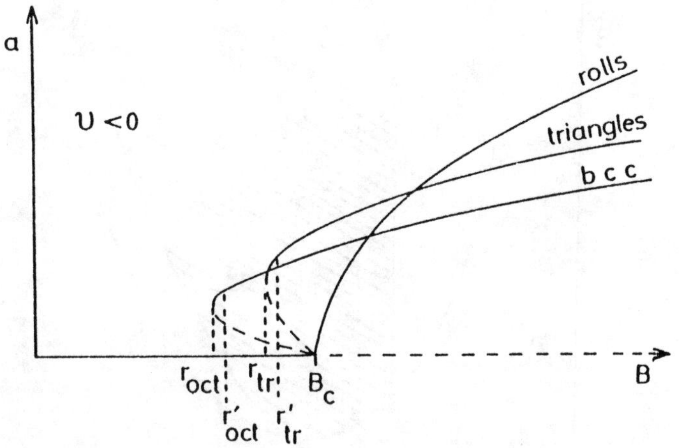

Figure 2. Bifurcation diagram near a patterning instability (a is the amplitude of the order parameter, r_i the stability limit, r'_i the metastability limit).

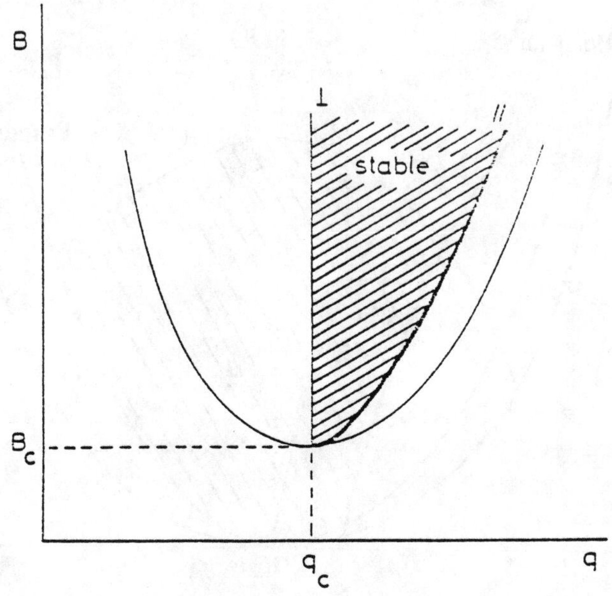

Figure 3a. Stability diagram of the layered structure in the isotropic case.

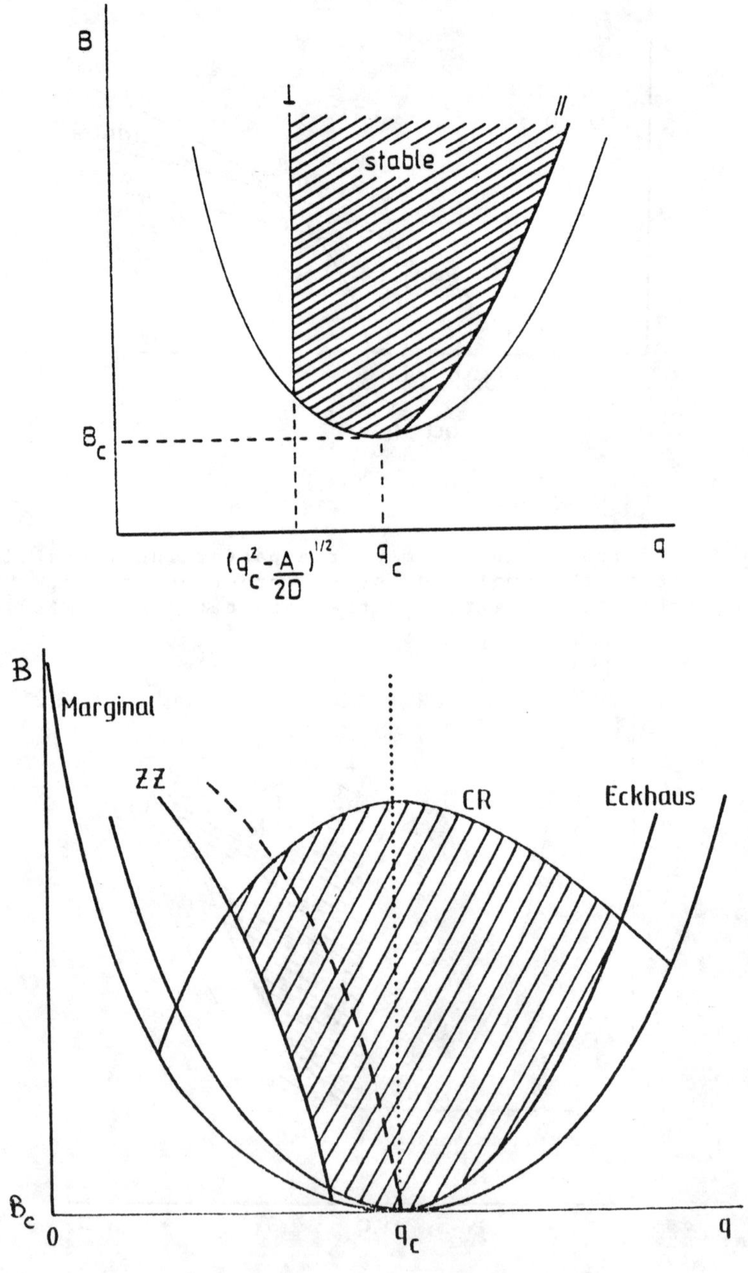

Figure 3. Stability diagram of the layered structure in
b) an anisotropic system
c) the anisotropic Sivashinski model.

similarities with the experimental results obtained in nematics heated from above in the presence of a horizontal magnetic field or in nematics submitted to an elliptical shear (10,12). Another point to be emphasized is that above threshold a range of wavevectors of length near critical- $(r_o - D(q^2 - q_c{}^2)^2 - Aq_y{}^2) > 0$ - have positive growth rates and may thus also lead to patterns. However only a part of these wavevectors may lead to physically realisable solutions. Effectively, by checking the stability of the corresponding layered structures for example, we find two types of instabilities in the isotropic model (9):

1) the zig-zag instability which induces a wavy distortion of the rolls and takes place when $q^2 < q_c{}^2$,

2) the Eckhaus instability which corresponds to a longitudinal compression of the structure and occurs when q lies in the range $q_c - (r_o/3\xi^2)^{1/2} < |q| < q_c + (r_o/3\xi^2)^{1/2}$, $\xi^2 = 4Dq_c^2$

Here also the anisotropy modifies these results. Namely the borderline of the zig-zag instability is shifted to lower values of q. The corresponding stability diagrams for structures periodic in one dimension in isotropic and anisotropic models are given if fig.3.

The patterning instabilities described in this section are also symmetry-breaking instabilities and the breakdown of translational and rotational symmetry leads to the nucleation of long range fluctuations in the ordered region of the parameter space. These fluctuations lead to defects in the patterns which are very similar to the dislocations and disclinations of equilibrium crystals.

To conclude this section let us emphasize the delicate balance between linear and nonlinear terms of the dynamics in the pattern selection processes. Moreover structures which appear to be stable in the deterministic analysis may however be strongly affected by the long range phase fluctuations induced by the breaking of translational and rotational symmetries and which lead to the spontaneous nucleation of defects in nonequilibrium patterns. Small systems will in general not be affected by such fluctuations while in large systems they are responsible for the "melting" of the structures, i.e. the destruction of translational and rotational order in the medium.

I will now show in the next sections how the concepts discussed above may be used to build a unified framework well adapted to the description and understanding of the formation of defect microstructures in nonequilibrium materials with dislocation patterning during fatigue experiments as specific example.

3. SPATIO-TEMPORAL ORDER AND DISLOCATION PATTERNS IN SOLIDS.

In this section, I would like to show how dynamical nonequilibrium concepts and reaction-diffusion equations may improve the understanding of the fundamental aspects of dislocation patterning in stressed materials. Before discussing the the principal aspects of a reaction-diffusion model developed for the description of self-organisation phenomena in fatigue experiments. Let me stress upon a few points in favor of the study of defect microstructures in the framework of nonlinear dynamics.

The macroscopic behavior of irradiated or mechanically or chemically driven materials is usually dictated by cooperative effects between defects like vacancies, interstitials, dislocations or voids. Moreover as such materials are driven away from thermal equilibrium, we cannot rely on traditional mechanical or thermodynamical approaches. Fortunately, elementary reactive interactions between defects and defect motion may be characterized experimentally and theoretically. As a result master equations or kinetic rate equations may be derived for the defect concentrations leading to nonlinear reaction-diffusion descriptions of their dynamics.

This is particularly interesting since such models are able to generate oscillatory or inhomogeneous solutions according to the value and variations of the control parameter and such behaviors are experimentally observed (13,14). Furthermore theoretical modelling of yielding, spatial or temporal variation of dislocation concentration, fracture kinetics,... in the framework of continuum mechanics (15) or irreversible thermodynamics (16) needs to incorporate the behavior of dislocation populations.

In addition to the different aspects of the instabilities discussed in the preceeding sections, specific properties of crystalline solids are expected to strongly influence the analysis. Effectively we deal here with strong anisotropies and the control parameter, say the stress or strain rates, control not only the dislocation creation and interaction rates but also the anisotropy itself via the velocity of mobile dislocations.. Moreover the main problems are related to the dynamics of the nucleation of the spatio-temporal structures since they affect the macroscopic properties of the samples during thei development. It may also happen that the failure of the sample occurs before the steady state is reached.

I will discuss here dislocation patterning in fatigue experiments on monocrystals of metallic compounds submitted to cyclic loading at controlled stress or strain rates. It is now well known that, according to the loading intensity, two types of dislocation patterns may appear in fatigued metals (13-16)(Fig.4). One is the vein structure of the matrix consisting in dense multipoles of primary edge dislocations separated by dislocation poor channels. The second one is ladder-like in copper monocrystals for example and corresponds to persistent slip bands(PSB). It develops in primary slip planes and consists of regularly spaced walls of

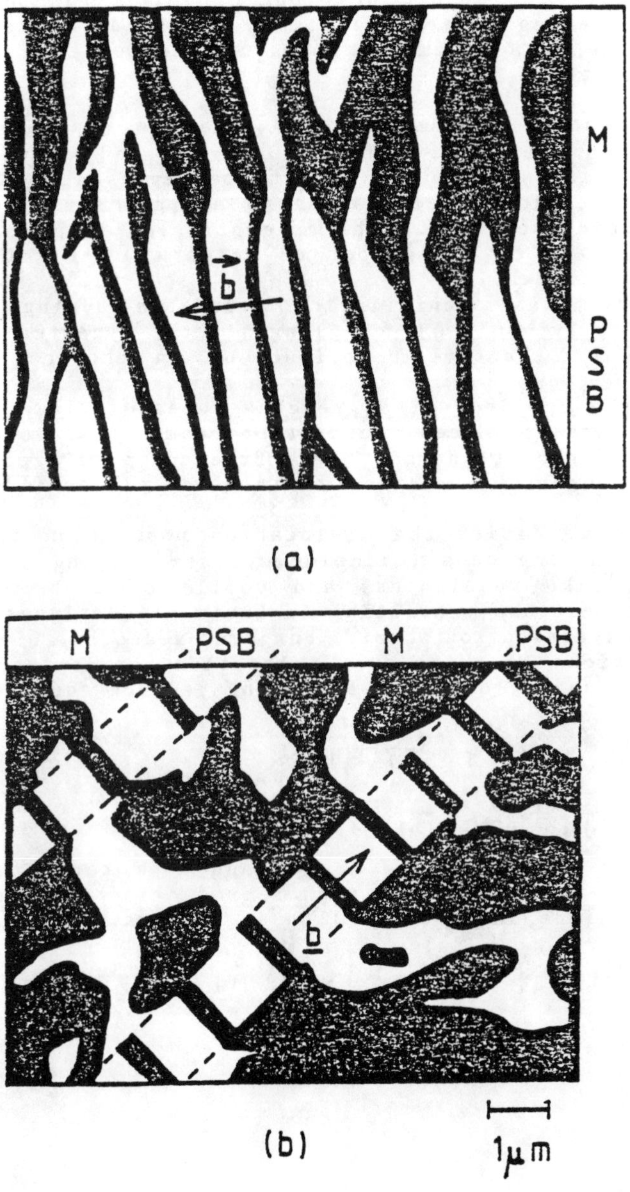

Figure 4. Vein and ladder-like structures in fatigued copper single crystals in the primary slip planes (a) and in the cross-slip plane (b).

high dislocation density. The experimental observations also lead to the conclusion that the dislocation patterns within the PSBs are the result of dynamic equilibrium between different processes such as creation, annihilation, pinning and motion. Hence a dynamical model may be built on the following elements:

1) The creation of dislocations under an applied stress is described by means of internal sources and the dislocation population is sufficiently high to be represented by a continuous dislocation concentration on a space scale larger than a few lattice spacings. The concentration is a local variable associated to the number of Burgers vectors originating in a space element.
2) The dynamical annihilation processes during glide are represented by nonlinear couplings in the dynamics which have the form of reactive contributions in chemical-like rate equations.
3) When the applied stress reaches threshold, i.e. when thermal or stress-induced activation dominates local energy barriers, dislocations may break free and move rapidly with a stress-dependent velocity in their glide planes (17).

Hence we divide the dislocation population in immobile or slowly moving ones (climbing, cross-slipping or pinned at obstacles like multipoles) and mobile or fast moving ones (free dislocations travelling in their glide planes according to the direction of their Burgers vectors). The local concentrations of these two dislocation populations, ρ_I and ρ_M, consequently obey the following balance equations:

$$\dot{\rho}_I + \mathrm{div}\,\vec{j}_I = \hat{c}_I(\rho_I, \rho_M)$$
$$\dot{\rho}_M + \mathrm{div}\,\vec{j}_M = \hat{c}_M(\rho_I, \rho_M) \qquad (25)$$

where $\hat{c}_I$ and $\hat{c}_M$ are source and sink terms taking into account the various mechanisms of production and annihilation of dislocations and are modelled as follows:

$$\hat{c}_I(\rho_I, \rho_M) = f(\rho_I) - b\rho_I + \rho_M \rho_I^2$$
$$\hat{c}_M(\rho_I, \rho_M) = b\rho_I - \rho_M \rho_I^2 \qquad (26)$$

where f(ρ_I) represents creation and annihilation of slow dislocations, b ρ_I the production rate of free dislocations and $\rho_M \rho_I^2$ corresponds to the pinning rate of free dislocations by immobile dipoles. b may be considered as proportional to the local rate of plastic deformation. A detailed analysis of the currents (18,19) leads to diffusive contributions to the kinetic equations.
Effectively the current associated to the slow dislocations may be written as:

$$\dot{j}_{I,i} = -\left[D_{ij}^{(0)} \nabla_j + D_{ijk}^{(1)} \nabla_j \nabla_k^2 \right] \rho_I \qquad (27)$$

Due to the attractive character of the elastic interactions between dislocations of opposite Burgers vectors, $D_{ij}^{(0)}$ may be negative in high density regimes while $D_{ijk}^{(1)}$ remains positive. Hence a diffusive instability may occur for high creation rates or high stress levels. By defining the x axis as parallel to the resolved shear stress we also have:

$$\vec{\dot{J}}_M = \dot{J}_{M,x} \vec{1}_x = - D_M \nabla_x \rho_M \qquad (28)$$

The diffusive nature of the flux $j_{m,x}$ is supported by the fact that the stress is periodically reversed and that positive and negative dislocations move in opposite directions. It turns then out that D_m is of the order of the thermal diffusivity below the yield point and increases rapidly to its maximum value above this threshold. Effectively it may be shown that (41):

$$D_M \simeq V^2 / \omega^2 \qquad (29)$$

where v may be related to the stress intensity via different relationships ranging from a power law to an exponential law of the type $v=v_0\exp(-(\tau_0/\tau)^m)$. Short range effects which occur at the atomic scale for example and which are neglected in this description may be introduced via noise terms leading to stochastic reaction-diffusion equations.

Among the various experimental situations describable by this model, let me just quote the results obtained in the simplest case, i.e. low initial densities in one slip plane. In this case the diffusion coefficients are positive and if xOy is the slip plane under consideration, the dynamical model may be reduced to:

$$\dot{\rho}_I = f(\rho_I) + D_I^x \nabla_{xx}^2 \rho_I + D_I^y \nabla_{yy}^2 \rho_I - b\rho_I + \rho_M \rho_I^2 \qquad (30)$$

$$\dot{\rho}_M = D_M^x \nabla_{xx}^2 \rho_M + b\rho_I - \rho_M \rho_I^2$$

where $D_I^x \gg D_I^y$.

The homogeneous steady state is given by:

$$f(\rho_I^o) = 0 \qquad \rho_M^o \rho_I^o = b \tag{31}$$

and becomes unstable versus homogeneous oscillations (Hopf bifurcation) when:

$$b \geqslant b_c^o = a + \rho_I^{o2} \qquad (a = -f'(\rho_I^o)) \tag{32}$$

or to inhomogeneous perturbations for:

$$b \geqslant b_c = [a^{1/2} + \rho_I^o (D_I^x / D_M^x)^{1/2}]^2 \tag{33}$$

b being an increasing function of the stress or strain rates, the patterning instability is reached before the Hopf bifurcation if:

$$(D_I^x / D_M^x)^{1/2} < a^{1/2} [(1 + (\rho_I^{o2}/a))^{1/2} - 1] / \rho_I^o \tag{34}$$

This condition is satisfied for a wide range of dislocation densities when $D_I^x \ll D_M^x$. This is the case for sudden increases of stress intensity beyond threshold. For slow increases through the threshold region this condition may not be satisfied since D_M^x is proportional to the square velocity of mobile dislocations. In the latter case oscillations in the dislocation populations should occur which could be related to the strain bursts observed in corresponding experimental situations (20).

When the patterning instability occurs first, the slow mode dynamics may be easily obtained showing that the preferred patterns beyond the dislocation freeing point correspond to modulations in the primary slip direction. Such dislocation patterns may be associated to the ladder-like structure of the PSBs.

4. EVOLUTION OF THE DISLOCATION POPULATIONS BEYOND THE PATTERNING INSTABILITY.

The nonlinear character of the dislocation dynamics does not allow in general analytic solutions for the two dislocation populations. However, near the instabilities, this dynamics may be reduced to normal forms or slow mode dynamics leading to amplitude equations for the unstable modes. Effectively, the existence of two time scales near the instability allows the elimination of the fast or stable modes. The asymptotic dynamics may then be expressed in terms of an order parameter like variable which is here a linear combination of the two dislocation populations and may be written, in the absence of diffusional instability (all the diffusion constants positive) and in a two-dimensional description (restricted to dislocation density variations in the slip and cross-slip directions) as (21) :

$$\dot{\sigma} = (r_0 - d_{\|}(q_c^2 + \nabla^2)^2 + d_{\perp}\nabla_y^2)\sigma - v\sigma^2 - u\sigma^3$$

$$(d_{\|} = D_I^*/b_c q_c^2 ,\ r_0 = (b - b_c)/b_c ,\ d_{\perp} = D_I^* D_M^*/b_c(D_I^* + D_M^*)) \quad (35)$$

with σ being the slow mode while the nonlinear coupling intensities u and v may be obtained explicitely in terms of the parameters of the model. In the isotropic case the only stable stationary solutions of this equation correspond to hexagonal or layered patterns while in highly anisotropic situations the cellular structures become unstable. Hence by increasing the stress or strain rate intensity well beyond threshold, the mobility of the free dislocations also increases in the primary slip direction and so does the anisotropy. Near threshold cellular and wall structures may coexist while beyond threshold layered structures with walls perpendicular to the slip direction are the only possible ones. Let us consider the latter case and on expressing the slow mode as :

$$\sigma = A \exp iq_c x + A^* \exp - iq_c x \quad (36)$$

with A=A(x,y,t) being a slowly varying function of its variables and A* its complex conjugate, we obtain the following anisotropic amplitude equation :

$$\dot{A} = [r_0 + \xi_x^2 \nabla_{xx}^2 + \xi_y^2 \nabla_{yy}^2]A - 3u\,|A|^2 A \quad (37)$$

where $\xi^2_x = 4q_c^2 d_x$, $\xi_y^2 = d_y$ and the terms with rapidly varying contributions have been neglected. Hence the preferred solution for steady state modulations in one direction with homogeneous amplitude in a medium with natural boundary conditions is then :

$$|A_0|^2 = r_0/3u \quad (38)$$

From this expression one may then go back to the dislocation densities and obtain their spatial variations. In the case of the successive activation of secondary slip directions, modulations in different directions may develop, leading to the formation of labyrinth structures (22-24).

Up to now the predictions of the one-dimensional aspects of the model corresponding to patterning in the primary slip plane only have been tested numerically. We found for example, for a sudden increase of the strain rate beyond threshold, the propagation of a wall structure from inhomogeneities in the bulk or at the boundaries as illustrated on fig.5 and 6.

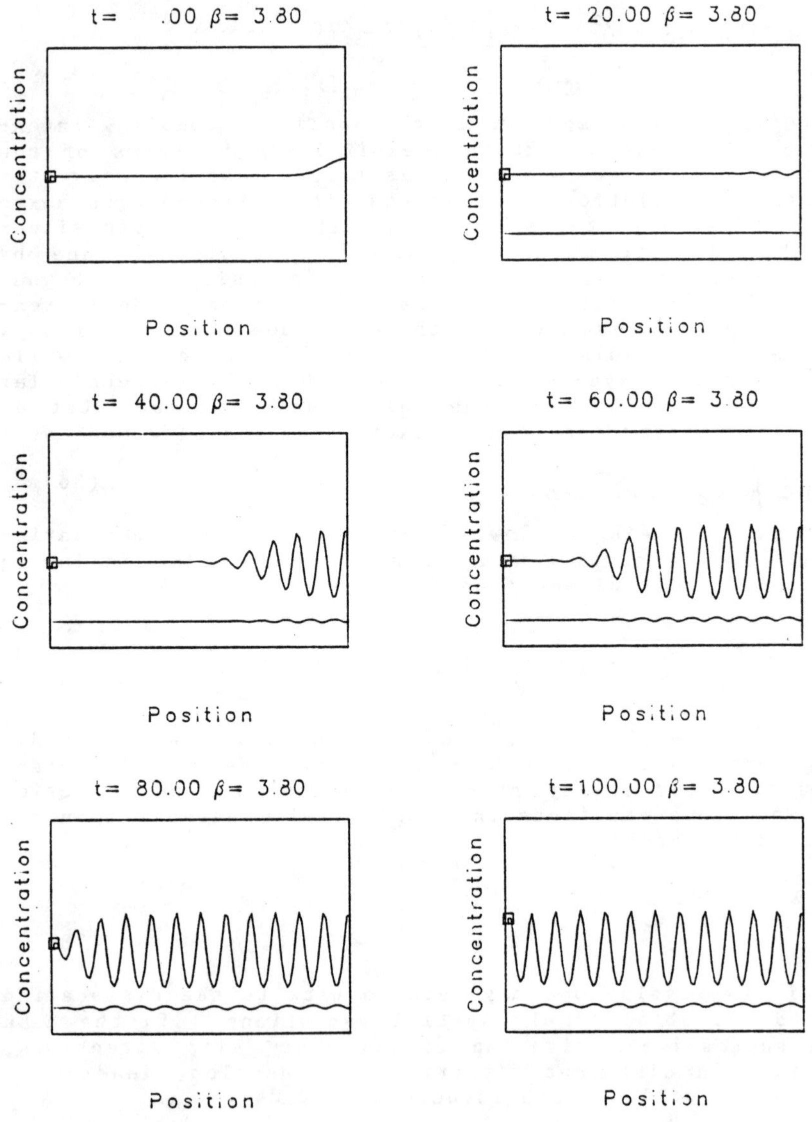

Figure 5. Spatio-temporal evolution of the dislocation populations after a sudden increase of the strain rate slightly betond threshold (D /D =100, =0.07) and computed with the complete reaction-diffusion dyna mics. The concentration of immobile dislocations is marked by a square on the concentration axis.

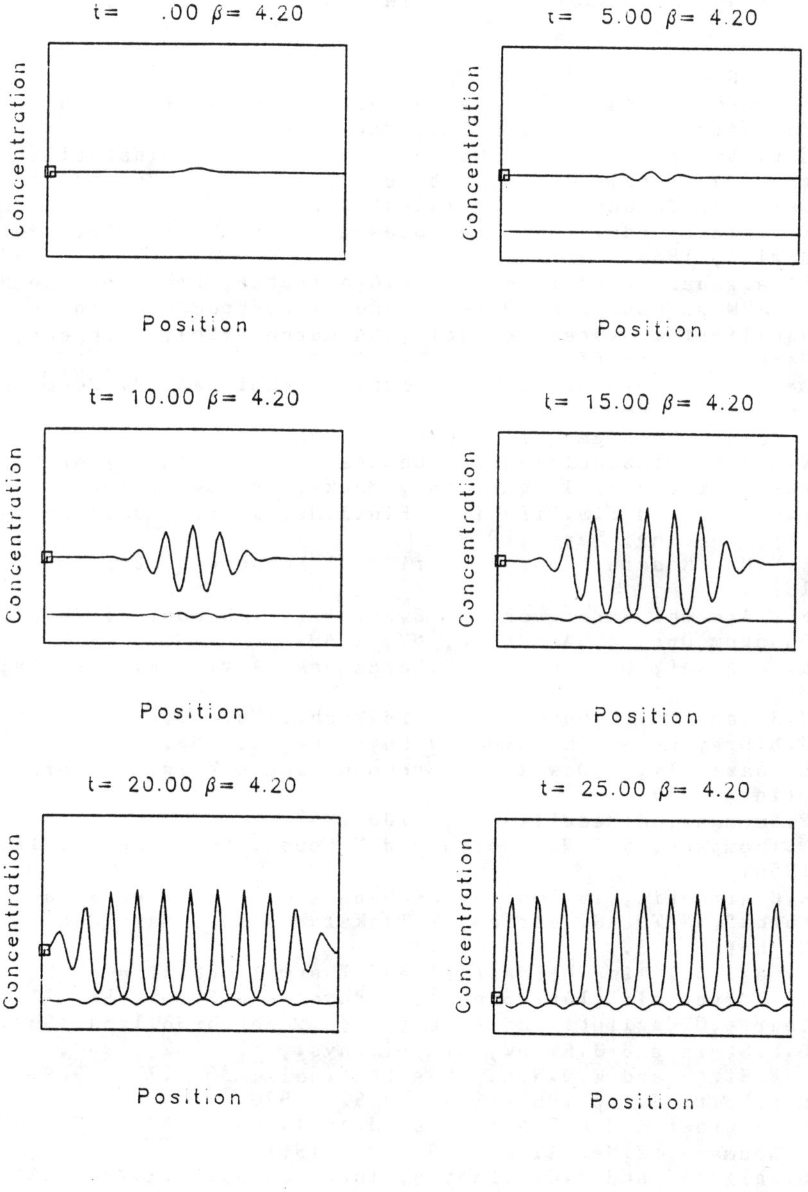

Figure 6. Spatio-temporal evolution of the dislocation populations like in fig.5 but for =0.18 and anothe initial condition corresponding to an inhomogeneity in the center of the sample.

ACKNOWLEDGMENTS

The collaboration of E.C.Aifantis and C.Schiller at various stages of this work and the constant support and interest of Professor I.Prigogine are gratefully acknowledged.

REFERENCES

1. G. Nicolis and I. Prigogine: Self-Organisation in Non Equilibrium Systems, Wiley, New York, 1977.
2. H.L. Swinney and J.P. Gollub : Hydrodynamic Instabilities and the Transition to Turbulence, Springer, Berlin, 1981. Cellular Structures and Instabilities, J.E. Wesfreid and S. Zaleski eds., Lecture Notes in Physics 210, Springer, Berlin, 1984.
3. D.Walgraef, C.Schiller and E.C.Aifantis, Proc. of the NATO ARW on Patterns, Defects and Microstructures in Non-equilibrium Systems (Austin,USA,March 1986),to appear, Martinus Nijhoff,
see the papers of W. Gerberich, L. Kubin and P. Neumann in this volume.
4. H.Haken, Synergetics, 2nd ed., Springer, Berlin,1978.
5. V.I.Arnold: Supplementary Chapters to the Theory of Ordinary Differential Equations, Nauka, Moscow, 1978.
6. L.Landau and E.M.Lifshitz: Fluid Mechanics, Addison-Wesley, Reading, Mass.,1959.
7. S.K.Ma: Modern Theory of Critical Phenomena, Benjamin, NY 1976.
8. F.T.Arecchi, Proc. of the E.Fermi Int. School of Physics Quantum Optics, Academic, NY, 1969.
9. D.Walgraef, G.Dewel and P.Borckmans, Adv.Chem.Phys. 49, 311,1983.
10. J.Salan and E.Guyon, J.Fluid.Mech.,126, 13,1983.
11. J.M.Dreyfus and E.Guyon, J.Physique,42, 283, 1981.
12. E.Guazzelli, G.Dewel, P.Borckmans and D.Walgraef, preprint, 1985.
13. P.Neumann, Z.Metallkde,58, 780,1967.
14. N.Thompson, N.J.Wadsworth and N.Louat, Phil.Mag.,1, 113, 1956.
15. E.C.Aifantis, in "On the Mechanics of Modulated Structures", NATO ASI series, T.Tsakalakos ed., Martinus Nijhoff,1983.
16. G.Caglioti, in "Mechanical and Thermal Behavior of Metallic Materials",Int. School of Physics E.Fermi, LXXXII course,G.Caglioti and A.Ferro eds., North Holland 1982.
17. D.L.Stein and J.R.Low, J.Appl.Phys., 31, 362, 1960. J.P.Hirth and W.D.Nix, Phys.Stat.Sol., 35, 177, 1969.
18. D.L.Holt, J.Appl.Phys., 41, 3197, 1970.
19. D.Walgraef and E.C.Aifantis, J.Appl.Phys., 58, 688, 1985.
20. P.Neumann, Z.Metallkde, 59, 927, 1968.
21. D.Walgraef and E.C.Aifantis, Int.J.Engng.Sc., 23, 1351 - 1365, 1985.
22. D.Walgraef and E.C.Aifantis, to appear Int.J.Engng.Sc.
23. N.Y.Jin and A.T.Winter, Acta Metall., 32, 1173, 1984.
24. F.Ackermann, L.P.Kubin, J.Lepinoux and H.Mughrabi, Acta Metall., 32, 715, 1984.

TRANSPORT PROPERTIES OF LASER-GENERATED NON-EQUILIBRIUM PLASMAS IN SEMICONDUCTORS

HENRY M. VAN DRIEL
DEPARTMENT OF PHYSICS AND ERINDALE COLLEGE
UNIVERSITY OF TORONTO
TORONTO, CANADA, M5S 1A7

1. INTRODUCTION

The absorption of high intensity laser pulses by semiconductors leads to the establishment of highly non-equilibrium electron-hole plasmas[1-8]. Because the presence of such plasmas can profoundly influence the properties of the semiconductors, there has long been fundamental interest in studying the characteristics of these ambipolar plasmas and exploiting these characteristics in such applications as laser-material processing[1,2], opto-electronic switching[9,10], and reflection switching[11], to name but a few. In those applications involving short, intense laser pulses, the transport properties of the plasmas are particularly instrumental in determining surface carrier density, energy deposition near the surface, etc. and a detailed knowledge of the tranport properties of the plasma is essential.

In this article I will attempt to outline some of the salient features associated with the generation and spatial and temporal evolution of high density ($N > 10^{20}$ cm^{-3}) and high energy plasmas which can exist near the surface of a semiconductor. At the outset, I wish to indicate that this is an extremely complex task to undertake in general since the interaction of intense laser radiation with semiconductors unleashes a myriad of physical processes in the carrier and lattice systems. Non-equilibrium effects in semiconductor plasmas have, of course, been known for a long time, particularly with respect to hot-carrier effects in electronic devices under external (low frequency) bias[12]. In that case, it is the electric field which supplies the energy to the carriers and maintains them out of thermal equilibrium with the lattice. In the case of laser-semiconductor interactions, energy is being supplied on a short time scale and with a large degree of spatial inhomogeneity to create and/or heat carriers, usually in the absence of a field. There are, therefore, some fundamental differences between transport and evolutionary processes which govern the two situations. For laser excitation, as the time scale on which information is sought decreases, the number of independent macroscopic variables which are needed to describe the plasma increases. Stated alternatively, non-equilibrium effects increase as the laser pulse width decreases and the number of effects that must be taken into account becomes large. Although the goal of science is to strive towards agreement between theory and experiment this becomes increasingly difficult since complete information is not attainable, at least not from the experiments conducted to date. In this article I will attempt to avoid detailed discussion of the many interesting and esoteric effects that occur on the very short time scale of femtoseconds but rather concern myself with non-equilibrium aspects in the supra-picosecond domain. As will be seen in what follows, this allows for a macroscopic description of non-equilibrium states based on three independent variables, carrier density (N), carrier temperature (T_e) and lattice temperature (T). The criteria by which to establish the validity of this description will be discussed. This is the approach taken to deal with the fundamentals of laser annealing as discussed in a recent review article[2]. Although the description is simplified, it nonetheless serves to deal with most situations of practical interest while concerning itself with the many-body effects of the carrier gas and the non-equilibrium state of the plasma.

I begin with a discussion of the time scales associated with the many processes which allow the initially excited semiconductor to return to its equilibrium state. This is followed by a phenomenological theory incorporating generation, heating, and transport of the carriers and heat generation and diffusion in the lattice. Explicit numerical results are presented for simulations involving the interaction of picosecond and nanosecond pulses with silicon. Finally, I offer some interesting experimental results which illustrate several key features of the transport processes particular to highly nonequilibrium plasmas.

2. FUNDAMENTALS OF LASER ENERGY DEPOSITION AND TRANSPORT IN SEMICONDUCTORS

2.1. Time Scales

In this section I will summarize the basic physical processes associated with the evolution of laser deposited energy in semiconductors. Ignoring spatial aspects for the moment, let me set the fundamental time scales for some of the important phenomena which are responsible for the initially excited semiconductor to reach its equilibrium state in which lattice temperature (T) alone can serve as the only independent variable. These processes are illustrated in the block diagram of Figure 1 and the illustration of microscopic processes in Figure 2.

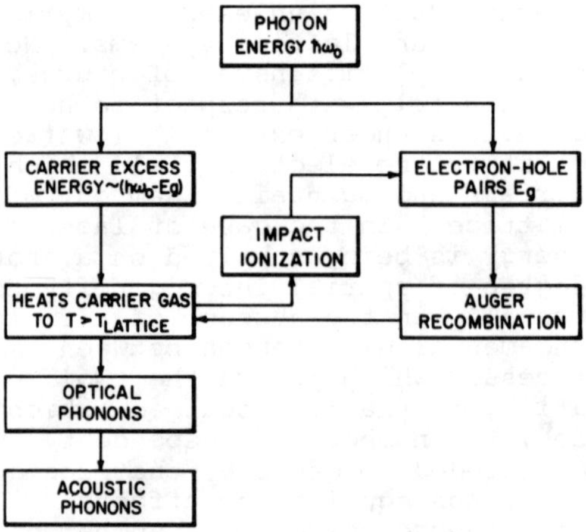

Fig.1. Schematic diagram of the evolution of laser deposited energy in the carrier and lattice systems.

The initial result of an inter-band absorption of a photon of energy $\hbar\omega_0$ is the generation of electrons and holes in a set of Bloch-states with a narrow range of energy and momentum[13,14]. This non-thermal distribu-

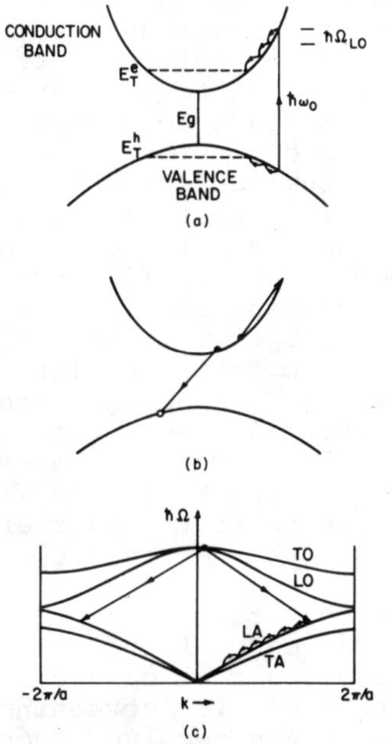

Fig.2. Schematic diagram indicating a) creation and thermalization of electron-hole pairs; b) auger recombination of electron-hole pairs and c) phonon scattering processes leading to thermalization of the lattice.

tion of carriers evolves on a 100 femtosecond, or less, time scale through carrier-carrier interactions into a thermal distribution with a well-defined common electron and hole temperature T_e. This can be thousands of degrees higher than the lattice temperature or even hundreds of degrees below the lattice temperature depending on the difference between the photon energy and the band gap, E_g.

On a picosecond time scale the carriers attempt to come to thermal equilibrium with the lattice, but more specifically with the longitudinal optic (LO) phonons to which they are most energetically coupled. This inter-system thermalization time, τ_E, has been estimated[15] to be between 0.5 and 10 picoseconds and depends on the initial lattice temperature, carrier temperature and density. In calculations presented later in this article, we will take it to be constant at 0.5 picoseconds so as to be conservative in estimating non-equilibrium effects. Upon receiving energy from the carriers the LO phonon population is enhanced and, in those situations where a phonon temperature can be defined, that temperature increases dramatically, even if the temperature of the other phonons is still low. This "hot phonon" effect can slow[16] the cooling of the carriers on a 1-10 picosecond time scale as the combined LO phonon-carrier system attempts to come to equilibrium with the other phonons in the lattice. In a time of the order of 10 picoseconds, as indicated in Figure 2c the LO phonons decay via lattice anharmonicity, releasing energy to zone boundary acoustic phonons which themselves interact with other phonons. Depending on wavevector and energy of the phonon state, thermal equilibrium may not be attained in times as long as a microsecond. Nonetheless, on the time scale of 10's of picoseconds most of the energy released from the carriers will be deposited in a near thermal equilibrium state of the lattice and there should be negligible error in assuming internal equilibrium as pertains to the matters of interest here.

During the time associated with carrier generation and internal thermalization, the equilibrium Fermi level is split into separate Fermi levels for the electrons and holes. The two subsystems will attempt to come to particle number equilibrium via recombination. This occurs on a time scale of picoseconds to milliseconds depending on the carrier density. At low densities ($N < 10^{17}$ cm^{-3}) recombination occurs through traps or defects at a rate proportional to N. In addition radiative band-band recombination can occur at rate proportional to N^2. In the former case, besides the reduction in carrier numbers which occurs, energy is delivered directly to the lattice. In the latter case light is generated which can lead to carrier creation in other parts of the sample. This leads to a redistribution of the carrier population or photo-diffusion[17]. At high carrier densities and in indirect gap semiconductors recombination is often dominated by an auger process, a three body interaction in which two carriers recombine, with a third carrier taking up the recombination energy. This is illustrated in Figure 2b. The effect of auger recombination is to reduce the carrier density, while increasing T_e. The inverse process, impact ionization, occurs when an energetic carrier creates an electron-hole pair, while itself losing energy. From requirements of momentum and energy conservation the initial energy of the impacting carrier must be $>3/2E_g$. Whereas, auger recombination occurs at a rate $\propto N(N^2 - N_0{}^2)$, where N_0 is the equilibrium carrier density for the temperature of the carrier gas, impact ionization occurs at a rate $\propto N$ with the coefficient being a highly sensitive function of T_e.

Because of the large N and T_e which can be established by picosecond laser pulses, it is possible that many of the rate coefficients which describe some of the processes discussed above can vary. For example, the subject of carrier screening in auger recombination and energy relaxation has received much attention[18] in many types of semiconductors, but a clear picture has yet to emerge mainly because of the difficulty of isolating individual processes from the many that occur simultaneously. (For example hot phonon effects have also been cited as important in the case of slowing of the carrier energy relaxation.) For the sake of simplicity, at least in the numerical simulations involving the evolution of plasmas in silicon on a greater than 10 picosecond time scale, we will ignore possible N and T_e dependence of the rate coefficients for lack of suitable, non-speculative information.

From the above discussion, in a spatially homogeneous material and on a timescale long compared to the internal thermalization time the properties of classical phonons and electrons can be given in terms of microscopic quasi-equilibrium distribution functions. For the electrons the distribution function can be written as

$$f_e^0\,(\vec{p}) = \left[\exp\left(\frac{E-\mu_e}{kT_e}\right) + 1\right]^{-1} \tag{1}$$

where μ_e is the Fermi level, T_e is the quasi-temperature, and E is the electron energy which depends on momentum, $\vec{p}$, through a dispersion relation. Here we will assume a parabolic dispersion relation, although at densities above 10^{20} cm^{-3} or temperatures in excess of 2000K this may not be strictly correct[19]. The local density of electrons can be found in the usual way by integrating over all the momentum states to yield the spatial density, N_e.

The distribution function for the holes is similarly given by:

relaxation time approximation. From a microscopic description I will proceed to a macroscopic description in the form of balance equations for particle number, particle energy and lattice energy. The key assumptions in this approach are:

1) The actual local carrier and phonon distribution functions are not far from the local equilibrium functions (characteristic of a spatially homogeneous situation) that they are trying to reach. Thus in the case of the carriers one assumes that it is possible to define local quasi-temperatures, $T_e(\vec{r})$, $T(\vec{r})$, and quasi-Fermi levels, $\mu_e(\vec{r})$, $\mu_h(\vec{r})$ that determine the carrier distribution functions in a small but macroscopic volume.

2) The mean free path that characterizes a particular transported quantity, such as particle number or energy, is small compared to the distance over which that quantity changes substantially. For example, one must have that the mean free path of carriers is small compared to $T_e|\nabla T_e|^{-1}$ or $N|\nabla N|^{-1}$. Otherwise it is not meaningful to speak of local equilibrium values of thermodynamic variables.

3) The size of the volume for which a local distribution can be defined must be large enough so that one can safely ignore statistical fluctuations in the values of variables in these systems.

4) The particles of the various subsystems are considered to be independent in the statistical mechanics and quantum mechanics sense. This implies that the overall distribution function of the system can be written as a product of single particle distribution functions. The particles need not be bare particles, but may be quasi-particles whose total energy may include a self-energy component related to the interaction of bare particles in the system. Thus, the energy of the electrons and holes, which includes a band-gap component can change with carrier number and phonon number (lattice temperature) since the band-gap varies[20,21] with both N and T.

Given the extremes in density, temperature, and gradients of these which occur, it is not obvious that all these criteria can be satisfied, although they are for most situations, including the numerical simulations presented here in which the absorption depths are greater than 0.1 μm. For shallower absorption depths and/or shorter pulses, such may not be the case.

We will now consider transport and other evolutionary processes in the electron (e) and hole (h) systems. Later, in a self-consistent manner we will combine the carrier transport equations with the phonon (lattice energy) transport equations.

Within the kinetic description, the evolution of the actual, local distribution function for electrons, f_e, is given by the Boltzmann equation as[22,23]

$$\left(\frac{\partial}{\partial t} + \vec{v}\cdot\frac{\partial}{\partial \vec{r}} + \vec{F}\cdot\frac{\partial}{\partial \vec{p}}\right) f_e(\vec{r},\vec{p}) = G_e(\vec{r},\vec{p}) - R_e(\vec{r},\vec{p}) - L_E - \frac{f_e - f^0_e}{\tau_e} \quad (6)$$

where $\vec{v}$ is the velocity of the electron of energy E and G_e and R_e are generation and recombination rates per unit volume of real and momentum space; $\vec{F}$ is the macroscopic force on the electron and L_E represents scattering events leading to energy loss to the lattice. In the relaxation time approximation scattering effects associated with momentum transfer are

$$f_h^0(\vec{p}) = \left[\exp\left(\frac{E-\mu_h}{kT_h}\right) + 1\right]^{-1} \quad (2)$$

On a picosecond time scale each of the subsystems can be described in terms of a common temperature.

$$T_e = T_h \quad (3)$$

but different Fermi levels

$$\mu_e \neq \mu_h. \quad (4)$$

The lattice phonon gas can be described in terms of a collection of distribution functions, $f_\Omega(\vec{p})$, where Ω is the frequency of a particular phonon mode. In keeping with the comments made earlier and for lack of a more easily justified assumption, we will assume that on a time scale greater than 10 picoseconds thermal equilibrium of all the phonons exists with a common temperature, T. All the lattice properties are therefore functions of the phonon density of states and Bose-Einstein distribution functions.

$$f_\Omega^0(\vec{p}) = \left[\exp\left(\frac{\hbar\Omega}{kT}\right) - 1\right]^{-1} \quad (5)$$

Because the carrier densities considered here are always much less than 1 per cent of the total electron density, we will assume no self-energy correction to the phonon energies via the carriers.

2.2. Spatial Aspects

All of the above remarks have concentrated on the purely temporal aspects of the evolution of the carrier-lattice system. Let us now address the issue of transport effects. Because of the inhomogeneous nature of the optical absorption process carrier density, lattice temperature and quasi-Fermi levels will vary in space as indicated in Figure 3. Particle and energy transport can therefore strongly determine the temporal evolution of the various subsystems. Because the temperatures and densities can vary significantly over distances of less than 1 μm, and because of the large gradients which are present, the transport problem is not easily handled in general. In what follows we will analyze the problem on the basis of the classical distribution functions (as indeed I have already assumed) and a kinetic theory approach based on the classical Boltzmann's equation in the

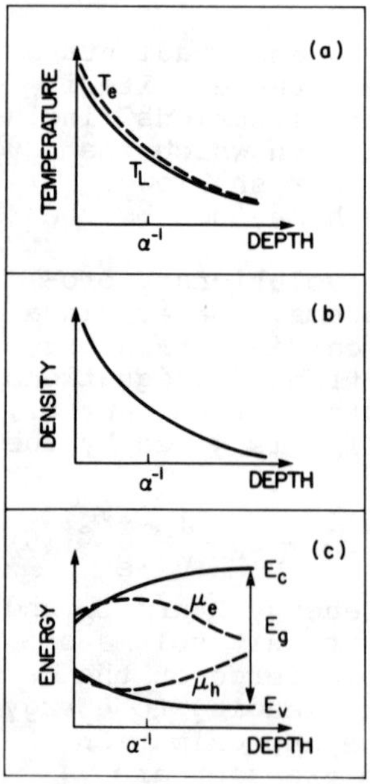

Fig.3. Schematic diagram indicating spatial profiles of a) temperature, b) density and c) electron and hole band energies and quasi-Fermi levels.

accounted for by the last term on the right with, τ_e being the momentum relaxation time. The vectors $\vec{v}$ and $\vec{F}$ are given by

$$\vec{v} = \frac{\partial E}{\partial \vec{p}}$$

and

$$\vec{F} = \vec{F}_0 - \vec{\nabla} S. \quad (7)$$

Here $\vec{F}_0$ is an external force and S is the self-energy of the electron due to interactions with other carriers or phonons. Note that E is the total energy of the carrier, including kinetic and self-energy contributions. In terms of the first assumption above and the quasi-equilibrium distribution functions, equation 6 can be rewritten as

$$\frac{\partial f_e}{\partial t} = -\frac{\partial f^0{}_e}{\partial \vec{p}} \cdot [T_e(E-\mu_e)\vec{\nabla}(\frac{1}{T_e}) - \vec{\nabla}\mu_e] + G_e(\vec{r},\vec{p}) - R_e(\vec{r},\vec{p}) - L_E - \frac{f_e - f^0{}_e}{\tau_e} \quad (8)$$

Note that because the carrier generation rate is sharply peaked in momentum space (state-filling), at high excitation intensities a quasi-equilibrium distribution function may not be maintained unless $\tau_e^{-1} \gg G_e$, R_e. However for the suprapicosecond pulses considered here and the range of intensities such pulses usually possess, the "kinks" in the distribution function will most likely be small. In addition, I will assume that carrier scattering events which lead to thermalization of the carrier distribution occur on a time scale much faster than τ_E, since otherwise the concept of a carrier-distribution function with a temperature T_e may not be possible.

An equation similar to equation 8 dictates the evolution of the hole distribution function, wherein f_e is replaced by f_h and μ_e is replaced by μ_h. For the phonons, the situation is only altered slightly since the chemical potential of the phonon gas is zero and phonon loss only occurs as a result of scattering. One then finds that the actual phonon distribution function $f_\Omega(\vec{r})$ evolves according to

$$\frac{\partial f_\Omega}{\partial t} = -\frac{\partial f^0{}_\Omega}{\partial \vec{p}} \cdot [TE_\Omega \vec{\nabla}(\frac{1}{T})] + G_\Omega(\vec{r},\vec{p}) - \frac{f_\Omega - f^0{}_\Omega}{\tau_\Omega}. \quad (9)$$

where τ_Ω is the scattering time of a phonon of frequency Ω and G_Ω is the microscopic generation rate which in general reflects interaction with the carriers as well as with other phonons. In the spirit of the discussion in Section 2.1 we will ignore the preferential coupling of carriers to specific phonons; this coupling would obviate against the existence of distribution functions as we have defined them as well as thermal equilibrium within the phonon system.

In what follows I will consider evolution of the system which is slow compared to the relevant scattering and generation processes. Memory effects in the distribution functions are therefore also negligible and hydrodynamic or balance equations can be obtained to yield a macroscopic description of the subsystems.

2.3. Carrier Number Balance Equation

Balance or continuity equations for the local electron density, $N_e(\vec{r})$, can be obtained by integrating both sides of the Boltzmann equation over momentum[23,24]. In performing these integrations we assume that the energy dependence of τ_e is given by $\tau_e \propto (E-E_c)^{\ell}$ where E_c is the energy of the conduction band edge and where, for carrier-phonon scattering, $\ell = -1/2$. (Note that the form for the Fermi-Dirac integrals that follow are explicity dependent on this particular value of ℓ). One then arrives at

$$\frac{\partial N_e(\vec{r})}{\partial t} = \frac{1}{q}\vec{\nabla}\cdot\vec{J}_e + G_e(\vec{r}) - R_e(\vec{r}) \tag{10}$$

where $\vec{J}_e$ is the electron current; R_e and G_e are the total recombination and generation rates of electrons per unit volume. For laser excitation $G_e(\vec{r})$ is determined by single photon absorption with possible contributions from two photon absorption. We then have

$$G_e(\vec{r}) = \frac{(1-\Gamma)\alpha I(\vec{r},t)}{\hbar\omega_0} + \frac{(1-\Gamma)^2\beta I^2(\vec{r},t)}{2\hbar\omega_0} + \delta_e(T_e)N_e \tag{11}$$

with Γ the reflectivity (which can depend on N_e and T), I(t) the intensity of the optical pulse, α the temperature dependent absorption coefficient for valence-conduction band transitions, β the two photon absorption cross-section and δ_e is the impact ionization coefficient for electrons. The attenuation of the beam is governed by

$$\frac{dI}{dx} = -\alpha I - \beta I^2 - \sigma_{FCA}N_e I \tag{12}$$

where x is the depth into the sample and σ_{FCA} is the free carrier absorption cross-section for electrons. If all three absorption mechanisms are significant it is not possible to obtain an analytic expression for $I(\vec{r},t)$.

The auger recombination rate for electrons can be written as

$$R_e = \gamma_e N_e (N_e N_h - N_0^2). \tag{13}$$

The current density of carriers is given by

$$\vec{J}_e = \frac{1}{q}\sigma_e\vec{\nabla}\mu_e - \sigma_e\rho_e\vec{\nabla}T_e \tag{14}$$

where -q is the charge of an electron and σ_e is the electron conductivity given by

$$\sigma_e = qN_e\mu^e{}_{MB}\frac{F_0(\eta_e)}{F_{1/2}(\eta_e)} \tag{15}$$

and $\mu^e{}_{MB}$ is the mobility of Maxwell-Boltzmann (non-degenerate) electrons; the mobility is proportional to τ_e. The reduced Fermi level is given by

$$\eta_e = \frac{(\mu_e - E_c)}{kT_e}. \tag{16}$$

and $F_j(\eta)$ is the Fermi integral of order j.

$$F_j(\eta) = \frac{1}{\Gamma(j+1)}\int_0^\infty z^j[\exp(z-\eta) + 1]^{-1}dz. \tag{17}$$

with $\Gamma(\cdot)$ being the gamma function. In what follows it is useful to remember that $F_j(\eta)/F_k(\eta) > 1$ if $j > k$ and $F_j(\eta)/F_k(\eta) \to 1$ for $\eta << -1$. The quantity ρ_e is the Peltier coefficient, given by:

$$\rho_e = \frac{k}{q}(\eta_e - \frac{2F_1(\eta_e)}{F_0(\eta_e)}) \tag{18}$$

A description of the hole density and current can be obtained in terms of the same formalism as for the electrons with N_e, μ_e, ρ_e, E_c, etc. replaced by N_h, μ_h, ρ_h, E_v, etc. with $\eta_h = (E_v - \mu_h)/kT_e$.

In a laser generated plasma, both types of carriers move together in the absence of any applied field but because of the difference in the electron and hole mobilities, an electric field, the Dember field, $\vec{E}_d$, develops due to charge separation. The Dember field (which might have been included in the Boltzmann equation at the outset) contributes to both the electron and hole currents which then become

$$\vec{J}_e = \frac{1}{q}\sigma_e\vec{\nabla}\mu_e - \sigma_e\rho_e\vec{\nabla}T_e + \sigma_e\vec{E}_d$$

$$\vec{J}_h = \frac{1}{q}\sigma_h\vec{\nabla}\mu_h - \sigma_h\rho_h\vec{\nabla}T_e + \sigma_h\vec{E}_d \tag{19}$$

Because large Dember fields would develop if the charge imbalance became too high, the two currents are approximately equal in magnitude but opposite in sign so that we can take

$$\vec{J}_e = -\vec{J}_h \tag{20}$$

Using this one obtains for $\vec{E}_d$

$$\vec{E}_d = \frac{1}{\sigma_e+\sigma_h}(-\frac{1}{q}\sigma_e\vec{\nabla}\mu_e + \sigma_e\rho_e\vec{\nabla}T_e - \frac{1}{q}\sigma_h\vec{\nabla}\mu_h + \sigma_h\rho_h\vec{\nabla}T_e) \tag{21}$$

and the particle current for electron and hole pairs becomes

$$\begin{aligned}\vec{J} &= -\frac{1}{q}\vec{J}_e = \frac{1}{q}\vec{J}_h \\ &= \frac{1}{q^2}\frac{\sigma_e\sigma_h}{\sigma_e+\sigma_h}[\vec{\nabla}(\mu_h-\mu_e) + q(\rho_e+\rho_h)\vec{\nabla}T_e]\end{aligned} \tag{22}$$

This can be rewritten in terms of N and T_e using the equality of electron and hole densities,

$$N_e = 2(\frac{m_e kT_e}{2\pi\hbar^2})^{3/2}F_{1/2}(\eta_e) = N_h = 2(\frac{m_h kT_e}{2\pi\hbar^2})^{3/2}F_{1/2}(\eta_h) = N \tag{23}$$

where m_e and m_h are the electron and hole density of states effective masses. This equation directly relates the electron and hole quasi-Fermi levels to the pair density N. Using this, equation 22 becomes,

$$\begin{aligned}\vec{J} = -D\Bigg[\vec{\nabla}N &+ \frac{N}{kT_e\left[\frac{F_{1/2}(\eta_e)}{F_{-1/2}(\eta_e)} + \frac{F_{1/2}(\eta_h)}{F_{-1/2}(\eta_h)}\right]}\vec{\nabla}E_g \\ &+ \frac{N}{T_e}\left[\frac{2\left[\frac{F_1(\eta_e)}{F_0(\eta_e)} + \frac{F_1(\eta_h)}{F_0(\eta_h)}\right]}{\left[\frac{F_{1/2}(\eta_e)}{F_{-1/2}(\eta_e)} + \frac{F_{1/2}(\eta_h)}{F_{-1/2}(\eta_h)}\right]} - \frac{3}{2}\right]\vec{\nabla}T_e\Bigg]\end{aligned} \tag{24}$$

where the band-gap energy is $E_g = E_c - E_v$ and the ambipolar diffusivity is given by

$$D = \frac{1}{q}\,\frac{kT_e(\sigma_e\sigma_h)}{\sigma_e + \sigma_h}\left[\frac{F_{1/2}(\eta_e)}{F_{-1/2}(\eta_e)} + \frac{F_{1/2}(\eta_h)}{F_{-1/2}(\eta_h)}\right]$$

$$= D_0 \text{ for } \eta_e,\ \eta_h << -1 \tag{25}$$

where D_0 is the ambipolar diffusivity for a Maxwell-Boltzmann distribution of carriers. As has been pointed out by us and others earlier[20,21,25], the significance of the self-energy correction to the particle current, through the ∇E_g term, is that normal diffusivity is inhibited and carrier confinement can occur. This is illustrated in Figure 3c, wherein a strong gradient of the band-gap can induce a positive (negative) gradient in the electron (hole) Fermi level, despite a negative gradient in the carrier density. The self-energy correction results from the dependence of the carrier energy on carrier density through exchange and correlation effects and on lattice temperature, through carrier-phonon interactions. The band-gap gradient can thus be rewritten as

$$\vec{\nabla} E_g = \frac{\partial E_g}{\partial T}\vec{\nabla} T + \frac{\partial E_g}{\partial N}\vec{\nabla} N. \tag{26}$$

Figure 4 shows the variation[21] of the diffusivity in the absence of temperature gradient effects. In this case one can define an effective diffusion coefficient as the coefficient of ∇N after combining equations 24 and 26. As the Figure shows, in the realm of Maxwell-Boltzmann statistics the diffusion coefficient decreases initially due to the self-energy corrections of carrier-carrier interactions, but as Figure 3c shows, as the band-gap increases away from the surface diffusion will be opposed. As degeneracy sets in at high carrier densities, the kinetic energy of the carrier (or Fermi pressure) becomes sufficiently high so as to enhance the diffusion process, and the self-energy effects become relatively less important. If, in addition, temperature gradients are also large, it is not possible to define an effective diffusion coefficient as a coefficient of ∇N alone, but inhibition of diffusion will be enhanced. From equation 24 it is easy to see that diffusion is defeated if the decrease in the band-gap near the surface is greater than the average carrier energy. In that case the carriers do not have sufficient kinetic energy to

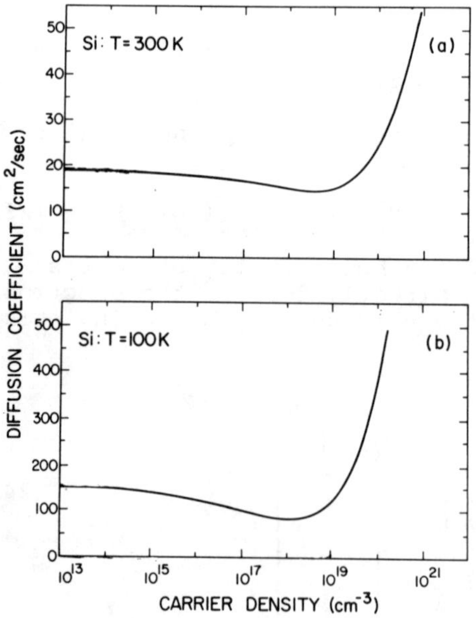

Fig.4. Variation of the effective diffusion coefficient with carrier density, illustrating many-body and degeneracy effects.

mount the potential barrier.

The balance equation for electron-hole pairs becomes,

$$\frac{\partial N}{\partial t} + \vec{\nabla}\cdot\vec{J} = G - R \tag{27}$$

where G and R are now given by

$$G(\vec{r}) = \frac{(1-\Gamma)\alpha I(\vec{r},t)}{\hbar\omega_0} + \frac{(1-\Gamma)\beta I^2(\vec{r},t)}{2\hbar\omega_0} + \delta(T_e)N \tag{28}$$

and $R = \gamma N(N^2 - N_0{}^2)$ where $\delta(= \delta_e + \delta_h)$ and γ $(= \gamma_e + \gamma_h)$ are the total coefficients of impact ionization and auger recombination for electrons and holes.

2.4. Carrier Energy Balance Equations

The equation for energy balance in the electron system is found by multiplying the Boltzmann equation for electrons by E and integrating over the electron distribution. A similar equation can be derived for the holes. For the ambipolar gas one can define a total energy density, Q, (kinetic energy plus band-gap energy per unit volume) in the electron-hole pairs as,

$$Q = NE_g + \frac{3}{2}NkT_e\left[\frac{F_{3/2}(\eta_e)}{F_{1/2}(\eta_e)} + \frac{F_{3/2}(\eta_h)}{F_{1/2}(\eta_h)}\right] \tag{29}$$

for which one has the energy balance equation,

$$\frac{\partial}{\partial t}Q + \vec{\nabla}\cdot\vec{J}_Q = S_Q - L_Q \tag{30}$$

Here J_Q is the total energy current, S_Q is the source of total energy and L_Q is a loss term for total energy in the carrier system. This equation can be converted into a differential equation for T_e by using equation 29 together with equation 27 for $\partial N/\partial t$. The total energy current is easily determined to be

$$\vec{J}_Q = (E_g + 2kT_e\left[\frac{F_1(\eta_e)}{F_0(\eta_e)} + \frac{F_1(\eta_h)}{F_0(\eta_h)}\right])\vec{J} - (\kappa_e + \kappa_h)\vec{\nabla}T_e \tag{31}$$

where κ_e and κ_h are the electronic components of the heat conductivity, given by

$$\kappa_e = k^2\sigma_e\frac{T_e}{q^2}\left[\frac{6F_2(\eta_e)F_0(\eta_e) - 4F_1(\eta_e)^2}{F_0(\eta_e)^2}\right] \tag{32}$$

and

$$\kappa_h = k^2\sigma_h\frac{T_e}{q^2}\left[\frac{6F_2(\eta_h)F_0(\eta_h) - 4F_1(\eta_h)^2}{F_0(\eta_h)^2}\right] \tag{33}$$

The source term for total energy in the carrier system is determined by the laser intensity and is given by

$$S_Q = (1-\Gamma)(\alpha + \sigma_{FCA}N)I(\vec{r},t) + (1-\Gamma)^2\beta I^2(\vec{r},t) \tag{34}$$

while the loss term is due to loss of kinetic energy through equilibration of the lattice and carrier temperatures and loss in energy of the carrier system due to band-gap shrinkage. Therefore we have:

$$L_Q = \frac{C_e(T_e - T)}{\tau_E} - N\frac{\partial E_g}{\partial t} \tag{35}$$

where C_e is the electronic specific heat at constant particle number given by

$$C_e = \frac{\partial Q}{\partial T_e}\Big|_N = \frac{Nk}{4}\left[15\left[\frac{F_{3/2}(\eta_h)}{F_{1/2}(\eta_h)} + \frac{F_{3/2}(\eta_e)}{F_{1/2}(\eta_e)}\right] - 9\left[\frac{F_{1/2}(\eta_e)}{F_{-1/2}(\eta_e)} + \frac{F_{1/2}(\eta_h)}{F_{-1/2}(\eta_h)}\right]\right] \tag{36}$$

Note that impact ionization and auger recombination do not represent source or loss terms for total carrier energy, since both cycle kinetic and ionization energy of the carriers. Nonetheless, auger recombination, for example, can speed the delivery of total energy to the lattice, since carrier ionization energy (E_g) is converted into carrier thermal energy, whose loss rate to the lattice is rapid.

2.5. Lattice Energy Balance Equations

Finally, the balance of energy in the lattice (phonon) system is arrived at by taking the Boltzmann equation for the phonon distribution function, equation 9, multiplying by phonon energy and integrating over the distributions. One then arrives[23] at

$$\frac{\partial}{\partial t}Q_L + \vec{\nabla}\cdot J_L = L_Q \tag{37}$$

where Q_L is the total energy density of the lattice and $\vec{J}_L$ is the lattice energy heat current given by

$$\vec{J}_L = -\kappa_L \vec{\nabla} T \tag{38}$$

where κ_L is the lattice thermal conductivity. By definition of the lattice heat capacity, C_L, equation 37 can be converted into

$$\frac{\partial T}{\partial t} = C_L \vec{\nabla}(\kappa_L \vec{\nabla} T) + L_Q \tag{39}$$

Equation 27, 30 and 39 completely determine N, T_e and T when solved together with the boundary conditions,

$$D\partial N/\partial x|_{x=0} = S(N(0,t) - N_0(0,t)),\ N(x\to\infty,t) = N_{eq} \simeq 10^{12}\,cm^{-3}\ \text{at}\ T = 300K$$

$$\partial T_e/\partial x|_{x=0} = 0,\ T_{e(x\to\infty,t)} = 300K$$

$$\kappa_L \partial T/\partial x|_{x=0} = -S(Q(0,t) - Q_0(0,t)),\ T(x\to\infty,t) = 300K \tag{40}$$

In the next section we consider explicit computer generated solutions to the balance equations for various picosecond and nanosecond pulses interacting with silicon. For simplicity, here I will assume that the surface recombination velocity is zero. The values for most of the other parameters used in the simulations are indicated in Table I. These parameters were deduced from experiments involving only modestly excited Si and are used here for lack of a better parameter set.

3. INTERACTION OF PICOSECOND AND NANOSECOND PULSES WITH SILICON

3.1. 0.53 μm Pulses

We first consider the interaction of 0.53μm pulses with Si for which the only significant absorption mechanism is single photon, interband absorption. As one benchmark to indicate that the equations are reasonable and consistent, we note that they predict a melting threshold of 300 mJ/cm^2 for 20 nanosecond pulses and 200 mJ/cm^2 for 20 picosecond pulses. Both of these values are within 10% of the experimental values[1,2,26]. In all honesty, however, the melt threshold at 0.53μm, for which the absorption depth is $\simeq$1μm is not particularly sensitive to the carrier dynamics, since the pulse widths are long compared to the carrier energy relaxation time and recombination time (see below). One makes negligible error in assuming instantaneous energy transfer of the laser deposited energy to the lattice in these cases. The small increase in fluence required of nanosecond pulses is attributed to lattice heat diffusion whose influence begins to be felt on a nanosecond time scale.

At fluences below the threshold of melting the carrier dynamics are more sensitive to details of the model and this becomes more apparent in the picosecond domain. Figure 5 shows the time evolution of the surface values of N, T_e, and T for F = 50 and 150 mJ/cm^2 for 20 picosecond pulses. Figure 6 illustrates the spatial and temporal evolution of N and T for these pulses at two different times. At the two values of F considered the shortest recombination time is less than the pulse width but not instantaneous so that the peak density is reached beyond the peak of the pulse (taken at t = 0). After the peak, the surface density decays due to recombination and diffusion. From Figure 5 it can be seen that two peaks occur in the surface value of T_e during laser excitation. On the leading edge of the pulse there is a rise in temperature because the intensity of the pulse and energy deposition is increasing more rapidly than N due to the ambient carriers already present. The surface carrier temperature peaks at $T_e \simeq 1400$K at t $\simeq$-40 picoseconds when N has reached a value of $\simeq 10^{14} cm^{-3}$. T_e then decreases since the pulse is heating an increasingly dense plasma which itself is losing energy to the lattice. T_e rises near t $\simeq$0 when the density reaches $\simeq 3\times10^{20} cm^{-3}$ and increasing auger recombination causes an increase in

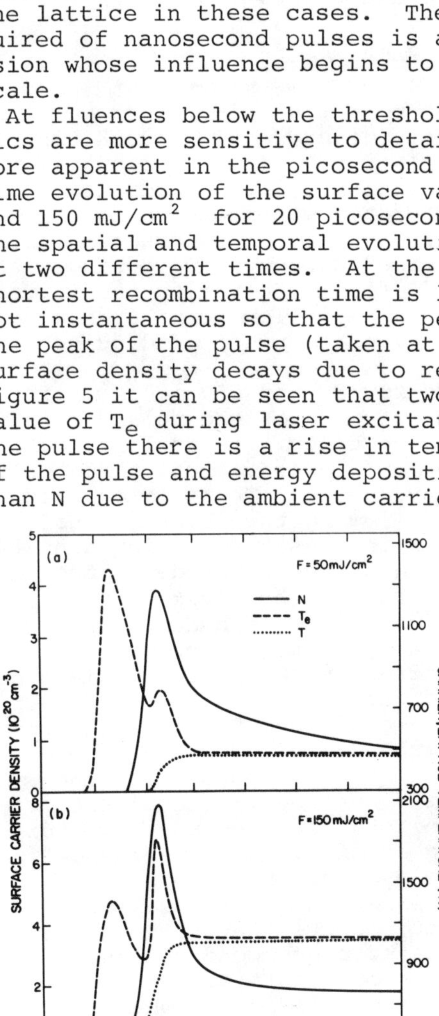

Fig.5. Carrier density (N), carrier temperature (T_e) and lattice temperature (T) surface values in Si as a function of time for 20 picosecond, 0.53 μm pulses of fluence: a) 50 and b) 150 mJ/cm^2 incident on intrinsic silicon.

T_e. The peak carrier temperature, is of course a function of the peak intensity and peak density as Figure 5 indicates. When the density drops, T_e begins dropping as well. Note that there remains a significant difference between T_e and T due to ongoing auger recombination. In general, although the energy relaxation time of the carriers can be much less than the width of the pulse, the lattice and carrier temperatures can still be different since the carrier system is energeticly pumped. Indeed, it was shown earlier[38], that for pulsed excitation, in the approximation that the energy relaxation time and carrier recombination time are much less than the pulse width so that N and T_e essentially follow the pulse profile, the difference between carrier and lattice temperature is given by

$$T_e - T = \frac{(\hbar\omega_0 - E_g)\alpha D^{1/2}}{3k(1+\alpha(D\tau_E)^{1/2})} \frac{\tau_E}{\tau_r^{1/2}} \tag{41}$$

where τ_r is the instantaneous recombination time. For N = 6 x $10^{20}cm^{-3}$ and $\tau_r \simeq 8$ picoseconds this gives $T_e - T \simeq 1000K$, in agreement with the numerical simulations. In nanosceond experiments[38], because of the lower intensities and densities encountered, the temperature difference was measured by thermionic emission to be $\simeq$200K, as also predicted by equation 41. Because of the rapid and complicated changes in T_e during picosecond pulses, it would be very difficult to verify the values experimentally. The values of T_e are too low to make impact ionization a significant process even for picosecond pulse excitation.

The surface lattice temperature for 50 and 150 mJ/cm^2 picosecond pulse excitation basically integrates the pulse as it also does for nanosecond pulses. On these time scales and for the initial absorption depths of $\simeq 1\mu m$, heat diffusion has only a minor effect on the surface temperature. There is a residual increase in lattice temperature following the pulse because of ongoing recombination. Indeed, as long as the carrier density remains high, a significant portion of the total deposited laser energy is stored as ionization energy of the plasma. This effect is more noticeable at lower intensities and densities. For example, for a 50 mJ/cm^2 picosecond pulses, at the peak of the pulse only 50% of the laser deposited energy is in the lattice while at t = 100 picoseconds the figure is 80%. For 150 mJ/cm^2 pulses the corresponding figures are 66% and 99% respectively. At the fluence required for melting, 85% of the laser energy is deposited in the lattice within 20 picosecond after the peak of the pulse although near the surface the local electronic energy is less than 5% of the total lattice energy, indicating that delayed auger heating is of little consequence in initiating melting. Similarly, delayed auger heating is of little consequence in nanosecond pulse interactions. Although the recombination times are longer, so are the pulse widths, and unless one uses rapidly truncated nanosecond pulses, the effect would not be observed. This contrasts with the situation observed for femtosecond pulse excitation[39], where significant lattice heating occurs through auger recombination after the pulse, albeit still on a picosecond time scale.

In Figure 6 we show the spatial profiles of N and T, at 15 and 200 picoseconds after the pulse maximum of 50 and 150mJ/cm^2,

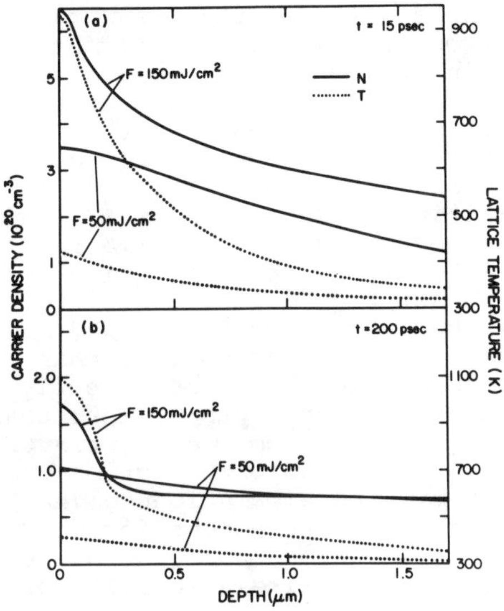

Fig.6. Spatial profiles of carrier density (N) and lattice temperature (T) following 50 and 150mJ/cm^2, 20 picosecond, 0.53μm pulses incident on silicon for time delays of a) 15 picoseconds, b) 200 picoseconds from the peak of the incident pulse.

20 picosecond pulses. The strong dependence of α on T introduces dramatic changes in the spatial profile of N and T with increasing F. The gradients in T and N increase during the pulse as seen from a comparison of the profiles at the two different times. This effect is aided by the large band-gap gradients which, as stated above, inhibit diffusion and, at high fluences, create carrier confinement. This causes more energy from auger recombination to be deposited near the surface which causes α to increase, causing more energy to be deposited near the surface, etc., steepening the profiles even further.

3.2. Interaction of 1.06 μm Picosecond Pulses with Si

For 1.06μm radiation α is only about 10 cm^{-1} at T = 300K but increases rapidly with increasing temperature. Higher intensities are therefore required to produce N, T_e and T values comparable to those of 0.53μm picosecond pulses. Indeed as Smirl and coworkers have recently determined[40] and as indicated in Fig. 7 the fluence required for melting ranges from 0.6 to 3J cm^{-2} for pulses with widths between 6 and 228 picoseconds. The fluence required, besides being very sensitive to the pulse width, does not scale as the room temperature absorption coefficient for the two wavelengths. Indeed simulations of the melting threshold indicate that two-photon absorption and free carrier absorption are very important. Figure 7 shows results of the numerical simulations with and without carrier diffusion and two photon absorption. Although it was indicated

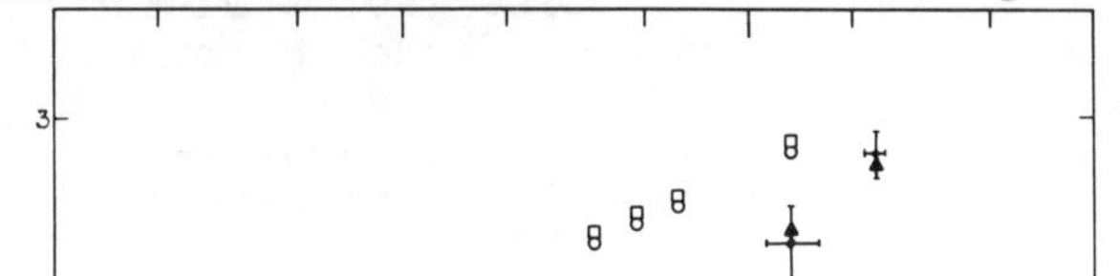

Fig.7. Fluence at 1.06μm required to melt Si as a function of pulsewidth. Result of numerical simulations discussed in the text are also indicated.

earlier[40] that diffusion might explain the dependence of fluence on pulse width, it can be seen from the Figure that diffusion plays a negligible role. The main source for the pulse width dependence is two photon absorption, which, since it varies as I^2 becomes more important for shorter pulses. On the other hand the main source of energy deposition is free carrier absorption. Without free carrier absorption it was found that the fluence required for melting exceeded 10J cm^{-2} at all pulsewidths. Clearly the carrier dynamics plays a role in determining energy deposition. The reason that diffusion doesn't is because of the large absorption depths associated with the different absorption mechanisms. Two photon and direct absorption never produce absorption depths less than a few microns, while free carrier absorption which is directly proportional to NT only gives an absorption depth of less than 1μm near the end of the pulses when $N \gg 10^{20} cm^{-3}$ and $T \simeq 1500K$. By this time there is little time for diffusion to act on the gradients in N that are established. The simulations give good agreement with the experimental results except for the very shortest pulse. Some of the assumptions related to the formalism presented in Section 2 are most likely invalid for the very short, intense pulses.

It is also interesting to note that at 1.06μm, due to the increase in α with T and the increase in free carrier absorption with NT, the lattice temperature, carrier temperature and carrier density become highly nonlinear functions of F for pulses of a given width. Indeed each of these parameters is subject to a "runaway" effect as the laser intensity increases. For example, for a 8.5 picosecond pulse (selected to match one of the experimentally used pulses), the peak carrier density achieved for a 0.5J cm^{-2} pulse is only $7 \times 10^{19} cm^{-3}$ while for a 0.9J cm^{-2} pulse the value is 5×10^{20} cm^{-3}.

Fig.8. Carrier density and lattice temperature profiles occuring at various times during an 8.5 picosecond pulse at a fluence just below the melting threshold.

In Figures 8 and 9 we show results of some of the numerical simulations illustrating the dynamics of the onset of melting. Figure 8 shows the spatial profiles and increasing gradients which develop in N and T with increasing time during an 8.5 picosecond pulse at a fluence just below the threshold fluence. The increasing gradients contribute to the runaway

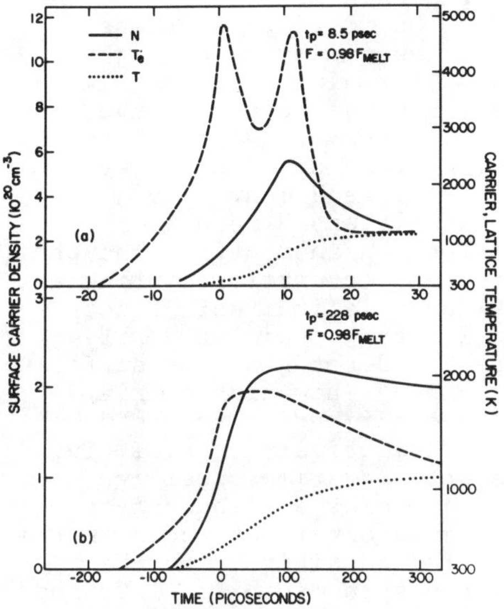

Fig.9. Surface carrier density, carrier temperature and lattice temperature occuring as a function of time for illumination of Si with 1.06μm pulses at a fluence just below threshold; a) 8.5 psec pulse and b) 228 psec pulse.

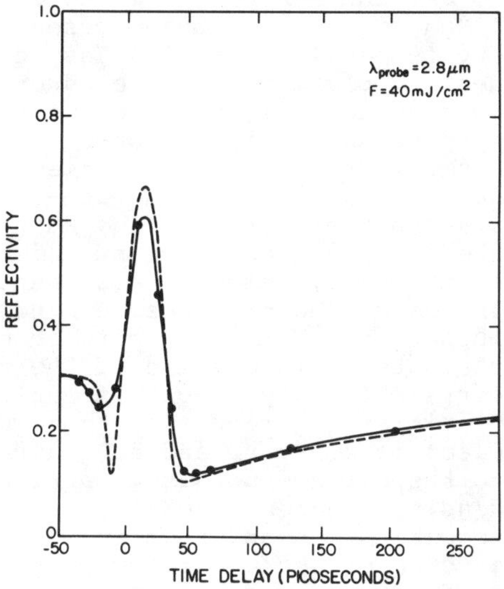

Fig.10. Silicon reflectivity at 2.8μm as a function of time during illumination by a 20 psec, 0.53μm pulse of fluence 40mJ/cm^2. The dots indicate data taken from van Driel et al[8]. The dashed curve indicates reflectivity, without taking into account the non-zero temporal profile of the pulse and non-infinite spatial profile of the carrier density. The solid curve indicates reflectivity taking into account these effects.

behavior of the surface values of N and T. Figure 9 shows the temporal behavior of N, T and T_e during 8.5 and 228 picosecond pulses at a fluence just below that of melting. Note that the carrier density achieved can be quite high, comparable to that associated with 0.53μm pulses. For 8.5 picosecond pulses one observes a two peak structure in T_e similar to that observed for 0.53μm excitation. Of note is the rapid increase in T_e to values where the possibility of impact ionization can occur, particularly for a reduced band-gap at high T. Indeed the inclusion of impact ionization is seen to drop the threshold for melting by about 30% at the shortest pulsewidths but has little effect at the longer pulsewidths. A curious effect occurs in the carrier temperature at an early stage in the pulse excitation process (not indicated in the Figure because of the scale). Because the photon energy is only slightly larger than the band-gap energy at 300K, new carriers which are created via single photon, direct absorption have a temperature lower than the lattice temperature. Indeed the overall temperature of the plasma drops to ≃250K before increasing due to further energy deposition by two photon absorption, free carrier absorption etc.

4. OPTICAL REFLECTIVITY AND CARRIER CONFINEMENT

Many ideas have been presented in the previous sections on the temporal and spatial aspects of the evolution of high density plasmas in semiconductors. It would be impossible to consider all the experiments which have dealt with the properties of highly non-equilibrium plasmas. Since the main topic is transport properties I will close by offering evidence of a truly unique non-equilibrium, and many body transport effect occuring at high densities - a reduction of the diffusion coefficient and carrier confinement. Although, in the past, evidence has been offered for carrier confinement in semiconductors, most of the evidence is indirect if not incorrect. To verify carrier confinement, one must verify the characteristic profile of the plasma and not just the surface carrier density. We have earlier shown[41] how high carrier densities contribute a strong Drude contribution to the dielectric constant and how infrared reflectivity is sensitive to the surface density, carrier scattering time, positive density gradients near the surface associated with surface recombination and negative density gradients that occur in the absence of surface recombination. The sensitivity to the gradients occurs because of the large skin depth for infrared light which can therefore probe not only the surface density but also the spatial inhomogeneity of the plasma. For a spatially inhomogeneous plasma the plasma resonance is broadened and this broadening influences the reflectivity. Evidence for carrier confinement and a reduction in the diffusion coefficient has earlier been presented for nanosecond laser induced plasmas in Si as probed by 10.6 and 5.3μm infrared radiation[41]. Here we show that more dramatic carrier confinement exists for picosecond pulse excitation.

Figure 10 shows the reflectivity at 2.8μm of silicon illuminated by a 40 mJ/cm^2, 0.53μm, 20 picosecond pulse[8]. The 2.8μm pulse, with a width of 20 picoseconds was generated from a 1.06μm fundamental pulse through stimulated Raman scattering in methane. The main features of the reflectivity can be understood in terms of the Drude model of the dielectric constant which at these short wavelengths is given by

$$\varepsilon = \varepsilon' + i\varepsilon'' \tag{42}$$

with

$$\varepsilon' = \varepsilon_0 - \frac{4\pi N q^2}{\omega^2 m_{op}} + 4\pi\chi \tag{43}$$

and

$$\varepsilon'' = \frac{4\pi\sigma}{\omega} \tag{44}$$

where $\varepsilon_0 = 11.8$, ω is the probe frequency, m_{op} is the optical effective mass ($\simeq 0.16 m_o$ in Si) and χ and σ are the susceptibility and conductivity associated with interconduction and intervalence band absorption[42]. Damping associated with momentum scattering is negligible since $\tau_e^{-1} \simeq 10^{13}$ $sec^{-1} << \omega$ at the wavelength considered. Plasmon resonances occur for those N and ω such that $\varepsilon' = 0$.

The reflectivity for an inhomogeneous plasma can be determined using the method of characteristic matrices or solving a multiple boundary value problem[41]. This was done in simulating the experiment which led to the results shown in Fig. 10. The character density characteristics are close to those shown in Figure 5a. The dashed curve shows the result of the numerical calculation without taking into effect the non-zero width of the probe pulse and spatial inhomogeneity effects (which in this particular case are small). The solid curve shows the results of a simulation which takes into account the 20 picosecond width of the probe pulse and the inhomogeneity of the plasma. Obviously temporal convolution effects are non-negligible. In any event, the reflectivity is seen to go through a minimum when the surface density, on the leading and trailing edge of the pump pulse, reaches $N \simeq 2.5\ cm^{-3}$ where $\varepsilon' = 1$. The slow recovery of reflectivity following the second minimum is attributed to the slowing down of auger recombination as the density decreases. The value of the reflectivity minimum is directly determined by ε''.

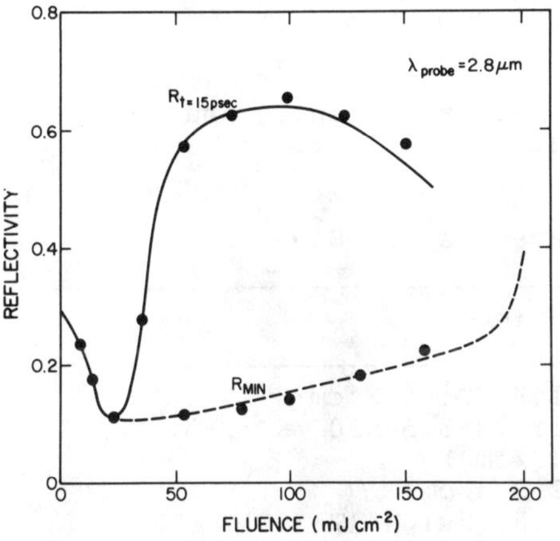

Fig.11. Reflectivity at 2.8μm of silicon as a function of fluence of a 20 psec, 0.53μm excitation pulse. The upper curve indicates data taken from Kurz and Rasolt[44] for the reflectivity at 15 psec after the peak of the pulse. The lower curve indicates data taken from Kurz and Bloembergen[43] on the minimum reflectivity following the pulse. The dots indicate results of theoretical calculations as outlined in the text.

When the pump fluence is increased[43] the value of the reflectivity minimum that occurs after the peak of the pulse rises as indicated in Figure 11. From Figure 6 it is seen that the carrier density gradient increases as a function of fluence due to increasing band gap reduction and carrier confinement which occurs with increasing lattice temperature. The steepening of the plasma density profile increases the inhomogeneous broadening of the plasmon resonance, leading to an increase in the reflectivity minimum. The dots in Figure 11 indicate results of the numerical simulation incorporating plasma inhomogeneity effects. Kurz and Bloembergen[43] interpreted the increase in the value of the reflectivity minimum as due to strong momentum damping of the plasma. However, this requires $\tau_e \simeq 5 \times 10^{-15}$ sec, a value which is anomalously low.

Further evidence of plasma confinement is indicated by the reflectivity[44] as a function of fluence at t = 15 picoseconds. As seen in Figure 11, the reflectivity begins decreasing with increasing fluence after approximately 100 mJ cm^{-2}. The dots indicate numerical simulations of the reflectivity based on carrier density profiles similar to those of Fig.6a. The reduction of the reflectivity can be explained by the steepening of the carrier density profile due to a reduction of the diffusion coefficient. Kurz and Rasolt have earlier offered an explanation for the reduction of the reflectivity due to the presence of a new recombination mechanism, such as plasmon recombination, which becomes important at very high T and N. Although such recombination may occur, it is not necessary to explain the data.

5. SUMMARY

I have attempted to outline the main physical effects and transport processes which occur in non-equilibrium laser generated plasmas. Obviously the story is not complete and only some of the salient features were considered. The emphasis has been on short-lived, high density and high temperature plasmas. There are also many interesting effects which occur at lower density plasmas which live in a cold lattice. Overall, the field is rich in effects and will continue to interest pure and applied physicists for some time.

6. ACKNOWLEDGEMENTS

Support for this work from the Natural Sciences and Engineering Research Council of Canada.

TABLE I. Optical and Thermal Parameters for Si.

Quantity	Symbol	Value
Thermal Conductivity[27]	κ_L	$1585T^{-1.23}$ W/cm·K
Specific Heat[27]	C_L	$1.978+3.54\times10^{-4}K^{-1}\cdot T-3.68K^2T^{-2}$ (J/cm^3)
Density	ρ	2.3 g/cm^3
Reflectivity(0.53μm)[28]	Γ	$0.37+5\times10^{-5}(T-300)$
Reflectivity(1.06μm)[28]	Γ	$0.30+5\times10^{-5}(T-300)$
Absorp.Coef.(0.53μm)[29]	α	$5.02\times10^3\exp(T/430)$ cm^{-1}
Absorp.Coef.(1.06μm)[30,31]	α	(see references)
FCA Cross-section(1.06μm)[32]	σ_{FCA}	$5\times10^{-18}(T/300K)$ cm^2
Latent Heat of Melting[27]	L	4000J/cm^3
Auger Recomb. Coef.[33]	γ	3.8×10^{-31} cm^6/sec
Impact Ionization Coef.[34]	δ	$3.6\times10^{10}\exp(-3/2E_g/kT_e)$ sec^{-1}
Ambipolar Diffusivity[35]	D_o	$18(300/T)$ cm^2/sec
Bandgap[36]	E_g	$1.16-7.02\times10^{-4}T^2/(T+1180K)$ -0.2×10^7cm $N^{1/3}$eV
Two Photon Absorp.Coef. (1.06μm)[37]	β	2 cm/GW

REFERENCES

1. See for example recent proceedings of the Material Research Society such as Energy Beam-Solid Interactions and Transient Thermal Processing, edited by J.C.C. and N.M. Johnson (Elsevier North Holland, New York, 1984), Mat. Res. Soc. Symp. Proc., 23, (1984) or Energy Beam-Solid Interactions and Transient Thermal Processing, edited by D.K. Biegelsen, G.A. Rozgonyi and C.V. Shank (Materials Research Society, Pittsburgh, U.S.A.), 35, (1985).
2. van Driel HM: "Physics of Pulsed Laser Processing of Semiconductors", in Semiconductors Probed by Ultrafast Laser Spectroscopy, Volume II, edited by R.R. Alfano (Academic Press, Orlando, U.S.A., 57 (1984).
3. Bloembergen N: in Laser Solid Interactions and Laser Processing, edited by S.D. Ferris, H.J. Leamy and J.M. Poate, (Academic Press, New York), 1, (1979).
4. Shank CV, Yen R and Hirlimann C: Phys. Rev. Lett.50, 454 (1983); Shank CV, Yen R and Hirlimann C: Phys. Rev. Lett.51, 900 (1983).
5. Van Vechten JA: J. Phys.41, C4.
6. Combescot M and Bok J: Phys. Rev. Lett.48, 1413 (1982); Bocking VR, Johnstone ES and van Driel HM: Phys. Rev. Lett.56, 205 (1986).
7. Boyd IW, Moss SC, Boggess TF and Smirl AL: App. Phys. Lett. 46, 366 (1985).
8. van Driel HM, Lompre LA and Bloembergen N: App. Phys. Lett. 44, 285 (1984).
9. Auston DH, Johnston AM, Smith PR and Bean JC: App. Phys. Lett. 37, 371 (1980).
10. Panchhi PS and van Driel HM: IEEE J. Quant. El, QE-22, 101, (1985).
11. Alcock AJ and Corkum PB: Can. J. Phys. 57, 1280 (1979).
12. Brunetti R and Jacoboni C: in Volume 1 of Semiconductors Probed by Ultrafast Laser Spectroscopy edited by R.R. Alfano (Academic Press, New York, 1984), 367.
13. Smirl AL, Boggess TF, Wherret BS, Perryman GP and Miller A: IEEE J. of Quantum Elec., QE-19, 609 (1983).
14. Tang CL and Erskine DJ: Phys. Rev. Lett. 51, 840 (1983); Oudar JL, Hulin D, Migus A, Antonetti A, and Alexadnre F: Phys. Rev. Lett.55, 2074 (1985); Knox WH, Hirlimann C, Miller DAB, Shah J, Chemla DS and Shank CV: Phys. Rev. Lett.56, 1191 (1986).
15. Conwell EM: High Field Transport in Semiconductors (Academic Press, New York, 1967).
16. van Driel HM: Phys. Rev. B19, 5928 (1979); Vasconcellos R and Luzzi R: Phys. Rev. B22, 6355 (1980); Kocevar P, Potz W and Porod W: Physica 117B and 118B, 220 (1983).
17. Epifanov MS, Bobrova EA and Galkin GN: Soviet Physics Semiconduct., 9, 1008 (1976).
18. Yoffa EJ: Phys. Rev. B21, 2415 (1980); Yoffa EJ: Phys. Rev. B23, 1909 (1981); Shah J: J. Physique (Paris) 42, C7, 445 (1981).
19. van Driel HM: App. Phys. Lett. 44, 617 (1984).
20. van Driel HM, Preston JS, and Gallant MIA: App. Phys. Lett. 40, 385 (1982).
21. Young JF and van Driel HM: Phys. Rev. B26, 2147(1982).
22. Klimontovich YL: Kinetic Theory of Nonideal Gases and Nonideal Plasmas, (Pergamon Press, New York, 1982); Mahler G and Fourikis A: J. of Lumin., 30, 18 (1985).

23. Kreuzer HJ: Nonequilibrium Thermodynamics and Its Statistical Fundations, (Clarendon Press, Oxford, 1981).
24. Drabble JR and Goldsmid HJ: Thermal Conduction in Semiconductors (Pergamon, New York, 1961), 203.
25. Van Vechten JA and Wautelet M: Phys. Rev. B23, 5551 (1981).
26. Lompre LA, Liu JM, Kurz H and Bloembergen N: App. Phys. Lett. 43, 168 (1983).
27. Wood RF and Giles GE: Phys. Rev. B23, 2923 (1981).
28. Jellison GE and Modine FA: Phys. Rev. B27, 7466 (1983).
29. Jellison GE and Modine FA: Appl. Phys. Lett. 41, 180 (1982).
30. Jellison GE and Lowndes DH: App. Phys. Lett. 41, 594 (1982).
31. MacFarlane GG, McLean TP, Quarrington JE and Roberts V: Phys. Rev. 111, 1245 (1958).
32. Svantesson KG: J. Phys. D12, 425 (1979).
33. Dwiezor J and Schmid W: App. Phys. Lett. 31, 346 (1977).
34. Sze SM: Physics of Semiconductor Devices, (Wiley, New York, 1969), 60.
35. See for example K. Seeger, Semiconductor Physics, (Springer-Verlag, New York, 1973).
36. Varshni YP: Physica 34, 149 (1967).
37. Boggess Jr. TF, Bohnert KM, Mansour K, Moss SC, Boyd IW and Smirl AL: IEEE J. of Quant. Elec. QE-22, 360 (1986).
38. Leung TLF and van Driel HM: App. Phys. Lett.45, 683 (1985).
39. Downer MC and Shank CV: Phys. Rev. Lett.56, 761 (1986).
40. Boyd IW, Moss SC, Boggess TF and Smirl AL: App. Phys. Lett. 45, 80 (1984).
41. Gallant MI and van Driel HM: Phys. Rev. B26, 2133 (1982); Preston JS and van Driel HM: Phys. Rev. B30, 1950 (1984).
42. Balkanski M, Aziza A and Amzallag E: Phys. Stat. Solidi 31, 323 (1969); Lambert LM: Phys. Stat. Solidi 11a, 461 (1972).
43. Kurz H and Bloembergen N: in Energy Beam-Solid Interactions and Transient Thermal Processing, edited by D. K. Biegelsen, G.A. Rozgonyi and C.V. Shank (Materials Research Society, Pittsburgh, U.S.A.), 35, 3, (1985).
44. Rasolt M and Kurz H: Phys. Rev. Lett.54, 722 (1985).

NONEQUILIBRIUM PHASES AND PHASE TRANSITIONS IN THE SURFACE MELT MORPHOLOGY OF LASER IRRADIATED SILICON

J. S. PRESTON, J. E. SIPE AND H. M. VAN DRIEL
DEPARTMENT OF PHYSICS
UNIVERSITY OF TORONTO
TORONTO, CANADA, M5S 1A7

1. INTRODUCTION

The increased use of lasers in material processing is directly related to the coherence of the laser beam. It is the spatial coherence property which allows one to direct and/or focus the modest intensity of laser beams onto small areas of semiconductors, metals or insulators, in such high temperature[1] processing applications as annealing and melting, or low temperature[2] processing applications, such as catalysis, etching or chemical vapor deposition. The temporal coherence or quasi-monochromaticity of the beam, permits one to initiate state specific effects. The coherence aspect of laser-material processing is often overlooked by many researchers who would wish to regard the laser as little more than a directed energy source or an ultrafast heat-gun. It has become apparent[3-5], however, that the coherence of the laser beam is responsible for development of surface patterns in the processed material. These patterns range from random morphologies to highly periodic structures whose spacing is related to the wavelength of light. It is clear that the development of these structures is related to the interference of the incident beam with surface scattered fields associated with surface polaritons or radiation remnants (lateral waves)[5]. The generation of surface periodic structures or "ripples" can also be viewed[4,6] as a stimulated surface scattering process or a stimulated Wood's anomaly. The specific types of patterns are determined[4] by material properties such as surface roughness and dielectric constant and beam properties such as wavelength, coherence, polarization and angle of incidence. Because of the interference effects there is often a significant variation in the surface fields and beam intensities leading to inhomogeneous melting, vapor deposition, etc. In some cases, such as in the production of grating structures, this effect is seen as desirable, but more often, it is regarded as deleterious and efforts are taken to suppress it, usually with a sacrifice of some other aspect of the process. It is clearly desirable to have a fundamental understanding of the influence of beam coherence and polarization on surface processing and to a large extent we have developed such an understanding over the past five years.

In this work we extend the understanding of laser induced structures by indicating several interesting non-equilibrium aspects of cw laser-material processing. In particular, we indicate that transverse instabilities can lead to several different types of surface morphologies or patterns consisting of co-existing regions of solid and melt induced on a silicon surface with various degrees of short range order. One can identify these structures as macroscopic surface structural phases and order-disorder "phase transitions" can be induced between these phases by adjusting the beam intensity and/or the spot size.

The development and maintenance of patterns under non-equilibrium conditions has been a synergetic subject of continuing interest in physics, chemistry, biology and sociology[7] for many years. In physical systems alone, pattern formation can be associated with such diverse mechanisms as dendritic crystal growth, laser mode competition and Rayleigh-Bernard instabilities. Although we illustrate the salient characteristics of laser-induced non-equilibrium phase transitions in connection with melting, similar types of effects might be expected to occur in other laser-material interaction processes. Indeed, there is most likely a whole class of structures which develop from an initially laterally homogeneous surface because of the transverse instabilities which characterize laser material interactions.

The basic source of the lateral instability can be understood as follows. At the onset of laser melting of Si, the bulk reflectivity increases substantially as liquid Si is metallic. Since a range of intensities exists for which the solid surface would be overheated while a molten surface would be undercooled, neither uniform phase is an acceptable steady state and a lateral phase separation occurs in which mixed states composed of solid and molten regions in alternate strips or irregular "lamellae" exist[8-12]. Details of the structures and the breaking of the initial translational symmetry are, of course, related to the interference process mentioned above. The arguments for a mixed phase can also be extended to considering the growth of optically thin molten regions for which the reflectivity changes continuously due to skin depth effects[13]. In the case of pulsed laser induced structures, the types of patterns that are formed are determined solely by the surface electrodynamics leading to interference[3,4]. Such is not the case during cw melting, and the patterns that evolve depend on the interplay between the spatially coherent deposition of energy via the laser beam, including interference effects, and the removal of that energy via heat diffusion. Here we indicate that with the new degree of freedom represented by heat loss through the material, both periodic and irregular microscopic patterns can be formed in steady state and morphological phase transitions between these states can be induced. At the same time we indicate that a geometric optics approach[10-12] is inadequate for an understanding of the patterns and that the detailed surface electrodynamics must be taken into account.

2. EXPERIMENTAL DETAILS

In our experiments, a linearly polarized 20W cw CO_2 laser operating in a single mode at $\lambda = 10.6$ μm was used to melt 2 μm thick films of silicon on a thick sapphire substrate. A schematic of the experimental set-up is shown in Figure 1. The

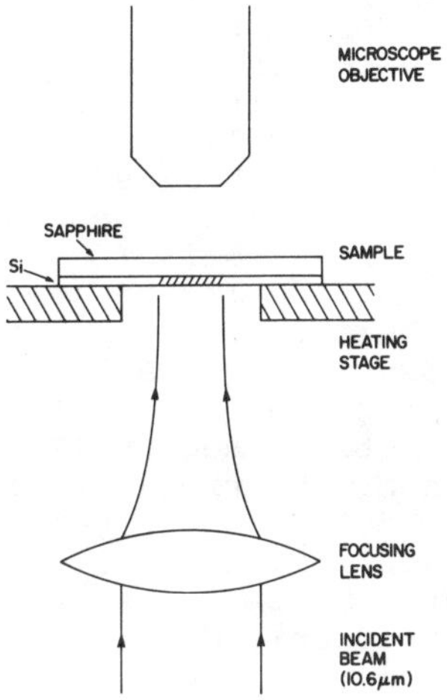

Fig.1. Schematic diagram of the experimental set-up

location and size of the irradiated spot was adjusted with a 10 cm focal length ZnSe lens, while the position and temperature of the samples were controlled using a movable resistive heating stage. The sample's own blackbody emissions near the melting point of 1685K were used to resolve the melt patterns.

3. NONEQUILIBRIUM PHASES

Photographs of patterns previously reported in the literature are shown in Figure 2. In the photographs the molten regions appear dark corresponding to the melt's lower emissivity. The irregular or lamellae patterns are indicated in Figure 2a and the periodic strips of solid and melt are indicated in Figure 2b. The strips orient themselves perpendicular to the polarization[4] and the spacing between adjacent solid strips is equal to the free space wavelength, λ. The grating-like structures occur at higher intensities than those for which lamellae structures are produced. This feature led us to consider experiments in which the melt morphology was varied as a function of laser intensity and the heat flow rate as determined by the spot-size. Simple estimates indicate that for our experimental conditions thermal conduction into the sapphire is the dominant heat loss mechanism and the two most salient experimental parameters in determining the microstructures are the incident intensity and the thermal conduction rate. In our experiments the rate of heat loss was adjusted by changing the laser spot size. (A simple dimensionality argument demonstrates that thermal conduction per unit area changes with the spot size.)

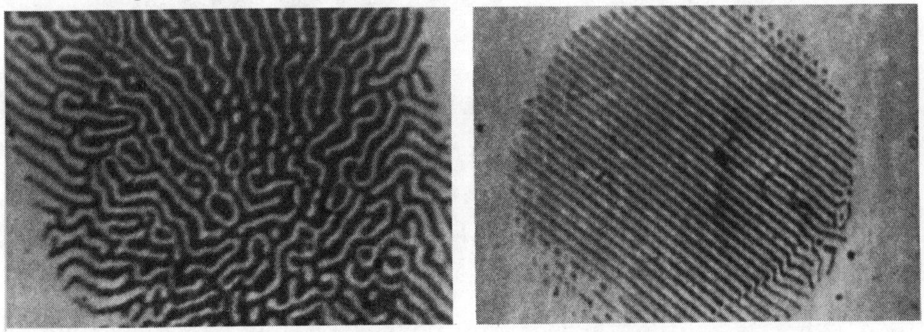

Fig.2. Photographs depicting different molten morphological structures formed; a) lamellae structures and b) fully periodic structures.

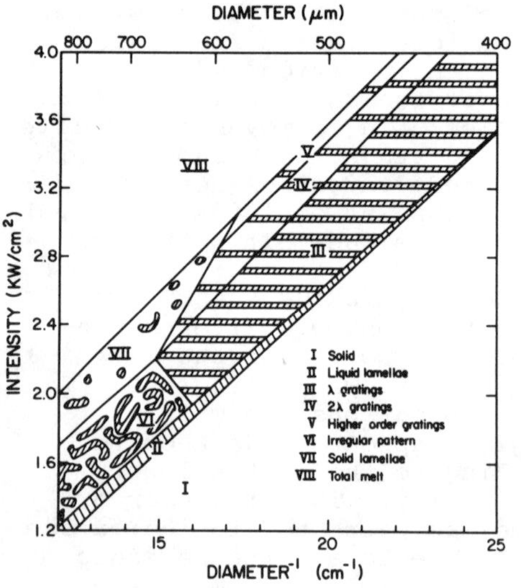

Fig.3. Nonequilibrium phase diagram for silicon under laser irradiation at 10.6 μm.

In Figure 3 we summarize in a stability or "phase" diagram the different nonequilibrium states associated with a given incident intensity and spot size. In general, for all spot sizes, as the laser intensity increases one proceeds from a uniform solid (Region I) to a uniform liquid (Region VIII). In between, the exact morphology is critically dependent on the intensity and spot diameter. Region II consists of an irregular distribution of isolated molten regions approximately 3 μm in diameter. Since the structure is observed only over a narrow range of intensities, and hence over only small geometric areas, Region II may not be a phase of the infinite system but rather an artifact of the spot size. In Region VI, the patterns of liquid and solid have an average spacing of approximately λ; however, the regions are irregularly arranged. The lamellae shown in Fig. 2a correspond to Region VI. Also within Region VI, the topology of the structure undergoes a rather subtle transition from isolated liquid regions to isolated solid regions as the intensity is increased. Region VII consists of isolated solid lamellae which have become separated by more than λ. The isolated lamellae then appear to move independently, interacting only if the separation becomes comparable to λ. In Regions III, IV and V periodic structures oriented perpendicular to the incident polarization are formed with spacing equal to λ, 2λ and 3λ respectively. Observations of the transition

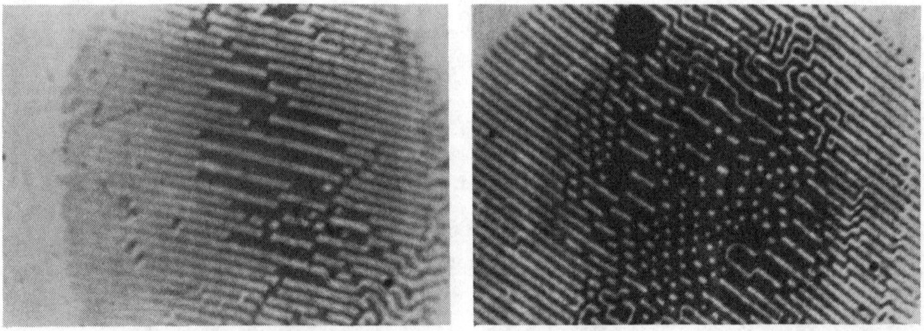

Fig.4. Photographs of structures corresponding to a) Regions III and IV and b) Regions III, IV and V.

from Region III to Region IV have revealed that alternate solid strips melt during the period doubling. This is indicated in Figure 4a, where in the center of the laser spot higher intensities occur. At yet higher intensities, in Region V, structures with 3λ or even 4λ spacing were observed but these structures were not stable and generally formed into a pattern consisting of a two dimensional array of solid regions as indicated in Figure 4b. In all cases, provided the experimental conditions don't change, the patterns noted are steady state patterns and there is no time evolution of the system. It is simple however, to change from one type of pattern to the other by varying the experimental parameters with the exception of the patterns noted below.

The phase diagram indicates that the grating-like structures are associated with relatively small spot sizes and hence large heat loss. This can be understood since, for the larger intensities required to initiate melting under these conditions, coherent energy deposition is more effective in ordering the surface. In addition, these structures are found to respond much more rapidly to variations of the polarization or intensity of the laser beam. This is not surprising since the timescale over which a system evolves is expected to be proportional to the ratio of the total energy to the energy flux. The disordered structures are associated with smaller heat loss and are inherently sluggish in response to variations in the experimental parameters. Moreover, in agreement with Nemanich et al[9], we have found that in some cases the disordered structures can be transformed into gratings by manipulating the laser spot. This is particularly true near the boundary between Regions III and VI in the phase diagram. This leads to the conjecture that the disordered structures observed in Regions VI and VII may not be true steady states, but rather the metastable "amorphous" phase associated with the "crystalline" grating phase.

Another trend apparent from Figure 3 is that the fraction of liquid increases with laser intensity as expected. However, this increase does not occur linearly with intensity but rather occurs in sporadic jumps as each new pattern emerges.

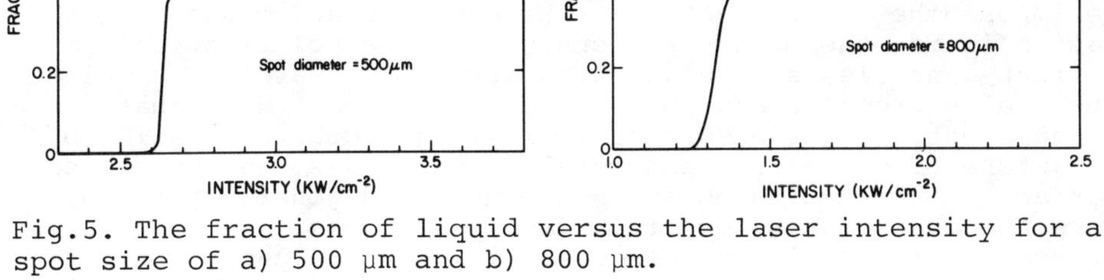

Fig.5. The fraction of liquid versus the laser intensity for a spot size of a) 500 μm and b) 800 μm.

We can thus identify the fraction of liquid as an order parameter which marks the onset of a transition between morphological phases. In Figure 5, the fraction of liquid is plotted as a function of laser intensity for two different spot sizes. At the smaller spot size wherein grating structures occur, the discrete jumps in Figure 5a occur solely as a result of the changes in the periodicity of the gratings. Changes in the width of the solid regions were found to be negligible, resulting in the striking plateaus. In the corresponding plot for the "amorphous" structures, Figure 5b, the fraction of liquid increases smoothly although some remnant of the plateaus remain.

4. THEORETICAL CONSIDERATIONS

Any fundamental understanding of this system's evolution requires a description of the coupling of the incident fields to the surface structures. In the geometric optics limit, relatively little energy is deposited in the melt since the fields are limited by the optical skin depth to a shallow region near the surface. If one assumes that the energy is deposited preferentially in the solid the surface morphology must then consist of rather unconventional boundaries between supercooled molten regions and superheated solid regions. While such a remarkable boundary might appear unphysical, Jackson and Kurtze[12] have argued that such a boundary between two unstable phases can in fact be stabilized by surface tension. However, for melt structures with size on the order of λ, the geometric limit is simply not valid.

As a prerequisite to understanding the steady states observed, we require a description of the coupling between the incident light and the melt patterns which meets the following criteria. First, since the length scales associated with the melt patterns are comparable to the incident wavelength, it is necessary to account for interference between the incident and scattered electrodynamic fields. Such interference can lead to energy deposition rates in the solid and molten regions dramatically different from the geometric optics limit. A self consistent approach is required since the large differences in the optical properties of solid and molten silicon render a perturbative treatment inadequate. Finally, we seek a description which retains more physical insight than numerically evaluating the fields and within which the energy deposition can be easily calculated.

In our treatment the response of the silicon film is described by three parameters, an average polarization density in the solid and surface current densities at the top and bottom of the metallic molten regions. To evaluate these parameters we impose the requirement that the electric and magnetic fields vanish inside the molten regions. Once the polarization and current densities are known it is straightforward to calculate the fields produced and the energy dissipated. Note that, because optically the surface is inhomogeneous, consisting of a mixture of dielectric and metallic materials, the generated surface field structures are not simple and generally do not have the characteristics of surface polaritons. While details of our approach will be discussed in a future paper, some results can be presented here for the case of the grating structures.

We first consider an existing grating structure and demonstrate that laser energy is deposited primarily in the metallic region contrary to the assumption of geometrical optics. We take a free standing film (no substrate) consisting of a solid-liquid grating of variable wavenumber κ with the sample 50% molten and illuminated by a laser beam of wavenumber $\tilde{\omega} = \omega/c$ where ω is the frequency of the beam. Figure 6 shows a cross section of the assumed melt-morphology for a grating of spacing λ. For simplicity, in this initial calculation we retain only the leading term of an expansion of the power spectrum of the liquid-liquid correlation function of the interface. In the case of linearly polarized light of field strength E_o normally incident on the grating the expressions for the magnitude of the top and bottom interface current densities, J^t, J^b, and the field, E, in the solid are given by:

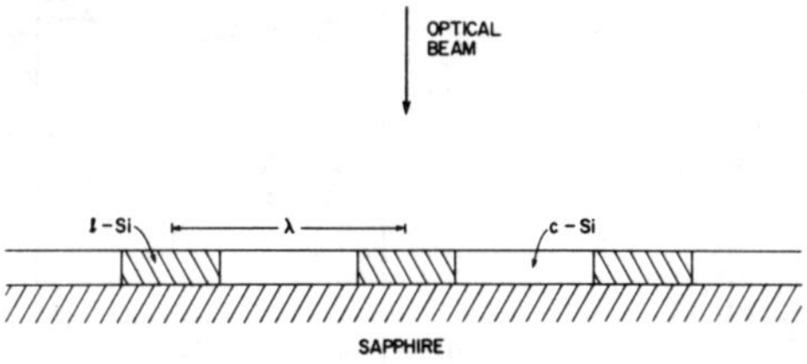

Fig.6. Schematic diagram of an induced solid-liquid grating in silicon-on-sapphire.

$$-i\omega J^t = \frac{c}{4\pi}\;\frac{3\tilde{\omega}+\sqrt{\tilde{\omega}^2-\kappa^2}}{\tilde{\omega}+\sqrt{\tilde{\omega}^2-\kappa^2}}\;E_o\;;\quad -i\omega J^b = \frac{c}{4\pi}\;\frac{\omega-\sqrt{\tilde{\omega}^2-\kappa^2}}{\tilde{\omega}+\sqrt{\tilde{\omega}^2-\kappa^2}}\;E_o$$

$$E = \frac{2\sqrt{\tilde{\omega}^2-\kappa^2}}{\tilde{\omega}+\sqrt{\tilde{\omega}^2-\kappa^2}}$$

in the case of the field oriented perpendicular ($\phi=\angle\vec{\kappa},\vec{E}=0°$) to the grating. Similar expressions exist for the field oriented parallel ($\phi = 90°$) to the grating. These expressions can be used to calculate the correction required to the geometric optics limit ($\kappa \to 0$) for the energy deposited in the molten and solid regions. This correction is shown in Figure 7 for the two field orientations. As κ approaches $\tilde{\omega}$, dramatic departures from the geometric optics limit occur. Indeed the energy deposited in the solid is a minimum for $\kappa = \tilde{\omega}$ in the case of light polarized perpendicular to the gratings. For the grating patterns observed ($\phi = 0°$) the energy deposited in the melt is dramatically enhanced, while the energy deposited in the solid regions is suppressed, leading to the energy being deposited primarily in the molten regions. Although the fields are confined to the surface of the molten regions, constructive interference between the scattered and incident fields greatly enhances the energy deposited. In the solid regions

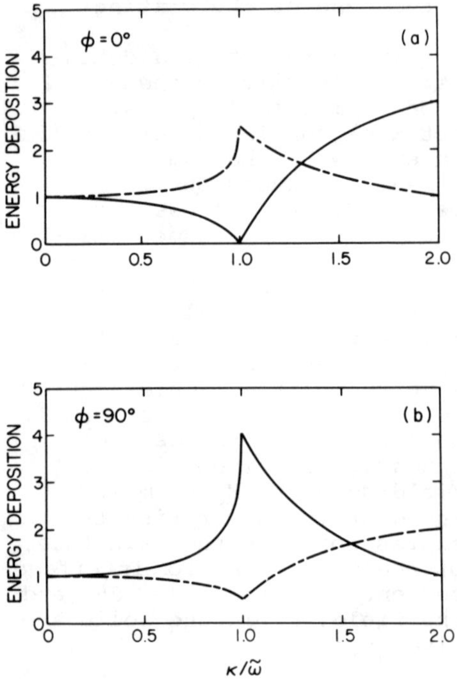

Fig.7. The ratio of the actual energy deposition to that of the geometric optics limit ($\kappa \to 0$) for a grating structure of wavevector κ oriented perpendicular ($\phi = 0°$) and parallel ($\phi = 90°$) to the polarization. The dashed and solid lines refer to the molten and solid regions respectively.

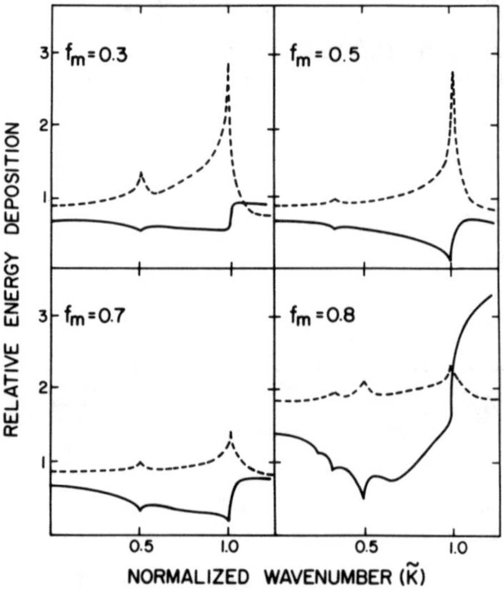

Fig.8. Relative energy deposition in the liquid (dashed curve) versus the solid as a function of normalized wavenumber for solid-liquid gratings with different fractions of the surface molten.

the fields interfere destructively leading to a phenomena which we have labelled interference shielding. Through this mechanism the melt is maintained at a higher temperature than the solid, leading to a conventional interface. In the case of the incident field being parallel to the grating structure ($\phi = 90°$) it is seen that the energy would be preferentially deposited in the solid and the grating structure would not be maintained. From Figure 7 then, we can understand both the orientation and the spacing of the preferred structures.

While inclusion of the substrate does not lead to any qualitative changes, retaining higher Fourier components of the induced current densities leads to new peaks in the energy deposition at $\tilde{\kappa} = \kappa/\tilde{\omega} = 1/n$, where n is any integer. These peaks are responsible for the formation of gratings at 2λ, 3λ, etc. In Figure 8, the corrections to the geometric optics limit is shown for 4 different fractions, f_m of the surface molten in the case of grating structures with variable $\tilde{\kappa}$. For these calculations the optical properties of the sapphire substrate and higher order terms in the Fourier expansion of the liquid-liquid correlation function were retained. In particular, the amount of shielding of the solid regions (dashed curves) is a sensitive function of f_m. For a grating with a spacing of λ, the shielding is most efficient when the surface is 50% molten. As f_m increases the shielding becomes less efficient until $f_m = 0.8$ when it is clear that the doubled grating is much more efficient at shielding the solid regions. The exact intensity at which the transition between the structures occurs is sensitive to details of the heat flow. Calculations are currently underway to demonstrate how the details of the heat flow in conjunction with the electrodynamics lead to the final selection of the steady state melt pattern observed.

5. ACKNOWLEDGEMENTS

We would like to acknowledge the generosity of D. Biegelsen and D. Guidotti for providing the silicon on sapphire samples used in this work. This work is supported by the Natural Sciences and Engineering Research Council of Canada, including a postgraduate scholarship to J.P.

REFERENCES

1. Laser and Electron-Beam Interactions with Solids, B. R. Appleton and G. K. Celler eds. (North Holland, New York, 1982); Laser-Solid Interactions and Transient Thermal Processing of Materials, J. Narayan, W. L. Brown and R. A. Lemons eds. (North Holland, New York, 1983).
2. Laser-Controlled Chemical Processing of Surface, A. W. Johnson, D. J. Ehrlich and H. R. Schlossberg eds. (North Holland, New York, 1984).
3. Young JF, Sipe JE, Gallant MI, Preston JS and van Driel HM in Laser and Electron-Beam Interactions with Solids, B. R. Appleton and G. K. Celler eds. (North Holland, New York, 1982) p. 233; Fauchet PM, Guosheng Z and Siegman AE in Laser-Solid Interactions and Transient Thermal Processing of Materials, J. Narayan, W. L. Brown and R. A. Lemons eds. (North Holland, New York, 1983), p. 205.
4. Sipe JE, Young JF, Preston JS and van Driel HM: Phys. Rev. B27, 1411 (1983); Young JF, Preston JS, van Driel HM and Sipe JE: Phys. Rev. B27, 1152 (1983); Young JF, Sipe JE and van Driel HM: Phys. Rev. B30, 2001 (1984).
5. Sipe JE, van Driel HM and Young JF: Can. J. Phys. 63, 104 (1985).
6. Brueck SRJ and Ehrlich DJ: Phys. Rev. Lett. 48, 1679 (1982).
7. Pattern Formation by Dynamic Systems and Pattern Recognition, H. Haken ed. (Springer-Verlag, New York, 1979).
8. Bosch MA and Lemons RA, Phys. Rev. Lett. 47, 1151 (1981).
9. Nemanich RJ, Biegelsen DK and Hawkins WG in Laser and Electron-Beam Interactions with Solids, B. R. Appleton and G. K. Celler eds. (North Holland, New York, 1982) p. 211.
10. Hawkins WG and Biegelsen DK: Appl. Phys. Lett. 42, 358 (1983).
11. Combescot M, Bok J and Benoit à la Guillaume C: Phys. Rev. B29, 6393 (1984).
12. Jackson KA and Kurtze DA, Jour. of Cryst. Growth 71, 385 (1985).

13. Sipe JE, Young JF and van Driel HM in Laser-Controlled Chemical Processing of Surface, A. W. Johnson, D. J. Ehrlich and H. R. Schlossberg eds. (North Holland, New York, 1984), p. 415.

ADSORPTION, DESORPTION, AND SURFACE REACTIONS

H.P. BONZEL

Institut für Grenzflächenforschung und Vakuumphysik
Kernforschungsanlage Jülich GmbH, Postfach 1913, D-5170 Jülich
Federal Republic of Germany

1. INTRODUCTION

The interaction of atoms and molecules with solid surfaces can be broken up into several fundamental processes, such as adsorption, surface diffusion, surface reaction, and desorption. The details of these fundamental processes have been investigated in numerous studies during recent years. Many of these investigations have been carried out on well defined single crystal surfaces utilizing various surface analytical tools and modern ultra-high vacuum technology. The use of single crystal surfaces is of particular importance in these studies in order to relate the physical characteristics of the fundamental processes mentioned above to specific adsorption sites.

The experimental investigations can be classified into two different groups. The first group consists of dynamic experiments concerned with the rates of adsorption and desorption of gases and molecules. The species involved are usually analyzed in the gas phase by mass spectrometry. The other type of experiment concerns the stationary adsorbed phase on the surface which is analyzed by various electron spectroscopies and additional scattering experiments involving atoms and ions. The goal of the latter experiments is to determine the physical properties of the adsorbate, such as the adsorption energy, adsorption site, the substrate molecule bond including bond length and molecule orientation, the ordering tendency of the adsorbate, its vibrational properties, and its propensity of dissociation. Of further interest are adsorbate-induced changes in substrate atom-atom bonding which may change the degree of relaxation between surface layers, eventually leading to surface reconstruction. In this brief review we will emphasize the dynamical aspects of the experiments and only give a few examples of investigations of the stationary adsorbed phase.

2. ADSORPTION

2.1. Kinetics of adsorption

The rate of adsorption of molecules on a solid surface depends on several parameters and is given by the following equation [1,2]:

$$r_a = \frac{dn}{dt} = J\ s(n,T) = s_0\ f(n)\ F_1 \exp\left(-\frac{E}{kT}\right) \qquad (1)$$

In this equation, n is the number of adsorbed molecules/cm^2, J the impingement flux (molecules/cm^{2}s), s(n,T) the sticking probability, s_0 the initial sticking coefficient, f(n) a function describing the coverage dependence of sticking, F_1 an entropy term, E the activation energy of adsorption, k the Boltzmann constant and T the absolute temperature of the gas molecules. At this point we can distinguish several different kinds of adsorption processes. Depending on the magnitude of E, one speaks of activated or non-activated adsorption (E=0). It is also important to know whether the adsorption will be molecular or dissociative. In the simplest case of molecular adsorption the site function f(n) will be $(1-n/n_0)$ where n_0 is the number of sites per cm^2. In that case, one speaks of Langmuir adsorption. The energetics of adsorption are illustrated in Fig.1 in simple terms by a one-dimensional adsorption potential [3]. The left-hand figure

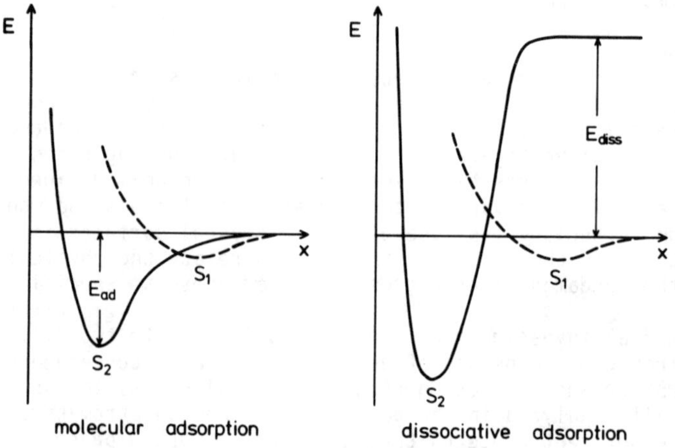

FIGURE 1: One-dimensional adsorption potential for molecular (left) and dissociative (right) adsorption. The activation barrier for adsorption results from the crossing point of the potentials for S_1 and S_2 above the x-axis (E >0).

shows the situation for molecular adsorption featuring a potential for a weakly held state, S_1, and a second potential for a more strongly held state, S_2 whose energy of adsorption is E_{ad}. The weakly held state S_1 is due to dispersion forces (physisorbed state) and is believed to be a precursor for the more tightly bound chemisorbed state S_2. There is no activation barrier for the transition from S_1 to S_2. The right-hand side of Fig. 1 shows the situation for a dissociative adsorption with a finite activation barrier between the first state S_1 and the tightly held dissociated state S_2. At large distances from the surface there is an energetic separation between the two potentials, the energy of dissociation of the molecule in the gas phase. Activated adsorption is generally only observed in context with dissociative adsorption. The first adsorbed state S_1 may be a weakly held precursor state to dissociation or a tightly bound, chemisorbed molecular state.

The measured quantity characterizing the rate of adsorption is the temperature and coverage dependent sticking coefficient defined by equation (1):

$$s(n,T) = \frac{1}{J}\frac{dn}{dt} \tag{2}$$

As the impingement flux J is known from gas kinetic theory

$$J = \frac{P}{\sqrt{2\pi m\, kT}} \tag{3}$$

the rate of adsorption dn/dt must be determined experimentally. There are several different techniques to measure dn/dt. In the first, the concentration of the adsorbate n is measured as a function of t by some kind of surface analytical technique. This results in the so-called uptake curve n(t). The sticking coefficient is then obtained by taking the derivative of this curve and dividing by the impingement flux. In the second approach the coverage is measured by a quantitative desorption experiment. The temperature of the crystal is raised, and the partial pressure P(t) of the desorbing gas is monitored as a function of temperature by a mass spectrometer or an ion gauge. As the experiment is performed in a continuously pumped system, the partial pressure is proportional to the desorption rate, and the area under the desorption trace is a measure of the coverage of the adsorbed species prior to the flash experiment (compare section 4.1.). The third approach is based on measuring the intensity of a molecular beam scattered at the surface [4]. At the beginning of the experiment the surface is clean and therefore adsorbs molecules with a relatively large rate. Thus, the intensity of the scattered beam is low. With increasing coverage the rate of adsorption decreases leading to an increase in the scattered intensity of the molecular beam. The sticking coefficient is therefore simply the difference between the incident intensity and the scattered intensity of the molecular beam divided by the incident intensity:

$$s(t) = \frac{I_0 - I(t)}{I_0} = \frac{I_\infty - I(t)}{I_\infty} \tag{4}$$

The incident intensity I_0 is usually replaced by the intensity of the scattered beam for a saturated surface, I_∞. It is also important that the scattered intensities are integrated over all angles. The coverage is calculated by integrating over the sticking coefficient.

In the following we discuss two examples of measurements of the rate of adsorption, one for reversible molecular adsorption and the other for dissociative adsorption. The first example concerns the adsorption of CO on Pt(111). In this case the uptake of CO on Pt was obtained by measuring the adsorbed amount in a flash desorption experiment [5] . Figure 2 shows desorption traces of CO from Pt(111) for various initial coverages. The adsorbed amount is determined quantitatively when the mass spectrometer is calibrated and the pumping speed of the system is known. The uptake curve of CO determined in this way is shown in Fig. 3. Here the coverage in molecules/cm^2 is plotted versus the total CO exposure in molecules/cm^2. The sticking coefficient is simply the derivative of this function which is shown in Fig. 4 for the adsorption temperature of 300 K. Apart from the initial rise at very low coverage which is not understood at this time, the sticking coefficient is roughly constant at 0.8 up to a coverage of about 0.2 and then decreases with increasing coverage until it reaches a very low value at a coverage of half a monolayer. The carbon monoxide molecules form an ordered layer at this coverage such that all ordinary adsorption sites are filled [6]. Fig. 4 shows for comparison also a measurement of the sticking coefficient by the molecular beam technique [7]. The substrate temperature is 310 K. Although the absolute values of the two measurements are quite comparable, the functional behaviour is somewhat different. The data of the molecular beam experiment obey quite well the Kisliuk model which we will describe very briefly.

The Kisliuk model [8,9] deals with a particular form of the site function f(n). It is based on a molecular adsorption potential as shown in Fig. 1. Two different precursor sites are distinguished, one located over

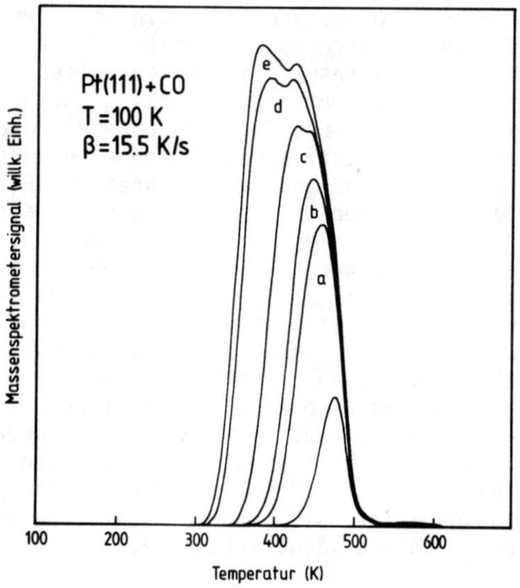

FIGURE 2: Desorption spectra of CO from Pt(111) for increasing initial coverages. CO was adsorbed at 100 K [5].

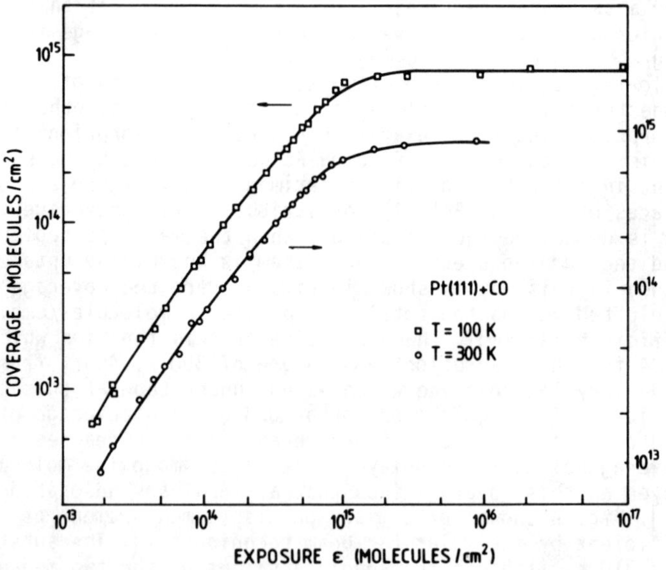

FIGURE 3: Plot of CO coverage versus CO exposure at 100 and 300 K [5].

an empty chemisorption site and the other over a filled chemisorption site. Free exchange between the two precursor sites via surface diffusion is allowed. Chemisorption can occur only from the precursor located over the

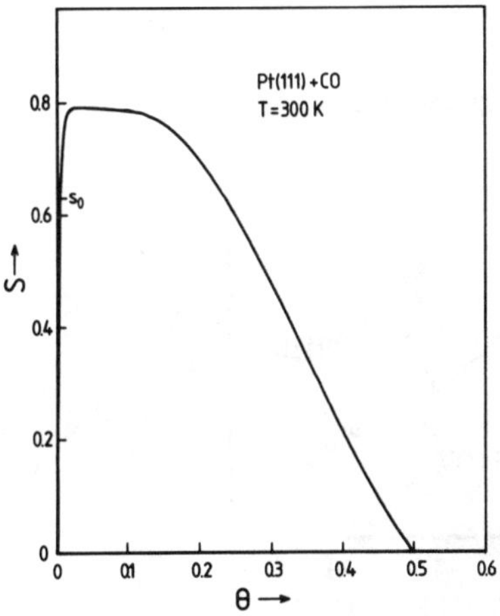

FIGURE 4: (a) Coverage dependence of sticking coefficient of CO on Pt(111) at 300 K [5]. (b) Same at 310 K but obtained by molecular beam technique [7].

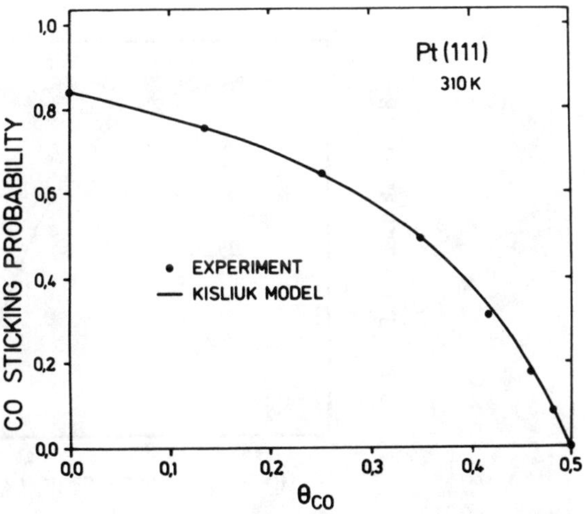

empty site. This model has had some success in explaining experimental data but this is perhaps not surprising because of a relatively large number of fitting parameters.

In the second example we present some data for the dissociative chemisorption of H_2 on several Ni surfaces. It was found that the adsorption of H_2 on Ni was structure-sensitve in the sense that the close-packed Ni(111) surface showed much less reactivity than the more open Ni(110) surface. The sticking coefficient for hydrogen on these two surfaces at a sample temper ature of 140 K is presented in Fig. 5 [10]. The data illustrate the strikingly different behavior of the sticking coefficient with coverage for these two surfaces. Whereas the initial sticking coefficient on Ni(111) is only 0.05, it is 0.96 for the Ni(110) surface. Furthermore, the sticking coefficient on Ni(111) decreases rapidly with increasing coverage towards very low values while it stays constant on Ni(110) up to coverages of about 0.5 and then decreases up to a coverage of 1.0. The data for both surfaces are well described by a modified Kisliuk model [10]. Such low sticking coefficients as observed for H_2 on Ni(111) are indicative of activated

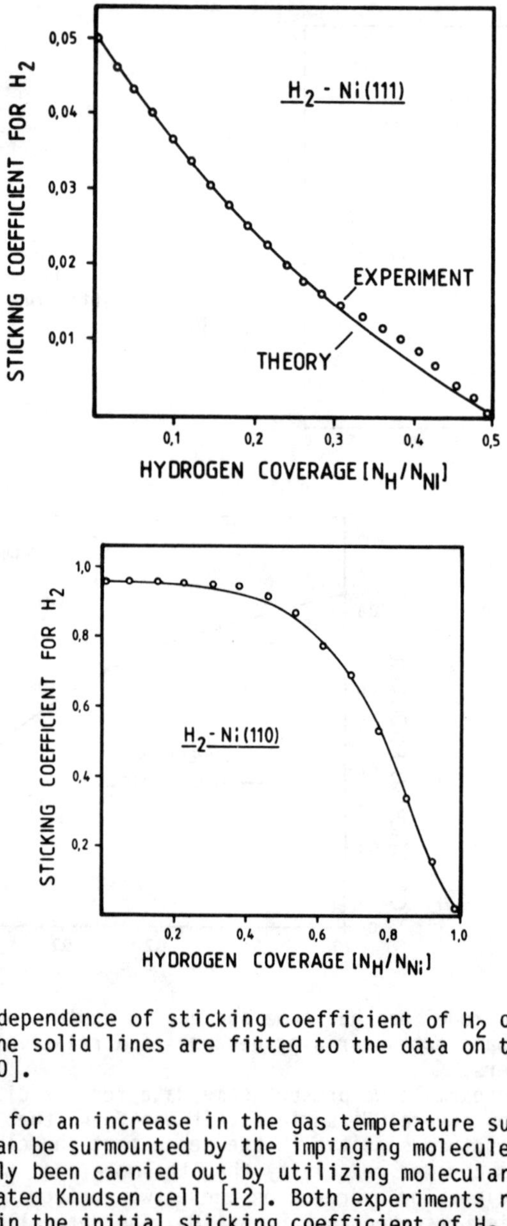

FIGURE 5: Coverage dependence of sticking coefficient of H_2 on Ni(111) and Ni(110) at 140 K. The solid lines are fitted to the data on the basis of an adsorption model [10].

adsorption and call for an increase in the gas temperature such that the activation energy can be surmounted by the impinging molecules. Such an experiment has actually been carried out by utilizing molecular beam techniques [11] or a heated Knudsen cell [12]. Both experiments report a considerable increase in the initial sticking coefficient of H_2 on Ni(111) with increasing gas temperature, as shown in Fig. 6 although the functional dependence is not the same. The same effect cannot be achieved by raising the surface temperature of the Ni crystal [12].

Structure sensitivity in adsorption is observed quite frequently. The simplest kind is the sensitivity of the rate of adsorption to the crystallographic surface orientation, as we saw in the example of H_2 on Ni. Mostly structure sensitivity is only pronounced with dissociative adsorption.

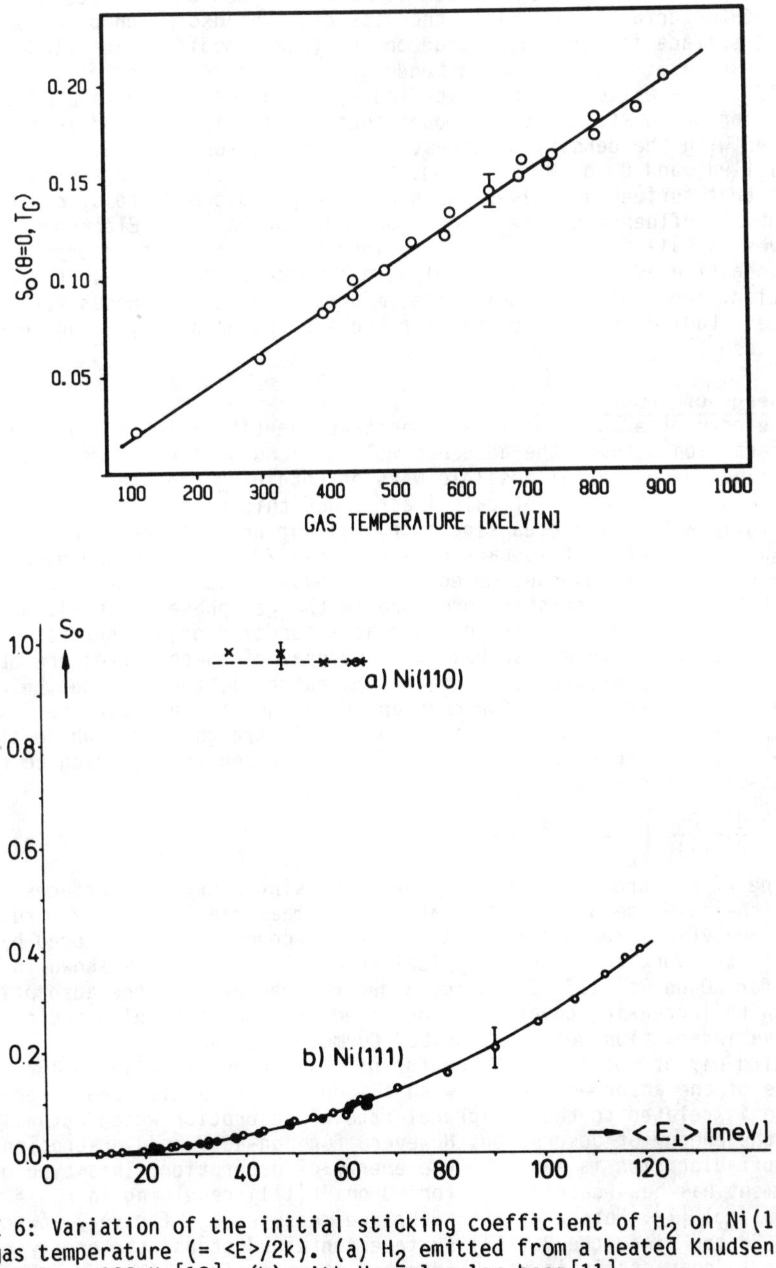

FIGURE 6: Variation of the initial sticking coefficient of H_2 on Ni(111) with gas temperature (= <E>/2k). (a) H_2 emitted from a heated Knudsen cell, Ni crystal at 190 K [12]. (b) with H_2 molecular beam[11].

Other well known examples are N_2 [13] and H_2 on W surfaces [14]. A more subtle case of structure sensitivity has to do with reconstructed surfaces. For example, the adsorption of O_2 on a non-reconstructed Pt(100) 1x1 surface is considerably faster at room temperature than on the reconstructed Pt(100) 5x20 surface [15]. Also, the dissociative adsorption of NO on the same 1x1 surface is more rapid than on the (5x20) modification [16]. In general, surface roughness has a tendency to increase the rate of adsorption [17]. This effect has been studied by using surfaces with a distribution of monatomic steps. It was found that the initial rate of adsorption increased with the density of steps, for example, for O_2 on Si [18], H_2 on Pt(111) [19], and O_2 on W(110) [20]. Finally it should be mentioned in this context that surface additives, such as foreign adsorbed atoms, can have a substantial influence on the rate of gas adsorption [21]. Electropositive additives (alkali metals) quite often increase the rate of adsorption while electronegative additives (S,P,C,O) have a tendency to decrease the rate of adsorption. These effects, which are important in heterogeneous catalysis, have been studied in great detail for the adsorption of CO, O_2 and other simple gases.

2.2. Energy of adsorption

The energy of adsorption is an important quantity which characterizes the interaction between the adsorbed molecule and surface atoms of the substrate. There are several possible ways of obtaining the energy of adsorption. One way is by means of calorimetry, but this technique does not work successfully with small-area single crystal surfaces. A second popular technique is by means of isobars or isotherms [22]. In this approach, the surface coverage of adsorbed molecules is measured as a function of the sample temperature at constant pressure in the gas phase (isobar). An alternate way is to measure the coverage as a function of pressure at constant sample temperature (isotherm). If several of these curves are obtained for different pressures (different temperatures), the data can be cross-plotted as log P versus 1/T for constant coverage of the adsorbate (isostere). This plot should result in a number of straight lines whose slopes are the isosteric heat of adsorption. The evaluation is according to the Clausius-Clapeyron equation, as follows:

$$\left.\frac{d\ln P}{d(1/T)}\right|_{\Theta} = -\frac{E_{ad}(n)}{k} \tag{5}$$

This type of measurement works very well for single crystal surfaces as long as the coverage of the adsorbate can be measured in-situ by a suitable surface analytical technique. Quite often the coverage is monitored by a change in the work function. A typical result for E_{ad} (n) is shown in Fig. 7 for CO on Pt(111) [6]. Here, a general decrease of the adsorption energy with increasing coverage is observed, a behaviour which indicates repulsive interaction between adsorbed CO molecules.

A third way of obtaining the energy of adsorption is by measuring the lifetime of the adsorbed species with a modulated molecular beam. Here, the lifetime is related to the reciprocal rate of desorption which actually yields the energy of desorption. However, for non-activated adsorption, the energy of adsorption is equal to the energy of desorption. This type of measurement has been carried out for CO on Pt(111) resulting in E_{ad} = 34.9 kcal/mol [7]. Note that the technique measures E_{ad} for very low coverages of CO and since the Pt(111) surface contained steps, the energy of adsorption is characteristic of CO adsorbed at step sites. Therefore E_{ad} determined here is higher than that in Fig. 7 for low CO coverage.

3. CHARACTERIZATION OF ADSORBED MOLECULES

As already outlined in the introduction, adsorbed molecules are investigated by a number of surface-spectroscopic and surface-scattering

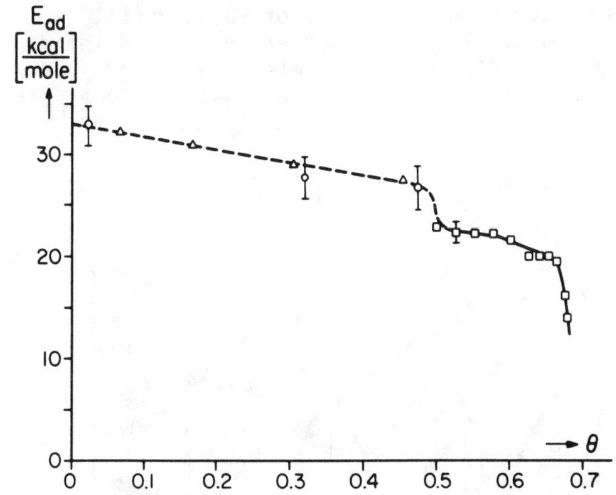

FIGURE 7: Adsorption energy of CO on Pt(111) as a function of CO coverage [6].

techniques. The general goal is a physical characterization of the adsorbed species as well as the adsorbate-substrate bonding. The physical properties are studied as a function of coverage and on different crystallographic surface orientations, such that the site specificity of adsorption can be elucidated.

Many adsorbed species have a tendency to form ordered layers on the surface. Depending on the kind of interaction between the adsorbed species, this ordering can occur at low or at high coverage. The most common tool to study ordering is low energy electron diffraction (LEED). For example, LEED studies of CO adsorption on Pt(111) have shown a whole series of ordered structures with increasing CO coverage [5].

When the energy of adsorption is very large, the presence of adsorbed species on the surface represents a major perturbation of the interatomic bonding near the surface. This alters bond distances as well as the separation between successive layers of the solid. In some cases the perturbation of the energetic equilibrium at the surface can be so dramatic that the structure of the surface changes (reconstruction). Several of these adsorbate-induced reconstructions of metal surfaces are known. Examples are O_2 and H_2 on Ni(110) [23,24] and Cu(110) [25] surfaces, and Na on Ag(110) [26] surfaces. Elaborate experimentation was necessary in each of these cases, in order to verify surface reconstruction and, in particular, to clarify the structure in each case.

Of great interest is the knowledge of the adsorption site in which the molecule resides on the surface. Several techniques are suitable for solving this problem. Particularly successful are vibrational spectroscopies where the measured vibrational frequencies of adsorbed molecules have been linked to the geometry the adsorption site. Symmetry selection rules and vibrational frequency shifts both serve to identify the adsorption site [27]. More about this problem will be said in detail in the lecture by Benedek (this volume).

A further question concerns the orientation of the molecule relative to the surface. Although vibrational spectra can also yield information on this question, other techniques such as the study of the near-edge structure of absorption spectra (NEXAFS) [28] or X-ray photoelectron diffraction (XPD) [29] are specifically suited to address this problem. As an example we show in Fig. 8 electron yield spectra of CO on Pt(111) [30]. Here, the partial electron yield is plotted versus energy of the incident photons which are absorbed to produce secondary electrons. The upper part of the figure shows the carbon K edge structure whereas the lower part shows the

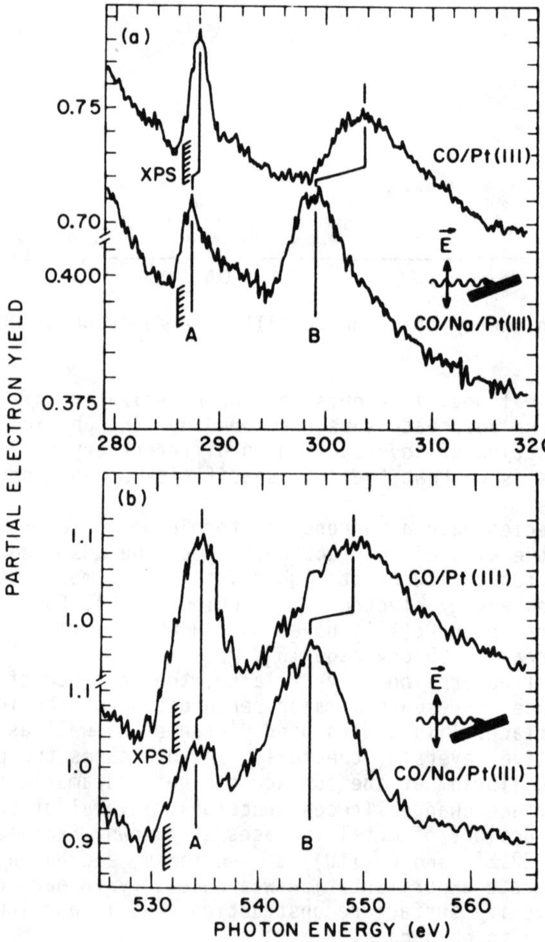

FIGURE 8: NEXAFS spectra for CO on clean and Na-covered Pt(111) recorded at 20° grazing X-ray incidence. Peak A is a π resonance and peak B a σ shape resonance. XPS denotes the absorption edge of the C1s(O1s) level. (a) Carbon K edge spectra. (b) Oxygen K edge spectra [30].

oxygen K edge structure. All spectra are characterized by a narrow π resonance and a broad σ resonance. The energetic separation between the π and σ resonances can be evaluated to yield the bond length of the adsorbed CO molecule. Furthermore, the polarization dependence of the π and σ resonance intensities tells the experimenter the orientation of the CO molecule relative to the surface. For example, the σ resonance is only observed when the electric field vector of the incident photon radiation is close to parallel to the CO bond axis. The result of the experiment for CO on clean Pt(111) is that CO molecules are perpendicularly adsorbed on the Pt surface, with the carbon atom closest to the surface. The same experiment was also carried out for a Na-covered Pt(111) surface. Fig. 8 shows in this case that the energetic separation between π and σ resonance has decreased by about 4 eV. This decrease corresponds to an increase in the CO bond length of 0.12 Å, which shows that coadsorbed alkali metals can drastically alter the intramolecular bond of the CO molecule, and therefore also the adsorption bond of CO.

Another way of obtaining the orientation of a molecule relative to a surface is by X-ray photoelectron diffraction. Here the emission of photoelectrons from strongly bound core levels is measured as a function of polar angle for a certain azimuth. Fig. 9 shows again as an example CO on clean Pt(111) [31]. The polar angle dependence of the C1s and O1s photoelectron emission is plotted in the upper part of the figure and shows a peak near normal in the C1s intensity. This peak is enhanced by taking the ratio of carbon to oxygen emission shown in the lower part of Fig. 9. The

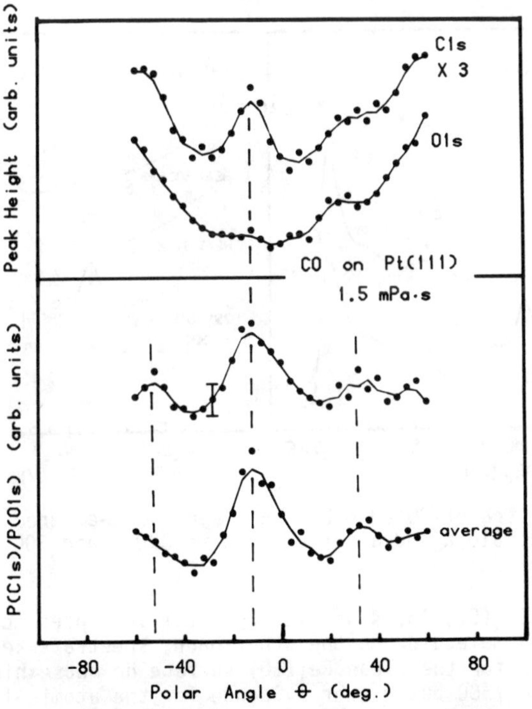

FIGURE 9: Polar angle dependence of C1s and O1s core level emission of CO adsorbed on Pt(111). Upper part shows intensities after background subtraction; lower part shows the ratio of peak heights; data labelled "average" correspond to the average of 3 independent measurements [31].

peak in the C1s emission occurs because the photoelectrons emitted from the carbon atom are strongly forward scattered by the oxygen atom. This peak then tells us two things: first, the CO molecule is adsorbed perpendicular to the surface, and secondly it is bonded with the carbon atom down. By comparison, the C1s/O1s dependence for CO on a K-covered Pt(111) surface shows a very similar dependence such that the mode of CO bonding does not seem to be altered by the presence of K atoms [31].

The electronic characterization of adsorbed molecules is usually carried out by photoelectron spectroscopy (UPS, XPS). The objective is to measure the binding energy of electrons in the valence band region (molecular orbitals) as well as the core level region. These binding energies are then compared to those of gas phase molecules such that energy shifts can be identified. These energy shifts often indicate molecular orbitals which are involved in the adsorption bond. The same technique is also extremely valuable for distinguishing molecular and dissociative adsorption. In many cases, it is quite easy on the basis of identified molecular orbital peaks to say whether the adsorbed species is in its original molecular form or whether it has dissociated on adsorption. The same kind of identification can be carried out by means of binding energy shifts of core levels. Fig. 10 shows as an example UPS spectra of CO adsorbed on clean and K covered Fe(110) at room temperature and also at elevated temperatures [32]. The room temperature spectra are characterized by the familiar molecular

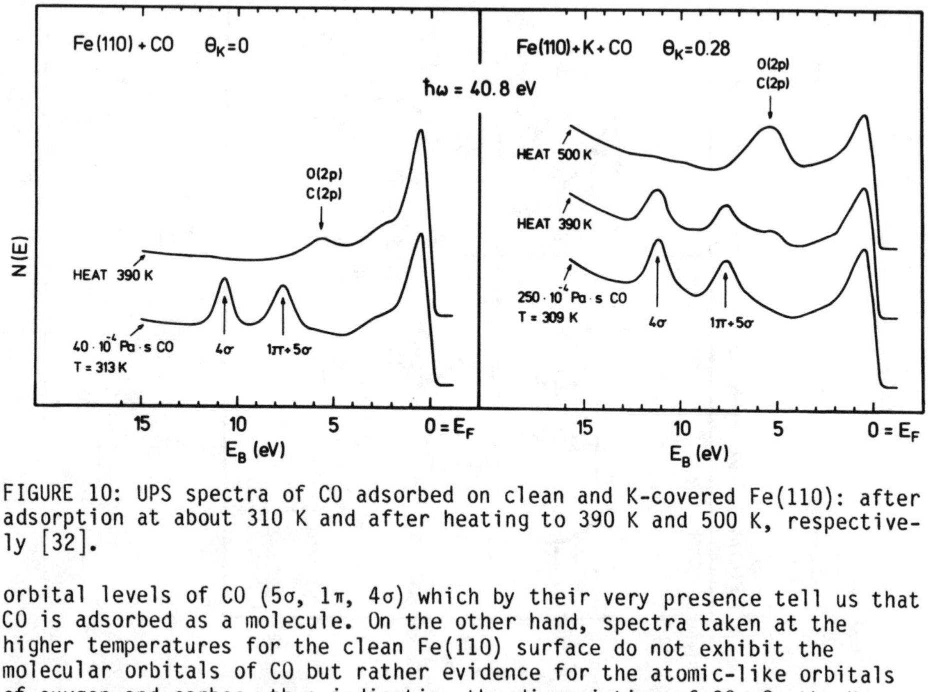

FIGURE 10: UPS spectra of CO adsorbed on clean and K-covered Fe(110): after adsorption at about 310 K and after heating to 390 K and 500 K, respectively [32].

orbital levels of CO (5σ, 1π, 4σ) which by their very presence tell us that CO is adsorbed as a molecule. On the other hand, spectra taken at the higher temperatures for the clean Fe(110) surface do not exhibit the molecular orbitals of CO but rather evidence for the atomic-like orbitals of oxygen and carbon, thus indicating the dissociation of CO. On the K-covered surface, on the other hand, heating to 390 K results in a mixture of molecular and dissociated CO as seen by the spectrum on the right hand side of the figure. From this we conclude that E_{ad}(CO) is increased by the presence of K. Furthermore, a comparison of the clean and K-covered Fe(110) surface shows that CO dissociation is more pronounced on the K-covered surface. Thus, the total amount of dissociated CO is larger on that surface, an observation which can be readily explained by the higher energy of adsorption of CO on the K-covered Fe surface.

A final remark concerns the rate of surface diffusion of adsorbed species. This process is particularly difficult to analyze because a spectroscopy with a very high lateral resolution is needed. One such technique is the field electron microscope with which the diffusion of adsorbed species across a metal surface has been studied [33]. Another very useful technique should be the newly developed scanning electron tunneling micro scope [34]. A completely different approach involves the scattering of helium beams from surfaces [35]. Here, the scattered beam intensity is a measure of the coverage of adsorbed molecules on flat parts of the surface. If molecules are adsorbed at low temperature, sufficiently low such that no surface diffusion takes place, the intensity of the scattered beam decreases. As the temperature is raised and the adsorbed species begin to diffuse to sites of higher adsorption energy (step sites, defect sites), the scattered helium beam intensity starts to increase again. Thus the increasing helium intensity is linked to the onset of surface diffusion of adsorbed species. Generally speaking, there is a great lack of information on surface diffusion data of adsorbed atoms and molecules on solid surfaces. Most recently, NMR has been applied to the measurement of surface diffusion of adsorbed molecules on metallic clusters [36].

4. DESORPTION

4.1. Rate of desorption

The desorption rate of atoms and molecules from surfaces is generally assumed to be described by the following equation [1,37]:

$$r_d = \frac{dn}{dt} = n^x \, k_{ox} \exp\left(-\frac{E_d(n)}{kT}\right) \tag{6}$$

In this equation x is the order of the desorption reaction, k_{ox} is the preexponential factor which includes an entropy term, and E_d is the energy of desorption which is generally coverage dependent. Strictly speaking, this equation cannot be applicable if the adsorbed species are residing in different adsorption sites with different adsorption energies. However, this difference is usually approximated by the coverage dependent adsorption energy. In a typical desorption experiment, the temperature of the substrate is raised in a linear fashion, such as

$$T(t) = T_0 + \beta t \tag{7}$$

where β is the heating rate. At a temperature where the desorption of the adsorbed species begins, the partial pressure of this particular species rises as shown in Fig. 2. If the experiment is carried out in a continuously pumped vacuum chamber, the pressure rise due to desorption will only be temporary because the amount of adsorbed gas is finite. Therefore the partial pressure of the desorbing gas as a function of temperature (time) will exhibit a maximum. As the partial pressure of the gas is directly proportional to the rate of desorption, the position of the maximum can be obtained from differentiating equation (6). The result is a condition given by the following equation [38]:

$$\frac{E_d}{kT_p^2} = x \, n^{x-1} \frac{k_{ox}}{\beta} \exp\left(-\frac{E_d}{kT_p}\right) \tag{8}$$

which relates the desorption energy to the temperature at the maximum, T_p. This equation is generally very useful for estimating E_d of the desorbing

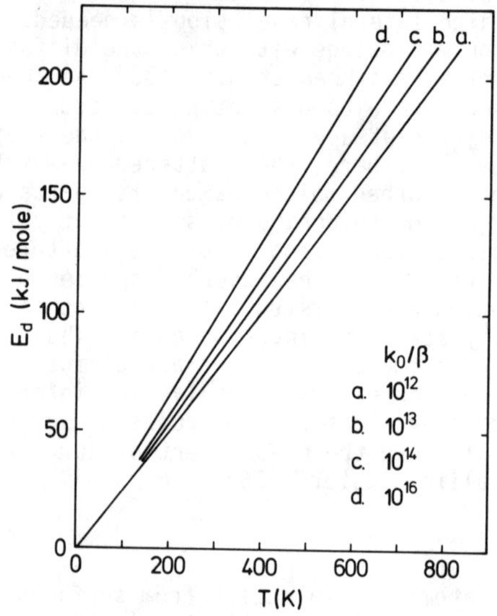

FIGURE 11: Plot of energy of desorption versus peak temperature of first order flash desorption traces, according to eq. (8) [38].

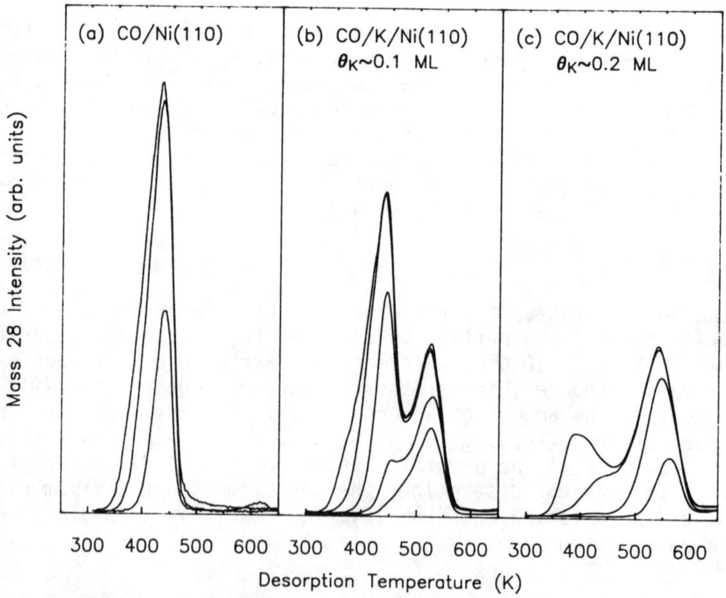

FIGURE 12: Thermal desorption spectra of CO from clean (a) and K-covered (b,c) Ni(110) surface. The adsorption temperature was 300 K, heating rate 13 K/s [39]

species. For a first order desorption reaction (x=1) and a known ratio of pre-exponential factor to heating rate the relationship in equation (8) can be evaluated quantitatively. This evaluation leads to the plot in Fig. 11 which shows the energy of desorption versus peak temperature for different values of k_{ox}/β. First order desorption is usually observed for a species that exhibits molecular (reversible) adsorption. Even if the pre-exponential factor k_o is not known, a change in T_p for different surface conditions (same β) can be taken as a measure of a change in E_d. A nice example is shown in Fig. 12 for CO desorption from a clean and K-covered Ni(110) surface [39]. Whereas the peak temperature for clean Ni(110) is 440 K, the spectra for K-covered Ni(110) exhibit two peaks, one characteristic of CO on clean Ni, and the other characteristic of K-influenced sites at a higher temperature of 525 K and even 560 K, depending on the K-coverage (Θ_K). Provided k_o does not change significantly, this increase in T_p of 85-120 K corresponds to an increase in E_d of approximately 20 - 40 kJ/mole according to Fig. 11. This effect illustrates the increase in the adsorption energy of CO due to coadsorbed K [21].

If decomposition or dissociation occurs during adsorption, the order of desorption is generally 2 or even higher. When the desorption order x is unknown, it is still possible to evaluate the energy of desorption from the peak temperature of the flash desorption trace if the heating rate β is varied. For this purpose we rewrite equation (8) such that [38]

$$\ln \frac{\beta}{T_p^2} = \ln \left[\frac{kxn^{x-1} k_{ox}}{E_d}\right] - \frac{E_d}{kT_p} \tag{9}$$

If desorption traces for equal initial coverages are now recorded with different heating rates, a plot of log (β/T_p^2) versus $1/T_p$ should yield a straight line whose slope can be evaluated to give the energy of desorption. This kind of evaluation has been used in a number of cases where adsorbed molecules decompose during desorption in a rather complicated fashion. A further advantage of the technique related to equation (9) comes from the fact that the pre-exponential factor k_{ox} does not have to be known.

Another important quantity that can be obtained from a flash desorption experiment is the relative or even absolute coverage of the adsorbate. Since the partial pressure of the desorbing species is directly proportional to the rate of desorption, the coverage can be obtained by integrating over the desorption trace:

$$n = \frac{S}{AkT} \int_0^\infty P(t)\, dt \tag{10}$$

In this equation, S is the pumping speed of the system, A the area of the sample and P(t) is the time-dependent partial pressure recorded during the experiment. Based on a modification of this equation (10), other more complicated schemes for integrating partial desorption traces have been worked out in order to obtain the coverage as a function of temperature for a given initial total coverage. From a complete set of these data, the coverage-dependent desorption energy and pre-exponential factor can, in principle, be calculated [37,40].

Most of the evaluation discussed here works well for reversible first order desorption. Equation (8) has also been applied to second order desorption as long as a single desorption peak is present. The situation becomes rather complex when multiple desorption states are present. Similar difficulties arise when desorption is coupled to decomposition of the desorbing species. In those cases, several different desorbing products have to be monitored simultaneously. The question is then whether desorption or decomposition is rate-determining. Several authors have dealt in great detail with questions of this kind [41,42].

4.2. Pre-exponential factor k_0

In order to evaluate the energy of desorption from the peak temperature of a flash desorption trace according to equation (8), a pre-exponential factor of 10^{13} s^{-1} was usually assumed. Independent measurements of the pre-exponential factor of desorbing molecules, such as CO and NO, showed, however, that it was considerably larger than $10^{13} s^{-1}$ and even coverage-dependent. Several authors reported pre-exponential factors exceeding $10^{16} s^{-1}$. A very detailed investigation of CO desorption from a Ru(001) surface by the technique of the heating rate variation resulted in pre-exponential factors between 10^{16} and 10^{19} s^{-1}, as shown in Fig.13 [43].

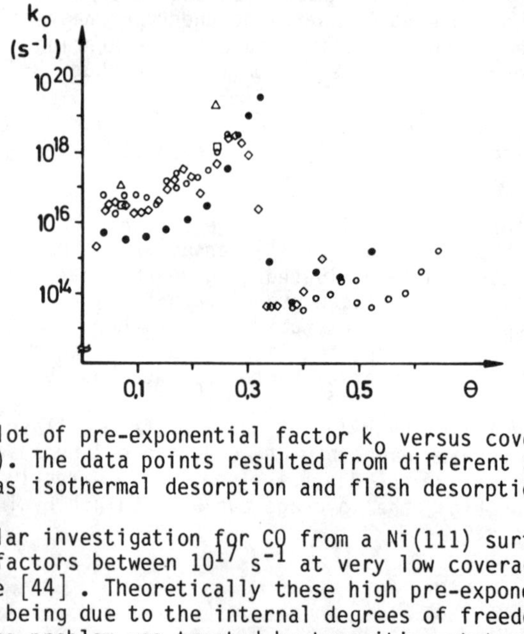

FIGURE 13: Plot of pre-exponential factor k_0 versus coverage for CO adsorbed on Ru(001). The data points resulted from different kinds of measurements, such as isothermal desorption and flash desorption [43].

Another similar investigation for CO from a Ni(111) surface reports pre-exponential factors between 10^{17} s^{-1} at very low coverage up to 10^{21} s^{-1} at high coverage [44] . Theoretically these high pre-exponential factors were explained as being due to the internal degrees of freedom of the desorbing molecules. The problem was treated by transition state theory where the pre-exponential factor was then proportional to the ratio of partition functions of the transition state and the adsorbed state. Translational, rotational and vibrational degrees of freedom of the CO molecule were taken into account yielding a pre-exponential factor of about $10^{18} s^{-1}$ at very low coverage [43]. The drawback of this simple theory is that it is not suitable for explaining the coverage dependence of the pre-exponential factor. A somewhat different approach starts from the equilibrium between gas phase and the adsorbed phase resulting in a modified Langmuir isotherm (Fowler equation) [44]

$$P = \frac{\alpha\Theta}{1-\alpha\Theta} \frac{kT}{h^3} (2\pi mkT)^{3/2} \exp\left(-\frac{E_d}{kT}\right) \qquad (11)$$

$\Theta = n/n_o$ is the relative coverage of the adsorbate, h = Planck's constant. Under equilibrium conditions the rate of desorption is equal to the rate of adsorption

$$r_d = r_a = \frac{P}{(2\pi mkT)^{1/2}} \; s(n,T) \tag{12}$$

and by inserting equation (6) and (11) into equation (12), one obtains an expression for the pre-exponential factor of desorption. One can see that this factor is coverage- and temperature-dependent. The same treatment can be extended to the desorption of molecules by including rotational and vibrational partition functions. Under those assumptions the pre-exponential factor is then given by the following equation:

$$k_o = \frac{\alpha}{n_o(1-\alpha\Theta)} \frac{kT}{h} \frac{2\pi mkT}{h^2} \frac{2I_m kT}{h^2} \; s(\Theta,T) \tag{13}$$

with I_m = moment of inertia. Equation (13) is used semi-empirically in the sense that the experimental data for $s(\Theta,T)$ are inserted. For a sticking coefficient s = 1 and small coverages the pre-exponential factor for CO can be calculated to be of the order of 10^{17} s^{-1} at 300 K, in good agreement with the experiment [44]. Recently another calculation of the pre-exponential factor for NH_3 on Ni(111) surface was carried out in the framework of transition state theory [45].

4.3. Angular and energy dependence

Thus far we have talked about integrated quantities of rates of adsorption and desorption of molecules and therefore ignored any possible angular dependencies of fluxes. Equally we have ignored the energy distribution of desorbing molecules and the question of the internal energy distributions of these molecules. In fact, asking questions about the angular and energy dependence of desorbing molecules, is the point where the real science of desorption begins. Under the assumption of thermodynamic equilibrium, the angular distribution of the flux, $F(\Theta,\varphi)$ of molecules leaving from a surface area element dA is given by the following expression (Knudsen) [46]:

$$F(\vartheta,\varphi)\, dA = \frac{n_v}{4\pi} \left(\frac{8kT}{\pi\, m}\right)^{1/2} \cos\vartheta \, dA \tag{14}$$

In equation (14) n_v is the concentration of molecules per unit volume (cm^{-3}), ϑ the polar angle and φ the azimuth. Whereas the probability of desorption does not depend on the azimuth, it does depend on the polar angle ϑ and is proportional to cos ϑ. This means that there is a maximum in the rate of desorption normal to surface.

The velocity (energy) dependence of molecules (flux) leaving the surface in equilibrium can also be given by a simple expression. The flux of molecules d^3f normalized to unit solid angle desorbing with velocities between v and v + dv are given by [46]:

$$\frac{d^3 f}{\sin\vartheta \; d\varphi \; d\vartheta} = n_v \left(\frac{m}{2\pi kT}\right)^{3/2} \exp\left(-\frac{mv^2}{2kT}\right) v^3 \cos\vartheta \, dv \tag{15}$$

This equation describes the Maxwell distribution. Of some importance is the root mean square of the velocity given by [46]

$$v_{rms} = \left(\frac{4kT}{m}\right)^{1/2} \tag{16}$$

This velocity is related to the mean kinetic energy of desorbing molecules:

$$\langle E \rangle = \frac{1}{2} m v_{rms}^2 = 2kT \tag{17}$$

The laws for angular and velocity distributions, equations (14) and (15), have been derived, under the assumption of equilibrium and microscopic reversibility, for all molecules leaving the surface. Here it is important to realize that the flux of molecules coming from the surface at equilibrium includes those species that were actually adsorbed and those that are scattered or reflected. When the sticking coefficient $s(T,\vartheta,n)$ is equal to unity, there are no reflected but only adsorbing molecules. In that case the desorption flux consists of desorbing molecules only, and eqs. (14) and (15) are a correct description for the angular and velocity distributions of the desorbing species. In the general case of $s(t,\vartheta,n) < 1$, however, there is at present no analytical expression for the description of the angle and velocity dependence of molecules desorbing from a surface. In this more complicated case one can derive a relationship between the desorption and adsorption probabilities which can be tested experimentally. The physical law of microscopic reversibility requires that desorption and adsorption should show the same angular dependence. Assuming thermodynamic equilibrium, one obtains the relationship [12]:

$$D(T,\vartheta,n) = s(T,\vartheta,n) \cos\vartheta \tag{18}$$

where $D(T,\vartheta,n)$ is the desorption probability. A test of this relationship under the conditions of thermodynamic equilibrium (adsorption and desorption taking place simultaneously) is practically impossible. Hence, as a compromise, experiments are performed under conditions of quasi-equilibrium, checking the probabilities of adsorption and desorption in separate experiments [12]. This procedure will be illustrated by examples.

There is a large difference in sticking probability for the adsorption of CO and H_2 on a Ni(111) surface [10]. Whereas CO adsorbs with an initial sticking coefficient near unity, that of H_2 is only 0.05. Therefore, the adsorption/desorption of these gases on Ni(111) was chosen to check eq. (18). The result for desorption of CO from a Ni(111) surface is shown in Fig. 14 [12]. Here, the desorption probability as well as the adsorption probability is plotted versus polar angle ϑ. One can see that the experimental data obey equation (14) very well and exhibit a cosine distribution. For comparison, Fig. 15 shows the adsorption and desorption probabilities versus polar angle for H_2 on Ni(111). In this case, the distribution deviates very substantially from a simple cosine law. It is fitted with a $\cos^x\vartheta$ function where $x = 4.5$. There is still agreement in the angular dependence of the normalized adsorption and desorption probabilities suggesting that the condition of quasi-equilibrium is fulfilled. For the same data it was also demonstrated that the desorption of H_2 did not obey the Maxwell law. The mean energy of desorbing H_2 molecules was considerably higher than the corresponding surface temperature of the Ni crystal [47].

The point in a comparison of CO and H_2 adsorption/desorption from a Ni(111) surface is that CO adsorbs reversibly and without any hint of an activation barrier for adsorption, while H_2 adsorbs dissociatively with considerable evidence for an activation barrier of adsorption. The angular velocity distribution of H_2 from a Ni(111) surface has also been studied in detail by time-of-flight mass spectroscopy [47]. These experiments were carried out for clean- and sulfur-covered surfaces. Typical energy

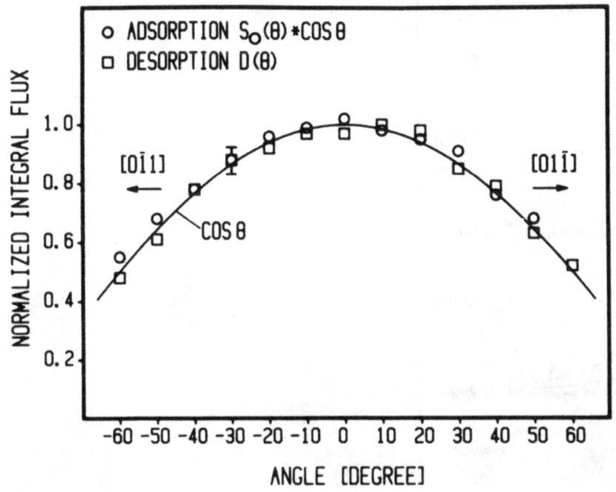

FIGURE 14: Angle dependence of the adsorption and desorption probability of CO on Ni(111) at 300 K [12].

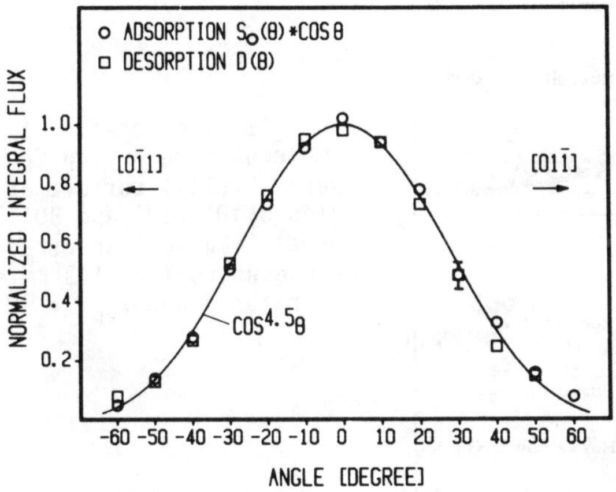

FIGURE 15: Angle dependence of the adsorption and desorption probability of H_2 on Ni(111) at 190 K [12]. The adsorbing gas has a temperature of 300 K (Knudsen cell).

distributions are shown in Fig. 16 for three different polar angles ϑ. In each case the temperature of the Ni crystal was 1143 K. The mean energy of each distribution is given as $\langle E \rangle/2k$ which deviates considerably from the surface temperature. For comparison, Maxwell distributions with the surface temperature and a temperature corresponding to $\langle E \rangle/2k$ are shown in each one of the figures. It is clear that the experimental distributions are non-Maxwellian. It turns out that molecules desorbing normal to the surface ($\vartheta = 0$) have the highest mean temperature, i.e. 1871 K, whereas those desorbing at larger polar angles have lower temperatures, e.g. at 80° only 771 K. The angular dependence of the mean energy of desorbing D_2 molecules from Ni(111) at a surface temperature of 1134 K is summarized in Fig. 17. The reason for the non-applicability of eq. (15) comes from the

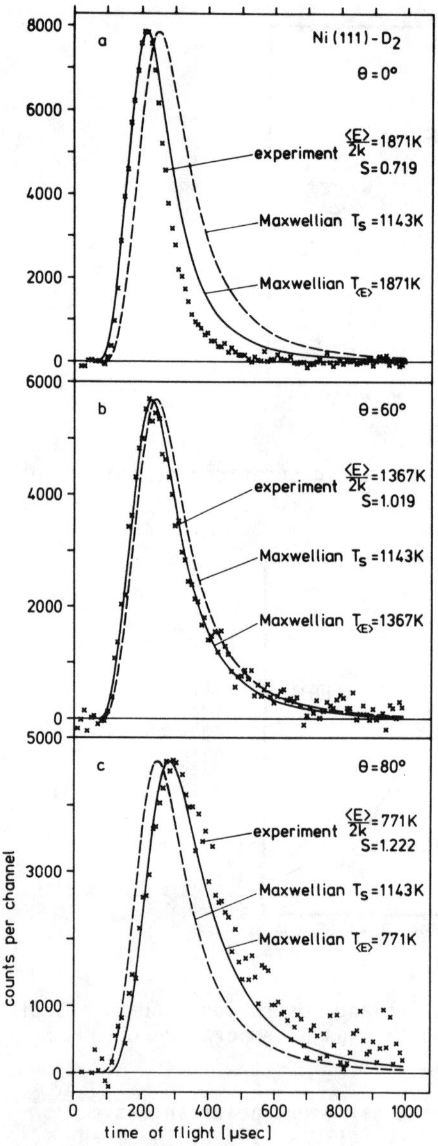

FIGURE 16: Time-of-flight spectra of D_2 molecules desorbing from a S-covered Ni(111) surface at polar angles of 0°, 60° and 80°. The curves represent Maxwell spectra corresponding to T = 1143 K (surface temperature) and $T_{<E>} = <E>/2k$ [47].

fact that there is an activation barrier for adsorption in this case. Even though the barrier is not known in any detail, one may infer that molecules desorbing have crossed this barrier and hence are more energetic than corresponding to the surface temperature. Since the barrier is a true surface property, molecules having crossed it no longer have an opportunity to lose energy to the solid. Hence they carry along the extra energy necessary to surmount the barrier as translational energy.

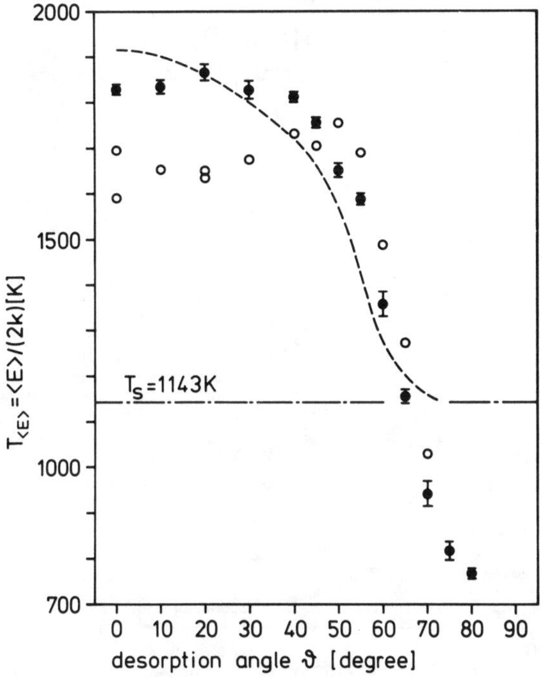

FIGURE 17: Polar angle dependence of mean energy of D_2 molecules desorbing from clean (o) and S-covered (•) Ni(111) [47].

Another interesting example in this connection concerns the desorption of D_2 from a Pd(100) surface [48]. Fig. 18 shows again a set of energy distribution curves of desorbing D_2 at a surface temperature of 500 K and for $\vartheta = 0^o$. The distributions shown represent the clean surface as well as surfaces with an increasing amount of adsorbed sulfur. The peculiar feature is now that the distribution for the clean Pd surface is Maxwellian whereas the distributions with a larger amount of sulfur on the surface are strongly non-Maxwellian. For example, the distribution for a sulfur coverage of 0.65 is characterized by a very narrow peak at high energy corresponding to a temperature of about 3 times the surface temperature. Hence one could conclude that D_2 desorption from clean Pd is not influenced by an activation barrier of adsorption, but that on sulfur-covered Pd it is. As a rationale for these experimental data, a one-dimensional potential energy diagram shown in Fig. 19 was proposed [46]. Since hydrogen dissolves easily in Pd, the energy diagram continues into the bulk of the crystal. On clean Pd the chemisorption state of hydrogen is characterized by an energy of about 20 kcal/mol. Hydrogen penetrating the bulk of a crystal and desorbing from the surface equilibrates in the chemisorption well, then recombines to form the H_2 molecule and desorbs. In the absence of an activation barrier of adsorption this desorption obeys the Knudsen and Maxwell distributions in equations (14) and (15). On the other hand, when sulfur is adsorbed on the Pd surface, the adsorption of hydrogen in its normal chemisorption state is poisoned, i.e. the chemisorption well in Fig. 19 is no longer accessible. Therefore, hydrogen coming from the bulk of the Pd, will

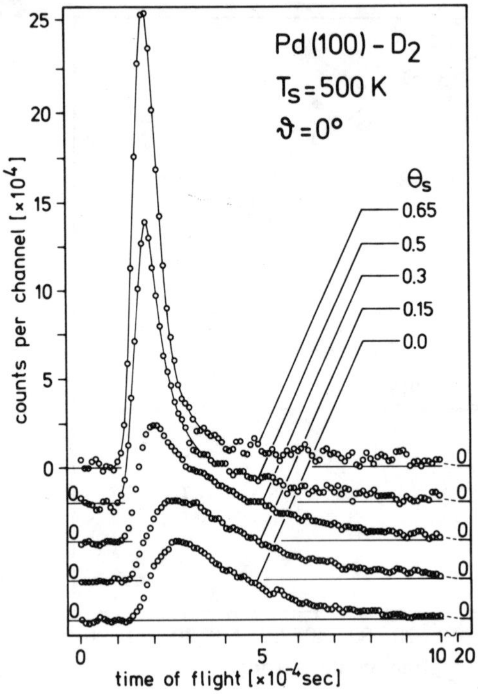

FIGURE 18: Time-of-flight spectra of D_2 molecule desorbing from clean and S-covered Pd(100) at 500 K (surface temperature). The sulfur coverage is given by Θ_s [48].

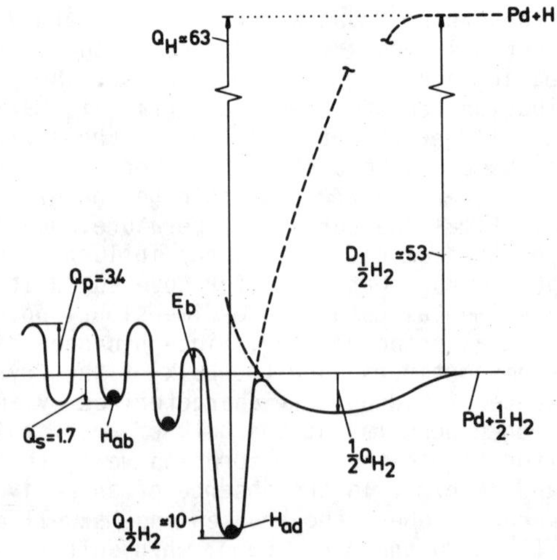

FIGURE 19: Energy diagram for H_2 on Pd, showing a molecular (H_2) and atomic H_{ad} adsorption state as well as bulk absorption states (H_{ab}). E_b last bulk barrier; Q_p permeation energy; Q_s solution energy (all energies in (kcal/mol)) [46].

equilibrate in the last well of the bulk potential and then recombine to form H_2 molecules near the top of the last barrier and then desorb. Because of the existence of this bulk potential barrier, E_b, the desorbing H_2 molecules will have again a high translational energy giving rise to the narrow peak in the energy distribution of Fig. 18. This example then underlines again the correlation of activated adsorption and the deviations from Maxwell and Knudsen distributions.

4.4. Internal energy of desorbing molecules

The naive point of view that the internal energy of desorbing molecules should correspond to the surface temperature of the substrate is generally incorrect. The reasoning is quite analogous to that in the previous section. Again it is useful to distinguish between non-activated adsorption on the one hand, and activated dissociative adsorption on the other hand. In the first case, the internal energies may correspond to the surface temperature, but in the second case considerable deviations can be anticipated. The principal reason for these deviations is the activation barrier of adsorption. The recombination of atoms into a molecule occurs near the top of the barrier, and any excess (or deficiency) in internal energy associated with this event will be carried over into the gas phase because the desorbing molecule has ceased to interact with the surface. Hence, the molecule might desorb in an excited vibrational or rotational state. The same occurs when desorbing molecules are formed as a product of a surface reaction, e.g. in the case of CO oxidation.

An interesting example is the desorption of N_2 from a polycrystalline Fe surface [49]. Here the vibrational state of desorbing N_2 molecules was measured by electron beam induced fluorescence. The vibrational energy of the molecule was then converted into an equivalent vibrational temperature. The experiments were carried out for surface temperatures between 1080 and 1400K. The vibrational temperature is shown as a function of the surface temperature in Fig. 20 for a clean Fe surface. Here the difference between

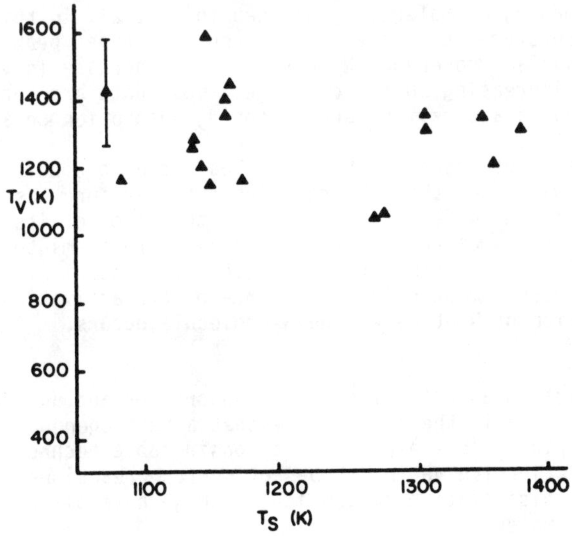

FIGURE 20: Vibrational temperature versus surface temperature for N_2 desorbing from clean Fe [49].

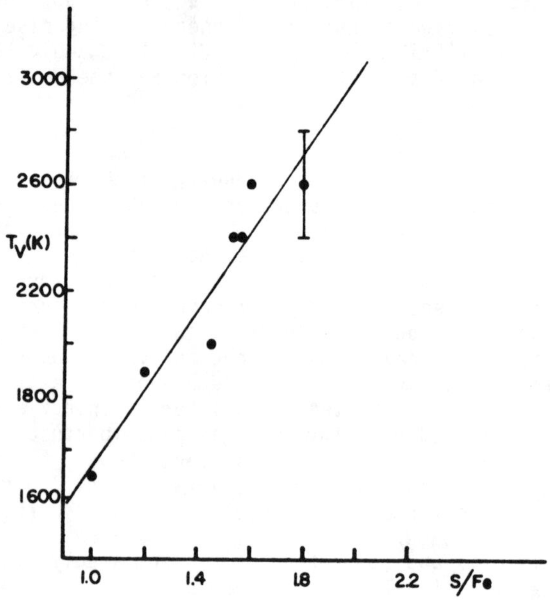

FIGURE 21: Vibrational temperature of desorbing N_2 molecules versus S/Fe Auger peak ratio for a surface temperature of Fe of 1150 K [49].

these two temperatures does not seem to be very significant. On the other hand, sulfur-contaminated Fe surfaces gave rise to very large vibrational temperatures of desorbing N_2 molecules, as seen in Fig. 21. In this figure, the vibrational temperature is plotted versus the S/Fe Auger peak ratio, i.e. versus sulfur surface coverage. Note the strong increase in vibrational temperature with increasing sulfur coverage. These data are rationalized in terms of an increased activation barrier for N_2 adsorption on a sulfur-covered Fe surface [49].

The rotational state of desorbing N_2 molecules was also studied by the same group [49]. It was found that the rotational temperature was about 400 ± 50 K independent of the surface temperature of the Fe or of its state of cleanliness. The low value of the rotational temperature seems to indicate that the recombination of N atoms into a N_2 molecule is sterically hindered. This steric hinderence must be a consequence of the activation barrier where the recombination of N atoms to the N_2 molecule occurs.

5. Surface reactions

One of the interesting aspects of studying adsorption and desorption of gases from solid surfaces is the possibility that a heterogeneous surface reaction might take place. This aspect has a considerable technological relevance. We can distinguish various types of surface reactions:

a) The first kind is dissociative adsorption which we have already learned about by a few examples.
b) The second type is dissociative desorption or decomposition. This means that on raising the temperature, the adsorbed species begin to decompose

just prior to desorption, sometimes concurrently with desorption. For example, the desorption and decomposition of formic acid which has been studied in great detail may occur as follows [50]:

$$HCOOH \begin{cases} \nearrow CO_2 + H_2 \\ \searrow CO + H_2O \end{cases} \tag{19}$$

It is a challenge for the surface scientist to find out when and why the decomposition occurs via the first or the second channel. Many other more complicated examples of this type have also been investigated [51].

c) A more significant example is a bimolecular heterogeneous reaction such as the oxidation of CO to CO_2:

$$CO_{ad} + O_{ad} \rightarrow CO_{2ad} \rightarrow CO_2 \tag{20}$$

This reaction occurs by a so-called Langmuir-Hinshelwood mechanism in which both reactants are coadsorbed on the surface [52]. In this particular example the reactants are CO and O which requires the dissociative adsorption of oxygen as a primary step. Many different investigations of this reaction have appeared in the literature [52].

d) In conjunction with this bimolecular heterogeneous reaction attempts have been undertaken to search for a mutual bond weakening/strengthening of the coadsorbed reactants, an effect, which might be interpretable as a precursor situation to an actual surface reaction. This search involved the investigation of stationary adsorbed phases by electron spectroscopies, for example, but proved to be unsuccessful in most cases.

The general experimental approach to studying surface reactions is similar to that for studying adsorption and desorption of gases. Experiments are usually carried out on well characterized clean single crystal or polycrystalline surfaces by utilizing surface spectroscopic techniques and working under ultra high vacuum conditions. Of course, the study of the steady-state surface reaction requires the presence of reasonably high partial pressures of the reactants inside the vacuum system. Under such conditions it is sometimes difficult to maintain and guarantee a high purity of the gas mixture. This problem can be overcome by using molecular beams of the reactant molecules. In some cases the reaction probability is extremely low, however, such that it can no longer be studied inside a vacuum system. A rather well-known example is the hydrogenation of CO on transition metal surfaces which occurs with a measurable rate at pressures of at least several mbar. In order to study such a reaction, a combination system is generally used featuring a high-pressure microreactor, a fast transfer system between reactor and vacuum, and a ultrahigh vacuum system for surface spectroscopy. The experimental results consist of reaction-kinetic data obtained by mass spectrometry or by a gas chromatograph and by surface analytical data.

The hydrogenation of CO on transition metal surfaces is an important example of a surface reaction because of its technological relevance [53]. It is also a very complicated reaction in terms of mechanism. The spectrum of products obtained from this reaction is extremely broad, ranging from simple methane to long-chain hydrocarbons, alcohols and other oxygen-containing molecules. The selectivity of this reaction depends on a large number of variables, such as temperature, reactant composition, gas flow, and surface composition. Surface additives such as alkali metals have a tremendous influence on the selectivity and reactivity of the catalyst [53]. Figure 22 shows as an example C1s and K2p spectra of a polycrystal

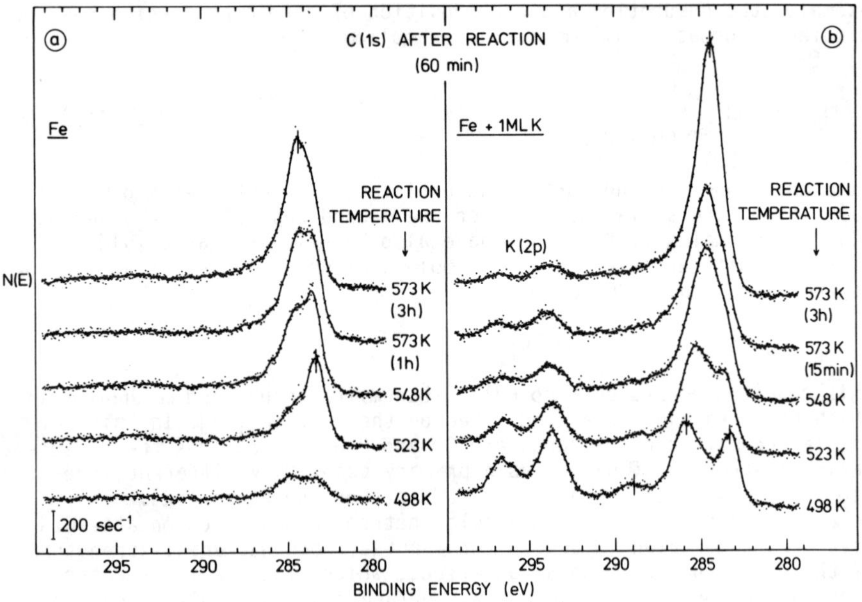

FIGURE 22: C1s and K2p XPS spectra after CO hydrogenation on Fe at various temperatures as indicated. The reaction time was 60 min except where otherwise indicated. Past (a) was initially clean Fe, part (b) Fe with one monolayer of K [54].

line Fe foil after CO hydrogenation at various temperatures. The reaction took place in a flow system at 1 bar total pressure and a partial pressure ratio CO/H_2 of 1:20, the reaction time in each case was 60 min [54]. The left hand side of the figure shows C1s spectra for the unpromoted (K-free) surface while the right hand side is characteristic of a surface with about one monolayer of pre-adsorbed K evaporated onto the clean Fe inside the UHV chamber. The C1s spectra as a function of reaction temperature for clean and K-covered Fe demonstrate the dependence of the surface composition of this catalyst on running conditions. In some cases there are several distinguishable C1s contributions which may be correlated with different carbon-containing adsorbed species. The identification of the various C1s peak contributions in Fig. 22 was accomplished by a comparison of binding energies with layers of known composition, such as simple carbidic (atomic) and graphitic carbon, adsorbed CO, or adsorbed hydrocarbon molecules [55]. The shape of the differentiated carbon Auger peak was also taken as an additional aid in identifying the carbonaceous species. With that knowledge the C1s peaks at 283.3 eV and 284.5 eV binding energy in Fig. 22 were attributed to carbidic and graphitic carbon, respectively. The additional peaks at 285.8 eV and 289.0 eV observed for the K-covered Fe were interpreted as being due to a polymerized surface carbon species ("Polymethylene") and a carbonate-type species, respectively. No detailed knowledge of these latter species (structure, stoichiometry, etc.) exists. Additional more sensitive analytical techniques are required in order to analyze these species in more detail and also to detect minority species which must be present as reaction intermediates. The remarkable feature in the results of Fig. 22 is that the surface carbon composition is indeed different for the clean and K-promoted Fe, and that the K causes the rate of carbon deposition to increase [56,57]. That could be shown by integrating the total C1s intensity as a function of reaction time.

ACKNOWLEDGEMENT

I am very grateful to Prof. George Comsa for clarifying discussions and critical comments on the manuscript.

REFERENCES

1. D.O. Hayward and B.M.W. Trapnell: Chemisorption (Butterworths, London 1964).
2. G. Ehrlich: J. Phys. Chem. Solids 1 (1956) 1.
3. J.E. Lennard-Jones: Trans.Faraday Soc. 28 (1932) 333.
4. D.A. King and M.G. Wells: Surface Sci. 29 (1972) 454.
5. H. Steininger, S. Lehwald and H. Ibach: Surface Sci. 123 (1982) 264.
6. G. Ertl, M. Neumann and K.M. Streit: Surface Sci. 64 (1977) 393.
7. C.T. Campbell, G. Ertl, H. Kuipers and J. Segner: Surface Sci. 107 (1981) 207.
8. P. Kisliuk: J. Phys. Chem. Solids 3 (1957) 95; and J. Phys. Chem. Solids 5 (1958) 78.
9. K. Schönhammer: Surface Sci. 83 (1979) L 633.
10. A. Winkler and K.D. Rendulic: Surface Sci. 118 (1982) 19
11. H.J. Robota, W. Vielhaber, M.C. Lin, J. Segner and G. Ertl: Surface Sci. 155 (1985) 101.
12. H.P. Steinrück, K.D. Rendulic and A. Winkler: Surface Sci. 154 (1985) 99.
13. S.P. Singh-Boparai, M. Bowker and D.A. King: Surface Sci. 53 (1975) 55.
14. P.W. Tamm and L.D. Schmidt: J. Chem. Phys. 54 (1971) 4775.
15. G. Pirug, G. Brodén and H.P. Bonzel: Proc. 7th Intl. Vac. Congr. and 3. Intl. Conf. on Solid Surf., p. 907, Vienna 1977.
16. H.P. Bonzel, G. Brodén, and G. Pirug: J. Catalysis 53 (1978) 96.
17. J. Perdereau and G. E. Rhead: Surface Sci. 24 (1971) 555; B. Lang, R.W. Joyner and G.A. Somorjai: Surface Sci. 30 (1972) 454.
18. H. Ibach, K. Horn, R. Dorn and H. Lüth: Surface Sci. 38 (1973) 433.
19. K. Christmann and G. Ertl: Surface Sci. 60 (1976) 365.
20. K. Besocke and K. Berger: Proc. 7. Intl. Vac. Congr. and 3. Intl. Conf. on Solid Surf., p. 893, Vienna 1977.
21. H.P. Bonzel: J. Vac. Sci. Technol. A2 (1984) 866.
22. J.C. Tracy and P.W. Palmberg: Surface Sci. 14 (1969) 274.
23. J.A. Van den Berg, L.K. Verheij and D.G. Armour, Surface Sci. 91 (1980) 218; H. Niehus and G. Comsa, Surface Sci. 151 (1985) L171.
24. K.H. Rieder and T. Engel, Phys. Rev. Letters 43 (1979) 373; K. Christmann, F. Chehab, V. Penka and G. Ertl, Surface Sci. 152/153 (1985) 356.
25. R.P.N. Bronckers and A.G.J. de Wit, Surface Sci. 112 (1981) 133; H. Niehus and G. Comsa, Surface Sci. 140 (1984) 18.
26. B.E. Hayden, K.C. Prince, P.J. Davie, G. Paolucci and A.M. Bradshaw: Solid State Comm. 48 (1983) 325.
27. H. Ibach and D.L. Mills: Electron Energy Loss Spectroscopy and Surface Vibrations (Academic Press, New York, 1982).
28. J. Stöhr and R. Jaeger: Phys. Rev. B 26 (1982) 4111.
29. K.A. Thompson and C.S. Fadley: J. Electr. Spectr. Rel. Phen. 33 (1984) 29.

30. F. Sette, J. Stöhr; E.B. Kollin, D.J. Dwyer, J.L. Gland, J.L. Robbins and A.L. Johson: Phys. Rev. Letters 54 (1985) 935.
31. D.A. Wesner, F.P. Coenen and H.P. Bonzel: Phys. Rev. B 33 (1986) 8837.
32. G. Brodén, G. Gafner and H.P. Bonzel: Surface Sci. 84 (1979) 295.
33. R. Gomer: Disc. Faraday Soc. 28 (1959) 23.
34. G. Binnig and H. Rohrer: Physica 127 B (1984) 37.
35. B. Poelsema, L.K. Verheij and G. Comsa: Phys. Rev. Letters 49 (1982) 1731.
36. C.P. Slichter: Bull. Am. Phys. Soc. 31 (1986) 534.
37. D.A. King: Surface Sci. 47 (1975) 384.
38. P.A. Redhead: Vacuum 12 (1962) 203.
39. L.J. Whitman and W. Ho: J. Chem. Phys. 83 (1985) 4808.
40. E. Bauer, F. Bonczek, H. Poppa and G. Todd: Surface Sci. 53 (1975) 87.
41. R. Gorte and L.D. Schmidt: Appl. Surf. Sci. 3 (1979) 381.
42. A. Cassuto and D.A. King: Surface Sci. 102 (1981) 388.
43. H. Pfnür, P. Feulner, H.A. Engelhardt and D. Menzel: Chem. Phys. Letters 59 (1978) 481; H. Pfnür and D. Menzel: J. Chem. Phys. 79 (1983) 2400.
44. H. Ibach, W. Erley and H. Wagner: Surface Sci. 92 (1980) 29.
45. A. Redondo, Y. Zeiri, J.J. Low and W.A. Goddard III: J. Chem. Phys. 79 (1983) 6410.
46. G. Comsa and R. David: Surface Sci. Reports 5 (1985) 145.
47. G. Comsa, R. David and B.J. Schumacher: Surface Sci. 85 (1979) 45.
48. G. Comsa, R. David and B.J. Schumacher: Surface Sci. 95 (1980) L210.
49. R.P. Thorman and S.L. Bernasek: J. Chem. Phys. 74 (1981) 6498.
50. J.L. Falcomer and R.J. Madix: Surface Sci. 48 (1975) 393.
51. R.J. Madix: in The Chemical Physics of Solid Surfaces and Heterogeneous Catalysis, Vol. 4, ed. by D.A. King and D.P. Woodruff (Elsevier, Amsterdam, 1982) chapters 1 and 4.
52. T. Engel and G. Ertl: in Ref. 51, chapter 3.
53. R.B. Anderson: in Catalysis, ed. by P.H. Emmett (Reinhold, New York, 1956), Vol. 4.
54. D.A. Wesner, F.P. Coenen and H.P. Bonzel: Langmuir 1 (1985) 478.
55. H.P. Bonzel and H.J. Krebs: Surface Sci. 91 (1980) 499.
56. H.P. Bonzel and H.J. Krebs: Surface Sci. 109 (1981) L527.
57. H.P. Bonzel, G. Brodén and H.J. Krebs: Appl. Surface Sci. 16 (1983) 373.

THEORY OF SPECTROSCOPY AND DYNAMICS IN LASER-IRRADIATED ADSPECIES-SURFACE SYSTEMS

THOMAS F. GEORGE AND DANIEL JELSKI
Departments of Chemistry and Physics & Astronomy
239 Fronczak Hall
State University of New York at Buffalo
Buffalo, New York 14260 USA

XI-YI HUANG
Department of Chemistry
University of Rochester
Rochester, New York 14260 USA

A. C. BERI
Systems Sciences Division
Computer Sciences Corporation
Beltsville, Maryland 20705 USA

1. INTRODUCTION

Recent progress in the theory of spectroscopy and dynamics in laser-irradiated adspecies-surface systems is discussed. This article is divided into three main sections. Beginning with Section 2 following this Introduction, a first-principles approach for describing vibrational excitation and relaxation of a laser-excited adatom is presented. A tractable approximation, known as the isomnesic (constant-memory) approximation is introduced in order to numerically deal with the long-time behavior of the dynamics. In Section 3 surface-dressed optical Bloch equations are solved for the resonance fluorescence spectrum of a laser-driven two-level atom near a flat metal surface and also near a rough metal surface modeled by a hemispheroid on a perfectly conducting plane. In Section 4 the problem of laser-induced period chemical vapor deposition is addressed, and a theoretical analysis of selected aspects of an experiment on the deposition of cadmium onto a silica surface is carried out, where the cadmium is produced by laser-induced dissociation of dimethyl cadmium gas above the surface.

2. VIBRATIONAL EXCITATION AND RELAXATION

We describe here a first-principles approach, which is divided into four parts. Part 2.1 is a brief introduction to the Zwanzig formalism for dealing with non-Markovian phenomena and approximations which make the method capable of describing long-range chemical processes with memory effects. Part 2.2 is an application of the theory to a fictitious system chosen to best illustrate effects peculiar to resonantly laser-excited vibrational manifolds, especially effects resulting from selective laser photochemistry. In Part 2.3 the theory is applied to a "real" system -- hydrated atomic hydrogen on the KCl(001) surface -- and the results are extended to chemical reaction times. Finally, in Part 2.4 a desorption channel is introduced, and important trends in the resulting desorption rate are discussed.

2.1. BASIC THEORY

The theoretical basis for the first-principles approach is the Heisenberg equation of motion

$$\dot{\rho}(t) = (i\hbar)^{-1}[H,\rho(t)] = -iL\rho(t) \quad , \tag{1}$$

where H is the Hamiltonian operator, L is the corresponding Liouvillian operator, and $\rho(t)$ is the density operator of the system. The Hamiltonian is partitioned into zeroth-order Hamiltonians for the solid and the adbond (the adspecies in the phonon-averaged field of all the lattice atoms) plus a perturbation representing the adbond-lattice interactions and another representing the adbond-laser interaction. Within the Born approximation and random-phase assumption, the equation of motion for an adatom-surface system leads to the generalized master equation (GME)[1]

$$\dot{P}_S(t) = \int_0^t dt' \, [\sum_{S'\neq S} K_{SS'}(t-t')P_{S'}(t') - \sum_{S'\neq S} K_{S'S}(t-t')P_S(t')] \quad , \tag{2}$$

which is an integrodifferential equation of the Volterra type. P_S is the occupation probability of the vibrational state $|s\rangle$ of the adbond, and the memory kernels $K_{SS'}$ incorporate the phonon (p) and the radiation (r) terms: $K_{SS'} = K^{(p)}_{SS'} + K^{(r)}_{SS'}$. For a Debye model of the solid, we have obtained the phonon portions $K^{(p)}_{SS'}$ in closed form,[2-4] and find them to be fairly complicated with a predominantly damped oscillatory behavior. Numerical solution of a Volterrra-type equation with a nonmonotonic kernel is very difficult.[5] In our case we were able to solve it only for $t \lesssim 5$ ps.[2] As a general rule, we have found that exact numerical solutions of the problem of vibration laser-adbond-surface energy transfer are possible only for times of the order of a few Debye periods (inverse of the Debye frequency, ω_D). This is hardly sufficient to describe chemical processes, so we have been studying possible techniques for overcoming this "non-Markovian bottleneck."

One such technique is based on a constant-memory approximation, i.e., isomnesic approximation (IA), to the memory kernel.[4] Briefly, the short temporal range of $K^{(p)}_{SS'}(t)$ suggests a delta-function representation if the long-time behavior (accurate beyond this temporal range) is of interest. Thus,

$$K_{SS'}(t) = \Omega_{SS'}\delta(t) + K^{(r)}_{SS'} \quad . \tag{3}$$

The second term on the right-hand side is time-independent. If it were a delta function also, we would obtain the Markovian approximation (MA). However, because of the constant terms $K^{(r)}_{SS'}$ we obtain instead a nonconvolution-type integrodifferential equation

$$\dot{P}_S(t) = \sum_{S'} K^{(r)}_{SS'} \int_0^t dt' \, P_{S'}(t') + \sum_{S'} \Omega_{SS'}P_{S'}(t) \quad , \tag{4}$$

which reduces to the following second-order differential equation (we note that the Markovian master equation is first order):

$$\ddot{P}_S(t) = \sum_{S'} K_{SS'}^{(t)} P_{S'}(t) + \sum_{S'} \Omega_{SS'}\dot{P}_{S'}(t) \quad . \tag{5}$$

We have found a convenient method for solving this based on Laplace transforms and the Heaviside expansion theorem.[6]

2.2. SELECTIVE PHOTOEXCITATION OF A PROTOTYPE SYSTEM

As a prototype system, we choose a mass 16 species on Ge which develops six bound states (S = 0,1,2,3,4,5) in the adbond potential. We also choose a situation where $|\omega_{10}| = \omega_D$ and $|\omega_{13}| = \omega_L$, where $\omega_{SS'}$ is the transition frequency between levels S and S' and ω_L is the laser frequency, so that the laser does not cause direct transitions between phonon states. The resulting population profiles are shown in Figure 1 for S = 0, 1 and 3 along with results of the exact numerical solution of the GME, Eq. (2). As expected, there is a loss in fine detail with the IA, but the average behavior is surprisingly well reproduced. Thus, for example, the crossing point where $P_1(t) = P_3(t)$ for the first time is at t = 1.7 ps in the IA as compared to t = 2.2 ps in the exact numerical solution. The relative behavior of P_0, P_1 and P_3 is also close to that obtained numerically.

Substantial chemical information is contained in the average adbond energy

$$\bar{E}(t) = \sum_{S} P_S(t)E_S \quad , \tag{6}$$

where

$$H_S|S\rangle = E_S|S\rangle \quad . \tag{7}$$

Figure 2 depicts $\bar{E}(t)$ obtained by solving the exact and isomnesic GME. We notice that the energy does not change monotonically, but consists of alternating influx and efflux. Figures 1 and 2 together display a feedback mechanism. Depletion of level S = 0 corresponds to energy transfer out of the phonon field, and we see such a depletion for $t \leq 5$ ps. This feedback is an important microscopic effect suggested in our earlier model calculation[7] but often missed in average statistical theories which examine only the long-time steady-state behavior of many-body systems. Figures 3 and 4, which show $P_S(t)$ and $\bar{E}(t)$ using the IA for 20 ps, further clarify the situation. The average energy content of the adbond is periodic, with an initial fast rise followed by a fall and subsequent smaller-amplitude damped oscillations. Between 0 and 4.8 ps, $P_0(t)$ decreases while $\bar{E}(t)$ increases; energy is transferred from the solid (and the laser, of course) to the adbond. However, subsequently between 4.8 and 7.7 ps, $P_0(t)$ increases while $\bar{E}(t)$ decreases, i.e., the phonon field is replenished at the expense of the adbond energy. But then again, the laser causes sufficient depletion of P_1 for phonon energy to get transferred back to the adbond, causing a depletion of P_0, which is energy feedback. The solutions of a Markovian version of this formulation

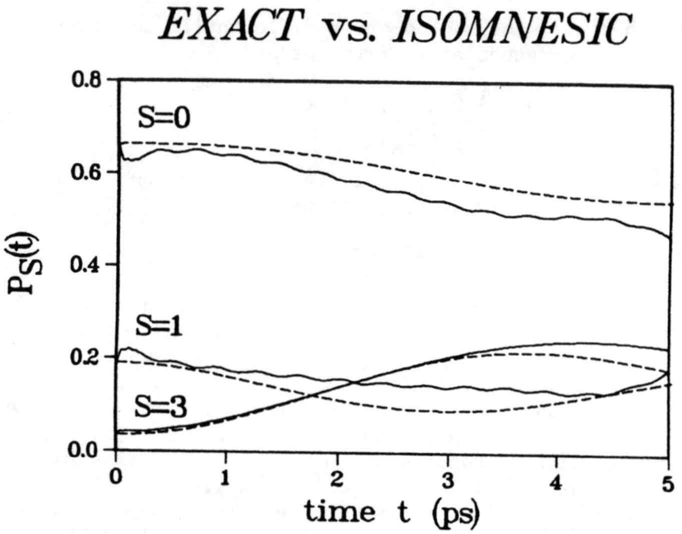

FIGURE 1. Probability profiles $P_S(t)$ for levels S = 0, 1 and 3 of the adbond. Solid lines: exact numerical solutions of the GME, Eq. (2); broken lines: closed-form solutions with the isomnesic approximation.

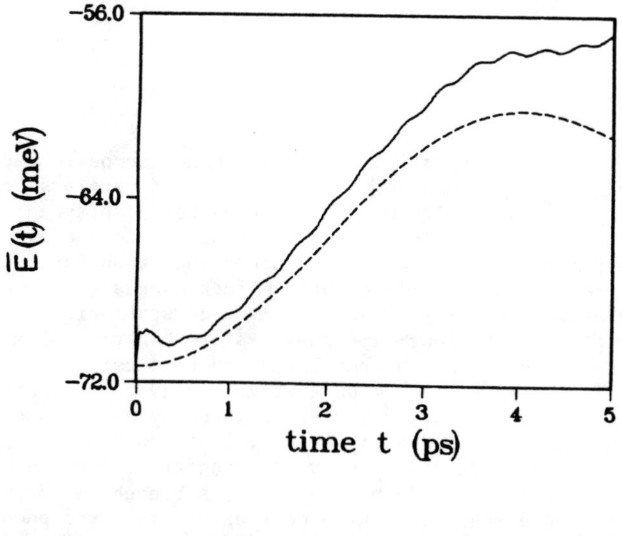

FIGURE 2. Average adbond energy $\bar{E}(t)$ obtained via the exact numerical solution of Eq. (1) (solid line) and closed-form solution with the IA (broken line).

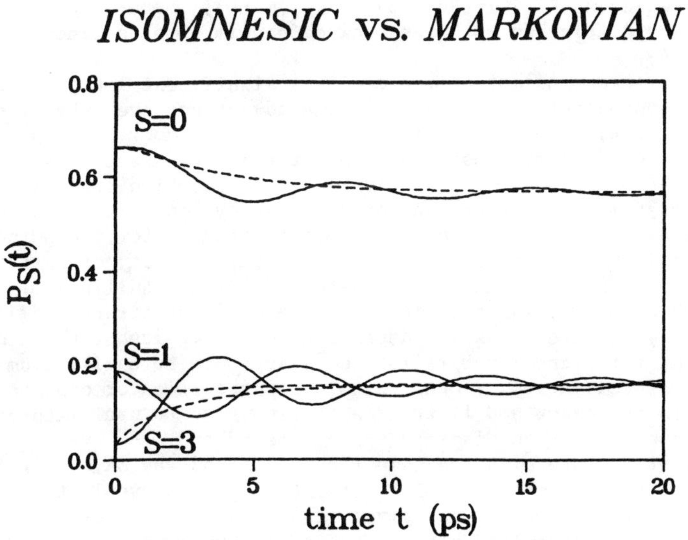

FIGURE 3. Comparison of $P_S(t)$ with the IA (solid lines) and the MA (τ = 1 ps; broken lines) for S = 0, 1 and 3.

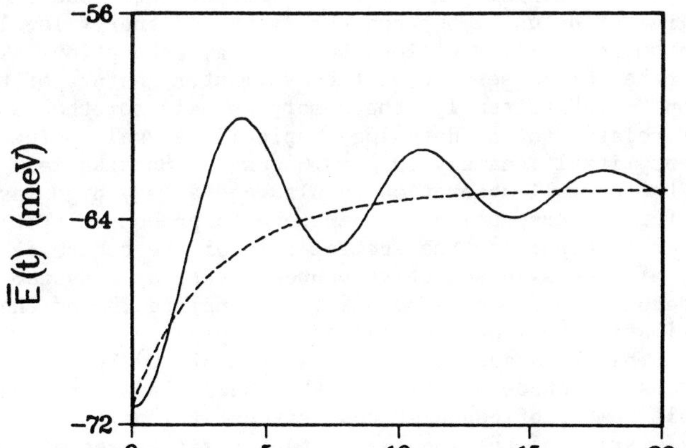

FIGURE 4. E(t) with the IA (solid line) and the MA (τ = 1 ps; broken line).

(obtained by replacing $K^{(r)}_{SS'}$ in Eq. (3) by $K^{(r)}_{SS'}\tau\delta(t)$, where τ is a time-range parameter) display no such feedback mechanism (broken lines in Figures 3 and 4).

The isomnesic approximation represents a significant improvement over the Markovian approximation and has definite advantages over the exact numerical treatment, since the IA is an analytic (nonnumerical) theory and is not subject to the time constraints inherent in the exact first-principles treatment. We are thus able to follow the time evolution of the adbond probability distribution for arbitrarily long times and to study the steady state. The Markovian limit is studied for the purpose of establishing a link between our new theory and the majority of past work on the chemical dynamics of many-body systems far from equilibrium. We find that such a (Markovian) treatment is inherently arbitrary and leads to qualitatively different results depending on the choice of the time-range parameter τ for the laser radiation.[4] Approach to equilibrium in the Markovian approximation is monotonic (a superposition of exponential changes), while the exact and IA treatments give rise to oscillatory behavior for the population distribution of the adbond. However, we do find that the steady state is the same for the IA and the MA. The important point here is that the IA is essentially as convenient and simple to work with as the MA, but gives results which are very faithful to the exact first-principles description of the phenomenon. We now proceed to applications of the above approach involving the IA.

2.3. LONG-TIME EFFECTS OF SELECTIVE PHOTOEXCITATION

Recent studies[8] have suggested that extremely high laser powers would be necessary to effectively desorb adspecies from a solid surface via a selective absorption mechanism. However, our recent work[3] has shown that a very small degree of detuning between the laser and energy levels of the system excludes the possibility of long-term energy absorption. The reason for this behavior is seen to be the oscillatory nature of the laser energy absorption "rate" (actually, the memory kernel) for the adbond, with a frequency related to the detuning, typically a small percentage of the laser frequency itself, namely 10^{10} - 10^{11} s^{-1}. Substantial cancellation in the overall absorption results and a very high laser power would be needed for the absorption to compete with phonon relaxation effects. In order to emphasize the special role of the coherence and monochromoticity of lasers in selective photoexcitation, a system with essentially no detuning has been chosen for the application of the isomnesic approximation to long time IR laser pumping.

Our formalism has been applied to the system $H(H_2O)/KCl(001)$,[9] where the primary reasons for choosing this are the shallow adsorption potential well and the small number of bound states (seven) it generates.[10] Figure 5 is a composite display of the function $\overline{E}(t)$ for six ranges of time differing by factors of 10 and for a laser intensity of 1 W/cm^2. It is seen that in going from a picosecond time range to a microsecond time range, the behavior of the system changes dramatically. The early behavior is dominated by fast energy transfer to phonon modes, completely overwhelming the slow laser pumping. The monotonic decay is representative of relaxation phenomena. For the 100 ps range, a false steady state seems to have been reached, but is in fact only an artifact of the large difference between the amplitudes of the laser and phonon kernels. The adbond energy starts to increase in the ns range, again

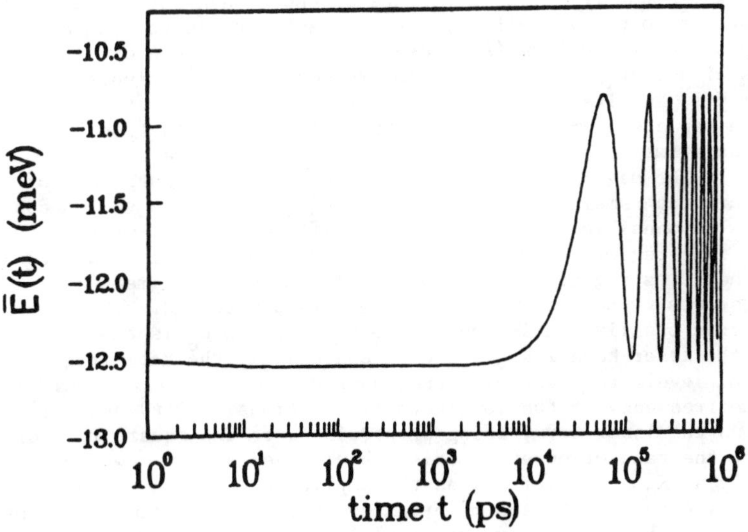

FIGURE 5. Ē(t) obtained by using the isomnesic approximation for time scales ranging from ps to μs for a laser intensity of 1 W/cm^2.

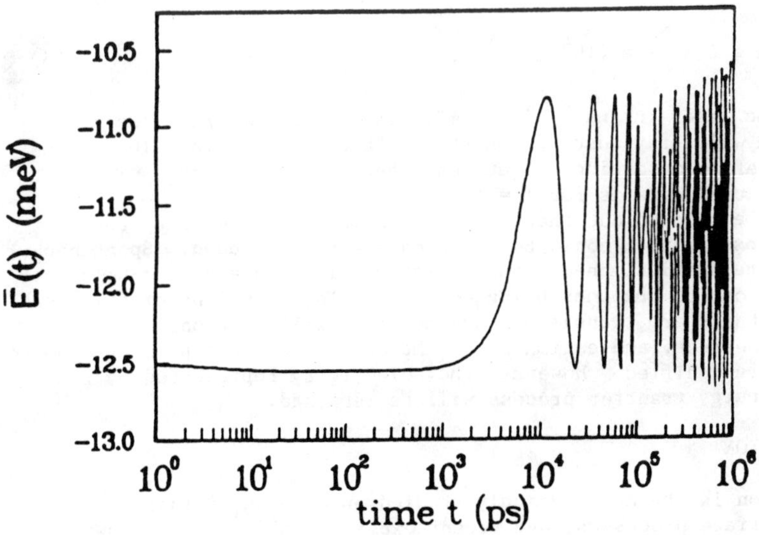

FIGURE 6. Same as Figure 5 for a laser intensity of 25 W/cm^2.

suggesting a possible steady state. However, this is a transitional period during which the relative importance of the phonon terms and the laser terms starts to change. The latter, in fact, originate from a probability profile of the form ($Ae^{\alpha t}\cos\beta t - \beta^{\alpha t}\sin\beta t$), where $\alpha + i\beta$ are complex roots of det $\underline{M}(s) = 0$, and $A + i\beta$ are complex coefficients obtained from $\underline{M}(s)$ and $\vec{T}(s)$, where the matrix $\underline{M}(s)$ and the vector $\vec{T}(s)$ are related to the Laplace transform of the isomnesic master equation.[9]

The phonon terms come from real roots and are therefore nonoscillatory. The magnitudes of the exponents α in the laser part of the solution are much smaller than corresponding exponents in the phonon term. This is responsible for the radically different nature of $\bar{E}(t)$ in the different time regimes. Thus, the seemingly monotonic rise in $\bar{E}(t)$ apparent in the 10 ns regime is seen to be only the early segment of a cycle which becomes evident in the 100 ns results and beyond. The frequency of this oscillation is seen to be $K_0^{1/2}$, where K_0 is the amplitude of the laser kernel $K_{SS'}$, and is essentially the Rabi frequency for the pair of levels resonant with the laser.

While the frequency of the long-term oscillations of P(t) and $\bar{E}(t)$ depends only on the laser term, being proportional to the square root of the intensity, the rate of onset of this behavior depends on α which varies linearly as K_0 and inversely as the amplitude Ω_0 of the phonon kernel. Thus, increasing the laser intensity from 1 to 25 W/cm^2 advances the onset time for the oscillatory state while increasing the frequency by a factor of 5 (as displayed in Figure 6).

Physically the results of Figures 5 and 6 can be interpreted in terms of desorption rates. Thus the time required for $|A|e^{\alpha t}$ and $|\beta|e^{\alpha t}$ to become large enough, so that the amplitude of E(t) in the oscillatory regime is substantially larger than the (t = 0) equilibrium Boltzmann value, is a reasonable measure of the desorption time to within an order of magnitude. This time is estimated by $\tau_D = \alpha^{-1}$. For our case,

$$\tau_D^{-1} = \alpha \cong 0.5\,\frac{K_0}{\Omega_0} \cong (10^4\ I)s^{-1} \quad , \tag{8}$$

where w is the laser intensity in W/cm^2. Thus, for a 1 W/cm^2 laser, a hydrogen atom will stay adsorbed on the KCl(H_2O) surface for ~10^{-4} s on the average before desorbing, or at least becoming very highly excited. The data mentioned here is for T = 150 K.

It must be pointed out that a number of mechanisms whereby the desorption cross section could be modified are not included. Spontaneous decay, phase relaxation, phonon-phonon interaction and electron-hole pair creation are some of the ones being considered in extensions of this work. As a result, the sharp transitions assumed here will be broadened by a detuning-type effect, and estimates of the desorption rate presented above will have to be modified. However, the selectivity represented by the dominant T_1 energy-transfer process will be retained.

2.4. DESORPTION

Desorption is the most commonly studied phenomenon of laser-stimulated surface processes, and recent experiments[11-13] have shown a strong dependence of the desorption cross section on the laser frequency. While energy transfer during IR-laser stimulated surface processes has been discussed for the above work in the absence of a desorption channel, selective absorption has been demonstrated for vanishing mismatch between the laser and adbond frequencies. The time-dependent energy profiles for the adbonds showed an initial strong increase followed by a decrease (transfer to bulk phonons) and, subsequently, another rise due to feedback from the phonons. We now present the results of an isomnesic formulation including desorption in the form of a transition from an excited state of the adsorption potential to the continuum.[14]

Let us consider a three-level system -- |S> = |0>, |1> and |3> -- one of whose transitions is resonant with the laser, and another equals the phonon Debye frequency. In the absence of desorption, application of the Laplace transform method to the GME leads to a third-order polynomial equation. With the inclusion of a depletive desorption channel, the GME reduces to a quartic. As expected, the <u>total</u> population for the adbond is depleted monotonically, and the desorption probability $P_D(t)$ grows correspondingly to a maximum value $P_T(0)$. The average adbond energy $\overline{E}(t)$ increases with time, but because of the competition between laser pumping and desorption, the detailed behavior of the energy profile is not as simple as that of $P_D(t)$. Depending on the magnitude of the bound-to-continuum transition rate R, $\overline{E}(t)$ has a monotonic (nonoscillatory), sigmoid profile, an oscillatory profile akin to the nondesorptive behavior, or a transitional profile. For 10^{-16} ps^{-1} $\leqq$ R $\leqq$ 1 ps^{-1}, both the cessation of slow oscillatory behavior and the onset of fast sigmoid behavior take place at higher laser powers for larger values of R. The trends observed are important from the point of view of selective laser photochemistry. Thus, for a given value of R, laser powers falling in the oscillatory region will cause a very slow accumulation of vibrational energy in the adbond, while laser powers falling in the transitional or sigmoid regions will excite the adbond quickly and cause almost instantaneous desorption.

3. RESONANCE FLUORESCENCE OF A TWO-LEVEL ATOM NEAR A METAL SURFACE

The lifetime of an excited molecule has experimentally been found to vary dramatically as a function of distance from a metal surface.[15] A number of researchers have examined this effect from the viewpoint of reflected-field theory, where the basic calculation is concerned with the interaction between an excited molecule and its own reflected radiation field. This theory generally provides good agreement with experiments.[15-26]

When an adatom is driven by strong resonant, driving coherent field, it creates for the atom or molecule an environment where the probability of stimulated emission can exceed that of spontaneous emission. Under this condition, dynamic ac-Stark splitting and nutational oscillation of the emitted light intensity become important parts of the laser-driven processes, such that interesting "resonance fluorescence" and other nonlinear optical phenomena can occur.[27] We have recently derived a set of surface-dressed optical Bloch equations,[21-26] by which we can examine the dynamics and relaxation of an adatom near a metal surface. By solving these equations, we have been able to evaluate the resonance fluorescence spectrum of a two-level atom near a metal surface, taking the following factors into account:

(i) the reflected electromagnetic field effects due to the presence of the interface
(ii) collision dephasing

(iii) surface-induced dephasing due to reflected photons
(iv) resonance excitation of surface plasmons
(v) random-phase fluctuations of the laser.

Besides these studies, we have been extending the flat-surface resonance fluorescence calculations to the rough surface case, which is modeled as a hemispheroidal protusion on a perfectly-conducting surface.[26] We discuss our progress below.

Let us begin with the case of a flat metal surface and the set of surface-dressed optical Bloch equations (SBE), which include the effect of surface-reflected photons, i.e., photons emitted by the laser-driven atom and reflected by the surface.[21,22] Within the rotating-wave approximation the SBE take the form

$$\frac{d}{dt}\begin{pmatrix}\hat{S}_{21}(t)\\ \hat{W}(t)\\ \hat{S}_{12}(t)\end{pmatrix} = \begin{pmatrix}-\tilde{\gamma}_2+i\Delta & i\Omega^-(t)/2 & 0\\ i\Omega^+(t) & -\gamma_1 & -i\Omega^-(t)\\ 0 & -i\Omega^+(t)/2 & -\tilde{\gamma}_2-i\Delta\end{pmatrix}\begin{pmatrix}\hat{S}_{21}(t)\\ \hat{W}(t)\\ \hat{S}_{12}(12)\end{pmatrix} - \gamma_1\begin{pmatrix}0\\ 1\\ 0\end{pmatrix}, \quad (9)$$

where $\hat{W} = |2\rangle\langle 2| - |1\rangle\langle 1|$ is the population inversion of the atom, $\hat{S}_{ij}$ is product of $\exp(i\omega_L t)$ and the transition operator $|i\rangle\langle j|$, $\Omega^{\pm}(t)$ is the time-dependent Rabi frequency, $\Delta = \omega_{21} - \omega_L$ is the detuning of the laser (ω_L) with respect to the two-level atom (ω_{21}), and the dephasing rate constant $\tilde{\gamma}_2$ is the sum of the radiative dephasing γ_2 and the surface-induced dephasing γ_s. The driving and reflected fields are treated semiclassically, and we assume the atom-surface distance to be large ($\gtrsim$ 30 nm). The relaxation time included in the Drude model for the complex dielectric constant of the metal medium represents the dissipation of electron gas which, together with surface plasmon resonances, influences the surface reflectivity and hence the behavior of the reflected field. Here we may neglect nonradiative transfer of energy from the excited atom to the metal. Effects of the laser bandwidth are included by means of a phase diffusion model for the driving field. In the weak-field or large-detuning limit, the power spectrum of scattered light has two peaks: one corresponding to Rayleigh scattering at the laser frequency ω_L and the other to fluorescence at the atomic transition frequency ω_{21}. For a sufficiently strong driving laser field, the spectrum exhibits three peaks: a central one at $\omega_0 = \omega_L$ (Rayleigh component), a left one at $\omega_- = 2\omega_L - \omega_{21} - \delta$ (three-photon component) and a right one at $\omega_+ = \omega_{21} + \delta$ (fluorescence component), where δ is the ac-Stark shift.

Results for the peak heights H_-, H_0 and H_+ are shown in Figure 7 for the case of a silver surface.[22] The key feature, which is a unique behavior due to the surface, is that for certain atom-surface distances H_- is larger than H_+, whereas in the pure gas-phase resonance fluorescence spectrum H_- is always less than H_+ (due to molecular collisions for positive detuning). Also, the population inversion of the adatom and the resonance fluorescence spectrum, as well as the surface-induced phase-decay constant of the adatom, show strong oscillatory behavior as a function of the adatom-surface distance.[21,22]

For a rough surface, let us consider a surface protusion modeled by a prolate hemispheroid on top of a plane.[28,29] We note that this model has been shown to be identical to a full spheroid in a vacuum,[30] so that our calculations can also be used for an ellipsoidal cluster. The two-level

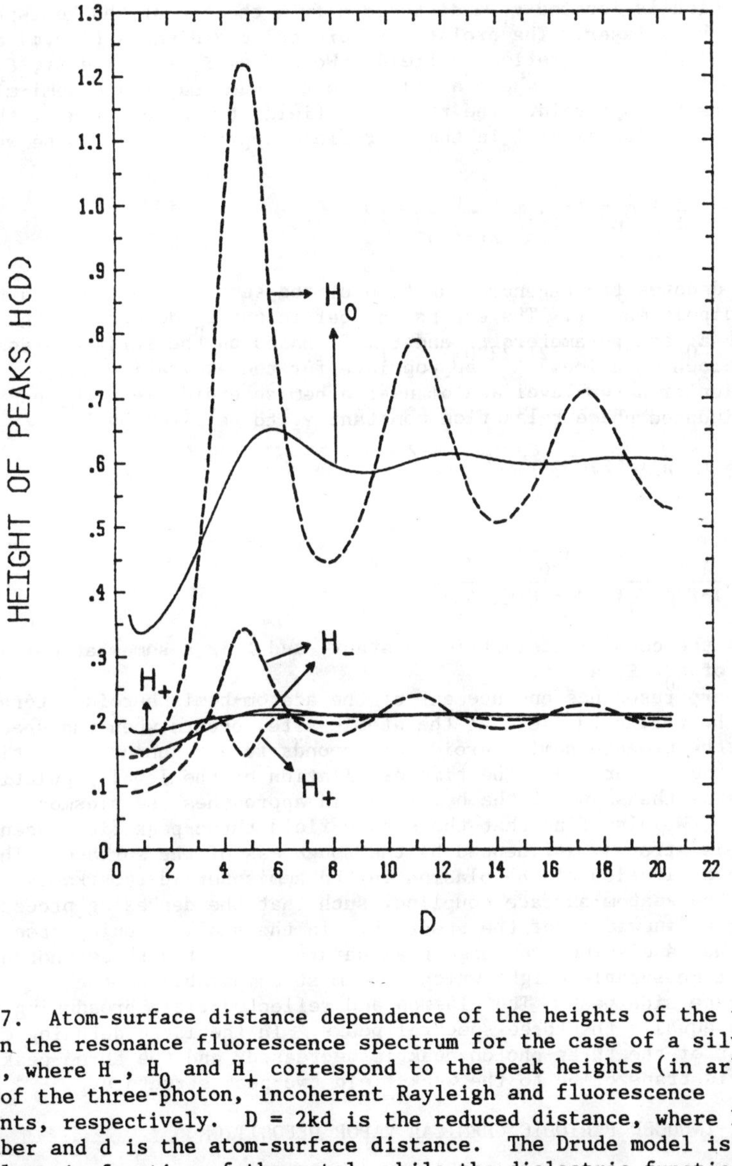

FIGURE 7. Atom-surface distance dependence of the heights of the three peaks in the resonance fluorescence spectrum for the case of a silver surface, where H_-, H_0 and H_+ correspond to the peak heights (in arbitrary units) of the three-photon, incoherent Rayleigh and fluorescence components, respectively. D = 2kd is the reduced distance, where k is the wavenumber and d is the atom-surface distance. The Drude model is used for the dielectric function of the metal, while the dielectric function of the gas medium is set equal to one. The detuning $\omega_{21} - \omega_L$ is 1 A and the Rabi frequency is 10 A, in the unit of Einstein's A coefficient. The solid curves are for the induced transition dipole of the atom oriented perpendicular to the surface, and the dashed curves are for the parallel case.

adatom, which is located at a distance d from the top of the hemispheroid, is driven by a laser. The prolated spheroidal coordinates (ξ,η,ϕ) are used to calculate the reflected field. We define $\xi_1 = (a + d)/f$, $\xi_0 = a/f$ and $f = (a^2 - b^2)^{1/2}$, where a and b are the semi-major and semi-minor axes of the hemispheroid. The reflected field at the position of the adatom (transition dipole) in the near-field approximation can be written as[26]

$$E_r = -\frac{1}{f}\sum_n C_n Q_n'(\xi_1) + \frac{\mu}{4(f\xi_1)^3} , \tag{10}$$

where Q_n denotes the Legendre function of the second kind and μ is the induced dipole moment. The expansion coefficient C_n depends on the laser amplitude E_0 and parameters ξ_0 and ξ_1.[12] Based on the surface-dressed optical Bloch equations[21,22] appropriate for the excitation and dissipation of a two-level adatom near a hemispheroid, we find the surface-induced phase relaxation constant γ_s to be given by[26]

$$\gamma_s = (2/\hbar)\mathrm{Im}(F) , \tag{11}$$

where

$$F = \frac{1}{1-\Gamma}\left[\frac{(1-\varepsilon)\xi_0 Q_1'(\xi_1)}{\varepsilon Q_1(\xi_0) - \xi_1 Q_1'(\xi_0)}\right] + \Gamma . \tag{12}$$

Here ε is the complex dielectric constant, and Γ is a somewhat complicated function of ξ_0, ξ_1 and ε.

A sharp resonance enhancement of the adatom-hemispheroid interaction, through the reflected field at the atomic site, occurs when the specific shape of the prolate hemispheroid corresponds to a resonant excitation of plasmons (see Figure 8). The time oscillation of the level population decreases as the shape of the hemispheroid approaches the plasmon resonance. We also find that the strong-field three-peak fluorescence spectrum is strongly influenced by the roughness of the surface. The resonance excitation of the plasmon in the hemispheroid remarkably enhances the adatom-surface coupling, such that the dephasing processes broaden the linewidths of the spectrum. In the small detuning case, the spectrum has a distinctive three-peak nature, where the three-photon side peak has a measurable height which is almost comparable to the fluorescence side peak. The plasmon and reflected-field broadening influence equally the three spectral peaks. In the large detuning case, the height of the three-photon peak is decreased, and the three-peak spectrum is transformed to the weak-field two-peak structure.

4. LASER-INDUCED PERIODIC CHEMICAL VAPOR DEPOSITION

We have been engaged in theoretical research to explain laser-induced periodic vapor deposition recently observed experimentally.[31] While some pieces of the puzzle remain unsolved, we think that we now have a sufficient understanding of the phenomena to warrant a more general discussion.

Briefly, the experiment[31] which we are attempting to model is as follows. A dilute organometallic gas ($Cd(CH_3)_2$ in this case) is placed above a silicon or silicon oxide surface. The system is illuminated with

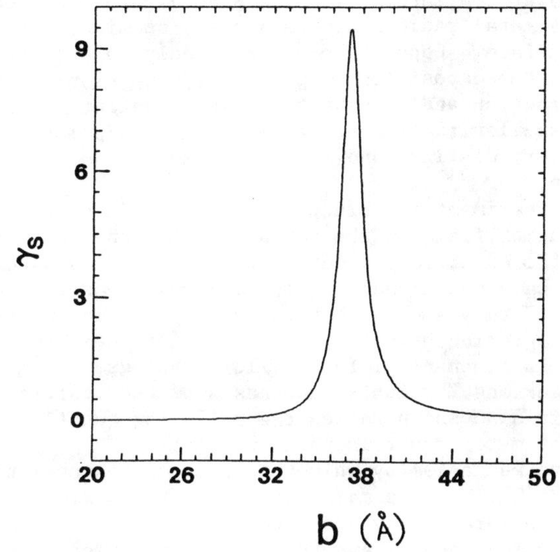

FIGURE 8. Rough surface-induced phase-decay constant γ_s as a function of the semi-minor axis b of the surface protusion, where the semi-major axis is fixed at 100 Å. γ_s is in the unit of Einstein's spontaneous decay constant A.

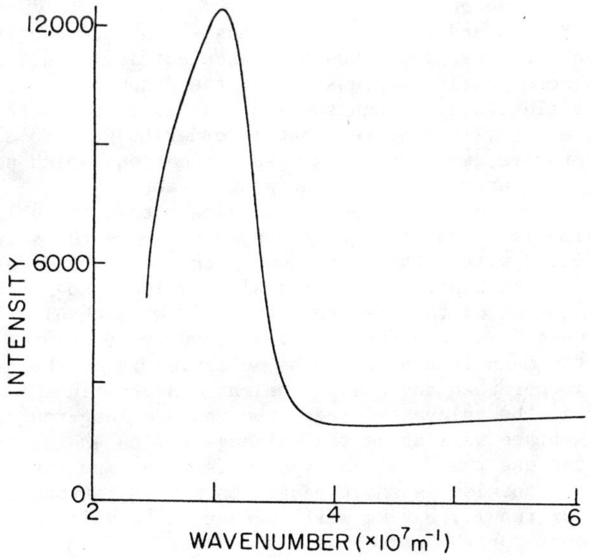

FIGURE 9. Graph of the plasmon field intensity as a function of grating wavenumber. The grating height is 25 nm, and 15 terms are kept when evaluating the modified Bessel functions.

a UV laser (257 nm) at low intensity (~ 10 W/cm^2). This dissociates the organometallic to a metal radical (in this case, cadmium), which is deposited on the surface. Deposition occurs evenly to a depth of about 200 nm, after which the deposition is periodic, with ripples developing in the p-polarized direction and growing to an added height of about 100 nm. In addition, much smaller ripples are formed in the s-polarized direction. This structure is very distinct and readily visible under a scanning electron micrograph.

The original experimenters, Brueck and Ehrlich,[31] postulated that the laser induces a plasmon field on the surface, where the varying field strength leads to the observed periodic deposition. They assumed the deposition rate to be proportional to the plasmon intensity, and using a first-order theory,[32] they showed that the plasmon field intensity is proportional to the grating height. It is therefore easy to conclude that the grating growth is exponential for shallow gratings.

We accept Brueck and Ehrlich's hypothesis as the starting point for our own work. Their discussion raises the following questions, some of which we shall be able to answer in this article:

1) What is the mechanism by which the plasmon enhances the deposition process? This is the major piece of the puzzle which remains unsolved, but we have more to say about this later.

2) What is the sequence of events in this experiment? Is the organometallic molecule dissociated in the gas phase, with deposition of the cadmium atom occuring later, or is the molecule physisorbed onto the surface, with dissociation occuring afterwards, perhaps as part of the chemisorption reaction of cadmium with the surface?

3) What is the dynamics of grating growth, and why does the grating stop growing? This last question is the one about which we now have the clearest understanding. We continue to work on the first two problems and shall discuss them in due course.

By accepting Brueck and Ehrlich's hypothesis that the grating growth is proportional to the strength of the plasmon electric field, then in order to derive a more precise expression for the dynamics, we must solve for the electric field above a rough surface. It is possible to do this numerically as accurately as required, but in order to gain some physical insight into the problem, we make two key approximations[33] which permit us to write an analytical expression for the growth rate:

1) We assume the Rayleigh hypothesis, namely that the solutions to the homogenous Helmholtz equation, which are exact above the selvedge region, are also valid within the selvedge region. This holds strictly only in the limit as the grating is infinitely shallow. How valid this approximation is depends on the precise nature of the mechanism by which the deposition occurs. We note that the gas pressure is about 1 torr, which means that the mean free path of the molecules is on the order of a millimeter, much longer than any grating height measured in the experiment. Thus if the relevant forces involved are long-range, most of the "action" takes place well above the selvedge region where our solution is exact, and we can use the Rayleigh hypothesis with impunity. On the other hand, if the mechanism is short range, then this introduces some uncertainty into our result, but we still expect to be able to qualitatively account for the phenomena.

2) Our second and more important approximation is the manner in which we truncate the Fresnel matrix. Toigo, Marvin and Celli[34] derived uncoupled equations for the electric field strength above a grating (given

the Rayleigh hypothesis), and their result has been used by many other researchers.[35,36] In principle, it is necessary to solve an infinite number of equations for an infinite number of unknowns, the unknowns being the amplitudes of the various order plasmons and/or Bragg reflections. Maradudin[34] has used this expression to derive the dispersion relation for a shallow grating. He did this by 1) assuming that the grating is shallow, and therefore only terms to first order in grating depth need be retained, and 2) that because the resonance condition is met, all field strengths are small relative to the resonance plasmon. This means that we can neglect all other reflections, and our infinite set of coupled equations reduces to two equations, easily soluble analytically.

Our contribution to this result is to notice that Maradudin's first assumption is unnecessary. If the resonance condition is met, then the resonance plasmon will be orders of magnitude larger than the other reflections, and hence we are justified in keeping all orders of the grating depth. This is true, of course, only as long as the resonance condition is met, but it implies that we can extend Maradudin's work to the deep grating case.

The implications of our approximation are best illustrated by studying Figure 9, where the x-axis is the grating wavenumber and the y-axis is the intensity of the plasmon field. Throughout we are insisting that the grating depth be 25 nm. Thus as the wavenumber becomes larger, the grating period becomes shorter, and the relative grating height becomes larger. Hence we see an increase in the plasmon field intensity, consistent with the previous result that the plasmon field increases linearly with grating height. The difficulty here is that this increase appears far from any resonance. Hence our approximation means that the two terms that we are keeping are on the same order of magnitude as the terms that we are dropping. This leads to a monotonically increasing function for the plasmon field strength, when in reality the curve would have more structure and would tend to approach some asymptotic value. In summary, for grating frequencies off resonance, our approximation over-estimates the plasmon field. Later we shall provide some measure for the degree of this error.

At resonance, on the other hand, the situation is different. Here the two terms that we are keeping are much larger than any others. Thus our approximation, while still tending to over-estimate the field strength, is much more accurate. Figure 10 illustrates the results of our method for various grating heights. Note the structure in the curves for gratings 15 nm and 20 nm deep. We suggest that this is physically reasonable since the resonance is very large, and the two terms we keep are quite complicated functions. On the other hand, our result for the 40 nm deep grating is suspect since there is no resonance whatsoever. In all cases we have kept 15 terms in evaluating the Bessel functions that appear in the equations.

Given the assumption that the grating growth rate is proportional to the plasmon field intensity, we are now in position to calculate the growth dynamics. Two problems are now presented. First, until we have a better knowledge of the mechanism by which the plasmon increases the growth, it is impossible to calculate the proportionality coefficient. Therefore, in our calculation shown in Figure 11, we have simply chosen the coefficient to match the time-scale of Brueck and Ehrlich's experiment. We are considering several possible models to explain this mechanism, and these will be discussed shortly.

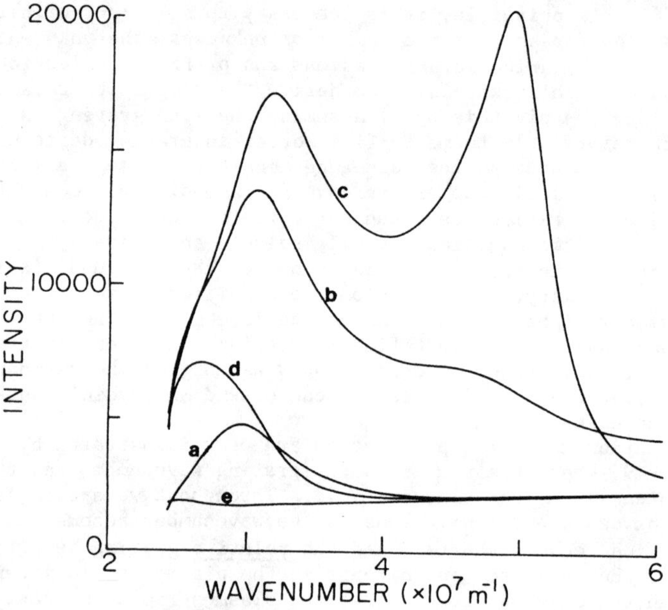

FIGURE 10. Same as Figure 9, except that the grating height is varied: (a) 5 nm, (b) 15 nm, (c) 20 nm, (d) 30 nm and (e) 40 nm.

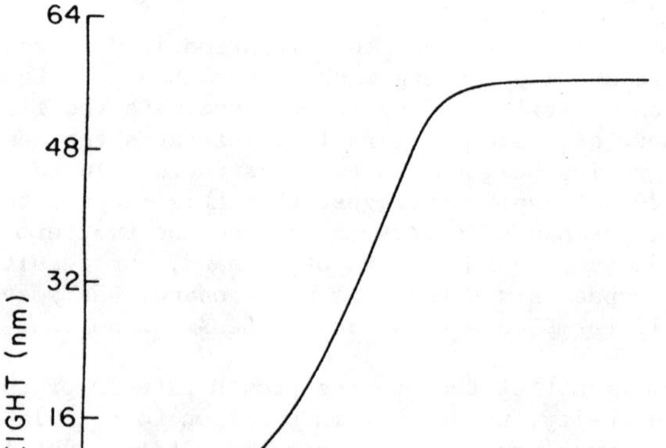

FIGURE 11. Grating height vs. time, calculated as described in the text.

The second problem is to determine when the grating stops growing. This happens when the resonance disappears altogether, and gratings of all frequencies grow at the same rate. Thus we have to subtract the "baseline" from the plasmon field intensity, namely the average intensity of all other plasmon frequencies. We have done this by treating the wavenumber of the light as the "base" wavenumber. This value is the left-most point of the curves shown in Figure 10. The reason for choosing this wavenumber is that it can never be at resonance since the laser-induced plasmons must oscillate at a wavenumber greater than that of the light. However, our approximation over-estimates the intensity of these off-resonant wavenumbers, so that we shall under-estimate the final peak height. This is shown in Figure 11, where the peak is shown growing to about 56 nm. Brueck and Ehrlich measure their peaks at about 100 nm. The discrepancy here gives us an indication of the accuracy of our approximations. If a method can be determined for calculating the "baseline" independently of our present model, which is obviously not valid for off-resonant wavenumbers, then we should be able to more accurately predict the maximum peak height.

Our discussion so far has revolved around the third of the three questions raised at the beginning of this article. We would now like to briefly discuss the other two. We have considered several possibilities for the mechanism by which the plasmon increases the deposition rate. The first, which at one point we considered most likely, is that the plasmon field induces a dipole in the adatom/molecule, which in turn attracts it to the surface. Unfortunately, the growth rate resulting from this mechanism is about 15 orders of magnitude too slow, and hence it is discarded.

A second possibility is that a charge transfer occurs between the surface and the cadmium atom, and hence the plasmon interacts with an ion and not a dipole. This appears to yield a growth rate on the right order of magnitude, but it remains unclear how the charge transfer takes place. It is conceivable that the plasmon field enhances the charge transfer process also, in which case a rate proportional to the square of the intensity is not impossible.

Yet another possibility is that the existence of the plasmon radically alters the shape of physisorption/chemisorption potential well. This raises the possibility that the growth is not proportional to the electric field intensity, but rather to the electron distribution along the surface. The very complexity of this problem forces us to exclude it from immediate consideration.

Finally, there is the dynamics of the dissociation process. We have reason to believe that this occurs in the gas phase, well above the surface.[37] If this is indeed true, then the problem becomes considerably simpler since the rate would depend only on the partial pressure of cadmium in the gas phase. We could then separate the problem into two parts: the gas-phase problem, i.e., the production of cadmium atoms, and the deposition problem. Depending on the pressure of the gas, one or the other will predominate in the total rate equation.

This concludes our discussion of recent progress made in the explanation of the periodic deposition phenomena. It only remains to make a few brief comments concerning the possible applications of such a model. An obvious question of interest would be how one could maximize either the grating height or growth rate. The development of a simple model will make it possible to predict ways to do this. Similarly, it may be possible to change the deposition pattern by changing the parameters of the experiment. The calculation of this phenomenon involves more than what we have discussed here.

ACKNOWLEDGMENTS

This research was supported by the Air Force Office of Scientific Research (AFSC), United States Air Force, under Contract F49620-86-C-0009, and the Office of Naval Research. The United States Government is authorized to reproduce and distribute reprints notwithstanding any copyright notation hereon.

REFERENCES

1. A. C. Beri and T. F. George, J. Chem. Phys. 78, 4288 (1983).
2. A. C. Beri, K. T. Lee and T. F. George, in Proceedings of the International Conference on Lasers '83 ed. by R. C. Powell (STS Press, McLean, Virginia, 1985), pp. 362-369.
3. A. C. Beri and T. F. George, J. Chem. Phys. 83, 2482 (1985).
4. A. C. Beri and T. F. George, Z. Phys. B 60, 73 (1985).
5. See, for example, C. T. H. Baker and G. F. Miller, Eds., Treatment of Integral Equations by Numerical Methods (Academic Press, New York, 1982); L. Collatz, The Numberical Treatment of Differential Equations (Springer-Verlag, New York, 1966), Chapt. VI.
6. E. Butkov, Mathematical Physics (Addison-Wesley, Reading, Massachusetts, 1968), Chapt. 5.
7. J. Lin, A. C. Beri, M. Hutchinson, W. C. Murphy and T. F. George, Phys. Lett. 79A, 233 (1980).
8. C. Jedrzejek, K. F. Freed, S. Efrima and H. Metiu, Surf. Sci. 109, 191 (1981).
9. A. C. Beri and T. F. George, J. Vac. Sci. Technol. B 3, 1529 (1985).
10. H. Frank, H. Hoinkes and H. Wilsch, Surf. Sci. 63, 121 (1977).
11. J. Heidberg, H. Stein and E. Riehl, Phys. Rev. Lett. 49, 666 (1982).
12. T. J. Chuang and H. Seki, Phys. Rev. Lett. 49, 392 (1982); T. J. Chuang, J. Chem. Phys. 76, 3828 (1982).
13. T. J. Chuang, Surf. Sci. Rep. 3, 1 (1983).
14. A. C. Beri and T. F. George, J. Chem. Phys., submitted.
15. See the review by R. R. Chance, A. Prock and R. Silbey, Adv. Chem. Phys. 37, 1 (1978).
16. M. R. Philpott, Chem. Phys. Lett. 19, 435 (1973).
17. H. Morawitz, Phys. Rev. 187, 1792 (1969).
18. R. R. Chance, A. Prock and R. Silbey, J. Chem. Phys. 60, 2744 (1974).
19. K. H. Drexhage, J. Lumin. 1/2, 693 (1970).
20. H. Kuhn, J. Chem. Phys. 53, 101 (1970).
21. X. Y. Huang, J. Lin and T. F. George, J. Chem. Phys. 80, 893 (1984).
22. X. Y. Huang and T. F. George, J. Phys. Chem. 88, 4801 (1984).
23. X. Y. Huang, K. C. Liu and T. F. George, in Laser-Controlled Chemical Processing of Surfaces, ed. by A. W. Johnson, D. J. Ehrlich and H. R. Scholossber (Elsevier, New York), Mat. Res. Soc. Symp. Prof. 29, 381 (1984).
24. X. Y. Huang, T. F. George and J. Lin, in Coherence and Quantum Optics V, ed. by L. Mandel and E. Wolf (Plenum, New York, 1984), pp. 685-693.

25. J. Lin, X. Y. Huang and T. F. George, Solid State Commun. 47, 63 (1983).
26. X. Y. Huang, K. T. Lee and T. F. George, J. Chem. Phys. 85, 567 (1986).
27. B. R. Mollow, Phys. Rev. 188, 1969 (1969).
28. J. Gersten and A. Nitzan, J. Chem. Phys. 73, 3023 (1980).
29. A. C. Pineda and D. Ronis, J. Chem. Phys. 83, 5330 (1985).
30. H. Metiu, Prog. Surf. Sci. 17, 153 (1984).
31. S. R. J. Brueck and D. J. Ehrlich, Phys. Rev. Lett. 48, 1678 (1982).
32. S. S. Jha, J. R. Kirtley and J. C. Tsang, Phys. Rev. B 22, 3973 (1980).
33. D. Jelski and T. F. George, J. Appl. Phys., submitted.
34. A. Marvin, F. Toigo and V. Celli, Phys. Rev. B 11, 2777 (1975).
35. A. A. Maradudin, in Surface Polaritons, ed. by V. M. Agranovich and D. L. Mills (North-Holland, Amsterdam, 1982), Chapt. 10.
36. D. Agassi and T. F. George, Phys. Rev. B 33, 2393 (1986); Surf. Sci. 172, 230 (1986).
37. P. T. Leung and T. F. George, J. Chem. Phys., in press.

MONTE CARLO SIMULATIONS OF SURFACE REACTIONS

Martine DUMONT* and R. DAGONNIER
Faculté des Sciences, Université de Mons, B-7000 BELGIUM.

1. INTRODUCTION

We present here Monte Carlo (MC) simulations of surface reactions. We consider reactions between gaseous species adsorbed on a catalytic solid surface. The reactions investigated occur in such a range of pressure and temperature that the surface remains almost entirely covered by an adsorbate which never exceeds a monolayer. The procedure of a computer experiment is very similar to that of a laboratory experiment designed to test a conceptual model within the framework of existing physical laws. However, contrary to a laboratory experiment, a MC simulation allows a complete control over the parameters that influence the operation of the model. The effect of a given physical factor on the behaviour of the model can be precisely *isolated* and systematically studied. A MC simulation can be used, therefore, to give an exhaustive account of the *logical* consequences issued from the specific assumptions, data, theory and mechanisms involved in the construction of a conceptual model.

A MC simulated surface reaction is a *synthetic* reaction based on a conceptual model which consists of :

(1) a set of *assumed* mechanisms for the "atomic" processes involved in the reaction e.g. molecular adsorption.

(2) a set of elementary physical laws (which represents accepted theoretical knowledge) e.g. Arrhenius formula for rate constants.

(3) a set of empirical relations and/or experimental data, which represents physical behaviour for which there is no adequate theory. e.g. activation energies.

The MC method is used whenever it is necessary to make either a logical decision or a numerical selection on a *probability* basis rather that on a deterministic basis. We make a difference between *reaction kinetics* and reaction dynamics. The reaction kinetics deals essentially with rates of chemical reactions, and the microscopic details (i.e. most of the physics of a reaction) is embedded in the rate constants. The details (the reaction dynamics) of the reaction are "summarized" in the well-known Arrhenius formula for the reaction rate.

We therefore understand why the MC method is by *definition* a "third mode" of investigation for surface reaction kinetics.

In surface reaction kinetics the various elementary mechanism involved in the reaction are described in terms of the *fractional coverages*

$$\theta_i = N_i/N$$

---

* Research Associate National Fund for Scientific Research (Belgium).

where N_i stands for the number of sites (per unit area) occupied by the species i; N is the number of active sites (per unit area) of the surface.

The fact that these coverages *are* the natural dynamical variables of the problem raises two main difficulties :
1) by definition, the coverages θ_i are *uniform* variables. Any description of a surface reaction in terms of θ_i implies a *random* distribution of all the different species adsorbed on the surface. In the case of a surface reaction inducing islands (clusters) in the adsorbate, such a description becomes problematic. We shall see later how MC simulations of such reactions may lead to solutions for this problem.
2) the coverages behave-*practically*-as "hidden" variables. In the case of surface reactions (contrary to homogeneous reactions) there is almost no *direct* method to follow the time and space evolution of these relevant concentrations.

In heterogeneous catalysis one can hardly follow the evolution of a surface reaction by using *indirect* methods i.e. mainly, by measuring the time variation of the different concentrations in the gas phase.

To illustrate this procedure let us consider the bimolecular reactions (our "guinea pig" reaction in the following) :

$$A + \frac{1}{2} B_2 \xrightarrow[\text{surface}]{\text{catalyst}} AB \tag{1.1}$$

e.g. the oxidation of CO on Pt

$$CO + 1/2\ O_2 \xrightarrow{Pt} CO \tag{1.2}$$

The reaction takes place in a gradientless reactor (of volume V) containing a catalytic surface of area Σ (i.e. $N\Sigma$ active sites). This ideal reactor is submitted to a flux q assumed constant ($q \equiv q_{in} = q_{out}$). The balance equations for the reactants A and B_2 are

$$V\frac{d}{dt}[A] = q\{[A]_{in}-[A]\}-q[AB] - V\frac{d}{dt}[AB] - N\Sigma\frac{d\theta_A}{dt} \tag{1.3.a}$$

$$V\frac{d}{dt}[B_2] = q\{[B_2]_{in}-[B_2]\}-q[AB] - V\frac{d}{dt}[AB] - N\Sigma\frac{d\theta_B}{dt} \tag{1.3.b}$$

where $[A]_{in}$,$[B_2]_{in}$ are the concentrations of A, B_2 in the feed stream; $[A]$,$[B_2]$ and $[AB]$ are the concentrations in the gas phase. These equations simply tell us that the rate of accumulation of a reactant in the gas phase equals the rate of this reactant flow in the system *less*

(i) the rate of this reactant flow out of the system
(ii) the reactant depletion due to the chemical reaction, and
(iii) the rate of accumulation of the reactant on the surface.

The measure on line of $[A]$,$[B_2]$ and $[AB]$ lead to a delicate determination of the *global* variation (if any) of θ_A and θ_B but not of the rate R_i of every mechanism involved in the reaction.

For instance, let us suppose that reaction (1.1) results from the competition of the following steps :

$$A + S \underset{-1}{\overset{1}{\rightleftarrows}} AS \tag{1.4.a}$$

$$B_2 + S_2 \underset{-2}{\overset{2}{\rightleftharpoons}} 2BS \qquad (1.4.b)$$

$$AS + BS \xrightarrow{3} AB + S_2 \qquad (1.4.c)$$

where S and S_2 denote an active site and an empty *dual* site (i.e. two nearest neighbour sites) respectively.

The kinetics of reaction (1.1) can only be tentatively described by means of a best fit procedure of model equations of the type :

$$\frac{d\theta_A}{dt} = R_1 - R_{-1} - R_3 = k_1 f_1(\theta_A,\theta_B) - k_{-1} f_{-1}(\theta_A,\theta_B) - k_3 f_3(\theta_A,\theta_B) \qquad (1.5.a)$$

$$\frac{d\theta_B}{dt} = R_2 - R_{-2} - R_3 = 2k_2 f_2(\theta_A,\theta_B) - 2k_{-2} f_{-2}(\theta_A,\theta_B) - k_3 f_3(\theta_A,\theta_B) \qquad (1.5.b)$$

where $f_i(\theta_A,\theta_B)$ represents the *coverage function* describing the rate R_i of the elementary mechanism i (see Eqs. (1.4.a-c)) and k_i, the corresponding rate constants.

In the case of a stable reaction, Eqs. (1.5a,b) lead to values for the reaction rate $R_3(\theta_A^o,\theta_B^o)$ where θ_A^o,θ_B^o should be the stable *steady state* (°) solutions of optimized Eqs (1.5a,b). In the case of an oscillatory reaction (or any transient situation) Eqs (1.5 a,b) lead to values for (i) reaction rate $R_3[\theta_A(t),\theta_B(t)]$ and (ii) the time variations of θ_A and θ_B. These quantities should be retrieved by a best fit of Eqs (1.5a,b). It should be emphasized that this procedure consists in a *simultaneous* best fit of *all* the different coverage functions involves in the model. This best fit procedure requires precise indications on the rate constants

$$k_i(T) = k_i' \, e^{-E_i/kT} \qquad (1.6)$$

We know that (i) the study of the adsorption/desorption process for each reactant, *separately* considered, gives values for $k_i(i=1,-1;2,-2)$. (ii) the measure of the production rate and its best fit description can lead to the reaction rate constants (e.g. k_3).

Nevertheless, the assignation of numerical values to the rate constants is often questionnable. For instance, how to be sure that

a) the rate of adsorption (desorption) of a single reactant has the same form when this process is competitive with a coadsorption, a reaction etc...?

b) the *apparent* activation energy of the production rate can be assigned to a given reaction mechanism (even unique!)?

Such difficulties plus the lack of a direct access to the *spatial* evolution of the coverages bring us to perform computer experiments.

The main advantages of MC simulations of surface reactions are the following :

(1) they give access directly to the kinetics of each elementary step involved in the reaction. One can investigate the effect of any relevant parameter on the rate expression of *each* of the steps *separately*. It is thus possible to test a given kinetic model for a reaction : one has to compare the solutions of the given kinetic equations with the results of MC simulations.

(2) they give access to the state of occupation of the surface *site by site*. They thus allow tests for the uniformity of the coverages. In the presence of islands one can picture the reaction in terms of cluster characteristics (distributions and averages of cluster sizes, perimeters,...) which, in turn, may tentatively be described in terms of coverages.

By means of MC simulations we can investigate for instance,

(i) the effect of rate "constants" depending on *local* coverages (e.g. the state of occupation of the nearest neighbours of each site considered) on the rate expressions of the elementary steps involved in the model.

(ii) the role of the surface (different sites, steps...) and its possible change of structure (appearing at a given value of a specific coverage) on the reaction.

(iii) laser assisted surface reactions, by tentatively following the thermal rate and the photorate of each step of the reaction etc...

The paper is organized as follows, our MC simulation programming is exposed chapter 2, the corresponding algorithms may be found in ref. [3]. Our results are presented chapter 3. In appendix B we report some elements of percolation theory.

2. MC SIMULATION OF A BIMOLECULAR SURFACE REACTION

In short, MC simulation of the surface reaction is performed as follows. A random rainfall models the arrival of the reactants on the surface. The time of arrival of each atom or molecule is given by an exponential law, the average of which is the average time interval between two collisions of the reactant with the surface (this time depends on the conditions of pressure and temperature of the reaction). The arrival site(s) is (are) randomly chosen with a uniform distribution. The atom (or molecule) adsorbs on the surface only if the chosen site(s) is (are) empty and the sticking probability favourable. Once adsorbed on the surface an atom (or molecule) can desorb or react with an atom of the other kind provided the space configuration is favourable to such a reaction. The choice between one possible future or another is based on the comparison of the respective times at which these different futures are programmed. The nearest event in time will be generated.

2.1. The reaction scheme

We restrict the kinetic scheme to the *direct* steps (1.4a-c). Other direct mechanisms, as well as their MFA rates (e.g. [1]) are reported in table 1. For sake of simplicity we do not consider here :

(i) *indirect* mechanisms which involve precursor states [2],

(ii) the migration (hopping) of the reactants at the surface.

Details for programming indirect events may be found in ref.[3].

N°: i	Step	Mechanism	MFA rate expression : R_i
1 -1	Adsorption/ desorption of A	$A + S \underset{-1}{\overset{1}{\rightleftharpoons}} AS$	$R_1 = k_1(1-\theta_A-\theta_B)$ $R_{-1} = k_{-1}\theta_A$
2 -2	Adsorption/ desorption of B_n	$B_n + nS \underset{-2}{\overset{2}{\rightleftharpoons}} n\ BS$	$R_2 = n\ k_2(1-\theta_A-\theta_B)^n$ $(n = 1,2)$ $R_{-2} = n\ k_{-2}\theta_B^n$
3	Langmuir-Hinshelwood reaction mechanism	$AS+BS \xrightarrow{3} AB+2S$	$R_3 = k_3\theta_A\theta_B$
3'		$AS+BS+nS \xrightarrow{3'} product+(n+2)S$	$R_{3'} = k_{3'}\theta_A\theta_B(1-\theta_A-\theta_B)^n$ $(n=1,2)$
4	Eley-Rideal (pick-up) reaction mechanism	$A+BS \xrightarrow{4} AB+S$	$R_4 = k_4\theta_B$
5		$B_n+nAS \xrightarrow{5} nAB+nS$	$R_5 = nk_5\theta_A^n$ $(n = 1,2)$
5'		$B_2+AS+S' \xrightarrow{5'} AB+BS'+S$	$R_{5'} = k_{5'}\theta_A(1-\theta_A-\theta_B)$

Table 1 : Elementary mechanisms and the corresponding MFA rate expressions

Other direct steps and hopping mechanisms can be easily included in our simulation along the lines developped here after.

The elementary processes (1.4.a-c) we consider for the bimolecular surface reaction (1.1) are detailed one by one. The reaction mechanism and the molecular adsorption-desorption of species A are treated in detail. For species B elementary processes, we indicate only the differences from the case of A. Hereafter the subscripts 1 and 2 refer to A (e.g. CO) and B (e.g.O) respectively.

i- Molecular adsorption of A

$$A + S \xrightarrow{k_1} AS \qquad (2.1.1)$$

An A molecule (or atom) adsorbs on a *single* empty site of the surface. The adsorption rate constant, which represents the number of A molecules striking the surface at an empty site and adsorbing per unit of time and surface, has the usual form [4]

$$k_i = \sigma_i p_i e^{-E_i/kT} (2\pi m_i kT)^{-1/2} \quad (i=1) \qquad (2.1.2)$$

where σ_i, p_i, E_i and m_i are the condensation coefficient (sticking probability), the partial pressure, the activation energy of the process and the molecular mass respectively. T is the gas temperature and k the usual Boltzmann constant.

ii- Molecular desorption of A.

This process is the inverse of step i : an adsorbed A atom (or molecule) desorbs back to the gas phase

$$AS \xrightarrow{k_{-1}} A + S \qquad (2.1.3)$$

The frequency of such desorptions per surface unit is written as follows

$$k_{-i} = k'_{-i}\, e^{-E_{-i}/kT} \quad (i=1) \qquad (2.1.4)$$

where E_{-i} and k'_{-i} represent respectively the activation energy for desorption and the frequency factor. k'_{-i} is of the order of a molecular vibration i.e. 10^{13} to $5.10^{14} s^{-1}$.

iii- Dissociative adsorption of B_2.

The adsorption of a B_2 molecule with dissociation on an empty *dual* site S_2 is represented by

$$B_2 + S_2 \xrightarrow{k_2} 2BS \qquad (2.1.5)$$

k_2 is given by Eq. (2.1.2).

iv- Associative desorption.

This is the inverse process of step (iii) : two neighbour B atoms desorb back to the gas and associate immediately to form a new B_2 molecule

$$2BS \xrightarrow{k_{-2}} B_2 + S_2 \qquad (2.1.6)$$

k_{-2} is given by Eq. (2.1.4).

v- Reaction mechanism :

We have limited our MC simulation of reaction (1.1) to the Langmuir-Hinshelwood 1 reaction mechanism between nearest neighbour (or adjacent) chemisorbed species :

$$AS + BS \xrightarrow{k_3} AB + S_2 \qquad (2.1.7)$$

k_3 denotes the reaction constant,

$$k_3 = k_3' \, e^{-E_3/kT} \qquad (2.1.8)$$

We assume that the product AB leaves the surface as soon as formed.

2.2. General hypothesis

Our MC simulation is based on the following assumption : gaseous reactants A and B_2 can competitively adsorb on active sites of the *same kind* and react to product AB molecules which leave the surface as soon as formed.

Our Monte Carlo programme consists thus in a parallel (simultaneous) simulation of the different elementary steps of the reaction :

i. competitive adsorption of reactants

ii. the reaction mechanism which takes place between *adjacent* adsorbed species A and B

iii. desorption of the reaction product AB and of the adsorbed reactants A and B.

2.3. Representation of the surface

We represent the catalyst surface by a plane square lattice of dimension NSITE x NSITE (e.g. NSITE=100). Each lattice site corresponds to a minimum of the surface potential well and is a possible fixation site for an atom or molecule. We consider all these lattice sites to be active sites of the same kind. Each site is localized with the help of two indices i and j. An array S [0:NSITE-1, 0: NSITE-1] indicates the state of occupation and evolution of any site. This array is defined in Table 2. We use periodic boundary conditions and start every simulation with an empty lattice (clean surface).

From now on, and for the sake of simplicity of notation, we designate by τ_i

$$\tau_i = \frac{1}{k_i} \frac{1}{(NSITE)^2} \qquad (2.3.1)$$

e.g. τ_1 can be interpreted as the average time interval between two successive collisions of a A atom with any site of the surface.

S(i,j)	Site (i,j) is occupied by...	programmed to...
0	nothing (empty site)	
1	A chemisorbed	desorb
5	B chemisorbed	desorb
9	A chemisorbed	react
10	B chemisorbed	react
11	isolated B	wait

Table 2 : Definition of the array S(i,j) caracterizing the state of occupation of each lattice site (i,j). (Some values i.e. 2,3 have been skipped to be used for the simulation of indirect processes [3]).

2.4. Molecular adsorption

The simulation of molecular adsorption (Cf. eq. (2.1.1)) can be visualized as a random rainfall of molecules A uniformly distributed on the surface. One has thus to answer the three following questions :

a- when does a A molecule hit the surface?
b- on which site?
c- once arrived on the surface does the molecule adsorb or immediately go back to the gas?

Let us examine now these points one by one.

a- As the arrival of a molecule A on the surface is independent of previous arrivals of other molecules A, i.e. is a Poisson process, the amount of time between two successive collisions on any point of the surface has an exponential distribution with mean τ_1 [5]. Arrival times of A molecules on any point of our surface have thus the form

$$t_o = t_{initial} \quad (2.4.1)$$

$$t_n = t_{n-1} + \Delta t \qquad n = 1,2,3,... \quad (2.4.2)$$

where the time interval t between two successive collisions of a A atom with the surface is pseudo-random number distributed according to an exponential law of mean τ_1. We generate this number by the usual logarithmic method [6]

$$\Delta t = -\tau_1 \ln U \quad (2.4.3)$$

where U is a pseudo-random number uniformly distributed on]0,1[. All elementary processes (1.4a-c) participating to reaction (1.1) are also Poisson processes and their realization time will be pseudo-random numbers generated with an exponential law of corresponding mean τ_i Eq (2.3.1).

b- the coordinates i and j of the adsorption site of the incoming molecule A are randomly generated from two pseudo-random numbers r_1 and r_2 uniformly distributed on the interval]0,1[as follows:

$$i = \text{integer part of } (r_1 * \text{NSITE}) \qquad (2.4.4a)$$

$$j = \text{integer part of } (r_2 * \text{NSITE}) \qquad (2.4.4b)$$

c- if the adsorption site is occupied this incident molecule goes back immediately to the gas. In the order case, the molecule adsorbs on the surface. The algorithm of the simulation of direct molecular adsorption is presented in [3] (Fig.2a). Let us remark that one may simplify the simulation procedure for molecular adsorption as follows. Instead of generating the arrival time with parameter τ_1 and choosing randomly the arrival site on the whole lattice, one may generate a successful arrival time with parameter $\tau_1 \, N_{free}/(\text{NSITE})^2$, see [3] (Fig.2b). It is easy to show the equivalence of these procedures for Poisson processes. Let us mention that in this way site by site information is not available, the simulation result being the global occupation rate. To get local information one should randomly choose an arrival site among the N_{free} free sites, which remains a lengthy operation.

From another point of view, [3] (Fig.2c) shows the algorithm of a simulation in which we consider a sticking probability σ_1 (see Eq. 2.1.2) as a function of the present state of occupation of contiguous sites, for instance.

$$\sigma_1(k,l) = \sigma_1^0 \, e^{-\mu n_{k,l}} \qquad (2.4.5)$$

where $n_{k,l}$ is the number of occupied neighbours of site (k,l), μ some appropriate constant and σ_1^0 the sticking probability in absence of occupied neighbours. One can treat in the same way a local dependance of the activation energy E_1 (see Eq. (2.1.2). This procedure can be easily extended to the coadsorption of two specie A and B.

2.5. Molecular adsorption-desorption

We simulate competitively the adsorption (Eq. (2.1.1.)) and desorption processes (Eq. (2.1.3)). The adsorption is treated as described here above. Desorption times are generated in the same way as adsorption times (Eq.(2.4.1) to (2.4.3)) relatively to the mean time interval τ_{-1} between two desorptions at any site of the surface.

The corresponding algorithm is presented in [3] (Fig.3). As adsorption and desorption are chronologically dependent we have to create and manage a list of the events preprogrammed for the future. This list contains, for each event, the time at which the event will occur and the affected site. The type of the preprogrammed event may be found in the corresponding S cell. The list of preprogrammed events is organized in such a way that it may be accessed in order of increasing time [7].

If the desorption is also considered as a Markovian process, it is not necessary to generate the desorption time at the moment of adsorption and to create a waiting list. After each event, it is possible to compare the time of the next successful adsorption (using as parameter $\tau_1 \, N_{free}/(\text{NSITE})^2$) and the time of the next desorption (using as parameter $\tau_{-1}N_{occ}/(\text{NSITE})^2$)

where N_{occ} is the number of occupied sites. The smallest time determines the next event to be considered. It is then only necessary to choose the affected site. In each case to avoid a tedious search for a site, one has to maintain a list of the occupied sites and a list of the free sites. Remark that these lists doe not need to be sorted and that one has to choose randomly a site in the list.

We could use a similar approach for the following more complicated processes. In the actual state of the programme, our approach is based on the generation of the possible futures of an adsorbed atom or molecule at the moment where it adsorbs. This way of proceeding allows us to consider more general laws of probability than the exponential law for the different elementary processes.

2.6. Dissociative adsorption of a B_2 molecule

For the simulation of dissociative adsorption of a molecule B_2, Eq (2.1.5), we have to specify, as in the molecular adsorption (section (2.4) the three following points

- when does an adsorption event occur?
- where?
- does the molecule adsorb on the surface or go back to the gas?

What mostly distinguishes dissociative (or atomic) adsorption from molecular adsorption is that the former requires two empty adjacent sites (i.e. one empty dual site S_2) and the latter only one empty site. We will insist in the following on the differences from the molecular case due to this characteristic requirement.

The adsorption time is as in the molecular case (Eq.(2.4.1) to (2.4.3)) a pseudo random number generated with an exponential law of mean τ_2. The first adsorption site is generated as in the molecular case (Eq.(2.4.4)). If this site is occupied the molecule goes back to the gas. We choose the second site randomly (uniform distribution) among the four nearest neighbours of the first site. If the second site is occupied the molecule goes back to the gas. The algorithm is presented in [3] (Fig.4).

2.7. Dissociative adsorption-associative desorption of a B_2-molecule (DA-AD) direct processes.

We simulate now, together with the dissociative adsorption treated here above, associative desorption Eq.(2.1.6). As in the case of molecular adsorption-desorption, once a molecule is adsorbed, its future desorption has to be generated. The difference is here that a desorption needs to occur via two neighbouring B atoms, these atoms perhaps originating from two different original B_2 molecules. Such a mechanism may lead to the isolation of B atoms on the surface. The definition of an isolated B atom is well illustrated Fig.1. The only future of an isolated atom is to stay on the surface and wait for a new neighbour. Isolation of an atom thus requires that the future previously assigned to this atom be suppressed from the waiting list.

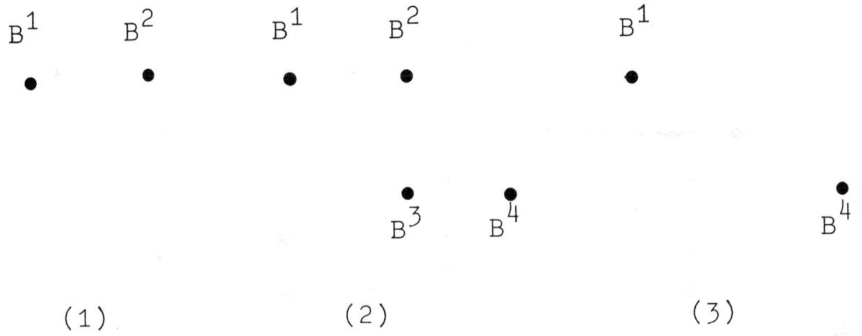

FIGURE 1.: Definition and formation of isolated B atom on the surface
(1) adsorption of the B^1-B^2 molecule. A desorption time t_1 is generated
(2) adsorption of the B^3-B^4 molecule at time $t<t_1$. A desorption time t_2 is generated ($t_2<t_1$) for the pair B^2-B^3.
(3) desorption of the B^2-B^3 molecule. B^1 and B^4 are isolated.

Another problem induced by associative desorption is the choice of the desorption which will occur if several different desorptions are possible. A desorption is uniformly chosen from all possible desorptions. Fig.2 gives a detailed illustration of the way in which we select the future of an adsorbing B_2 molecule.

The programmation of the future of the two B atoms of an adsorbing B_2 molecule may involve the participation of one or two neighbour atoms. This participation may cause the deprogrammation of the future of these neighbour atoms and by degrees, to the deprogrammation-reprogrammation of a large part of the lattice. To avoid a too heavy reprogrammation we adopt the following arbitrary rule. We do not allow any reprogrammation involving a B atom already programmed for a desorption with the exception the one or two neighbours of the adsorbing B_2 molecule participating to the desorption of its contributing atoms. The algorithms of direct DA-AD are presented in [3] (Fig.8-10).

2.8. Competitive adsorption-desorption

We are now able to simulate competitive adsorption-desorption : molecular adsorption-desorption of A molecules or atoms and dissociative adsorption-associative desorption of B_2 molecules. It is no more than a parallel simulation of case 2.5 and 2.7 hereabove.

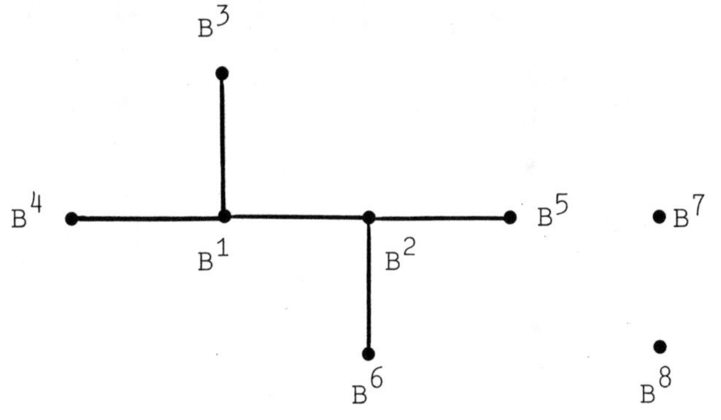

FIGURE 2. : Determination of the future of an adsorbing pair B^1-B^2. B^1 has three different possibilities of desorption (including the initial pair B^1-B^2), B^2 has two. The initial pair is retained once. For example are may select (subroutine FUTB, Fig.5 App. A) the desorptions B^1-B^3 and B^2-B^5. If B^5 was programmed to desorbe with B^7, B^7 has to be isolated and eventually reprogrammed to desorbe with another isolated atom (for example B^8, if B^8 is not itself programmed for another desorption) (see subroutine ISO,[3] (Fig.10)).

2.9. Competitive direct adsorptions-desorptions with reaction

We treat here simultaneously a surface reaction (Langmuir-Hinshelwood mechanism) (Eq.(1.4c)) together with competitive direct adsorptions-desorptions i.e. the elementary processes Eqs.(1.4a and b).

The reaction mechanism is similar to the direct associative desorption but involves two molecules of different kind. It offers a new possibility for the future of an adsorbed molecule A or B. In case of favourable configuration an adsorbed molecule can either desorb or react with a molecule of the other kind and leave the surface as product of the reaction. The choice between a possible reaction or desorption is based on a time comparison method. We use this time comparison method each time we are confronted with a choice between two (or more) possible events of *different* nature.

The selection between two possible futures is based on the comparison of their realization time. The one with the smaller time is retained. It is easy to show for Poisson processes the equivalence of this procedure with a selection based on the comparison of a uniformly distributed random number with the probability of one of the considered mechanism.

Let us point out that in the case of a choice between two (or more) events of the same type (see section 2.6 e.g.) this time comparison method would truncate the time distribution by promoting short times. The resulting time distribution will no longer be of the initial type (exponential in our case) and will no longer have the expected average value.

As in the case of associative desorption the reaction mechanism may lead to the isolation of atoms or molecules and to the reprogrammation of the future of atoms or molecules. Isolations due to the reaction mechanism appear one by one in contrast to those generated in pair by associative desorptions. Due to the reaction mechanism, reprogrammation of the future of A molecules may occur (see Fig. 3 for example). This was not possible when considering only desorption mechanisms.

FIGURE 3. : example of reprogrammation of A molecule.
a) A is programmed to desorbed at time t_1. At $t_2<t_1$ a B^1-B^2 molecules adsorbes next to A. The atom B^1 has two possible futures : either a reaction with A or a desorption with B^2. Suppose the time comparison method leads to the programmation of the reaction A-B^1 : the future already programmed for A has to be changed. Next step is to look if B^2 is isolated or can be programmed for a reaction or a desorption (see [3] routine ISOB, (Fig.14)).
b) A^1 and B are programmed to react and leave the surface at time t_1. At $t_2<t_1$ an atom A^2 adsorbes next to B and the reaction A^2-B is programmed. The already programmed future of A^1 has to be suppressed and a new one has to be simulated (see [3], subroutine ISOA (Fig.20)).

In principle the reprogrammation procedure is not limited to the neighbouring sites of a reacting pair A-B or a desorbing pair B-B and should be persued from site to site. In order to avoir reprogrammations of large parts of the lattice (for time limitation reasons) we proceed as follows :

i- the reprogrammation of the future of an atom or a molecule can occur only in case of mechanisms of different kind.

ii- a reprogrammation may occur only in favour of a mechanism whose mean time is smaller than the mean time of the mechanism already programmed. For example if the mean time for reaction τ_3 is smaller than the mean time for desorption τ_{-1}, an A molecule initially programmed for a desorption can be reprogrammed for a reaction but we shall never consider the inverse possibility. If two different mechanisms have same order mean times we shall not allow any reprogrammation.

The different algorithms and routines for the simulation of direct coadsorption with reaction are presented in [3] (Fig. 16-20).

3. SOME RESULTS

We present some of our results by going from simple cases to more complicated situations. When possible, the comparison with theoretical predictions will be made. By this way we build confidence in the use of MC methods for simulating the kinetics of surface reactions.

3.1. Molecular adsorption/desorption of A.

The MC simulation of the adsorption

$$A + S \xrightarrow{1} AS \tag{3.1.1}$$

leads to the equation

$$\frac{\Delta N_A}{\Delta t} = k_1(N-N_A) \tag{3.1.2}$$

optimized by

$$\frac{d\theta_A}{dt} = k_1(1-\theta_A) \tag{3.1.3}$$

It should be kept in mind that, in kinetics, any rate is not an individual rate but an average rate (over a *large* number of particles). Strictly speaking the variables θ are discrete ($\Delta\theta = 1/N$) and one cannot take the limit $\Delta t \to 0$ because in this limit the variations of θ are discontinuous. Nevertheless one can write differential equations like (3.1.3) by extending the results for a finite zone to a macroscopic zone. Therefore, in our best fit procedure we disregard the values of coverages θ_A, θ_B appearing early in time as we start with an empty surface.

By construction, the MC simulation of a molecular adsorption is equivalent to the MC simulation of a simple site percolation (see appendix A for definitions and similarities). No wonder that we find for the *critical* coverages θ_A^C the standard value (of p_c for a square lattice [8])

$$\theta_A^C = 0.593 \tag{3.1.4}$$

θ_A^C is definied as the coverage where, for the first time, a cluster of A particles percolates through the lattice. Moreover we retrieve for the cluster numbers C_S (defined in Eq. (a.1), App.A) the exact expressions given in [9] up to S=17.

The simulation of the adsorption/desorption of A

$$A + S \underset{-1}{\overset{1}{\rightleftharpoons}} AS \tag{3.1.5}$$

leads to the kinetic equation

$$\frac{d\theta_A}{dt} = k_1(1-\theta_A)-k_{-1}\theta_A \tag{3.1.6}$$

Moreover, we retrieve, for the steady state (°) coverage, the well-known Langmuir isotherm

$$\theta_A^o = \frac{k_1}{k_1+k_{-1}} = \frac{a_A p_A}{1+a_A p_A} \qquad (3.1.7)$$

with

$$a_A = \sigma_A/(2\pi M_A RT)^{1/2} k_{-1} \qquad (3.1.8)$$

We can also simulate the adsorption/desorption of A with an activation energy for the desorption (i.e. the heat of adsorption) E_{-1} decreasing with the *local* coverage θ_A^{loc} (e.g. the NN occupation of each site considered) as

$$E_{-1} = -E_{-1}^o \ln \theta_A^{loc} \qquad E_{-1} \gg T \qquad (3.1.9)$$

with, for instance

$$\theta_A^{loc} = \theta_A^{NN} = n_A/Z \qquad (3.1.10)$$

where n_A and Z are the number of the NN occupied by A and the coordination number of the lattice respectively. By this means we can investigate the validity of Freundlich isotherms i.e. steady states of the theoretical form (e.g. see [10])

$$\theta_A^o = \left(\frac{k_1}{k_{-1}}\right)^m \frac{m\pi}{\sin(m\pi)} , \quad 1 > m = T/E_{-1}^o > 0 \qquad (3.1.11)$$

i.e.

$$\lim_{m \to o} \theta_A^o \propto p_A^m \qquad (3.1.12)$$

3.2. Dissociative adsorption/associative desorption of B_2

The MC simulation of the dissociative adsorption :

$$B_2 + S_2 \xrightarrow{2} 2BS \qquad (3.2.1)$$

leads to the equation

$$\frac{d\theta_B}{dt} = 2k_2(1-\theta_B)^2 \qquad (3.2.2)$$

with an exponent γ_2 close to its MFA value (cf. Table 1)

$$\gamma_2 = 2.1 \pm 0.1 \qquad (3.2.3)$$

The average steady state coverage of empty *isolated* sites is (see also [11]) :

$$\theta_S^o = 0.0915 \qquad (3.2.4)$$

Now, if one simulates the adsorption/desorption of B_2

$$B_2 + S_2 \underset{-2}{\overset{2}{\rightleftharpoons}} 2BS \qquad (3.2.5)$$

with, for instance,

$$k_2 = 5.10^7 s^{-1} \quad , \quad 5.10^6 s^{-1} \leqslant k_{-2} \leqslant 10^7 s^{-1} \tag{3.2.6}$$

we find the kinetic equation :

$$\frac{d\theta_B}{dt} = R_2 - R_{-2} = 2k_2(1-\theta_B)^{\gamma_2} - 2k_{-2}^{eff.} \cdot \theta_B^2 \tag{3.2.7}$$

Some comments about this equation arise

(i) The value of γ_2 depends on the value of k_{-2} as follows:

$$\gamma_2 = 2 - \beta_2 k_{-2} \tag{3.2.8}$$

where (in the range considered here)

$$\beta_2 = (1.6 \pm 0.3) 10^{-8} s \tag{3.2.9}$$

Thus γ_2 decreases when k_{-2} increases. The more efficient is the associative desorption the more dual sites (per time unit) are freed too allow the dissociative adsorption of B_2.

(ii) In a MC experiment, we assign the value of k_{-2} to the inverse of the mean time spent by two atoms adsorbed in a favourable configuration (i.e. in NN position) before desorbing together. In a MC experiment, *like in the real situation* some atoms remain *isolated* during some *delay time* before being in such an associative desorption configuration (see section 2.7). In consequence there appears an *effective* rate constant k_{-2}^{eff} which is

(1) an average over all configuration (favourable and unfavourable) involved in a long sequence of mechanisms. K_{-2}^{eff} is the rate constant one reaches by optimization of results obtained in lab. experiments.

(2) smaller than k_{-2}; the discrepancy increasing as the pressure decreases.

Moreover if we set

$$k_{-2}^{eff} = \frac{1}{\langle \tau_{-2} \rangle_{eff.}} = \alpha k_{-2} \tag{3.2.10}$$

where $\langle \tau_{-2} \rangle_{eff.}$ can be measured during the simulation; we find that, for a given value of k_2, the average delay time increases slightly with k_{-2}. For instance for $k_2 = 5.10^7\ s^{-1}$ we find that

$$\begin{aligned} \alpha &= 0.440 \qquad \text{for } k_{-2} = 10^7 s^{-1} \\ \alpha &= 0.487 \qquad \text{for } k_{-2} = 5.10^6 s^{-1} \end{aligned} \tag{3.2.11}$$

Nevertheless, if we neglect this small effect we should rewrite Eq. (3.2.8) as

$$\gamma_2 = (1 - \tilde{\beta}_2\ k_{-2}^{eff}) \tag{3.2.12}$$

with

$$\tilde{\beta}_2 = (3.4 \pm 0.6) 10^{-8}\ s \tag{3.2.13}$$

Notice that the steady state coverage θ_B^o has a Langmuir isotherm for (see Eqs (3.1.7)

$$\theta_B^o = (a_B p_{B_2})^n/[1+(a_B p_{B_2})^n] \tag{3.2.14}$$

but with

$$n > n_{MFA} = 1/2 \tag{3.2.15}$$

a_B is similar to a_A, Eq.(3.1.8).

Here agin MC simulation of adsorption/desorption of B_2 with a heat of adsorption E_{-2} decreasing with the local coverage θ_B^{loc} as (see Eq.(3.1.9) for analogy)

$$E_{-2} = -E_{-2}^o \ln \theta_B^{loc} \quad , \quad E_{-2} \gg T \tag{3.2.16}$$

should provide a test for the corresponding Freundlich isotherm i.e.

$$\lim_{m\to o} \theta_B^o \propto p_{B_2}^{m/2} \quad , \quad m = T/E_{-1}^o \tag{3.2.17}$$

Let us notice that θ_B^{loc} can be approximated by (see (Eq.3.1.10))

$$\theta_B^{NN} = \frac{n_B^{(1)}+n_B^{(2)}}{Z} \tag{3.2.18}$$

where $n_B^{(i)}$ is the NN number of the site i (partner in the dual site considered) occupied by a B atom.

Finally let us mention that a test for the simulation of the processes (3.2.5) consists of the MC experiment of an *isotopic exchange*. Consider the random rainfall of an isotopic mixture of B_2 and B_2^* in equal proportion. Suppose that these isotopes have identical features with respect to the chemisorption and the desorption. If the MC simulation of processes (3.2.5) is correct *and* the resulting distribution of the particles on the surface is strictly random, one should observes a *complete* isotopic exchange. Namely, the adsorption of an equimolecular beam is *followed* (once the adsorbate is prepared) by the desorption of B_2, B_2^* and BB^* in the relative proportions 1/4, 1/4 and 1/2 respectively. Noteless to say that this result should also be obtained in the stable steady state resulting of the balance between adsorption and desorption.

3.3. Competitive coadsorption

We simulate the following competitive adsorption :

$$A + S \xrightarrow{1} AS \tag{3.3.1}$$

$$B_2 + S_2 \xrightarrow{2} 2BS \tag{3.3.2}$$

at ambient pressure, namely for

$$p_A/p_{B_2} \approx p_A/p_{atm} \approx 1\% \qquad T \lesssim 400K \tag{3.3.3}$$

i.e. more precisely with

$$k_1 = 2.94\ 10^6 s^{-1} \quad , \quad k_2 = 4.92\ 10^6 s^{-1} \qquad (3.3.4)$$

The kinetic equations obtained are

$$\frac{d\theta_A}{dt} = k_1(1-\theta_A-\theta_B), \quad \frac{d\theta_B}{dt} = 2k_2(1-\theta_A-\theta_B)^{\gamma_2} \qquad (3.3.5)$$

with

$$\gamma_2 = 2.1 \pm 0.1 \qquad (3.3.6)$$

The steady state coverages are

$$\theta_A^o = 0.44 \qquad \theta_B^o = 0.56 \qquad (3.3.7)$$

The results are similar to the MFA ones. Indeed when we analyse the adsorbate in terms of clusters characteristics we find for the *ramification parameters* η (defined in Eq. (a.6), App.A) the following values

$$\eta_A \approx \eta_B \approx 3.5 \qquad (3.3.8)$$

which are close to their MFA values $\eta = 4$.

3.4. Surface reaction (without desorption)

Let us consider the reaction (1.1) and restrict the kinetic scheme to the following steps

$$A + S \xrightarrow{1} AS \qquad (3.4.1)$$

$$B_2 + S_2 \xrightarrow{2} 2BS \qquad (3.4.2)$$

$$AS + BS \xrightarrow{3} AB + S_2 \qquad (3.4.3)$$

To compare this situation with the previous one (see section (3.3)) let us choose the same value for k_1 and k_2 (see Eq. (3.3.4). The values of k_1 and k_2 being fixed we select values of k_3 such that the simulated kinetics lead to steady state *mixed* adsorbates. Indicative values of k_3 may be found by referring to the MFA kinetic equations (see Table 1)

$$\frac{d\theta_A}{dt} = R_1 - R_3 = k_1(1-\theta_A-\theta_B) - k_3\theta_A\theta_B \qquad (3.4.4)$$

$$\frac{d\theta_B}{dt} = R_2 - R_3 = 2k_2(1-\theta_A-\theta_B)^{\gamma_2} - k_3\theta_A\theta_B \qquad (3.4.5)$$

with

$$\gamma_2^{MFA} = 2 \qquad (3.4.6)$$

From these equations one readily sees that the system reaches a steady state adsorbate (richer in B) :

$$\theta_A^o + \theta_B^o = 1 - (k_1/2k_2)^{1/(\gamma_2-1)} \qquad (3.4.7)$$

provided

$$k_3 \geqslant 4k_1(k_1/2k_2)^{1/(\gamma_2-1)}/[1-(k_1/2k_2)^{1/(\gamma_2-1)}]^2 \qquad (3.4.8)$$

i.e.

$$k_3 \geqslant 7.14\ 10^6\ s^{-1}$$

(using Eq.(3.3.4) and $\gamma_2 = 2$)).

In fact our MC experiments lead to stable steady state adsorbates (richer in B) for

$$k_3 \geqslant k_3^{min} = 5.5\ 10^6\ s^{-1} \qquad (3.4.9)$$

More important, the competition of the kinetic steps (3.4.1-3) induces an important clustering of the adsorbate. The best fit kinetic equations issuing from our experiments have the following form [11]

$$\frac{d\theta_A}{dt} = R_1 - R_3 = k_1(1-\theta_A-\theta_B) - \alpha k_3\theta_A(1-\theta_A)\theta_B(1-\theta_B) \qquad (3.4.10)$$

$$\frac{d\theta_B}{dt} = R_2 - R_3 = 2k_2(1-\theta_A-\theta_B)^{\gamma_2} - \alpha k_3\theta_A(1-\theta_A)\theta_B(1-\theta_B) \qquad (3.4.11)$$

with

$$\gamma_2 = 2-5.10^{-8}k_3 \quad ; \quad 10^7 \geqslant k_3 \geqslant k_3^{min} \qquad (3.4.12)$$

$$\alpha = 2.3 + 1.7\ 10^{-3}\ (k_3-k_3^{min})^{1/2} \qquad (3.4.13)$$

As first conclusion when k_3 increases the total coverage, Eq.(3.4.7), increases and becomes richer in B (see Table 3). Here again the more efficient is the reaction the more dual sites (per time unit) are freed to allow the dissociative adsorption of B_2. Let us recall that the Langmuir-Hinshelwood reaction mechanism looks like an associative desorption and is MC simulated as such (see section 2.9). Compare Eq. (3.4.12) and 3.2.8) for analogy.

The analysis of our results in terms of cluster features shows that the reaction rate R_3 is proportional to the product of the total *dual* perimeters B_A and B_B (defined in Eq. (a.2), App. A) i.e.

$$R_3 = \alpha' k_3 B_A B_B \qquad (3.4.14)$$

This reaction rate R_3 is proportional neither to the product of the coverages (MFA) nor to the product of the total "usual" perimeters (see App.A).

The explicit form (see Eq. (3.4.11)) :

$$R_3 = \alpha k_3\theta_A(1-\theta_A)\theta_B(1-\theta_B) \qquad (3.4.15)$$

shows that, within the parameter range considered, the clusters (even when percolating) are so much ramified that the MFA expression of the total dual perimeters remains qualitatively correct (see App. A, Eq. (a.5) and Table 3).

k_3	θ_A^o MFA	θ_B^o MFA	θ_A^o MC	θ_B^o MC	η_A	η_B
$5.7\ 10^6$	0.0881	0.6130	0.1135	0.7080	3.5	2.8
1.10^7	0.0470	0.6541	0.0360	0.8491	3.6	2.7

TABLE 3 : Steady state coverages and the corresponding coefficients η, Eq. (a.6), for two values of k_3.

Finally some questions remain open. How to interpret the reaction rate as being proportional to the product of the total dual perimeters? How will this picture be modified in the presence of rate "constants" k_i depending on the local coverages?

3.5. Surface reaction (with desorption of A)

Now we simulate the kinetic scheme

$$A+S \underset{-1}{\overset{1}{\rightleftharpoons}} AS \tag{3.5.1}$$

$$B_2 + S_2 \xrightarrow{2} 2BS \tag{3.5.2}$$

$$AS+BS \xrightarrow{3} AB + S_2 \tag{3.5.3}$$

e.g. with

$$k_1 = 2.32\ 10^6\ s^{-1}$$

$$k_2 = 1.57\ 10^6\ s^{-1}$$

$$k_3 = 1.65\ 10^3\ s^{-1}$$

$$k_{-1} = 6.74\ 10^3\ s^{-1}$$

i.e. with $E_3 = 5.5.$ kcal $\qquad E_{-1} = 25$ kcal; $\quad T \approx 550K$

$$k'_3 = 2.5\ 10^5 \qquad k'_{-1} = 5.10^{13}$$

The best fit kinetic equations are

$$\frac{d\theta_A}{dt} = k_1(1-\theta_A-\theta_B) - k_{-1}\theta_A^{\gamma_{-1}} - \alpha k_3\theta_A(1-\theta_A)\theta_B(1-\theta_B) \tag{3.5.4}$$

$$\frac{d\theta_B}{dt} = 2k_2(1-\theta_A\theta_B)^{\gamma_2} - \alpha k_3\theta_A(1-\theta_A)\theta_B(1-\theta_B) \tag{3.5.5}$$

with

$$\gamma_2 = 1.45 \quad , \quad \gamma_{-1} = 1.37 \quad , \quad \alpha = 8.04$$

$$\theta_A^o = 0.1945 \qquad \theta_B^o = 0.8052$$

3.6. MC test of a kinetic model. An example

Takoudis, Schmidt and Aris (TSA) [13] consider the surface reaction

$$A + B \longrightarrow \text{product} \tag{3.6.1}$$

involving the following steps :

$$A + S \underset{-1}{\overset{1}{\rightleftharpoons}} AS \tag{3.6.2}$$

$$B + S \rightleftharpoons BS \tag{3.6.3}$$

$$AS + BS\ 2S \xrightarrow{3} 4S + \text{products} \tag{3.6.4}$$

In TSA model, the lumped sequence of site exchange and reaction steps is "summarized" in a Langmuir-Hinshelwood mechanism that requires two adjacent vacant sites.

The authors study the following dimensionless kinetic equations ($\tau \equiv k_3 t$; $\alpha_i = k_i/k_3$)

$$\frac{d\theta_A}{d\tau} = \alpha_1(1-\theta_A-\theta_B) - \alpha_{-1}\theta_A - \theta_A\theta_B(1-\theta_A-\theta_B)^2 \tag{3.6.5}$$

$$\frac{d\theta_B}{d\tau} = \alpha_2(1-\theta_A-\theta_B) - \alpha_{-2}\theta_B - \theta_A\theta_B(1-\theta_A-\theta_B)^2 \tag{3.6.6}$$

and for

$$\alpha_1 = 0.016 \quad , \quad \alpha_{-1} = 0.001 \quad ; \quad \alpha_2 = 0.0278 \quad ; \quad \alpha_{-2} = 0.002 \tag{3.6.7}$$

obtain sustained oscillations characterized by a period $t_p \sim 0.08\tau$, *mean* values

$$\bar{\theta}_A = 0.1945 \quad , \quad \bar{\theta}_B = 0.7025 \quad ; \quad \bar{R}_3 \sim 1.5\ 10^{-3} \tag{3.6.8}$$

and amplitudes

$$\Delta\theta_A \sim \Delta\theta_B \quad 0.02 \quad ; \quad \Delta R_3 \sim 5.10^{-4} \tag{3.6.9}$$

For the MC simulation of this model [14] we use the dimensionless times (see Eqs (3.6.7)) :

$$\tau_i = 1/\alpha_i \tag{3.6.10}$$

and consider the reaction step, Eq. (3.6.4), in two ways :

(i) case 2/3 : the two vacant sites are assigned to be NN of the adsorbed and reacting molecule B

(ii) case 2/6 : the two vacant sites are assigned to be NN of the adsorbed pair AB.

We compared the time evolution of the simulated reaction rate R_3^{MC} with the rate $R_3^{calc.}=\theta_A\theta_B(1-\theta_A-\theta_B)^2$ calculated with the values of θ_A and θ_B issued from the simulation.

The main results are :

(1) In the case 2/3, the reaction vanishes in the long time limit ($\theta_A^o = 0.445$, $\theta_B^o = 0.554$). The production is essentially governed by the desorption and therefore practically non-detectable.

(2) The case 2/6 seems more favourable to the TSA model. In the long time limit we found (compare to Eqs (3.6.8))

$$\theta_A^o = 0.1959 \quad , \quad \theta_B^o = 0.7832 \quad , \quad R_3^{MC} \approx R_3^{calc.} \leqslant 10^{-4} \tag{3.6.11}$$

However we have not been able to detect oscillations in the "stable" (but fluctuating) regime of the reaction.

3.7. On coverage dependent activation energy. An example

We simulate the reaction (1.1) within the kinetic scheme (3.4.1-3) but with adsorption rate "constants" depending of the *local* coverage (here the NN occupation of each site considered, (see Eqs.(3.1.10).(3.2.18)). Namely, we choose ($\tau=k_3t;\alpha_i=k_i/k_3$)

$$\alpha_1 \equiv k_1/k_3 = \alpha_1^o \exp\left[-\mu\,(\theta_A^{NN} - \theta_B^{NN})\right] \tag{3.7.1}$$

$$\alpha_2 \equiv k_2/k_3 = \alpha_2^o \exp\left[+\mu\,(\theta_A^{NN} - \theta_B^{NN})\right] \tag{3.7.2}$$

i.e. we consider the coadsorption of two reactants opposing in their redox (electron donor-acceptor) properties. Working again at atmospheric pressure and for $T \approx 400$ K, we obtain the following kinetic equations (e.g. for $\mu = 0.1$).

$$\frac{d\theta_A}{dt} = \tilde{\alpha}_1 e^{-\mu(\theta_A-\theta_B)}(1-\theta_A-\theta_B)-(\theta_A\theta_B)^{1/2}e^{-\mu(\theta_A-\theta_B)} \tag{3.7.3}$$

$$\frac{d\theta_B}{dt} = 2\tilde{\alpha}_2\, e^{\mu(\theta_A-\theta_B)}(1-\theta_A-\theta_B)^2-(\theta_A\theta_B)^{1/2}e^{-\mu(\theta_A-\theta_B)} \tag{3.7.4}$$

with

$$\tilde{\alpha}_i \gtrsim \alpha_i \qquad (i=1,2) \tag{3.7.5}$$

We observe that

(i) the local coverage effects (see Eqs. (3.7.1,2)) cut up the clusters into small pieces *and* reduce the delay time mentionned above.

(ii) this local coverage dependancy is averaged into the adsorption rates.

(iii) more surprising, the reaction rate appears as an exponential function of the coverages.

It should be added that these results are preliminary, simulations at lower pressure and using higher values for μ are underway.

4. ON LASER ASSISTED SURFACE REACTIONS

We hardly begin our investigations on the simulated effects of a *low power* laser pulsed irradiation on surface reactions.

In order to explain how we proceed to simulate both the thermal and the photonic effects of a laser pulse on a surface reaction, let us start with the reaction, Eq. (1.1), governed only by the mechanisms, Eq. (3.4.1-3). The basic assumptions are the following :

(1) When the laser is switched on (e.g. at t=0) the coverage is in a given *unperturbed* stable steady state (prealably simulated).

(2) The photon flux supplied by the laser is uniformly distributed on the surface. Namely the spatial pulsewidth σ is larger than the linear size of the surface considered. The *"surface temperature"* $T_s(t)$ is uniform and assumed to be coverage-independent.

(3) The energy of the laser effectively used in "photonic" processes all along the experiment is negligible with respect to the total energy thermally converted into the sample.

(4) The thermal effect of the laser will be induced by *all* the mechanisms which are $T_s(t)$-dependent. Desorption, reaction, hopping,... mechanisms are of this type. We assume that direct adsorption mechanisms depend only on the (fixed) gas temperature. Thus in our simple example only k_3 is $T_s(t)$-dependent.

(5) The photonic effect of the laser will be induced *selectively* by one (or more) elementary mechanism(s). A given mechanism, when perturbed by photons, undergoes a decrease of its activation energy. In our simple kinetic scheme we choose, for instance, the reaction mechanism (3.4.3) to be affected by the photons i.e. its activation energy E_3^o becomes

$$E_3 = E_3^o(1-\alpha) \qquad 1 \geqslant \alpha > 0 \qquad (4.1)$$

Let us first consider the simulation of the *thermal effect* according to our basic assumptions. First we solve a *heat equation* modelling the experimental set-up we want to represent (e.g. see ref.[15] for approximate situations). We thus dispose of numerical values for the "surface temperature" $T_s(t)$ induced by the heating program supplied by the laser.

We have chosen two ways to insert the thermal effect on the reaction mechanism (for instance). Let us consider a reaction configuration implemented at a time t_o. Then we program the future of this reaction either, simply, at the time

$$t_1 = t_o + \Delta t_1\{\tau_3\} \text{ , } \tau_3 = \tau_3\{T_s(t_o)\} \qquad (4.2)$$

or, using more computer time, at the time

$$t_2 = t_o+\Delta t_2\{\tau_3\}, \ \tau_3=\tau_3\{T_s(t_o+\tau_3\{T_s(t_o)\})\} \qquad (4.3)$$

where clearly

$$\Delta t_2 \gtrless \Delta t_1 \quad \text{if} \quad T_s(t_o) \lesseqgtr T_s(t_o+\tau_3\{T_s(t_o)\}) \qquad (4.4)$$

If we denote by b the time pulselength we conclude that if $\tau_3(t{=}o) \ll b(=10^{-6}$s in our case) procedure (4.2) is sufficient. Noteless to say that in presence of desorption mechanisms the game becomes more tricky as the procedure (4.3) should be used.

Now we consider the simulation of the photonic effect according to our basic assumptions. To fix the ideas let us assume an *average* photon flux Φ (per site of the surface and per time pulse length)

$$\phi = 10^3 \text{ photons } (\text{site}^{-1}, b^{-1}) \qquad (4.5)$$

i.e. on average, a site is visited by a photon every 10^{-9}s.

Let us first remark that if the following condition is satisfied :

$$\phi \gg \frac{b}{<\tau_{ads}>} \qquad (4.6)$$

where $<\tau_{ads}> = 2/(k_1+k_2)$, each particle which adsorbs during the time pulse length *may* absorb at least one photon.

Now we insert the photonic effect on the reaction mechanism (for instance) as follows. We first define a treshold flux ϕ_c (per site)

$$\phi_c = f\ \alpha\ E_3^o\ \frac{\lambda}{hc} \qquad (4.7)$$

Where $\alpha\ E_3^o$ represents the decrease of the activation energy due to the photonic effect (see Eq. (4.1)). The efficacity factor f is simply (see Eq. (4.6) for analogy)

$$f = b/\tau_3(t{=}o) \qquad (4.8)$$

Therefore, throughout the time interval during which

$$\phi(t) \geqslant \phi_c \qquad (4.9)$$

we insert the photonic effect on the reaction mechanism (for instance) as follows. If a reaction configuration is implemented at a time t_o during this period, we program the future of this reaction either, at the time

$$t_1' = t_o + \Delta t_1'\{\tau_3\}\ ,\ \tau_3 = \tau_3\{E_3,T_s(t_o)\} \qquad (4.10)$$

or, at the time

$$t_2' = t_o + \Delta t_2'\{\tau_3\}\ ,\ \tau_3 = \tau_3\{E_3,T_s\{t_o+\tau_3\{T_s(t_o)\}\} \qquad (4.11)$$

Outside the above mentionned period the reaction future is assumed to be "photon-independent" and programmed along the rule (4.2) or (4.3).

Fig.4 and Table 4 [16] show some results obtained from such a simulation. These "exploratory" results concern the reaction, Eq.(1.2), governed by the mechanisms, Eq.(3.4.1-3) with $A=CO$, $B_2=O_2$. The thermal and photonic effects of the laser pulse on these mechanisms have been simulated along the procedures illustrated by Eqs. (4.2) and (4.10).

The *unperturbed* characteristic times are $\tau_1^o = 3.4\ 10^{-7}$ s $\tau_2^o = 2.10^{-7}$s ($E_2^o = 1250$ K), $\tau_3^o = 1.82\ 10^{-7}$ ($E_3^o = 4000$K). A gaussian dye laser pulse ($\lambda = 6000$ Å, $\phi \sim 8$ photon site^{-1}b^{-1}, $b=10^{-6}$s) is switched on at $t_o=10^{-7}$s, when $T_S(t_o) = 350$K. The threshold flux $\phi_c \sim 0.23$ photon site^{-1} (use Eqs.(4.7-8) with $\alpha=0.25$).

As shown in Table 4 the system is not yet in its stable steady state at t_o; we reach this state at $t \lesssim t_o + b$.

Fig.4 shows that when the reaction is sufficiently fast in its unperturbed regime, the totally laser-activated reaction rate R_3^{tot}. drops below its unperturbed value R_3^o near the end of the pulse. At the beginning of the pulse the reaction becomes so fast that the "minority" reactant (here A) is instantaneously removed from the surface. The reaction *delay time* (see section 3.2) increases drastically under the laser influence. Indeed, at the end of the pulse (at $t \sim t_o + b$), we have respectively

$$\langle\tau_3^o\rangle_{eff.}/\tau_3^o \lesssim 10; \quad \langle\tau_3^{tot}\rangle_{eff.}/\tau_3^{tot} \gtrsim 10^3 \qquad (4.12)$$

We thus conclude that, when assisting a "naturally" fast reaction, the laser does not increase significantly the time averaged turnover frequency (R_3/N). We are investigating how a kinetic scheme completed with desorption and hopping mechanisms modifies our simulation results.

t	θ_A^o	θ_B^o	θ_A^{th}	θ_B^{th}	θ_A^{tot}	θ_B^{tot}
10^{-7}	0.16	0.40	0.16	0.40	0.16	0.40
10^{-6}	0.19	0.56	0.008	0.69	0.005	0.73
	R_3^o	R_3^{th}	R_3^{tot}	τ_3^o	τ_3^{th}	τ_3^{tot}
10^{-7}		$6.1\ 10^9$			$1.83\ 10^{-7}$	
10^{-6}	$7.5\ 10^9$	$9.4\ 10^9$	$5.3\ 10^9$	$1.82\ 10^{-7}$	$3.59\ 10^{-9}$	$5.1\ 10^{-10}$

TABLE 4. Coverages, reaction rates R_3 and reaction times τ_3 in the unperturbed situation (°), the "thermal-laser" case (th) and when the laser acts like a thermal *and* a photonic source (tot). These values correspond to the beginning and near the end of the pulse respectively.

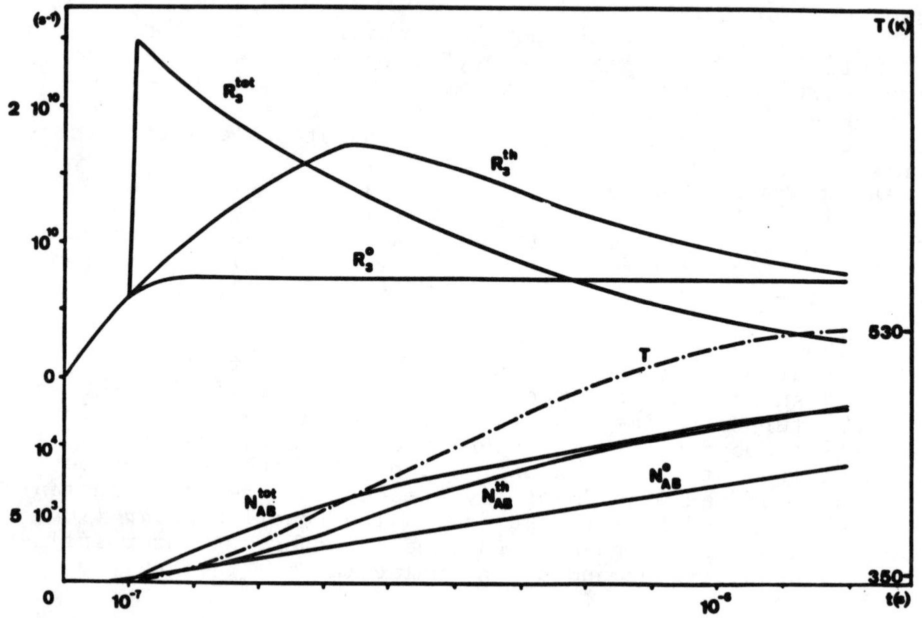

FIGURE 4. Time evolution of the production N_{AB}, the reaction rate R_3 (solid lines) and the surface temperature T (dashed-dotted line) during the laser pulse duration. The superscripts (o), (th) and (tot) are defined in Table 4.

5. CONCLUSIONS

We hope that the examples of MC simulated surface reactions reported here show how these methods are usefull to investigate the kinetics of such reaction. Procedures to simulate reactions in more than one monolayer are underway.

The most severe restriction in this field of investigations concerns the ratio of computing time to real time which can be prohibitive. Cellular automata, exclusively made to perform MC simulations of surface reactions on a large scale, should solve this crucial problem and develop a fair future for this field or research.

Acknowledgements

We wish to thank Prof. P. Dufour, M. Bouillon, A. Mercier, J.C. Minor, P.M. Petropoulos, M. Poriaux and J. Ravez for their assistance with the various programs. This work is partly supported by the IRIS Program sponsored by the Belgian Ministry for Science Policy.

APPENDIX A : ON THE SITE PERCOLATION PROBLEM [6]

In the site percolation problem, every site of the lattice is either occupied (with a probability p) or empty (with a probability (1-p)) entirely randomly, independent of whether its neighbours are occupied or empty. The whole problem is defined, for a given lattice, by the single parameter p. A cluster is a group of occupied sites connected by nearest-neighbour distances.

For an infinite lattice there exists a sharply defined percolation p_c i.e. a critical point where for the first time an infinite network percolates through the lattice :

For $p > p_c$, one percolating network exists, for $p < p_c$ no percolating network exists.

The surface reaction, Eq.(1.1), appears as a three-component reactive percolation problem.

The probability to have a A-occupied site (a B-occupied site) is θ_A (θ_B) or an empty site is $(1-\theta_A-\theta_B)$ (e.g. see (a.9) below).

To analyse the kinetics of the reaction in terms of cluster properties we need the following definitions :

(i) A s(σ) cluster is a group of s(σ) sites occupied by nearest-neighbour (NN) particles A(B). s(σ) is the size of the cluster. If C_s(C_σ) denotes the number of s(σ) clusters (*per site*), we have e.g. for the species A

$$\theta_A = \frac{1}{N}\sum_s n_s s = \sum_s C_s s = N_s\langle s\rangle/N = \sum_s n_s\theta_s \qquad (a.1)$$

where n_s(n_σ) is the number of s(σ) clusters present on the surface; N_s(N_σ) is the *total* number of clusters of average size $\langle s\rangle$($\langle\sigma\rangle$). θ_s(θ_σ) represents the fraction of the surface covered by a *single* s(σ)-cluster.

(ii) The perimeter t_s(τ_σ) of an s(σ) cluster is the number of empty sites *or* B(A) occupied sites that are NN to the A(B) occupied cluster sites.

(iii) We define the *dual* perimeter b_s(β_σ) of a s(σ) cluster as the number of NN connections between A(B) occupied cluster sites and perimeter sites. If B_A(B_B) denotes the *total* dual perimeter (*per site*) for species A(B), we have for the species B for instance,

$$B_B = \sum_\sigma C_\sigma\beta_\sigma = N_\sigma\langle\beta\rangle/N = \theta_B\langle\beta\rangle/\langle\sigma\rangle$$
$$= (N_{BS} + N_{BA})/N = 2(\theta_{BS} + \theta_{AB}) \qquad (a.2)$$

where N_{BS} is the number of B semi-occupied dual sites (per unit area), etc... and $\theta_{BS} = N_{BS}/2N$, etc...

Let us remark that if one uses the exact sum rule (for a square lattice)

$$\theta_{BB} + \frac{1}{2}\theta_{BA} + \frac{1}{2}\theta_{BS} = \theta_B \qquad (a.3)$$

one sees that Eq.(a.2), for instance, reads

$$B_B = 4(\theta_B - \theta_{BB}) \qquad (a.4)$$

Therefore, the MFA expression for the total dual perimeter B_B is

$$B_B^{MFA} = 4\theta_B(1-\theta_B) \qquad (a.5)$$

Consequently, we see that the determination of the *ramification parameter* η defined as follows

$$\frac{\langle\beta\rangle}{\langle\sigma\rangle} = \eta_B(1-\theta_B) \ , \ \frac{\langle b\rangle}{\langle s\rangle} = \eta_A(1-\theta_A) \qquad (a.6)$$

helps us to specify how much the clusters are ramified. The more η approaches its MFA value $\eta=4$, the more the clusters are ramified.

To specify the cluster distributions in a given adsorbate one should keep in mind the following remarks.

If one assumes that *all* the s-clusters (with the same size) are randomly distributed the probability P_s for a site to belong to an s-cluster is

$$P_s = s\frac{n_s}{N} = sC_s = N_s\theta_s \qquad (a.7)$$

Thus in the case of the adsorption of A, for instance, we have the exact sum rule :

$$1 - p + p\,P_\infty + \sum_s s\,C_s = 1 \qquad (a.8)$$

i.e. the evident result

$$\bar{p} = \sum_s sC_s = \sum_s n_s\theta_s = \theta_A \ , \ \theta_A<\theta_A^c \qquad (a.9)$$

It must be emphasized that clusters of same size $s(\sigma)$ can have different perimeters $t(\tau)$. Therefore the number $C_s(p)$ of s clusters can be calculated only if one knows the number g_{st} of *"animals"* i.e. of geometrically different cluster configurations corresponding to the same size s *and* perimeter t. This number g_{st} is independent of p; and we have, for instance, the exact result

$$C_s = \sum_t g_{st}\, p^s(1-p)^t \qquad (a.10)$$

which reduces the calculation of $C_s(p)$ to a purely geometric counting. See for instance [9] where the values of C_s are tabulated (up to s=17) for a square lattice.

To specify how the distribution of the small clusters (induced by the surface reaction) departs from a random distribution we can evaluate the following parameter all along the reaction; e.g. for the species A

$$\chi_A^2 = \sum_{s=1}^{s_{max}} [C_s^{sim} - C_s^{th}]^2 / C_s^{th} \qquad (a.11)$$

where C_s^{sim} is the number of s clusters (per site) issued from the simulation while C_s^{th} corresponds to this number calculated using Eq.(a.10) (i.e. up to s=17 at the present time).

REFERENCES

1) Clark A. : *The theory of Adsorption and Catalysis* (Academic Press, New-York, 1970).
2) Cassutto A. and King D. : Surface Sci., 102, 388 (1981).
3) Dumont M. and Dufour P. : Computer Phys. Commun. 41, 1 (1986).
4) Hayward D.O. and Trapnell B.M.W. : *Chemisorption* (Butterworths, London, 1964).
5) Feller W. : *An introduction to probability theory and its applications* (Wiley, New-York, 1966) p.446.
6) Knuth D.E. : *The art of computer programming*, Vol. II (Addison-Wesley, Reading, 1969) p.113.
7) Bouillon M. : private communication.
8) Stauffer D. : Physics Reports 54, 1 (1979).
9) Flammang A. : Z. Physik B28, 47 (1977).
10) Hu S.M. : J. Appl. Phys. 55, 4095 (1984).
11) Rossington D.R. and Borks R. : Surface Sci. 3, 202 (1965).
12) Dumont M., Poriaux M. and Dagonnier R. : Surface Sci. 169, L307 (1986).
13) Takoudis C.G., Schmidt L.D. and Aris R. : Surface Sci. 105 325 (1981).
14) Bouillon M., Dagonnier R., Dufour P. and Dumont M. : Surface Sci. 115, L111 (1982).
15) Ready J.F. : *Effect of high power laser radiation* (Academic Press, New-York, 1971).
16) Petropoulos P.M. : to appear.

MECHANISMS OF LASER-INDUCED DESORPTION FROM INSULATORS AND COMPOUND SEMICONDUCTORS

Noriaki Itoh

Department of Physics, Faculty of Science, Furocho, Chikusaku, Nagoya University, Nagoya 464, Japan

Abstract: Mechanisms of the atomic processes induced by electronic excitation are described and applied to explain the photo- and laser-induced desorption and interface reaction processes. It is emphasized that electron-lattice coupling plays a major role in inducing localization and forming an anti-bonding potential curve of the type assumed in the Menzel-Gomer-Readhead mechanism. It is pointed out that in some specific solids such as alkali halides and silver halides, formation of excitons near the surface results in atomic processes, while, in many others, interaction of two excitation events (two holes) is responsible for atomic processes. Several mechanisms of formation of two hole states are described.

1. Introduction

Primary interaction of photons with a non-metallic wide band-gap crystalline solids causes fundamental electronic excitation, which produces free electrons, free holes and excitons. All of these primary products have wave functions extending over the whole crystal volume. There are a few possible modes of the recombination of such electronic excited states: (i) the radiative recombination of free excitons, (ii) radiative or non-radiative recombination of excitons after being self-trapped, (iii) self-trapping of a hole followed by trapping of an electron to become a self-trapped exciton and (iv) trapping of either an electron or a hole by a recombination center (including a surface state) followed by trapping of a carrier of the opposite sign. Conversion of a pair of an electron and a hole to an exciton and vice versa may also occur. For desorption to be induced as a result of electronic excitation, the electronic excitation energy is to be converted to the energy for an atomic process that ejects an atom from the surface. Conversion of such electronic excitation energy for defect formation in the bulk of alkali halides has been well established (1).

Evidently no energy is available for atomic processes if excited states are subjected to radiative recombination. Non-radiative recombination for free excitons, which possesses a large energy and does not couple with the lattice strongly, is highly improbable because of a low non-radiative transition probability (2). Thus any atomic process induced by electronic excitation may be concerned with localization of extended wave functions: self-trapping

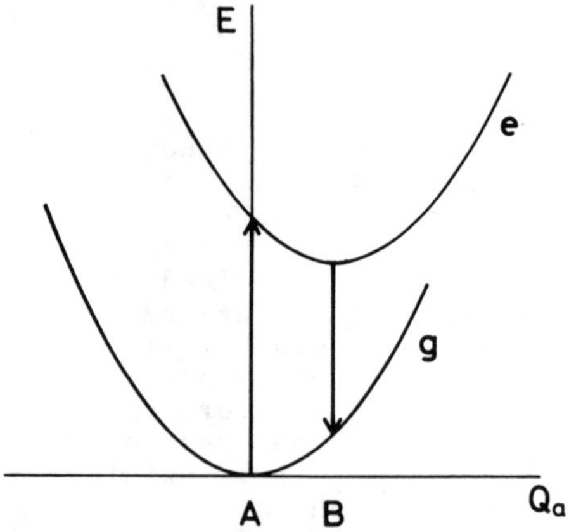

Fig. 1. A configuration coordinates curve for the ground state g and an excited state e, describing the radiative and non-radiative recombination. The upward vertical arrow denotes optical absorption and the downward arrow luminescence.

of an exciton or of a hole or trapping of an electron or of a hole by a recombination center (3). The self-trapping occurs only in materials in which the electron-lattice coupling is strong (4, 5).

The recombination process is generally described in terms of adiabatic potential curve as shown in Fig. 1, which describes the energy of an exciton or a electron-hole pair at a defect site as a function of configuration coordinates Q_a. At the ground state (g) the system takes configuration A. Upon optical excitation, which occurs vertically in the figure, the self-trapping of an exciton or a hole relaxes the lattice to the configuration B. At the configuration A the wavefunctions of the exciton and of the electron-hole pair are extended, while they are localized at the configuration B. Furthermore several local modes of lattice vibration are formed at the configuration B. The abscissa Q_a denotes the local mode that interacts most strongly with the electronic excitation. The excited state at the configuration B leads to the original configuration A after either radiative or non-radiative recombination. Upon non-radiative recombination, the electronic energy is converted to the energy of the local phonons Q_a. Thus Q_a is called accepting mode.

After an atomic process is induced the configuration should be converted to a configuration different from A. Two possible mechanisms have been suggested to explain the atomic processes in general including desorption and defect formation: the local heating (phonon kicking) mechanism and local excitation mechanism (3). These mechanisms are illustrated in Fig. 2 which shows a two dimensional configuration coordinates surface. Here Q_R shows the reaction coordinate that leads to the final product of the atomic process. In the local heating mechanism, the energy imparted to the accepting phonon mode is transferred to the

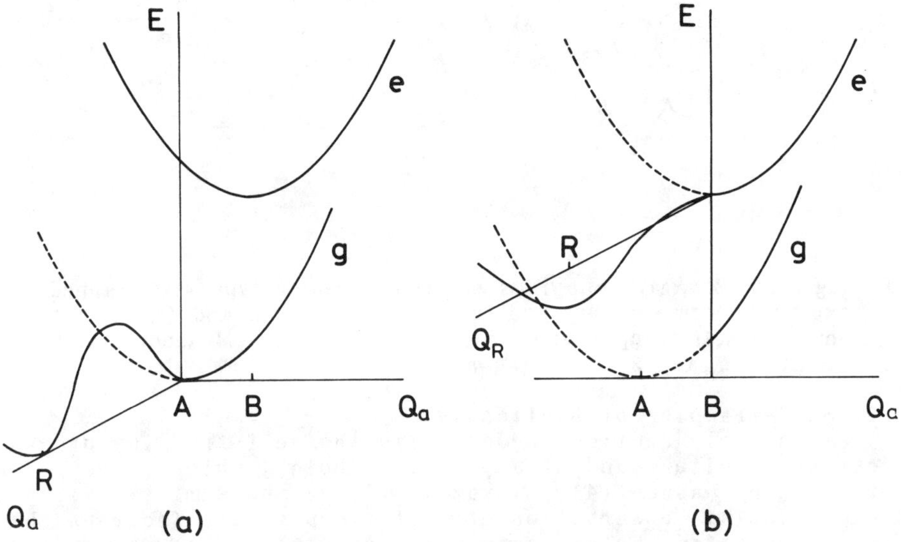

Fig. 2. Description of the local heating (phonon kicking) mechanism and (b) the local excitation mechanism. Q_a and Q_R are the accepting mode and the reaction mode, respectively. R denotes the configuration of the reaction products.

reaction mode that brings the configuration from A to R. In the local excitation mechanism one of the excited states has anti-bonding adiabatic potential curve with respect to the reaction mode and hence electronic excitation to the excited state leads to reaction. The instability at the configuration B against the coordinate Q_R is called adiabatic instability.

The purpose of the paper is to survey the laser-induced desorption for insulators and compound semiconductors and to interpret the results in terms of the electron-lattice coupling processes described above. In 2, we give a brief description of the electron-phonon coupling that leads to self-trapping. We survey in 3 experimental aspects of several types of photon-induced desorption including photo- and laser-induced desorption and interface reactions. Relevant ion-induced processes are included also. Finally in 4, we discuss the mechanism of several modes of photon-induced desorption.

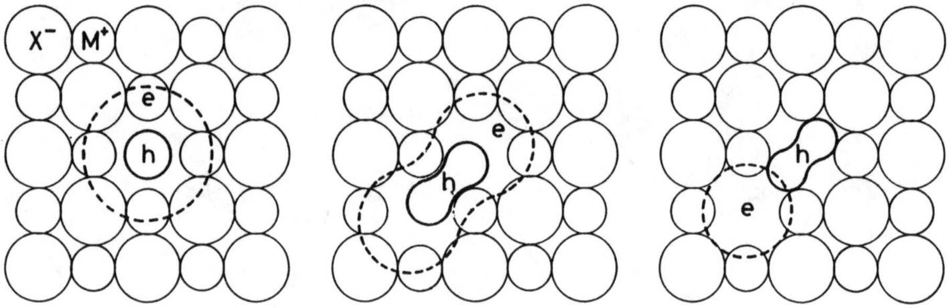

Fig. 3. Schematic diagrams of (a) one-center type self-trapped exciton, (b) two-center type self-trapped exciton and (c) off-centered self-trapped exciton in alkali halides. M denotes an alkali atom and X a halogen atom.

2. Self-trapping of Excitons

In this section we deal with the self-trapping process of an exciton and of a positive hole, which have been treated by Rashba (4), Toyozawa (5, 6) and Sumi (7, 8). No experimental observation of self-trapping of electrons has been reported. The necessary condition for the self-trapping of an exciton, as given by Rashba and Toyozawa, is that a half of the width of the exciton band is smaller than the lattice relaxation energy E_{LR}. Similar criterion is given for a positive hole.

The self-trapping process appears to be not a single process but consists of a initial stage where the state of an exciton coupled with phonons is generated and of the final stage where relaxation to the self-trapping configuration is induced (5, 7, 8). The initial stage determines the potential barrier for the self-trapping of excitons while the second stage the magnitude of E_{LR}. A crucial factor for the initial stage is the interaction energy of an exciton with phonons.

Here rather than concentrating on the theoretical description, we describe the self-trapping process of alkali halides for illustration. Since the top of the valence band in alkali halides consists of the p-orbitals of the halogen ions, a free exciton may be described in terms of a hole localized on a halogen ion, forming a halogen atom, with an electron trapped by the coulomb potential formed by surrounding alkali ions, The initial stage of the self-trapping is governed by two interaction energies, one with site-symmetrical phonons, associated with changes in alkali-halogen distance and the other with off-diagonal phonons, associated with the change in halogen-halogen distance. Coupling with site-symmetrical phonons lead to the self-trapped excitons of one-center type (Fig. 3(a)) and coupling with off-diagonal phonons to the self-trapped excitons of two-center type (Fig. 3(b)). In the former the accepting mode is the site-symmetrical distortion of the surrounding lattice while in the latter it is the halogen-halogen separation. The former interaction energy is approximately proportional to the inverse of the reduced mass of the alkali ion and the halogen ion and the latter interaction energy is approximately to the inverse of the mass of the halogen ion. Thus one-center self-trapping is favored in heavy alkali halides such as RbI, in which slightly modified one-center self-trapped excitons have been found (10). For the one-center self-trapped exciton, major contribution to E_{LR} comes from the Jahn-Teller energy and for the two-center self-trapped excitons the major contribution to E_{LR} comes from the bonding energy of the X_2^- molecular ion (X denotes a halogen atom.).

The two-center self-trapping is of particular interest for desorption because of its inherent instability. Itoh and Saidoh (11) have pointed out that the coulomb repulsion between the electron and X_2^- molecular ion will repel the X_2^- to the nearest neighboring halogen site, as shown in Fig. 4, unless the Pauli repulsion between X_2 and the neighbboring alkali ions makes the configuration stable. According to them the two center self-trapped excitons of this configuration is unstable and displaced to the configuration shown in Fig. 3(c) at a certain excited state where the Pauli repulsion between the X_2^--molecular ion and the neighboring alkali ion is weakened. Thus here the reaction mode is the translational displacement of the X_2^- molecular ion. The adiabatic potential curve at this excited state is shown schematically in Fig. 4.

Toyozawa (5, 12) has pointed out that the pseudo Jahn-Teller effect due to the accidental degeneracy between the $1s\sigma_g$ and $2p\pi_u$ orbital of the electron composing a two-center self-trapped exciton produces an adiabatic instability in the lowest excited state. Recently Song et al. (13, 14) have shown that, because of this interaction, the two-center self-trapped exciton of the configuration shown in Fig. 3(c) or Fig. 4(a) does not exist and is converted to that of the configuration $(F\text{-}H)_{nn}$, shown in Fig. 3(c), which is the nearest neighboring pair of an F center (a vacancy with an electron) and an H center (an interstitial atom). According to this result, the adiabatic potential curve shown in Fig. 4(c) is for the lowest excited state of the self-trapped excitons. Williams et al. (15) have described several features of defect formation processes in alkali halides according to this interpretation.

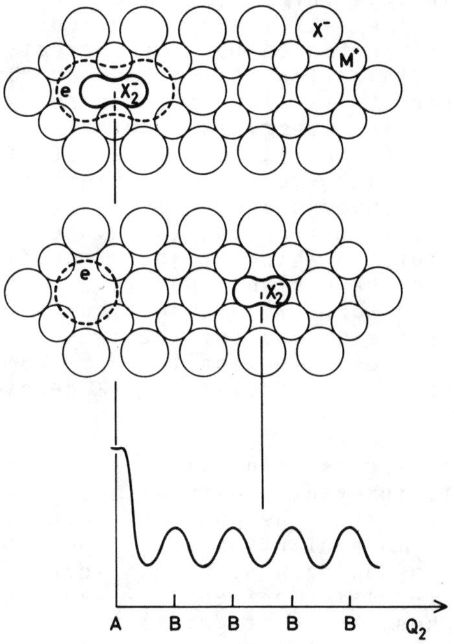

Fig. 4. A schematic diagram showing the adiabatic potential for the translational motion of the X_2^- ion composing a self-trapped exciton in alkali halides. The top curve denotes the two center self-trapped exciton and the middle curve the configuration where the halogen vacancy (the F center) and the interstitial halogen atom (the H center) are separated. The bottom curve shows the adiabatic potential curve.

Irradiation of alkali halides with an electron pulse produces the optical absorption bands due to the F centers and the H centers beside those due to the self-trapped excitons. A substantial fraction of the F band and H band is annihilated within a millisecond. Thus these optical absorption bands have been ascribed to the close pair of an F center and an H center, which has a separation slightly larger than that of $(F\text{-}H)_{nn}$.

The concentration ratio of the close F-H pairs to the self-trapped excitons generated by an electron pulse is close to unity at liquid helium temperature, indicating that the efficiency of the separation of F-H pair is relatively high.* According to the Itoh and Saidoh model the separation takes place at a highly excited state of the self-trapped exciton, while according to Toyozawa's model, it occurs at the lowest state of the self-trapped excitons. In any case, defect formation is according to the local excitation mechanism and it is clear that the configuration of the two-center self-trapped excitons is not stable and hence induces defect production, if it is formed. Similar displacement of the X_2^- molecular ion may occur if the self-trapped hole traps an electron. It has been shown that the self-trapped excitons having configurations of the type of Fig 3(c) are generated in some other materials as described later.

3. Survey of Experimental Results of Photo-Induced Desorption

In this section we survey the experimental results of laser-induced desorption. Because of the close correlation between the laser-induced desorption and desorption induced by photons, low energy (keV) electrons and energetic (MeV) heavy ions, we do not restrict ourselves to the laser-induced desorption but discuss all types of desorption induced by electronic excitation. Further details of the survey of the experimental results are published elsewhere (16).

3.1 Types of electronic excitation

It is convenient to distinguish three types of electronic excitation that may cause desorption depending on materials: (i) low density valence-electron excitation, (ii) core excitation and (iii) high-density valence-electron excitation. The laser-induced desorption belongs to the third type.

The desorption induced by valence electron excitation is characterized by following factors: (i) desorption is induced by low-intensity photon beams near the band-gap energy as verified for alkali halides and silver halides (17, 18) and (ii) the yield of the electron-induced desorption is almost as high as the yield of electronic excitation. The electron induced desorption yield of several materials is listed in Table I. It is clear that the yield for Ne is as high as that for alkali halides and is 71 eV per desorbing atom. This value is only about twice of the average energy needed for an excitation and hence desorption is ascribed to the valence-electron excitation. For other materials listed in Table I, it is not yet clear whether the desorption is induced by valence-electron excitation or by other mechanisms.

Photo-induced or electron-induced desorption of ions from metal oxides or from adsorbates on metals often show a distinct threshold energy near the core threshold (19, 20), indicating that core excitation is responsible for ion ejection. Desorption by core excitation occurs in a variety of adsorbate-metal systems, including ionic (21) and covalent (22) systems. For NaCl, in which excitation of valence-electron produce desorption, Na^+ ions and excited Na atoms start to be desorbed when the energy of incident particles crosses the Na 2p threshold (23). As we discuss later the mechanism of desorption of ground state Na atoms from NaCl and that of excited Na atoms and of Na^+ ions are different.

TABLE 1. Electron-Induced Desorption Yield of Condensed Gas for 2 keV electrons.

Substance	Energy (keV)	Yield (molecules per electron)
KI[a]	5	14.4
Ne[b]	2	28
CO[c]	2	1.75
D_2[d]	2	8
N_2[e]	2	1.2
O_2[e]	2	2.4

a. reference 34
b. J. Shou, P. Borgesen, O. Ellegard and H. Sorensen, to be published.
c. J. Shou, O. Ellegard, P. Borgesen and H. Sorensen, Desorpftion Induced by Electronic Transition, eds. W. Brenig and D. Menzel (Springer Verlag, Berlin, 1985) p.170
d. R. W. Ollerehead, J. Bottiger, J. A. Davies, J. L'ecuyer, H. K. Haugen and N. Matsunami, Radiat. Eff. 49, 203 (1982)
e. O. Ellegard, J. Shou, H. Sorensen and P. Borgesen, Surf. Sci. 167, 474 (1986)

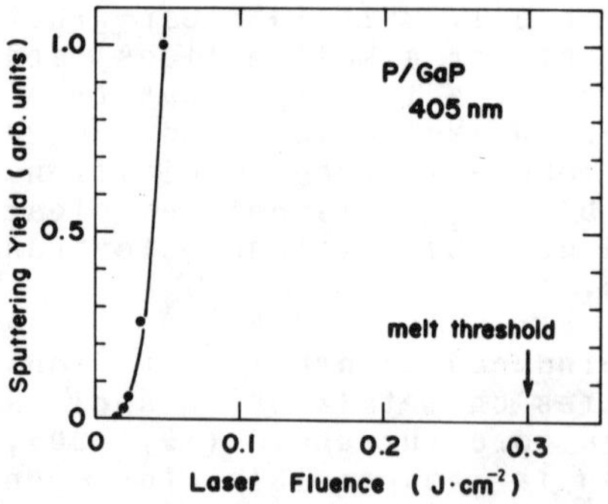

Fig. 5. Relation between laser-induced sputtering yield and laser fluence for P atoms from GaP. The melt threshold is shown by the arrow.

Laser-induced desorption discussed in this paper concerns desorption induced by laser pulses of a fluence much lower than the melt threshold. The most significant feature of the laser-induced desorption is its non-linear response to the fluence (24-26), as shown in Fig. 5. For GaP, it has been shown that desorption is induced by a laser fluence of about 1/10 of the melt threshold (27). Besides GaP laser-induced sputtering has been observed for other compound semiconductors, such as CdSe (24), InP (26), CdS and GaAs (29). Laser-induced desorption of ZnO (25, 30), Al_2O_3 (31-33), MgO, $MgAl_2O_4$ and SiO_2 (32) have been also observed.

3.2 Velocity distribution of ejected particles

Typical velocity distribution of particles ejected by three types of electronic excitation described in 3.1 is compared in Fig. 6, 7 and 8. Fig. 6 and 7 show the energy distribution of Br atoms desorbed from KBr by electron irradiation (34) and of P atoms desorbed from GaP by laser irradiation (25), respectively. Fig. 8 compares the energy distribution of CO atoms at the ground state and at the excited state desorbed from a CO layer on Ru metal (35). Clearly the kinetic energy of atoms at the ground state desorbed by valence-electron excitation is below 1 eV, while that of excited atoms and also ions desorbed by core excitation is distributed above 1 eV. As seen from Fig. 7, laser irradiation produces atoms below 1 eV. Similar results have been observed also for ZnO (30), CdS (36) and GaAs (29).

Mass analysis measurements have shown that most of the component atoms are ejected by laser irradiation, except for experiments on GaP (25), where only P atoms are reported to be desorbed. Since the melting point of metals composing compound semiconductors is relatively low, both components will be ejected even if the primary process imparts kinetic energy to one of the component species, as it is the case for alkali halides. The result that only P is desorbed from GaP at a low laser fluence (25) suggest strongly that the non-metallic atoms are ejected primarily.

3.3 Laser induced surface modification

In addition to laser-induced desorption, studies of laser-induced surface modification and gas-surface reaction of semiconductors and insulators have been carried out extensively. The detection method of surface modification having a sensitivity of a monolayer is LEED. It has been shown that the reconstructed surface of InP is modified by laser irradiation of a fluence near the desorption threshold (37). Studies of luminescence, UPS and Hall effect measurements of CdS (38), GaAs (39), InP (40) and CdTe (41) indicate that metal-enriched surface layers are formed. These results are consistent to the result that non-metallic species is ejected primarily. It has been shown that SiO_2 is reduced to Si (42) by laser irradiation. Laser-induced ablation of Al_2O_3 has been observed with an electron microscope (32, 33) and the result shows no indication of thermal evaporation.

The arguments above indicate that the laser-irradiation produces decomposition in a variety of materials. The desorption is considered to be a result of this decomposition. If impurities are adsorbed on the surface,

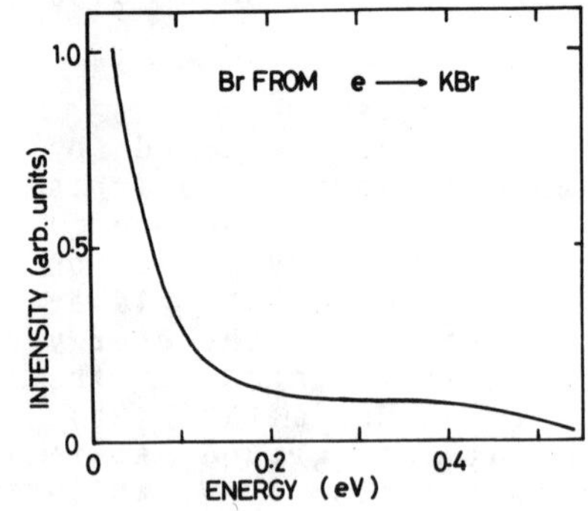

Fig. 6. Energy distribution of Br desorbed from KBr by 540 eV electrons (data taken from reference (34).

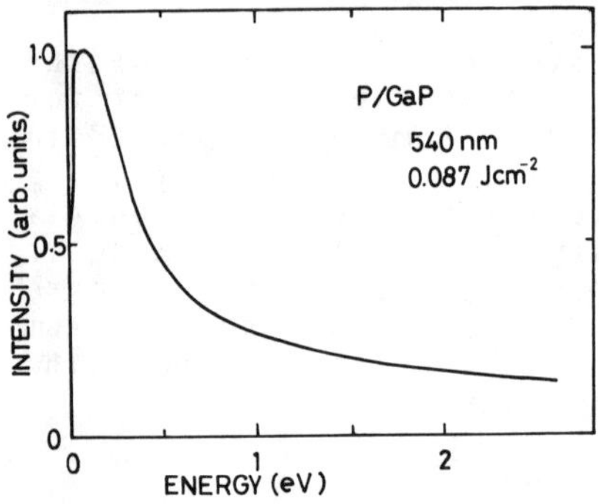

Fig. 7. Energy distribution of P desorbed from GaP by a laser pulse (data taken from reference (25)).

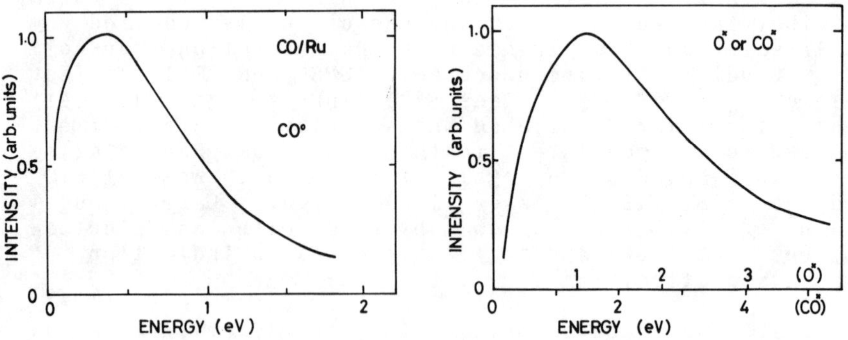

Fig. 8. Energy distribution of (a) CO and (b) excited CO desorbed from CO adsorbate on Ru (001) (reference 35).

reaction between the decomposition products and the adsorbates will take place. The laser-induced oxidation of GaAs (43, 44) is considered to be induced by this mechanism.

4. Mechanism of Laser-Induced Sputtering and Surface Modification

4.1 Localization

It has been argued that the one of the necessary conditions for the electronic excitation energy to be converted to the atomic energy to induced desorption in crystalline solids is the localization of electronic excitation energy. The localization is required for electronic excitation energy to be imparted to a (few) desorbing atom(s) and to form anti-bonding potential curve that breaks the bond of (a) component atom(s). For core excitation localization that satisfies both of above requirements have been established (45): the localization of two holes that induce coulomb repulsion larger than the width of the valence band can have a finite lifetime longer than the inverse of the characteristic lattice frequency. The anti-bonding potential curve for desorbing atoms may be formed in ionic solids because of the coulombic interaction and in covalent solids because of lack of covalent electrons.

The nature of localization for desorption by valence-electron excitation (for 1h1e or 1h2e where h denotes a hole e an electron (42)) is less clear. The Menzel-Gomer-Readhead (MGR) mechanism (46, 47) assumes presence of an anti-bonding potential surface but its nature and its correlation with localization is not clear. A modification of the MGR mechanism by Antoniewicz (48) has the same difficulty. The mechanism of dissociative attachment (22) may be applicable for molecular adsorbates on metals but has difficulty for crystalline solids since the attached electron does not form a localized state, namely no self-trapped electron is stable.

We propose that the localization for valence electron excitation (1h1e) is induced by self-trapping in a wide range of materials. The contribution of the self-trapping has been suggested for a few specific cases (27, 49, 50). Here we suggest that the same concept is applied to many other cases and that adiabatic instability is induced on the site of the self-trapped exciton. Only in the materials, such as alkali halides and silver halides valence-electron excitation at a low density induces desorption. For alkali halides, since self-trapping of excitons occurs in the bulk, it is likely that it occurs at the surfaces, where softer phonon modes are abundant. For silver halides, it has been shown that self-trapped excitons are formed in the bulk of AgCl but not of AgBr, but desorption is induced both in AgCl and in AgBr. We suggest (16) that self-trapping of excitons occur at the surface of both solids in a way similar to

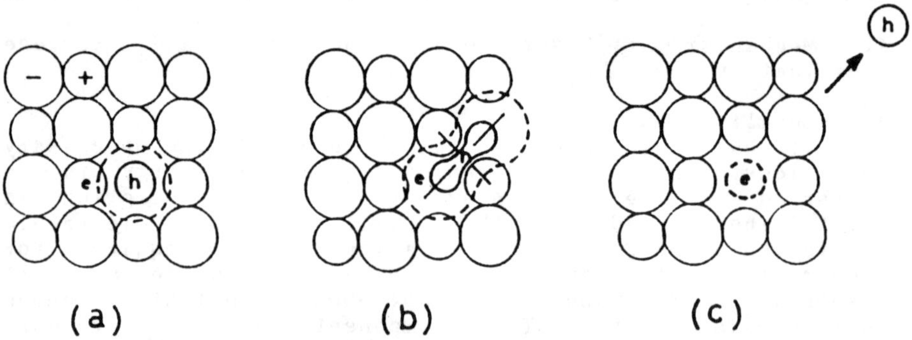

Fig. 9. Model of desorption of halogen atom from alkali halides: (a) free exciton, (b) two-center self-trapped exciton (c) ejection of a halogen atom.

the adiabatic instability described above. In this configuration because of a larger positive Madelung potential at the lattice point in the second layer than in the top surface layer, the X_2^- molecular ion will be polarized so that the electron density in the halogen inside is larger (Fig. 8(b)). Thus the final consequence of the reaction is formation of an F center and emission of an halogen atom (Fig. 9(c)).

The mechanism proposed above is similar to so called H center migration mechanism (54-56), in which it is assumed that the dynamic migration of the H centers is initiated at highly excited states of the self-trapped excitons. According to the recent investigation by Song et al. described above, it is likely that the H center migration is initiated at the lowest excited state of the self-trapped excitons.

So far we have discussed the origin of the anti-bonding adiabatic potential surface in alkali halides. We note that similar repulsive nature of the self-trapped exciton has been observed for monovalent ionic system, such as CaF_2, where it is shown that trapping of an electron by a self-trapped hole (a F_2^- molecular ion occupying two lattice points) displace the F_2^- molecular ion into an interstitial position (57). For rare gas solids it has been shown that the self-trapped exciton in Ne, which has a structure (Ne_2^++e), similar to that of the two-center self-trapped exciton in alkali halides, nucleate a self-trapped exciton bubble (58). The formation of the bubble is considered to be initiated by the translational displacement of Ne_2^+ from the site where the electron is trapped. Thus it appears that the anti-bonding adiabatic potential curve described in Fig. 4 is formed for the self-trapped excitons in the bulk alkali halides because of the difference in the nature of the valence band in the bulk and that at the surfaces.

We now apply the same concept to the laser-induced desorption, namely to the case of dense valence-electron excitation. To consider the localization in this case, it is convenient to divide materials into two categories (51): one in which self-trapping of excitons is induced and the other in which no self-trapping is induced. Since the self-trapped excitons has a relatively long lifetime before they are finally annihilated, the chance that a self-trapped exciton is excited again or another self-trapped exciton is generated in its proximity is high under dense valence-electron excitation. This produces two-hole state (2h2e), which may result also from core excitation. We propose that surface processes for materials, in which excitons are self-trapped, such as SiO_2 arise from this type of localization. For materials in which no self-trapping is induced, a new mechanism for localization under high density electron-hole plasma near the surfaces should be invoked. Itoh and Nakayama (51, 52) suggested that under dense electronic excitation two holes are attractive by virtue of lattice relaxation near the surfaces. Such localization is a manifestation of the Anderson negative U (53) and may be possible under dense electronic excitation, where coulomb repulsion is screened.

4. 2 Origin of adiabatic instability

In 4.1 we have argued that the self-trapping is a step for desorption. The next crucial problem is existence of the anti-bonding potential curve that ejects an atom from the lattice point near the surface out of the surface.

As described in 2, the two-center type self-trapped excitons in alkali halides has an anti-bonding adiabatic potential curve or adiabatic instability. In the bulk, a pair consisting of an F center and an H center is evolved from a self-trapped exciton through a series of replacement of the type:

$$X_2^-(A) + X^-(B) \rightarrow X^-(A) + X_2^-(B), \qquad (1)$$

where A and B denote neighboring lattice points. The replacement sequence (1) is the mechanism known as the thermal migration of the interstitial halogen atom. For the defect formation process, the replacement sequence (1) may be induced by the kinetic energy imparted to the molecular ion due to the anti-bonding adiabatic potential curve.

In the following arguments we assume that a self-trapped exciton is produced also near the surface. Alternatively, we may assume that a self trapped hole is formed near the surface which subsequently traps an electron, as shown in Fig. 9(a). In this configuration the X_2^- molecular ion is repelled towards the surface because of of Ne. We expect that in this type of materials the sputtering yield is relatively high as indeed observed in alkali halides and Ne.

For materials in which the self-trapped excitons near the surfaces are mostly subjected to radiative or non-radiative recombination, without forming anti-bonding adiabatic potential surface to induce desorption, the desorption yield induced by valence-electronic excitation at a low excitation density should be essentially zero. Even in these materials, we suggest that the sputtering is induced if the excitation density is high enough to cause excitation of a valence electron associated with the self-trapped exciton. The probability of such re-excitation of the self-trapped excitons under laser irradiation may be higher if the lifetime of the self-trapped excitons is longer. The lifetime is not the crucial factor for high density electronic excitation in an ion track. This suggestion is that the anti-bonding adiabatic potential surface is formed after the re-excitation even if it is not formed for the self-trapped exciton. This is plausible since the number of electrons contributing to the bond in the vicinity of the self-trapped exciton is reduced. The same type of anti-bonding adiabatic potential surface may be formed when two hole self-trapping is induced by the negative-U interaction (52).

We remark on the difference between the anti-bonding potential surface formed after core excitation through an Auger process and that formed by the self-trapping of excitons. In the former case the localization is a non-adiabatic process formed within 10^{-15} s while the lattice is rigid. In this case particles may be ejected before the two-hole localized state couples with the lattice or before the lattice around the two holes is deformed to minimize the total energy. Therefore ejected atoms will acquire relatively a large kinetic energy, as indeed observed. On the other hand the self-trapping itself is an adiabatic process and hence particle emission from the self-trapped site would follow an adiabatic potential surface against the reaction mode. Since the reaction mode is not necessarily a normal mode, the electronic potential energy will be distributed into a variety of phonon modes. Such a reaction is called unimolecular reaction and may be treated using the RRKM theory (59). The calculated velocity distribution (52, 60) is significantly smaller than that for core excitation and is in the range as that observed for valence-electron excitation.

Another important feature of the self-trapping mechanism suggested in the present paper is that the kinetic energy is imparted in one of the sublattices. The selection comes from the nature of the orbital composing the valence band. For example, as in alkali halides, if the top of the halogen orbital is composed of the halogen orbital, self-trapping of holes and hence of excitons distorts mainly the halogen sublattice. Thus the electronic excitation energy is converted to the kinetic energy of the halogen sublattice, ejecting a halogen atom from the lattice point and eventually leading to decomposition. If there are adsorbates on the surface, the dissociated atoms may react with the adsorbate, causing a gas-surface reaction.

5. Conclusion

In this paper we emphasized that the two crucial factors for desorption: localization and adiabatic instability. It is suggested that localization for valence-electron excitation is originated from self-trapping of excitons. The adiabatic instability of the self-trapped excitons in alkali halides and Ne have been well established. In materials, in which excitons are self-trapped but the self-trapped excitons have no adiabatic instability, we suggest that the adiabatic instability for reaction is introduced if a valence electron associated with the self-trapped excitons is re-excited. The re-excitation of the self-trapped excitons results in formation of two-holes close to each other. Therefore the bond is weakened locally and consequently an anti-bonding adiabatic potential curve on the reaction coordinate is formed.

The model presented in this paper is on the basis of the electron-lattice coupling in solids. Thus the desorption is considered to be induced under a strong coupling with phonons contrary to the desorption induced by core excitation. The general difference in the velocity distribution of particles emitted by core excitation and valence-electron excitation may be attributed to this difference.

According to the model described in this paper, the laser-induced desorption in materials in which excitons are self-trapped takes place by a mechanism similar to the desorption or defect formation in alkali halides. Further details of the possible desorption processes in specific materials are described separately (16). The model is now considered to be highly plausible, since the self-trapped excitons have been recognized to be formed in several solids, such as SiO_2 (61, 62) and Al_2O_3 (63). For materials in which excitons are not self-trapped, localization other than self-trapping has to be invoked. The negative-U interaction may be a plausible process. Further theoretical works are needed to elucidate this interesting new problem.

REFERENCES

1. See for example, N. Itoh, Adv. Phys. 31, 492 (1982) and N. Itoh and K. Tanimura, Radiat. Eff. to be published and R. T. Williams, Semiconductors and Insulators 30, 251 (1978)

2. R. Engelman, Non-Radiative Decay of Ions and Molecules in Solids (North-Holland, Amsterdam, 1979)

3. A. M. Stoneham, Rep. Prog. Phys. 44, 1251 (1981)

4. E. I. Rashba, Excitons, eds. E. I Rashba and M. D. Sturge (North-Holland, Amsterdam, 1982) p.543

5. Y. Toyosawa, Vacuum Ultraviolet Radiation Physics, eds. E. E. Koch, R. Haensel and C. Kuns (Pergamon Vieweg, Braunschweig, 1974) p.317

6. Y. Toyozaya, Physica B116, 7 (1983)

7. H. Sumi, J. Phys. Soc. Jpn. 53, 3498 (1984)

8. H. Sumi, J. Phys. Soc. Jpn. 53, 3512 (1984)

9. K. S. Song, J. Phys. Soc. Jpn. 26, 1131 (1969)

10. T. Eshita, K. Tanimura, N. Itoh and H. Nishimura, J. Phys. Soc. Jpn. 54, 4418 (1985)

11. N. Itoh and M. Saidoh, J. Phys. (France) 34, C-9, 101 (1973)

12. Y. Toyozawa, J. Phys. Soc. Jpn. 44, 482 (1978)

13. G. Brunet, C. H. Leung and K. S. Song, Solid State Commun. 53, 607 (1985)

14. C. H. Leung, G. Brunet, K. S. Song, J. Phys. C. 18, 4459 (1985)

15. R. T. Williams, K. S. Song, W. L. Faust and C. H. Leung, Phys. Rev. to be published

16. N. Itoh, Proceedings of the 6 th International Workshop on Ion-Surface Collisions; to be published in Nucl. Instrum. and Methods

17. H. Kanzaki and T. Mori, Semiconductors and Insulators 5, 401 (1983)

18. H. Kanzaki and T. Mori, Phys. Rev. B. 29, 3573 (1984)

19. M. L. Knotek, Desorption Induced by Electronic Transitions, eds. N. H. Tolk, M. M. Traum, J. C. Tully and T. E. Madey (Springer Verlag, Berlin, 1983) p.139

20. D. Menzel, Nucl. Instrum. and Methods B13, 507 (1986)

21. M. L. Knotek and P. J. Feibelman, Phys. Rev. Lett. 40, 964 (1978)

22. D. E. Ramaker, Desorption Induced by Electronic Transitions, eds. W. Brenig and D. Menzel (Springer Verlag, Berlin, 1985) p.10

23. N. H. Tolk, M. M. Traum, J. S. Kraus, T. R. Pian and W. E. Collins, Phys. Rev. Lett. 49, 812 (1982)

24. T. Nakayama, H. Ichikawa and N. Itoh, Surf. Sci. 123, L693 (1982)

25. N. Nakayama, Surf. Sci. 133, 101 (1983)

26. J. M. Moison and M. Bensoussan, J. Vac. Sci. Technol. 21, 315 (1982)

27. T. Nakayama and N. Itoh, Desorption Induced by Electronic Transition eds. W. Brenig and D. Menzel (Springer Verlag, Berlin, 1985) p.237

28. A. Namiki, K. Watabe, H. Fukano, S. Nishigaki and T. Noda, Surf. Sci. 128, L243 (1983)

29. A. Namiki, T. Kawai, Y. Yasuda and T. Nakamura, Jpn. J. Appl. Phys. 24, 270 (1985)

30. T. Nakayama, N. Itoh, T. Kawai, K. Hashimoto and T. Sakata, Radiat. Eff. Lett. 67, 129 (1982)

31. R. W. Dreyfus, R. E. Walkup and R. Kelly, Radiat. Eff. to be published

32. R. Kelly, J. J. Coumò, P. A. Leary, J. E. Rothenberg, B. E. Braren, C. F. Aliotta, Nucl. Instrum. and Methods B9 329 (1985)

33. J. E. Rotherberg, R. Kelly, Nucl. Instrum. and Methods B1, 291 (1984)

34. H. Overeijinder, M. Szymonski, A. Haring and A. E. de Vries, Radiat. Eff. 36, 63 (1978)

35. P. Feuler, Desorption Induced by Electronic Transitions, eds. W. Brenig and D. Menzel (Springer Verlag, Berlin, 1985) p.142

36. A. Namiki, H. Fukano, T. Kawai. Y. Yasuda and T. Nakamura, J. Phys. Soc. Jpn. 54, 3162 (1985)

37. J. M. Moison and M. Bensoussan, Surf. Sci. 126, 294 (1983)

38. M. S. Brodin, N. A. Davydova and I. Yu. Shablii, Phys. Statua. Solidi. B115, 641 (1983)

39. W. Wesch, E. Wendler, G. Gotz, K. Unger, H. Roppischer and Chr. Resagk, Phys. Statua. Solidi. B130, 539 (1985)

40. J. D. Obestar and B. G. Streetman, Surf. Sci. 108, L479 (1981)

41. G. P. Yablonski, Sov. Phys.-Solid. 26, 607 (1984)

42. C. Fiori, R. A. B. Devine, Phys. Rev. Lett. 52 2081 (1984)

43. C. Cohen, Siejka, M. Berti, A. V. Drigo, G. G. Bentini, D. Pribat and E. Jannitti, J. Appl. Phys. 55, 4081 (1984)

44. W. G. Petro, I. Hino, S. Eglash, I. Lindau, C. Y. Su and W. E. Spicer, J. Vac. Sci. Technol. B1, 469 (1983)

45. D. E. Ramaker, J. Vac. Sci. Technol. A1 1137 (1983)

46. D. Menzel and R. Gomer, J. Chem. Phys. 41, 3311 (1964)

47. P. Readhead, Can. J. Phys. 42, 886 (1964)

48. P. Antoniewicz, Phys. Rev B21, 3811 (1980)

49. R. E. Johnson, W. L. Brown, Nucl. Instrum. and Methods 209/210, 469 (1983)

50. M. Wautelet, Surface. Sci. 133, L437 (1983)

51. N. Itoh and T. Nakayama, Nucl. Instrum. and Methods B13, 550 (1986)

52. N. Itoh and T. Nakayama, Phys. Lett. A92, 471 (1982)

53. P. W. Anderson, Phys. Rev. Lett. 34, 953 (1975)

54. N. Itoh, Nucl. Instrum. and Methods 132, 201 (1976)

55. P. C. Townsend, Sputtering by Particle Bombardment vol II ed. R. Behrisch (Springer Verlag, Berlin, 1983) p.147

56. M. Szymonski, Proceeding of the Symposium on Sputtering eds. P. Varga, G. Betz and F. P. Viehbock (Perchtoldsdorf, Vienna, 1980) p. 761

57. R. T. Williams, M. N. Kapler, W. Hayes and J. P. Stott, Phys. Rev. B15, 725 (1977)

58. M. Ueta, H. Kanzaki, K. Kobayashi, Y. Toyozawa and E. Hanamura, Excitonic Processes in Solids (Springer Verlag, Berlin, (1986)

59. K. J. Laidler, Theories of Chemical Reaction Rates (Robert E. Krieger, Huntington, 1979)

60. A. R. Ziv, B. V. King, S. H. Lin and I. S. T. Trong, Nucl. Instrum. and Methods 218, 742 (1983)

61. K. Tanimura, T. Tanaka and N. Itoh, Phys. Rev. Letters. 51, 423 (1983)

62. W. Hayes, M. J. Kane, O. Salminen, R. L. Wood and S. P. Doherry, J. Phys. C. 17, 2943 (1984)

63. C. Itoh, T. Tanimura and N. Itoh, J. Phys. C to be published

GAS-SURFACE INTERACTIONS STIMULATED BY LASER RADIATION: BASES AND APPLICATIONS

T. J. Chuang*
IBM Almaden Research Center
San Jose, California 95120-6099, U.S.A.

1. INTRODUCTION

Lasers have been extensively used in recent years to induce or enhance chemical interactions at gas-solid interfaces. The rapid expansion of the field which can be generally referred as laser-induced gas-surface chemistry is driven by both the basic scientific interest and the potential technological applications in material processing and fabrication of new materials. In the basic area, there are fundamental questions concerning energy transformation from photons into chemical energies, the kinetics and dynamic processes involving energy acquisition, storage and decay that need to be understood. Because of their brightness and narrow bandwidths, and because they can be generated as ultrashort pulses, laser beams are particularly suitable for such studies. For instance, a small set of atoms or molecules can be selectively excited by a short light pulse and the dynamic interactions between the excited species and surface atoms including the change and redistribution of the translational and internal energies of the system can be followed by the time-resolved optical spectroscopy or mass spectrometry. The utilization of lasers in conjunction with other surface analytical methods has led to tremendous progress in our understanding of surface structures and dynamics. In the applied area, related in particular to material processing, the laser photochemical techniques have shown great promises in producing well-defined microstructures for electronic devices. Other potential applications such as photocatalytic reactions and energy conversions will undoubtedly be further developed in the future. These basic and applied aspects of the laser surface chemistry have been a subject of a few reviews appeared recently /1-3/. Here, a brief review is presented on the important surface processes stimulated by laser radiation, with particular emphases on recent studies on laser-stimulated desorption, photochemical etching, photodeposition and molecular dynamics on surfaces. For more general considerations and thorough discussions, one should refer to prior reviews /1-3/.

2. BASIC LASER-STIMULATED PROCESSES

A molecule-surface reaction usually involves adsorption, dissociative chemisorption, adsorbate-adsorbate and adsorbate-solid reaction, product formation and desorption of surface species. Photon beams can influence and in many cases enhance these processes via photoexcitation of gaseous species, the adsorbates and/or solid atoms. The modes of excitation include the vibrational and electronic states of the gaseous species and the adsorbate-surface complexes, and the electronic as well as the thermal activation of the substrate. Both the single and the multiple photon absorption may be involved in the excitation process. A noted example to demonstrate the photoenhanced chemisorption and reaction due to vibrational excitation is the interaction of SF_6 with silicon /4/. It was shown that although the molecules were inert to Si at 25°C, they could be chemically activated by multiple photon absorption of CO_2 laser photons either in the gas phase or in the adsorbed state. A systematic study was performed including the determination of reaction yields as a function of the laser wavelength, the laser intensity and the gas pressure as well as the reaction products. Reactive radicals generated by multiple infrared photon dissociation to react with surfaces resulted in chemical etching of solids were also observed on Si /4/ and SiO_2 /5/. IR multiple photon excitation was also used to decompose SiH_4 molecules for deposition of Si on quartz and glass substrates /6/.

Likewise, photodissociation to produce radicals can be accomplished by single photon absorption in the ultraviolet (UV) and visible spectral regions. The radicals can often react with surfaces which are inert to the parent molecules. Many surface etching and vapor deposition reactions are activated by such a photochemical process. The prominent examples are Si/Cl_2 /7/ and SiO_2/Cl_2 /8/ surface etching reactions enhanced by an Ar^+ laser, and metal deposition on quartz substrates by UV photolysis of $Al(CH_3)_3$ and $Cd(CH_3)_2$ /9/. Electronic excitation of solid substrates can also effectively enhance the surface reaction, most notably the band gap excitation of oxide and semiconductor surfaces. Photoenhanced oxidation of Si /10/ and GaAs /11/, and etching of Si by XeF_2 /12/ are noted examples. Electronic excitation effects other than the thermal effects on metal surfaces have not been so clearly elucidated because of the strong interaction between the adsorbed species and metal atoms, and the rapid electronic decay processes. There are, however, increasing evidences that nonthermal photochemical processes play important roles in certain gas-solid systems where the solids absorb the laser photons.

An important effect of laser irradiation on solid surfaces involves the desorption of adsorbed species, namely, the laser-stimulated desorption. This can be accomplished by means of surface heating or resonant vibrational and electronic excitation of the adsorbate-surface complexes. The subject has been reviewed recently /13/. Briefly, in laser-induced thermal desorption (LITD), surface heating can result from the direct lattice phonon excitation or the initial substrate electronic excitation, e.g. via the electron hole pairs, which subsequently decays to phonon bath before desorption occurs. The unique features of LITD are the rapid and the large range of heating rates allowing one to examine the desorption kinetics over a wide dynamic range. In addition, operation with short laser pulses can facilitate time- resolved studies and provide a means to desorb surface intermediate species not detectable by other methods. Indeed, LITD has been successfully employed to study the desorption kinetics of CO from a Fe(100) surface /14/ and the surface diffusion of CO adsorbed on Cu(100) /15/ and H_2 adsorbed on Ni(100) /16/. It was used to measure desorption energies of H_2 and CO from stainless steel surfaces /17/. The technique in conjunction with short laser pulses was also applied to determine the reaction intermediates present on surfaces. For example, the determination of the SiF_x $(x < 4)$ products in the Si etching reaction with XeF_2 /18/ and the product of methanol decomposition on Ni surfaces /19/ was obtained with the LITD method.

In resonantly stimulated photodesorption, one can use IR lasers to excite surface vibrational modes (IRPD) or UV-visible lasers to promote electronic excitation (UVPD). The phenomenon of IRPD has been investigated under UHV conditions by Heidberg, Chuang and their coworkers /20,21/. Some experimental results for NH_3 and ND_3 molecules adsorbed on a Cu(100) surface /22,23/ are shown in Figs. 1 and 2. In this system, the high frequency N-H and N-D stretching modes were excited with a tunable pulsed IR laser in the 2.5-4.2 um region, while the low frequency bending mode was excited with a pulsed CO_2 laser /24/. The systems were characterized with XPS, both the conventional and the laser-induced thermal desorption, and in the case of C_5H_5N/Ag film /21/, also with the surface-enhanced Raman scattering. Time-of-flight (TOF) mass spectrometry was employed in ammonia photodesorption studies. It is observed that the IRPD quantum yields are generally very low and the desorption rates are determined by the IR absorption cross sections. There are, however, no clear vibrational mode enhancement or isotope selectivity in photodesorption yields /22,23/. In a recent theoretical calculation performed for the NH_3/Cu(100) system by Hussla et al. /23/ involving phonon-mediated and electron hole pair damping mechanisms, a reasonably good agreement between the calculated desorption yield as a function of laser fluence with the experimentally measured yields was obtained when a reasonable set of parameters including the energy damping rates in the 10^{11} /sec range were used. The results of this calculation are also shown in Fig. 2. The same calsulation, however, showed that the resultant desorption temperature (T_d)could reach 200 K or higher in contradiction with the observed T_d = 90 K. Clearly, if the local temperatures due to the energy decay from the excited vibrational modes were so high, then the desorption yields due to the thermal effects should be much higher than those observed experimentally. This discrepancy had led the authors to suggest /22,23/ that in IRPD, the desorbed molecules might still be internally excited when they desorbed into the gas phase. Other theoretical treatments concerning IRPD include most noteably the work by George /25/, Gortel and Kreuzer /26/, Jedrzejek /27/, Janda /28/, Lin /29/ and their coworkers.

Photodesorption studies involving electonic excitaion of adsorbate-surface complexes by UV and visible lasers on well-characterized surfaces are still relatively few. In a recent

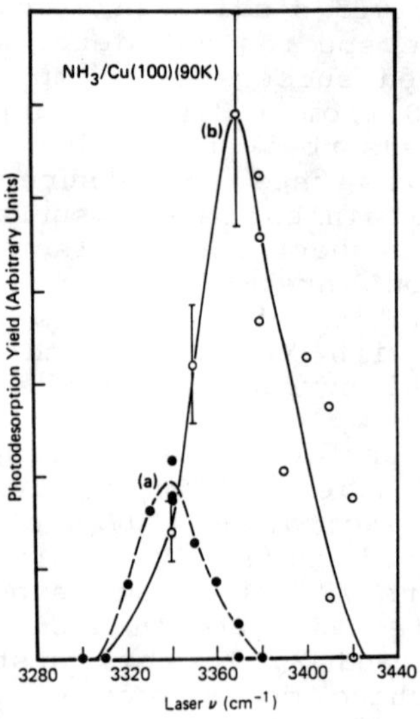

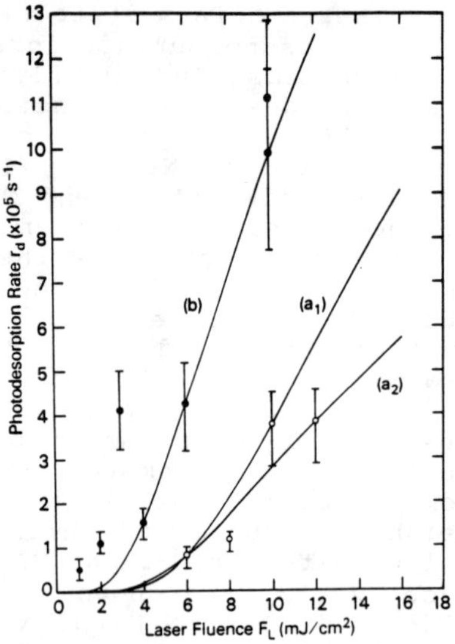

Fig. 1 IR photodesorption yields of NH_3 on Cu(100) at 90 K as a function of laser frequency: (a) surface coverage about a monolayer, $\theta = 1$ and laser intensity, $I = 10\ mJ/cm^2$; (b) multilayer coverage, $\theta = 3.4$ and $I = 10\ mJ/cm^2$. The laser is p-polarized and at 75° angle of incidence. Each data point is an average of mass spectrometer signals due to 20 laser pulses, (data according to Ref. /22/).

Fig. 2 IR photodesorption yields of NH_3/Cu(100) at 90 K as a function of laser fluence: (a) surface coverage $\theta = 1$, $\nu = 3340\ cm^{-1}$; (b) $\theta = 3.4$, $\nu = 3370\ cm^{-1}$, sampling average of 20 laser pulses. The dots and circles are experimental data and the solid curves are results of a model calculation according to Ref. /23/ based on the phonon and electronic damping mechanisms.

experiment by Bourdon et al. /30/, it was observed that CH_3Br molecules adsorbed on a LiF crystal could be photodesorbed by an excimer laser at 222 nm or 308 nm. The result was interpreted in terms of the UV absorption by the color centers of the substrate causing a shock wave to pass through the crystal and thereby inducing the ejection of the adsorbed molecules. It was further found that adsorbed CH_3Br could be photodissociated by the single-photon absorption of the 222 nm light to produce CH_3 radicals which subsequently desorbed from the substrate. In contrast to the photodesorbed CH_3Br molecules which possessed rather low kinetic energies, the photogenerated CH_3 radicals desorbd with relatively high velocities. Apparently, in the antibonding state reached by the photoexcitation part of the electronic energy after breaking of the chemical bond was transformed into the translational energy of the desorbing radicals. Studies of the CH_3 radical yields as a function of the 222 nm laser intensity and the surface coverage confirmed the photochemical effects on the LiF surface. Similar electronically excited photofragmentation and desorption have also been observed in CH_2I_2-Al_2O_3 system /31/. Photodesorption of neutrals and ions from surfaces of molecular solids or molecular films deposited on metal substrates irradiated by UV or visible lasers were investigated /32,33/. Quite often, nonthermal photodesorption characteristics were observed. In addition, there were reports of CO_2 and H_2O desorption from oxide semiconductors induced by band gap excitation with the UV-visible light /34/.

3. DYNAMIC PROCESSES

On a solid surface, a given chemical reaction of interest initiated by either the molecular or the solid absorption of photons must compete not only with the energy transfer between different molecules and among different degrees of freedom, but also with other surface processes such as desorption and energy relaxation to lattice phonons. Thus, in order to understand the surface chemical kinetics, we have to know the various energy transfer and decay rates. The electronic energy relaxation process was studied by Whitmore et al. /35/, who used a UV laser to excite pyrazine on top of an Ag(111) surface and monitored the optical emission. The phosphorescence lifetime of a triplet state of the molecule was measured as a function of the molecule-metal separation with Ar as a spacer between 10 Å and 400 Å. It was observed that the triplet lifetime decreased monotonically with the dipole-surface distance in quantitative agreement with the classical electromagnetic theory based on the mechanism of electronic energy transfer to surface plasmons. With this approach, the electronic lifetime of the molecule in direct contact with the metal could not be determined because of the diminished emission intensity. A siminlar method was adopted by Wokaun et al. /36/ to study the electronic energy transfer from a dye molecule to Ag island films separated with a thin SiO_x layer. A non-monotonic dependence of the luminescence intensity on the thickness of the spacer layer was observed due to two competing processes: The electromagnetic enhancement of absorption and emission versus the radiationless energy damping to the metal. But, no information for the dye molecules directly adsorbed on Ag could be obtained. The fluorescence properties of a dye molecule adsorbed on glass surfaces and Ag islands were also investigated by Leitner et al. /37/ with picosecond laser technique. It was found that on glass surfaces, the fluorescence intensity and the lifetime were reduced if the intermolecular distance was decreased to a critical point. This was essentially a concentration quenching effect due to dimerization often observed in the liquid phase. On the Ag islands, however, the fluorescence lifetime was drastically shortened due to electromagnetic interaction between the dye molecules and plasmons in the metal particles, modified by a damping influence of dye dimers.

Vibrational deactivation due to gaseous collisions with solid surfaces was investigated by Misewich et al. /38/ using an IR laser excited fluorescence quenching technique. It was observed that the vibrational deactivation probabilities per surface collision were very high, e.g. 0.16, 0.20 and 0.22 for an excited $CO_2(001)$ colliding with a silver, a nickel and a stainless steel surface, respectively. Namely, it required only 5 gas-surface collisions to damp the excited state. Similar high deactivation efficiency was found for excited CO ($v = 1$) and $CO_2(101)$ molecules colliding with an Ag film. Surface trapping followed by electron-hole pair formation was proposed to be the dominant mechanism for the effective decay of the vibrational energy. In a separate study /39/, an IR laser was used to excite a NO molecular beam before it was scattered from a LiF surface. The NO molecules were excited into $v = 1$ and $J = 3/2$ state, direct-inelastically scattered from the surface and its internal state distribution was analyzed by resonantly enhanced two-photon ionization. It was found that the vibrational survival probability for the scattered NO ($v = 1$) molecules could be very high and highly rotationally excited states could be produced in the surface scattering process. The experimental finding was in agreement with the result of a stochastic trajectory calculation performed by Lucchese and Tully /40/ on the same system. It was shown that when trapping of NO on the LiF surface was negligible, vibrational energy accommodation coefficient to the solid surface should be less than 1 %. Furthermore, the NO vibrational mode was most strongly coupled to the surface phonons with a much weaker coupling to the rotational and translational modes.

A direct measurement of vibrational energy relaxation rate for molecules adsorbed on dielectric surfaces was carried out by Heilweil et al. /41/ using the picosecond IR laser spectroscopic method. Here a picosecond IR pulse was used to excite some of OH groups adsorbed on silica particles, a probe pulse after some delay was employed to determine the time required for the system to return to its original state. The relaxation time for the excited O-H stretching vibration near 3000 cm^{-1} was about 204 psec for the system in vacuum. It was suggested that the relative slow relaxation rate was controlled by the vibrational coupling to lower frequency modes of the SiO_2-OH system. With large differences in vibrational frequencies, such vibrational to vibrational (V-V) energy transfer rates were expected to be slow. Further experiments showed the comparable relaxation rates for the OD groups, but a substantially shorter relaxation time for the surface NH_2 group /42/. These results clearly indicated that the observed IR line width of 8 cm^{-1} for the OH stretching mode was not due to lifetime broadening but rather due to the vibrational dephasing or inhomogeneous broadening or both. Naturally, the relative long vibrational relaxation time has important implications for surface photochemistry.

4. PHOTOCHEMICAL ETCHING AND VAPOR DEPOSITION

A surface etching reaction can be enhanced by photoexcitation in gas phase and/or in the adsorbed phase. For practical purposes, the most effective etching reactions are achieved by direct photon irradiation on solid surfaces exposed to reactive gases. The subject has been reviewed recently /2,18,43/ and a brief summary of recent studies is given in Table 1. Mechanistically, the characteristics of the surface etching reactions can be divided into three classes, namely, the spontaneous etching, the passive and the diffusive reaction systems. In a spontaneous etching system, the reaction can take place in the absence of radiation and a photon beam can enhance the reaction rate. In a passive reaction system, the reaction results in the formation of a thin passivated layer and in a diffusive reaction, the reactant not only can be adsorbed on the surface but also can penetrate quite deeply into the bulk of the solid. In these two classes of reactions, surface etching does not occur without thermal activation or external radiation.

An example for the spontaneous etching system is silicon exposed to XeF_2. Its reaction rate can be greatly enhanced by visible and IR lasers /2,12,18,44/. A short visible light pulse can induce desorption of surface intermediate species such as SiF and SiF_2. Time-of-flight (TOF) measurements showed that the translational temperatures of the photodesorbed SiF_x ($x \leq 3$) could be substantially lower than the calculated surface temperature raised by the laser pulse /18/. Figure 3 shows the TOF signals when a fluorinated Si surface was irradiated by 532 nm light pulses at the fluence F = 0.05 J/cm^2 (6 nsec pulse width). The translational temperatures of the photodesorbed SiF_x species were in the 400 - 560 K range. If the laser fluence was raised above 0.1 J/cm^2, plasma formation giving rise to a characteristic optical emission from the surface and desorption of Si^+ and SiF^+ ions were detected. In this case, the kinetic energies of the desorbed particles were very high ($\geq$ 0.4 eV). In any event, less F-coordinated species and even Si^- atoms were detected to be the dominant desorption species when excited by a UV or visible laser pulse. This is in great contrast to the dark (thermal) reaction /45/ and the cw Ar^+ laser enhanced reaction /12/, in which SiF_4 was found to be the major reaction product. A field-assisted F^- anion penetration into the subsurface region of Si was suggested /46/ to be the mechanism for Si-fluorine reaction. In analogy, a field-enhanced effect could also be important in the photon-enhanced etching reaction, although a mechanism involving the

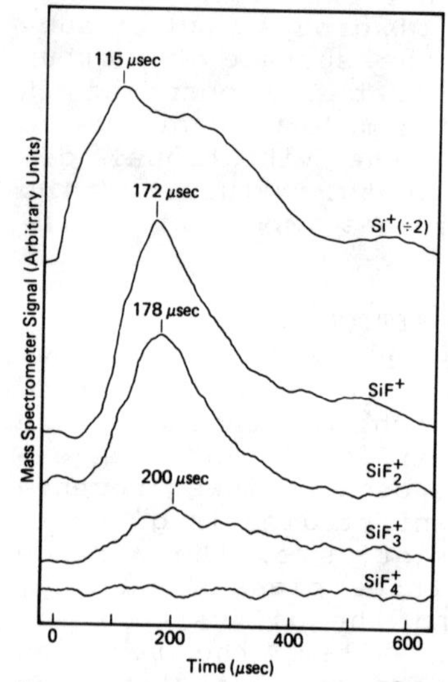

Fig. 3 Time-of-flight signals of SiF_x species detected by the mass spectrometer from a Si surface exposed to XeF_2 at 300 K, irradiated by 532 nm laser pulses at $I = 9\ MW/cm^2$ ($0.06\ J/cm^2$), (data according to Ref. /18/).

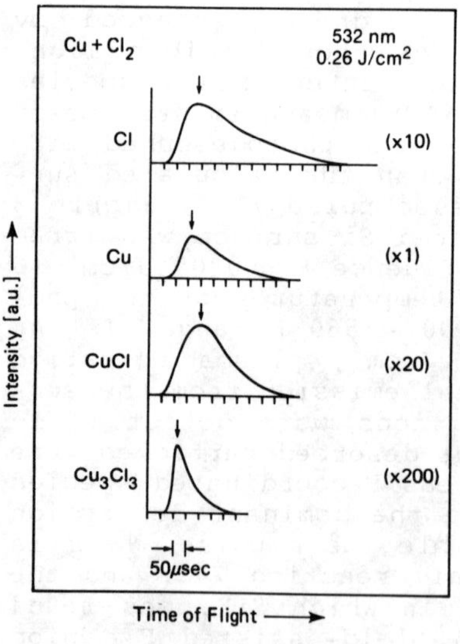

Fig. 4 Time-of-flight mass spectrometric signals of laser desorbed atoms and molecules from a Cu surface exposed to Cl_2 with 1500 L of exposure and laser radiation at 532 nm and $0.26\ J/cm^2$, (data according to Ref. /50/).

direct participation of electrons and holes generated by the photon irradiation was also proposed /12/.

Unlike XeF_2, Cl_2 does not etch a pure Si spontaneously at 25°C in the absence of radiation even though Cl_2 can dissociatively chemisorb on Si. The reaction rate decreases drastically after the formation of a thin chemisorption layer. It is thus a passive reaction system. Irradiation by UV lasers on silicon can induce efficient etching of the solid /7,47,48/. It was further found /47/ that the n-type Si could even be spontaneously etched by Cl atoms generated by Cl_2 photolysis. Also, the photoinduced etch rate on the n-type Si decreased with increasing sheet resistance, whereas the p-type material showed the opposite effect. The results were interpreted also in terms of the field-assisted mechanism /46/. Namely, the reaction rate could increase with the electron density in the conduction band which led to the formation of Cl^- anions. The anions could subsequently penetrate into the subsurface region of Si and thereby facilitate the formation of etching products. This lattice penetration was assisted by the surface electric field established across the chlorinated surface layer.

The diffusive reaction can be best illustrated in the reaction of Ag and Cu with Cl_2 /48-50/. In these systems, Cl^- anions penetrate quite deeply into the bulk of the metals. Chlorine penetration can weaken the metal-metal bonds even at relatively low Cl_2 exposures, and can thus facilitate the efficient etching of the metal by UV or visible laser pulses. When a Cu surface containing some chlorine, e.g. with the average surface Cl concentration in the form of $CuCl_x$ /49/ of about 20 % (x = 0.2), was irradiated by a UV or visible light pulse, desorption of Cu_3Cl_3, Cu_2Cl_3, Cu_2Cl, etc. molecular species and Cu atoms, in addition to some Cl were detected /50/. Typical TOF spectra for Cu exposed to Cl_2 with 1500 Langmuirs of exposure obtained by Sesselmann et al. /49/ with 532 nm laser pulses are shown in Fig. 4. The product mass distributions induced by both 532 nm and 355 nm pulses are dislayed in Fig. 5. The velocities of the photodesorbed particles were generally very high and could not be fitted into Boltzmann distributions. The mass distribution of desorbed species also differed drastically from that obtained by the thermal evaporation of a solid CuCl. In fact, laser irradiation on a solid CuCl produced a very different time-of-flight spectrum as shown in Fig. 6. Furthermore, plasma formation and ion emission were always detected along with the ejection of neutral particles from the laser-irradiated solid CuCl /50/. Apparently for a solid CuCl, photon irradiation can promote "photon-sputtering" involving the formation of a dense electron-hole plasma. As suggested by Itoh and Nakayama /51/ in other systems, multiple-hole states near the surface could be generated in a dense e-hole plasma. The nonradiative e-hole recombinations could energize surface atoms and induce desorption of particles with kinetic energies much higher than that expected from the surface temperature. Overall, however, the photochemical etching behavior of the Cu/Cl_2 system is very different from the photoetching or photoablation of a solid

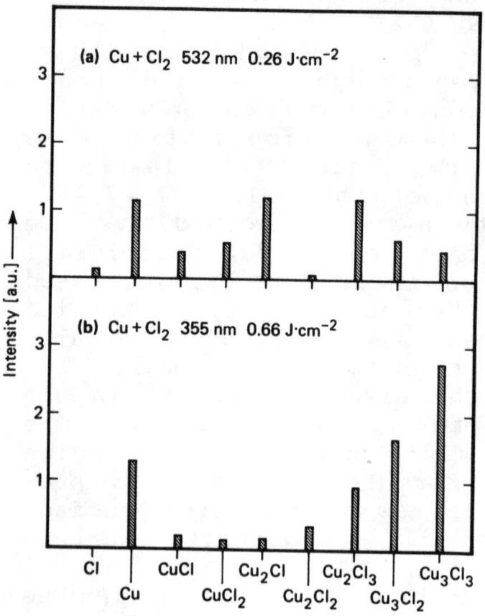

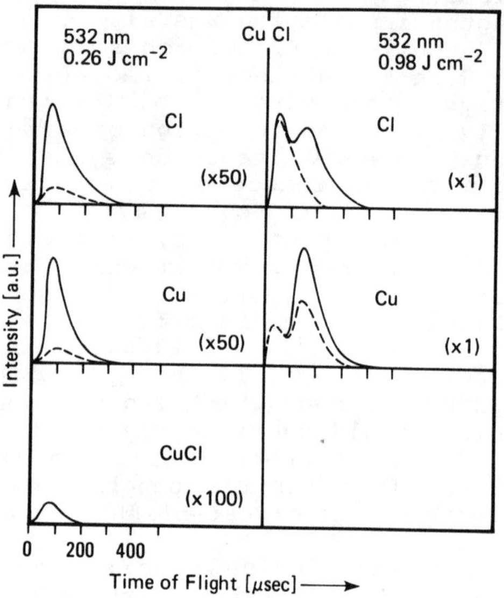

Fig. 5 Relative intensities of the mass distributions of photodesorbed particles from chlorinated Cu surfaces at (a) 532 nm and (b) 355 nm radiation. The intensities are corrected for the mass spectrometer transmission function. At 355 nm the Cl mass signal is not displayed, because 355 nm radiation can lead to dissociation of gaseous Cl_2 which may enhance the Cl signal, (data according to Ref. /50/).

Fig. 6 Time-of-flight mass spectrometric signals of laser desorbed atoms and molecules from a solid CuCl surface irradiated by 532 nm laser pulses at two different fluences. The dashed curves indicate the portion of genuine positive ions detected by the mass spectrometer with the ionizer switched off, (data according to Ref. /50/).

Table 1

Summary of recent studies on laser-enhanced chemical etching of solids by gases

Solid	Gas	Laser	Optical Configuration	Etch Rate or Etch Yield/Pulse	Spot Size/ Resolution	References
Ge	Br_2	Ar (cw)	focused //	$\sim 10^{-3}$ μm/sec	(NA)	(56)
Ge	Br_2	Dye (pulsed)	focused ⊥	0.2 Å/pulse	(NA)	(57)
GaAs InP	CH_3Br CH_3Cl CF_3I	Ar (cw)	focused ⊥	10^{-3} μm/sec	1.5 μm	(58)
GaAs	HCl/He plasma	Ar (cw)	unfocused ⊥	6×10^{-4} μm/sec	~100 μm	(59)
GaAs	Cl_2,CCl_4	Ar (cw)	focused ⊥	~2 μm/sec	~1 μm	(66)
Si	Cl_2	XeCl (pulsed)	unfocused ⊥	(NA)	~1 μm	(47)
Si	Br_2	Ar (cw)	focused //	$\leq 10^{-4}$ μm/sec	(NA)	(60)
Si	Cl_2 HCl	Ar (cw)	focused ⊥	0.02 ~ 10 μm/sec	5 μm	(7)
Si	SF_6	CO_2 (pulsed)	unfocused ⊥	1.2 Å/pulse	(NA)	(4)
Si	SF_6	CO_2 (pulsed)	focused //	0.05 Å/pulse	(NA)	(4)
Si	XeF_2	CO_2 (pulsed)	unfocused ⊥	0.6 Å/pulse	(NA)	(44)
Si	XeF_2	Ar (cw)	unfocused ⊥	6×10^{-4} μm/sec	(NA)	(12)
Si	COF_2	ArF (pulsed)	unfocused ⊥	0.13 Å/pulse	(NA)	(61)
SiO_2	HF	CO_2 (cw)	focused ⊥	0.05 μm/sec	~100 μm	(62)

Table 1 (continued)

SiO_2	CF_3Br CDF_3	CO_2 (pulsed)	focused //	0.3 Å/pulse	∿ 200 μm	(5)
SiO_2	Cl_2	Ar (cw)	focused ⊥	3×10^{-4} μm/sec	50 ∿ 80 μm	(8)
SiO_2	NF_3/H_2	ArF (pulsed)	unfocused ⊥	0.02 Å/pulse	(NA)	(63)
SiO_2	CF_2Br_2 C_2F_4	KrF (pulsed)	unfocused ⊥	0.6 Å/pulse	(NA)	(64)
SiO_2	CF_2Cl_2	KrF (pulsed)	unfocused ⊥	0.13 Å/pulse	(NA)	(65)
Mo,Ti	NF_3	ArF (pulsed)	unfocused ⊥	0.3 Å/pulse	(NA)	(65)
Al	Cl_2	XeCl (pulsed)	unfocused ⊥	30 Å/pulse	(NA)	(48)
Ag,Mo	Cl_2	N_2 (pulsed)	focused ⊥	3 Å/pulse	(NA)	(48)
Cu	Cl_2	Nd:YAG (0.53+0.35μm) (pulsed)	focused ⊥	(NA)	(NA)	(50)
Ta	SF_6	CO_2 (pulsed)	unfocused ⊥	2.6 Å/pulse	(NA)	(44)
Ta	XeF_2	CO_2 (pulsed)	unfocused ⊥	0.4 Å/pulse	(NA)	(44)
Te	XeF_2	CO_2 (pulsed)	unfocused ⊥	10 Å/pulse	0.5 μm	(8)
Ni(80%) Fe(20%)	SF_6,CF_4 CCl_4,Cl_2	Ar (cw)	focused ⊥	0.1 μm/sec	80 μm	(43)
Ni,Fe	Cl_2	N_2 (pulsed)	focused ⊥	2 Å/pulse	(NA)	(48)
TiC- Al_2O_3	CCl_4 SF_6	Ar (cw) Nd:YAG (pulsed)	focused ⊥ focused ⊥	0.1 μm/sec 5×10^{-3} μm/pulse	10-100 μm 10-100 μm	(43)

Notes: (1) // and ⊥, the laser beam incident either parallel (//) or perpendicular (⊥) to the solid surfaces.

(2) NA, data either not available or unspecified.

CuCl. Both systems exhibit electronic excitation effects, although the exact mechanism for the metal-chlorine system remains to be further elucidated.

The subject of laser-induced chemical vapor desorption (LCVD) has also been reviewed recently /3,52/. Basically a laser beam is used to decompose gaseous or condensed molecules and induce part of the photofragments to be deposit on a solid surface. By this approach, many metals, insulators and semiconductors have been deposited on various types of substrates. The photochemistry involves the unimolecular dissociation and bi- or tri-molecular reactions. The deposition reaction can be initiated pyrolytically or photolytically or a combination of both. While for the pyrolytic LCVD, the reaction kinetics is controlled by the local heating of the substrate and the newly formed film, direct photodecomposition of molecules in the gaseous and/or adsorbed phases is required for the photolytically activated LCVD, at least in the initial phase of the deposition. Although there are extensive LCVD studies on many chemical systems in the last five years, the microscopic mechanisms and the nucleation processes, whether it is initiated by the photothermal or the photolytical means are not at all clear, in particular, in the initial stage of the film growth.

In the case of strongly photon-absorbing substrates where the pyrolytic effect dominates, the molecules are presumably thermally decomposed near the hot spot of the substrate produced by the laser beam. Some photofragments desorb into the gas phase while some atoms are left on the surface. The free atoms subsequently form clusters which provide nucleation centers for further film growth. Because of the laser radiation, there is strong temperature gradients on the surface. When the film growth begins to take place, the surface temperature distribution changes rapidly with time due to the changing optical and thermal properties of the interface and the deposited substance. Detailed knowledge of the thin film structure and its optical and thermal properties are not available and thus the initial photothermal chemistry is difficult to understand. In the longer deposition period when a steady growth of the film or the rod is established, thermal modeling of the deposition rate becomes easier /52/. For the optically transparent substrate on which the initial deposition is produced by molecular photolysis, the situation is further complicated by the fact that once a metal or semiconductor cluster is formed on the surface it can greatly change the electronic properties of the physisorbed or chemisorbed molecules. The strong adsorbate-metal interactions can cause spectral shift, broadening of the dissociation continuum and reduction of the optical absorption strength. Indeed, the nucleation and cluster formation can modify the electromagnetic fields on the metal particles and promote the photdeposited films to develop periodic structures /53,54/. In the case of photolytically deposited Cd films from $Cd(CH_3)_2$ /53/, a ripple structure with the spatial period less than the laser wavelength was observed. This structure was interpreted to result from the stimulated light scattering into surface plasmons. In addition to the microscopic regular topography, periodicity in chemical composition and crystallinity was found in the photodeposited copper-carbon films /54/. Periodic structures were also detected in pyrolytic deposition of silicon /52,55/. Obviously this and other interesting aspects of surface photochemistry in microscopic dimensions deserve closer investigations.

5. APPLICATIONS TO MATERIAL PROCESSING

In chemical etching of solids for microfabrication of solid state devices, the current trend is definitely toward dry processing. Thus, we have seen an increasing use of plasma and reactive-ion-etching (RIE) replacing the conventional wet techniques. Processing by plasma and RIE requires fine-line lithography. The method of etching involves basically the use of a photoresist mask on a solid, and a glow discharge to generate chemically active species for reaction with the exposed surfaces. Although the general concept is rather simple, the complexity of the glow discharge and the sensitive dependence of the etching behavior upon instrument parameters have rendered detailed understanding of the surface chemistry in the etching process difficult to obtain. In many instances the achievable etch rates are slow and obtainable profiles are unacceptable for device applications. Also, the problems associated with the resist masks are sometimes difficult to overcome. Furthermore, as the devices get smaller in dimension, the charge-particle-induced radiation damages can become difficult to avoid. Such undesirable radiation effects can be particularly serious in RIE processing. In contrast, laser etching is a direct, maskless process. The radiation damage to the solid is expected to be negligible.

While plasma-assisted etching techniques undoubtedly will remain as essential and dominant techniques in dry processing technology in the foreseeable future, laser chemical etching is certainly emerging as a powerful supplement to these current methods. Moreover, the laser approach should be considered as an alternative when and where the plasma techniques become inadequate. Thus, it seems useful to compare the prominent features of the two etching methods. For clarity, this comparison is listed in Table 2.

Likewise, the expectation of laser-induced chemical vapor deposition (LCVD) for direct fabrication of microchips and new materials is very high. In comparison with existing techniques such as thermal CVD, ion sputtering, plasma deposition and electron-beam evaporation, the laser method can offer certain advantages in many cases. For instance, high deposition rates can usually be achieved at low substrate temperatures. Also, the process parameters are relatively simple and easy to control. At the present time, however, the most promising applications for LCVD appear to be in the area of circuit repairs and corrections. Naturally, the laser-induced chemical etching can also be used for repairing microcircuits. In combination, LCVD and photoetching indeed can become powerful tools for selective modification of micropatterns. For more general applications, however, it should be recognized that there are

Table 2

Comparison of laser- and plasma-assisted etching techniques

	Plasma etching	Laser etching
Resist mask	needed	not needed
Etch rate per unit area	slow e.g.,<< 1 μm/sec for Si	generally fast e.g., > 50 μm/sec for Si
Throughput	generally high many wafers processed per batch	low step-and-repeat processing may be required; single wafer processing possible in near future
Etching selectivity	high	generally very high
Resolution	width < 1 μm	< 1 μm
Substrate temperature	generally at 25°C	high localized temperature due to laser irradiation
Etching mechanisms	Reactive radicals charged particles ion bombardment	photolysis solid photoexcitation thermal activation
Radiation damage	sometimes serious	expected to be negligible
General applications	semiconductors certain metals and insulators organic polymers	semiconductors metals and insulators organic polymers
Special applications		3-D tooling circuit repair customized fabrications

still fundamental problems and practical limitations that need to be resolved, in particular, concerning the total process rates and yields. Nevertheless, we should realize that the radiation-enhanced chemical techniques are still in early stages of development. With the advent of more powerful and more reliable laser systems, these techniques will undoubtedly be further improved and utilized in manufacturing laboratories.

6. SUMMARY

In this report, the basic processes important in laser-induced molecule-surface interactions have been outlined. Recent advances particularly in the fileds of laser-stimulated desorption, photochemical etching, photodeposition and surface molecular dynamics promoted and monitored by laser beams have been discussed. Specific examples are given to illustrate the photon-enhanced chemisorption, adsorbate-adsorbate and adsorbate-substrate reactions, product formation and desorption of surface species. The surface interaction dynamics associated with the electronic and vibrational excitation, and subsequent energy transfer and relaxation processes in competition with chemical reactions are presented with available experimental data. Like the extensive laser photochemical investigations in homogeneous systems in the last two decades, the applications of lasers to gas-surface studies can yield tremendous insight into the interfacial reaction mechanisms and dynamics. Such studies also offer the exciting possibility for technological innovations. In particular, the techniques based on the laser-induced gas-surface chemistry for processing electronic materials have been well demonstrated recently. Here, their salient features and unique capabilities as well as their limitations are pointed out and compared with existing methods. With the advent of more powerful and versatile laser systems, these photochemical techniques will be further developed and new applications in other areas such as chemical syntheses of new materials, photocatalysis and energy utilizations will be realized. In short, it is a scientific discipline full of opportunities for fruitful studies from both basic and applied points of views.

(*) An Alexander von Humboldt-Senior U.S. Scientist Awardee, presently on leave (9/85-8/86) at the University of Munich in Munich and the Fritz-Haber-Institut in Berlin, F.R.Germany.

REFERENCES

1. T.J. Chuang, Surf. Sci. Reports 3, 1-105 (1983); also Surf. Sci. (1986, in press).
2. T.J. Chuang, J. Vac. Sci. Technol. 21, 798 (1982).
3. D.J. Ehrlich and J.Y. Tsao, J. Vac. Sci. Technol. B1, 969 (1983); R.M. Osgood,Jr., Ann. Rev. Phys. Chem. 34, 77 (1983).
4. T.J. Chuang, J. Chem. Phys. 74, 1453 (1981) and 72, 6303 (1980).
5. J.I. Steinfeld, T.G. Anderson, C. Reiser, D.R. Denison, L.D. Hartsough and J.R. Hollahan, J. Electrochem. Soc. 127, 514 (1980).
6. M. Hanabusa, A. Namiki and K. Yoshihara, Appl. Phys. Lett. 35, 626 (1979).
7. D.J. Ehrlich, R.M. Osgood,Jr., and T.F. Deutsch, Appl. Phys. Lett. 38, 1018 (1981).
8. T.J. Chuang, IBM J. Res. Develop. 26, 145 (1982).
9. D.J. Ehrlich, R.M. Osgood,Jr., and T.F. Deutsch, Appl. Phys. Lett. 38, 946 (1981).
10. S.A. Schafer and S.A. Lyon, J. Vac. Sci. Technol. 21, 422 (1982), and 19, 494 (1981); E.M. Young and W.A. Tiller, Appl. Phys. Lett. 42, 63 (1983).
11. M. Matsuura, M. Ishida, A. Suzuki and K. Hara, Jap. J. Appl. Phys. 20, L726 (1981); W.G. Petro, I. Hino, S. Eglash, I. Lindau, C.Y. Su and W.E. Spicer, J. Vac. Sci. Technol. 21, 405 (1982).
12. F.A. Houle, J. Chem. Phys. 79, 4237 (1983) and 80, 4851 (1984).
13. T.J. Chuang, J. Vac. Sci. Technol. B3, 1408 (1985).
14. G. Wedler and H. Ruhmann, Surf. Sci. 121, 464 (1982).
15. R. Viswanathan, D.R. Burgess,Jr., P.C. Stair and E. Weitz, J. Vac. Sci. Technol. 20, 605 (1982).
16. S.M. George, A.M. DeSantolo and R.B. Hall, Surf. Sci. 159, L425 (1985).
17. J.A. Tagle and A. Pospieszczyk, Appl. Surf. Sci. 17, 189 (1983).
18. T.J. Chuang, I. Hussla and W. Sesselmann, in Laser Processing and Diagnostics, edited by D. Bäuerle (Springer, Heidelberg, 1984), p. 300.
19. R.B. Hall and A.M. DeSantolo, Surf. Sci. 137, 421 (1984); also R.B. Hall, A.M. DeSantolo and S.J. Bares, Surf. Sci. 161, L533 (1985).
20. J. Heidberg, H. Stein, E. Riehl and A. Nestmann, Z Phys. Chem. (N.F.) 121, 145 (1980); J. Heidberg, H. Stein and E. Riehl, Z. Szilagyi and H. Weiss, Surf. Sci. 158, 553 (1985); J. Heidberg, H. Stein and E. Riehl, Phys. Rev. Lett. 49, 666 (1982).
21. T.J. Chuang and H. Seki, Phys. Rev. Lett. 49, 382 (1982); H. Seki and T.J. Chuang, Solid State Commun. 44, 473 (1982); T.J. Chuang, J. Electr. Spectr. Relat. Phenom. 29, 125 (1983).
22. T.J. Chuang, H. Seki and I. Hussla, Surf. Sci. 158, 525 (1985).
23. I. Hussla, H. Seki, T.J. Chuang, Z.W. Gortel, H.J. Kreuzer and P. Piercy, Phys. Rev. B32, 3489 (1985).

24. I. Hussla and T.J. Chuang, Ber. Bunsenges. Phys. Chem. 89, 294 (1985).
25. J. Lin and T.F. George, Surf. Sci. 115, 569 (1982); J. Chem. Phys. 78, 5197 (1983); A.C. Beri and T.F. George, J. Chem. Phys. 78, 4288 (1983); ibid., 83, 2482 (1985).
26. Z.W. Gortel, H.J. Kreuzer, P. Piercy and R. Teshima, Phys. Rev. B28, 2119 (1983); ibid., B27, 5066 (1983); H.J. Kreuzer and Z.W. Gortel, Phys. Rev. B29, 6926 (1984).
27. C. Jedrzejek, K.F. Freed, S. Efrima and H. Metiu, Surf. Sci. 109, 191 (1981); C. Jedrzejek, J. Vac. Sci. Technol. B3, 1431 (1985).
28. F.G. Celii, M.P. Casassa and K.C. Janda, Surf. Sci. 141, 169 (1984).
29. G.S. Wu, B. Fain, A.R. Ziv and S.H. Lin, Surf. Sci. 147, 537 (1984); B. Fain and S.H. Lin, Chem. Phys. Lett. 114, 497 (1985).
30. E.B.D. Bourdon, J.P. Cowin, I. Harrison, J.C. Polanyi, J. Segner, C.D. Stanners and P.A. Young, J. Phys. Chem. 88, 6100 (1984).
31. K. Domen and T.J. Chuang, (to be published) .
32. V.S. Letokhov, V.G. Movshev and S.V. Chekalin, Sov. Phys. JETP 54, 257 (1981); V.S. Antonov, V.S. Letokhov and A.N. Shibanov, Appl. Phys. 25, 71 (1981).
33. N. Nishi, H. Shinohara and T. Okuyama, J. Chem. Phys. 80, 3898 (1984).
34. T. Kawai and T. Sakata, Chem. Phys. Lett. 69, 33 (1980); N. Van Hieu and D. Lichtman, Surf. Sci. 103, 535 (1981).
35. P.M. Whitmore, H.J. Robota and C.B. Harris, J. Chem. Phys. 77, 1560 (1982); A.P. Alivisatos, D.H. Waldeck and C.B. Harris, J. Chem. Phys. 82, 541 (1985).
36. A. Wokaun, H.-P. Lutz, A.P. King, U.P. Wild and R.R. Ernst J. Chem. Phys. 79, 509 (1983).
37. A. Leitner, M.E. Lippitsch, S. Draxler, M. Riegler and F.R. Aussenegg, Appl. Phys. B36, 105 (1985).
38. J. Misewich, C.N. Plum, G. Blyholder, P.L. Houston and R.P. Merrill, J. Chem. Phys. 78, 4245 (1983), also, V.A. Apkarian, R. Hamers, P.L. Houston and J. Misewich, in Dynamics on Surfaces, edited by B. Pullmann, J. Jortner, A. Nitzan and B. Gerber (D. Reidel, Dordrecht, 1984), p. 135.
39. J. Misewich, H. Zacharias and M.M.T. Loy, J. Vac. Sci. Technol. B3, 1474 (1985); also, Phys. Rev. Lett. 55, 1919 (1985).
40. R.R. Lucchese and J.C. Tully, J. Chem. Phys. 80, 3451 (1984).
41. E.J. Heilweil, M.P. Casassa, R.R. Cavanagh and J.C. Stephenson, J. Chem. Phys. 82, 5216 (1985); ibid., 81, 2856 (1984).
42. E.J. Heilweil, M.P. Casassa, R.R. Cavanagh and J.C. Stephenson, J. Vac. Sci. Technol. B3, 1471 (1985).
43. T.J. Chuang, Mat. Res. Soc. Symp. Proc. 17, 45 (1983); ibid., 29, 185 (1984).
44. T.J. Chuang, J. Chem. Phys. 74, 1461 (1981); also J. Vac. Sci. Technol. 18, 638 (1981).

45. Y.Y. Tu, T.J. Chuang and H.F. Winters, Phys. Rev. B23, 823 (1981); H.F. Winters and F.A. Houle, J. Appl. Phys. 54, 1218 (1983).
46. H.F. Winters, J.W. Coburn and T.J. Chuang, J. Vac. Sci. Technol. B1, 469 (1983).
47. T. Arikado, M. Sekine, H. Okano and Y. Horiike, Mat. Res. Soc. Symp. Proc. 29, 167 (1984).
48. W. Sesselmann and T.J. Chuang, J. Vac. Sci. Technol. B3, 1507 (1985); also Surf. Sci. 162, 1007 (1985).
49. W. Sesselmann and T.J. Chuang, Surf. Sci. (1986, to be published).
50. W. Sesselmann, E.E. Marinero and T.J. Chuang, Appl. Phys. A (to be published).
51. N. Itoh and T. Nakayama, Phys. Lett. A92, 471 (1982); T. Nakayama, M. Okigawa and N. Itoh, Nucl. Instr. Meth. Phys. Res. B1, 301 (1984).
52. D. Bäuerle, in Laser Processing and Diagnostics, edited by D. Bäuerle (Springer, Heidelberg, 1984), p. 166.
53. S.R.J. Brueck and D.J. Ehrlich, Phys. Rev. Lett. 48, 1678 (1982); R.M. Osgood,Jr. and D.J. Ehrlich, Opt. Lett. 7, 385 (1982).
54. R.J. Wilson and F.A. Houle, Phys. Rev. Lett. 55, 2184 (1985).
55. Y.C. Du, U. Kempfer, K. Piglmayer, D. Bäuerle and U.M. Titulaer, Appl. Phys. A39, 1 (1986).
56. I.M. Beterov, V.P. Chebotaev, N.I. Yushina and B.Ya. Yurshin, Sov. J. Quantum Electron. 8, 1310 (1978).
57. G.P. Davis, C.A. Moore and R.A. Gottschc, J. Appl. Phys. 56, 1808 (1984).
58. D.J. Ehrlich, R.M. Osgood,Jr., and T.F. Deutsch, Appl. Phys. Lett. 36, 698 (1980).
59. C.I.H. Ashby, Appl. Phys. Lett. 45, 892 (1984); 46, 752 (1985).
60. L.L. Sveshnikova, V.I. Donin and S.M. Repinskii, Sov. Tech. Phys. Lett. 3, 223 (1977).
61. G.L. Loper and M.D. Tabat, Appl. Phys. Lett. 46, 654 (1985).
62. K. Daree and W. Kaiser, Glass Technol. 18, 19 (1977).
63. S. Yokoyama, Y. Yamakage and M. Hirose, Appl. Phys. Lett. 47, 389 (1985); M. Hirose, S. Yokoyama and Y. Yamakage, J. Vac. Sci. Technol. B3, 1445 (1985).
64. J.H. Brannon, J. Phys. Chem. 90, 1784 (1986).
65. G.L. Loper and M.D. Tabat, SPIE 459, 121 (1984).
66. M. Takai, H. Nakai, J. Tsuchimoto, J. Tokuda, T. Minamisono, K. Gamo and S. Namba, in Laser Processing and Diagnostics, ed. by D. Bäuerle (Springer, Heidelberg, 1984), p. 315.

PHOTOCHEMISTRY OF TRANSITION METAL COMPLEXES

H. Schröder, B. Rager, S. Metev
MPI für Quantenoptik, 8046 Garching, FRG

N. Rösch, H. Jörg
Lehrstuhl für Theoretische Chemie, Technische Universität München, 8046 Garching, FRG

1. INTRODUCTION

There is increasing interest in understanding the photoinduced decay channels of simple organometallic complexes, in particular those of metal carbonyls. These complexes are noted for their high photochemical instability since the energy of an absorbed photon in the near uv or uv is sufficient to break several of the weak metal-ligand bonds. Possible applications are therefore wide spread, ranging from the favourable production of free atoms and atomic clusters, their utility as convenient model systems in photocatalysis, to their advantageous properties for laser induced metal deposition.

There is a number of important review articles devoted to the photochemistry of transition metal cluster carbonyls [1]. In those systems, the metal atom is more tightly bound due to multiple coordination and electronically excited states of the ligand-field may be accessible [2,3]. Therefore, fluorescence or phosphorescence detection is feasible for monitoring photoinduced decay processes. Comparatively little is known about the homocentric carbonyls. The apparent reason for the lack of quantitative information concerning the photophysical pathways is the absence of convenient spectroscopic signature by which state specific excitation and relaxation events can be monitored [4]. Metal carbonyls and simple organometallics generally do not fluoresce, and their electronic absorption bands and those of their smaller fragments, seldom show resolved vibronic features in either the gaseous or condensed phase [5,6]. Even uv/vis laser based spectroscopic techniques have not proven successful up to now since the only species which can be detected selectively are most frequently the central metal atoms as a result of several poorly defined excitation steps.

In the present manuscript we shall first summarize current results related to the laser photolysis of volatile transition metal carbonyls. We shall primarily focus on $Fe(CO)_5$ and $Cr(CO)_6$ since most reports have dealt with these complexes. A complete literature survey will not be given. However, it should be possible for the reader to get a comprehensive view by means of the literature cited in the specified references. One outstanding conclusion shall be anticipated, namely the creation of electronically excited fragment states that retain the excess energy from dissipating into vibrational and translational degrees of freedom and thus rationalize the process of multiple photon decomposition.

In the second part, the electronic configuration of metal carbonyls and metal carbonyl fragments is discussed with particular reference to $Ni(CO)_4$. This is done since $Ni(CO)_3^*$ was the first case where emission arising from a metal carbonyl fragment has been observed [7]. We should mention that visible luminescence after laser excitation of other tetrahedral transition metal complexes like OsO_4 [8], CrO_2Cl_2 [9], $Pt(PF_3)_4$ [10] is a well documented phenomenon which still waits for a convincing explanation. Finally, we shall discuss our results concerning the KrF laser induced multiphoton ionization (MPI) of $Ni(CO)_4$. Here we were able to determine, for the first time, the absolute number of generated ions at sufficiently low laser fluences and could thus derive an absolute excitation cross section; in this case, for a third order process yielding predominantly Ni^+.

2. LITERATURE SURVEY

2.1. THERMOCHEMISTRY AND DISSOCIATIVE ATTACHMENT OF ELECTRONS

Mononuclear metal carbonyls are noted for their prompt decay upon electronic excitation, yielding at least one CO moiety with a quantum yield close to unity. To understand this mechanism in more detail, the bond dissociation energies for the individual species involved must be known. One major difficulty often preventing successful studies of organometallic bond energies is the general surface reactivity of these compounds which decompose catalytically in conventional experiments. This is precisely the reason why they are very attractive for the purpose of metal deposition. To prevent surface reactions, G.P. Smith et al. [11] have recently developed a pulsed laser technique which provides rapid indirect thermal heating and cooling to study homogeneous gas-phase decomposition. Their final results concerning the first dissociation bond energies are summarized in Table I together with the mean bond energies D_{av}.

Compound	$H^o_{298}(1)$		D_{av}	
$Cr(CO)_6$	36.8	[11]	26	[13]
$Mo(CO)_6$	40.5	[11]	36	[13]
$W(CO)_6$	46.0	[11]	43	[13]
$Fe(CO)_5$	41.5	[11]	28	[13]
$Ni(CO)_4$	25 ± 2	[12]	30 ± 3	[12]
	22	[14]	35	[13]

TABLE I. Metal carbonyl bond dissociation energies, kcal/mol

A complementary technique providing the bond energies for all the fragments has been demonstrated by W.C. Lineberger et al. [12] for the case of $Ni(CO)_4$. They measured the argon ion laser induced photoelectron spectra of the various nickel carbonyl ions $Ni(CO)_n^-$, n = 1 - 3, and they evaluated the following data:

	D $[Ni(CO)_{n-1} - CO]$
$Ni(CO)_4$	25 ± 2
$Ni(CO)_3$	13 ± 10
$Ni(CO)_2$	54 ± 15
$Ni(CO)$	29 ± 15

TABLE II. Fragment bond dissociation energies, kcal/mol

These authors and others [15] also pointed out that $Ni(CO)_4$ is not bound with respect to dissociation of CO. The electron capture cross section for this process is often extremely high, up to 200 Å² for thermalized electrons. This is due to the fact that the reaction

$$M(CO)_n + e \longrightarrow M(CO)_{n-1}^- + CO$$

is approximately thermoneutral, with the electron affinity of $M(CO)_{n-1}$ being close to or even exceeding the dissociation energy $D[M(CO)_{n-1}\text{-}CO]$. Electron capture may sometimes be the leading mechanism to initiate a metal deposition [16,17]. Comparing the energies required for the first bond scission with the average dissociation energy we realize that in all cases, with the exception of $Ni(CO)_4$, the first CO moiety is the most difficult to dissociate. This means that photodissociation is selfaccelerating upon constant irradiation. Only in the case of $Ni(CO)_4$ there is a severe bottle-neck on the way to the free metal atom.

2.2. LASER PHOTODISSOCIATION OF $Fe(CO)_5$ AND $Cr(CO)_6$

Until now, the dynamics of transition metal carbonyl decay processes have been most intensively investigated for $Fe(CO)_5$ [4,18-24] and $Cr(CO)_6$ [18,25-30]. These initial studies revealed that even mild excitation of metal carbonyls can cause complete decarbonylation with a quantum yield for the metal release close to unity [18]. The identity of the metal ion was faithfully demonstrated by means of mass spectrometry in conjunction with a molecular beam [4,18,26], by means of resonance enhanced multiphoton ionization (MPI) using a tunable dye laser [4,18,19,23,25] or by time-of-flight measurements of the emitted photoelectrons [23,24,26].

Usually only one pulsed tunable laser is utilized which has to dissociate the complex by an n-photon process and then to ionize the nascent atom in an resonantly enhanced m-photon process. Therefore, a conclusion concerning the fragmentation path is sometimes difficult to derive. Besides the resonant path, one also observes ions produced at laser frequencies which can not be assigned to an atomic transition. This part of the ion signal has not conclusively been interpreted and it is certainly different in nature for the individual molecules. It is generally observed that the off-resonant ion signal increases with increasing laser intensity [4] and also when the primary photons have a higher energy. Multiphoton ionization within the molecular ladder [4], power broadening of atomic transitions [26] and direct (not resonantly enhanced) ionization from an electronically excited atom (up to 3 eV in the case of iron [24]) have been proposed as viable mechanisms. From the evidence that larger molecular ions are by far the minority of the overall ion yield it has been concluded that dissociation for all fragments (except for the smallest one [26]) proceeds always faster than autoionization.

In particular, J.T. Yardley et al. [20,21] and E.R. Grant et al. [4], have discussed the kinetics of $Fe(CO)_5$ photofragmentation in excellent detail. They provide ample evidence that laser induced decomposition of $Fe(CO)_5$ is highly non-statistical with ultrafast dissociation and energy retention in the remaining metal complex. One possibility proposed is a sequence of electronically excited states $Fe(CO)_5^*$, $Fe(CO)_4^*$, +.... Adequate theoretical treatment of this problem will require construction of accurate excited state potential energy surfaces.

Another intriguing result has been reported by G.J. Fisanick et al. [26]. These authors studied the MPI spectroscopy of the series $Cr(CO)_6$, $Cr(CO)_3C_6H_6$ and $Cr(C_6H_6)_2$ and could not detect any MPI signal from the nascent photoproduct C_6H_6. Therefore, they proposed the elimination of a transformed benzene species in the dissociation of $Cr(CO)_3C_6H_6$.

2.3. FRAGMENT SPECTROSCOPY

Now we want to focus on those studies which mainly aim at the identification of neutral transient fragments. Such experiments have classically been performed in solid matrices [31] where typically only the monodecarbonylated fragment, e.g. $Fe(CO)_4$, could be detected, by means of infrared or Möß-bauer spectroscopy [22]. This was taken as a further indication for a sequential dissociation event, because an electronically excited $Fe(CO)_4^*$ species, capable of further decomposition, is rapidly quenched by interaction with the environment. In fact, the potential dissociation path can only be displayed under quasi-isolated molecule conditions.

Yardley et al. [21] have measured the primary photochemical fragmentation pattern of $Fe(CO)_5$ in the gas phase, resulting from one photon absorption at 352, 248 and 193 nm. They used a fragment trapping technique at an excess of PF_3 partial pressure and detected species such as $Fe(CO)_2(PF)_3$, $Fe(CO)_3(PF_3)_2$ and $Fe(CO)_4PF_3$ by means of gas chromatography mass spectrometry. It is well known that the vibrational frequency of CO depends on the fragment coordination number [12]. This means that one can discriminate between $M-(CO)_n$ and $M-(CO)_{n-1}$ in the infrared spectrum. This method has been applied recently by T.R. Fletcher and R.N. Rosenfeld [27] and by E. Weitz et al. [28] to investigate photofragmentation of $Cr(CO)_6$ upon excitation at 249 and 351 nm. The primary photoproducts which could be detected were CO, $Cr(CO)_5$, and, to a lesser extent, $Cr(CO)_4$.

In particular, Fletcher and Rosenfeld [27] succeeded in monitoring the back-reaction yielding the parent molecule [$Cr(CO)_5$ + CO ---> $Cr(CO)_6$, $k=(1.2\pm0.2)\times10^6$ $torr^{-1}s^{-1}$] as well as the formation of clusters [$Cr(CO)_4$ + $Cr(CO)_6$ ---> $Cr_2(CO)_{10}$, $k=1.8\times10^7$ $torr^{-1}s^{-1}$]. The reaction $Fe(CO)_4$ + CO ---> $Fe(CO)_5$, however, occurs almost three orders of magnitude more slowly. This difference in reactivity may be due to the spin multiplicity change associated with the recombination reaction. This kind of information is valuable for the proper design of gas mixtures for metal depositions, since one can thus estimate the interfering homogeneous cluster formation at a given density of primary radicals. An alternative method capable of producing similar data has been published be W.H. Breckenridge and G.M. Stewart [30]. They measured transient absorption in the visible spectral range and assigned an isolated diffuse band, centered around 620 nm, to the absorption of $Cr(CO)_5$. In the next section we will comment on this dramatic shift of the absorption contour between $Cr(CO)_6$ and $Cr(CO)_5$, in the light of our results for $Ni(CO)_4$.

3. ELECTRONIC STRUCTURE OF TRANSITIONS METAL CARBONYLS AND OF THEIR FRAGMENTS

The electronic structure of transition metal complexes derives its characteristic features to a considerable extent from the nearly degenerate set of rather localized metal d orbitals. It was therefore quite natural to start a theoretical treatment by focusing on the central metal ion borrowing from the well established quantum theory of atoms. The resulting formalism of crystal field theory was very successful in rationalizing magnetic properties and visible optical spectra of classical metal complexes [32]. However, true chemical problems, such as questions concerning structure and reactivity, could be tackled profitably only after the development of molecular orbital (MO) theory [33]. R. Hoffmann and his coworkers have shaped this theory into a standard tool of inorganic chemistry by skillfully applying it to a large and fascinating variety of organometallic systems [34,35]. These compounds, with their strongly interacting organic ligands lie far beyond the realm of crystal field theory where the assumption of a weak influence of the surrounding "field" on the metal ion constitutes a cornerstone of the theory. The virtues of MO theory become strikingly evident in the photochemistry of organometallic compounds [5] where excited states are probed that contain metal and ligand contributions to a varying degree. Despite impressive recent advances in the quantum chemistry of transition metal compounds [36], MO theory still remains the only reliable approach which is able to provide a simple, consistent picture of the ground and excited states.

3.1. MOLECULAR ORBITALS OF TETRAHEDRAL METAL COMPLEXES

Organometallic compounds display a wide variety of electronically excited states. Let us illustrate possible excitations by constructing the MO interaction diagram of a tetrahedral 3d metal complex (see fig. 1). The set of valence orbitals available to the metal atom comprises the 3d levels with their well known directional properties and the rather diffuse 4s and 4p orbitals which extend far into the ligand region of space. With carbonyls in mind, we assume (filled) σ donor orbitals and rather low-lying (empty) π acceptor levels on the ligands.

The σ levels, generating symmetry adapted orbitals of type a_1 and t_2 [37], will interact with metal 3d and 4s orbitals, thus forming four metal-ligand bonds. The degenerate metal 3d orbitals, as a consequence thereof, split into a nonbonding e and a metalligand antibonding t_2 level. This σ interaction entails ligand-to-metal charge donation which will be balanced by π backdonation in conformity to Pauling's electroneutrality principle. Thereby, some of the ligand π^* levels which span orbitals of e, t_1 and t_2 character, will interact with the e and t_2 metal 3d levels.

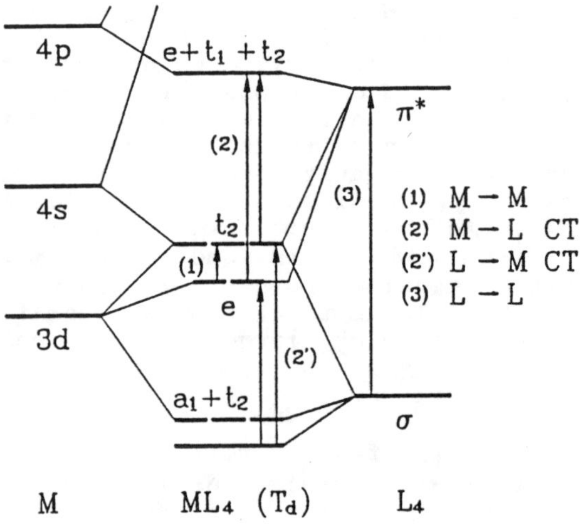

FIGURE 1. Simplified molecular orbital diagram for a tetrahedral transition metal complex indicating some electronic transitions in an organometallic compound

A peculiarity of tetrahedral geometry is the symmetry-allowed 4p-3d metal orbital mixing in the t_2 manifold possibly leading to increased intensity of metal-to-metal excitations (see (1) in fig. 1). Such so-called ligand field transitions are missing in the spectrum of $Ni(CO)_4$ with its d^{10} configuration completely filling the metal levels e and t_2 (see fig. 2). Next in energy are chargetransfer excitations which may be of metal-to-ligand (2) or ligand-to-metal (2') character. Again, the latter are missing in $Ni(CO)_4$. The third kind of excitations shown in fig. 1 give rise to appropriately modified intraligand bands in the spectrum. They involve only ligand derived molecular orbitals (3). The discussion so far has been restricted to complexes with only one metal atom, more complicated excitations will occur in cluster compounds containing two and more metal centers [5].

3.2. QUANTITATIVE DESCRIPTION OF THE GROUND STATE

Presently, the electronic structure of transition metal compounds is a field of very active research in quantum chemistry [36]. This is in part a consequence of the enormous increase of computer power which has become available during the last years and which allows the utilization of ever more sophisticated methods. However, almost all studies aiming at a quantitative description have so far been restricted to the ground state. For transition metal compounds, even this holds a formidable challenge due to the very distinct properties of the metal valence orbitals. For an accurate prediction of the metal-ligand bond lengths and the corresponding force constants, a balanced description of the localized 3d and the diffuse 4s and 4p orbitals is mandatory.

It is now well known that the Hartree-Fock (HF) approximation, the basic ab initio method [38], fails to maintain this balance since it is unable to properly account for the correlated motion of the localized d electrons. Instead, it favours the occupation of metal s and p levels which lead to an erroneous repulsive interaction with the filled σ donor orbitals of the ligands. Consequently, the Hartree-Fock method dramatically overestimates metal-carbon bond distances in organometallic compounds. As an example, we present in table III results for $Ni(CO)_4$ which is one of the most thoroughly studied transition metal compounds. Inclusion of configuration interaction (CI) leads to improved results, although at the expense of a much larger computational effort.

Method	d_{Ni-C} (A)	d_{C-O} (A)	$f_{NiC,NiC}$ (N/m)	$f_{CO,CO}$ (N/m)	$f_{NiC,CO}$ (N/m)	D_o (eV)
HF	1.921[a]	-	-	-	-	-1.977[b]
CI[c]	1.881	-	-	-	-	1.14
Experiment[d]	1.825	1.122	235.5	1823.3	23.5	6.12
LCGTO-Xα[e]	1.805	1.137	300.0	1800.0	24.0	8.57

[a] Ref.(39) [b] Ref.(40) [c] Ref.(41)
[d] Ref.(42,43) [e] Ref.(44)

TABLE III. Theoretical and experimental spectroscopic properties for the ground state of $Ni(CO)_4$

Density functional theory [45] provides an alternative approach to electronic structure, quite different in flavour and computational requirements. The corresponding SCF problem with one-electron equations containing a local potential, are solved in the local density (LD) approximation. The Xα method may be regarded as a variation of this scheme. Expansions employing familiar Gaussian-type basis sets yield an efficient and accurate computational procedure (LCGTO-Xα [46]) that has been applied to a variety of organometallic systems [44,47,48]. From these results, one may conclude that LD methods overestimate metal-carbon bonding resulting in too short bond lengths, too large force constants and too large binding energies (see table III). However, the error is much smaller than that of HF calculations and quite uniform [47,48]. The overbinding has been traced [44] to the LD approximation which overestimates the d-electron exchange [49]. Both methodologies, ab initio and local density, display their strengths and weaknesses already in the treatment of the bare transition metal atoms, the corresponding molecular properties only mellow their traits [44]. Somewhat pointedly, one may characterize the present situation in molecular electronic structure theory [36] as a renaissance of atomic theory.

3.3. STUDIES ON EXCITED STATES: PHOTOFRAGMENTATION OF NICKELTETRACARBONYL

Ab initio investigations on electronic spectra of transition metal compounds are still scarce (e.g. [50,51]) because the computational effort is quite demanding, a CI treatment [38] being mandatory. CI techniques have also been employed at the semiempirical level to transition metal complexes [52] including $Ni(CO)_4$ [53,54]. Although LD methods are formally restricted to the lowest state of each symmetry manifold [45], useful information may also be obtained on excited states [7]. As an example, let us discuss the molecule nickel tetracarbonyl [7,55].

Experiments show that $Ni(CO)_4$ looses one CO ligand after uv irradiation yielding intense luminescence in the visible [7] (see also part C). The emission spectrum consists of one rather broad (≈ 0.8 eV) band centered around 1.9 eV (650 nm). To rationalize these findings, consider the MO energy diagram in fig. 2 showing the one-electron energies for carbon monoxide, nickel tetracarbonyl and nickel tricarbonyl as obtained from LCGTO-Xα calculations [7,55]. In the latter fragment, the three CO ligands had been fixed in a pyramidal conformation with tetrahedral bond angles. One identifies two major contributions to the bonding of the fourth CO unit to the $Ni(CO)_3$ fragment. First, the 5σ orbital of CO (i.e. essentially a lone pair at carbon [35]) interacts with the $13a_1$ orbital of $Ni(CO)_3$ which is strongly pointing towards the

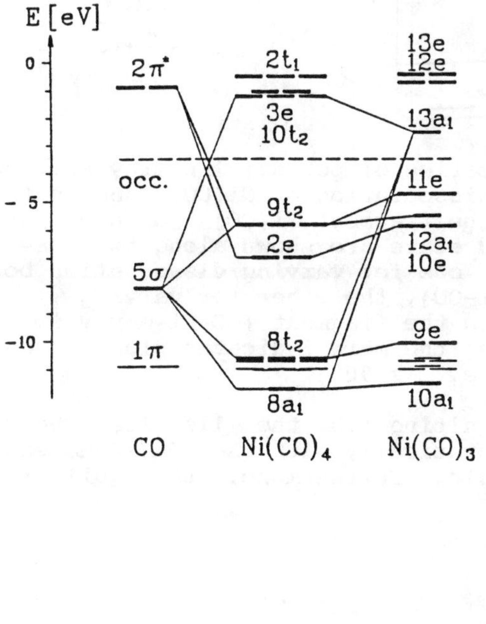

FIGURE 2. LCGTO-Xα orbital energies for $Ni(CO)_4$. Also shown are the levels of the photodissociation fragments $Ni(CO)_3$ and CO as well as their orbital interactions. The levels above the dashed line are unoccupied.

empty coordination site [56]. Thereby, the 5σ orbital lowers its energy and concomitantly pushes the $13a_1$ level upwards. Secondly, the occupied metal 3d levels are lowered due to additional backbonding (10e, $12a_1$, and 11e in $Ni(CO)_3$; 2e and $9t_2$ in $Ni(CO)_4$, see fig. 2).

Electronic excitations of $Ni(CO)_4$ can only take place across the gap $9t_2 \rightarrow 10t_2$ along path (2) of fig.1. We will therefore focus our attention on this lowest transition. The highest occupied molecular orbital (HOMO), 11e, and the lowest unoccupied molecular orbital (LUMO), $13a_1$, of the fragment $Ni(CO)_3$ correlate with the corresponding levels $9t_2$ and $10t_2$ of the parent molecule. The HOMO-LUMO energy difference of $Ni(CO)_3$ lies in the visible range and thus an electronic relaxation $13a_1 \rightarrow 11e$ would be able to explain the observed emission. This transition is predicted to display pronounced charge-transfer character since the $13a_1$ orbital is localized mainly on the CO ligand framework [55].

To confirm the assignment of the emission to the fragment LCGTO-Xα, total energy calculations have been performed [7,55] tracing the dissociation of one Ni-CO bond in $Ni(CO)_4$ and the nuclear rearrangement in $Ni(CO)_3$. The results for the ground and first excited (singlet) state are presented schematically in fig. 3. It turns out that the first excited

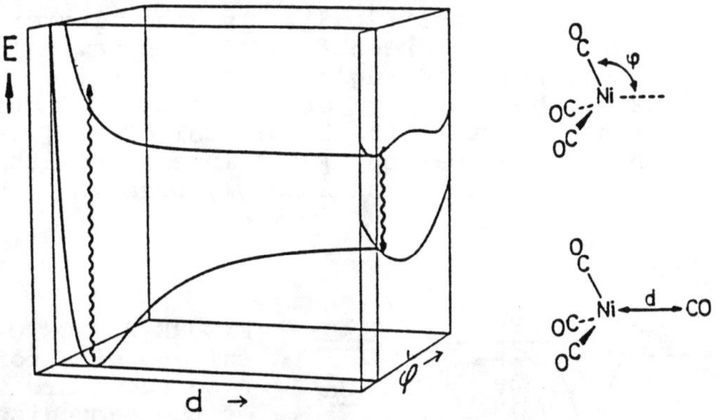

FIGURE 3. Schematic presentation of potential energy surfaces relevant to the dissociation of $Ni(CO)_4$ and to the luminescence of the fragment $Ni(CO)_3$. Ground state and first excited state are shown along two one-dimensional cuts, one for varying dissociation bond length d ($(CO)_3Ni$-CO), the other for varying bending angle in the fragment (CO-Ni-empty fourth coordination site; the mark indicates planar fragment geometry, i.e. = 90°).

state of $Ni(CO)_3$, ^1E, resulting from the $11e \rightarrow 13a_1$ charge-transfer, correlates asymptotically with the HOMO-LUMO excitation of the parent molecule. Furthermore, the equilibrium geometry of the excited state is pyramidal, in contrast to the planar conformation of the ground state. Thus, Franck-Condon factors for the corresponding transition will be very unfavourable, rationalizing the observed rather long lifetime of about 15 μs and the width of the experimental spectrum [7]. Note that the lifetime of excited $Ni(CO)_4$ with respect to dissociation is estimated to be several orders of magnitude smaller (see part C), justifying the assignment of the emission to the fragment.

It is tempting to speculate whether a similar mechanism, as proposed here for $Ni(CO)_4$, is at work in the photodissociation of other transition metal carbonyls. It is quite conceivable that the absorption around 620 nm observed after photodissociation of $Cr(CO)_6$ [30] involves an orbital of the fragment $Cr(CO)_5$ which is lowered into the HOMO-LUMO gap of the hexacarbonyl due to the lacking interaction with the missing sixth ligand [56]. In the case of $Fe(CO)_5$, there is a direct correlation between an excited state and the triplet ground state of the corresponding fragment $Fe(CO)_4$. Therefore, no related emission is to be expected. However, charge-transfer initiated photodissociation may provide an alternative to the mechanism based on Fe 3d(e)->3d(a_1) excitation which has been postulated [50]. Further theoretical work is certainly necessary to explain the luminescence of complexes which have no occupied d orbitals, such as OsO_4 and CrO_2Cl_2. In these cases, the nature of the dissociating ligands is not as clear-cut as for the carbonyl complexes where stable, closed-shell ligands are released.

4. UV LASER PHOTOLYSIS OF $Ni(CO)_4$

The tetrahedral molecule $Ni(CO)_4$ has the lowest coordination number of all volatile carbonyl complexes. To our knowledge its photochemical decay channels have experimentally been studied only in the classical work by A.B. Callear [57], although valuable information concerning the electronic properties of $Ni(CO)_4$ are now available and the assortment of possible photofragments is surveyable. The experimental technique we used to investigate the visible luminescence, resulting from dissociative excitation of tetrahedral transition metal complexes, has already been described in detail [8]. In addition to the luminescence studies, we have carried out MPI experiments with the KrF (248 nm, 5 eV) and the XeCl (308 nm, 4 eV) laser since those lasers are frequently used for metal depositions.

4.1. EXPERIMENTAL PROCEDURE

In the first series of experiments we have studied the interaction of the laser radiation with a pulsed molecular beam ($Ar/Ni(CO)_4$: 9/1; $p_{tot} \approx 3$ bar). A skimmer placed at a suitable distance was used to collimate the beam. Its temporal and spatial characteristics could be determined with a fast ion gauge. In this beam apparatus, we have also carried out a number of scattering experiments including photoinduced reactive metal deposition. Those results have been reported elsewhere [58]. Here, we shall concentrate on the MPI of $Ni(CO)_4$. For this purpose, the molecular beam was excited with a well defined focused laser beam, 22 cm downstream from the nozzle source. The $Ni(CO)_4$ density at this distance was $\approx 2x10^{12}$ cm^{-3}. The trigger of the laser pulse was adjusted to interact with the maximum density of the molecular pulse. An axially mounted quadrupole mass spectrometer (UTI 100), with the ionization source switched off, served to monitor the laser generated ions. The interaction region was placed $\approx$ 2 cm in front of the entrance diaphragm of the mass spectrometer.

Multiphoton ionization and dissociation of $Ni(CO)_4$ at rather low uv laser intensities ($>10^5$ W/cm^2) has been studied in a 50 cm long cell equipped with suitable detection electronics as illustrated in fig. 4. The two collector electrodes had a distance of 15 mm and the unfocused parallel uv beam (r = 0.3 mm) was aligned along the axis of the sample cell. The partial pressure of $Ni(CO)_4$ was accurately determined from the corresponding absorption losses (σ_{KrF} = $3.0x10^{-17}$ cm^2). The entrance and transmitted laser energies were also measured with high precision. In this way, it was possible to determine absolutely the number of one-photon excited $Ni(CO)_4$ molecules and the number of multiphoton generated ions. The ionic fragments, in turn, were measured with the pulsed Townsend discharge technique [59], which shall be briefly explained.

If the interaction of the uv laser radiation with the $Ni(CO)_4$ molecules, free charge carriers (electrons or ions)

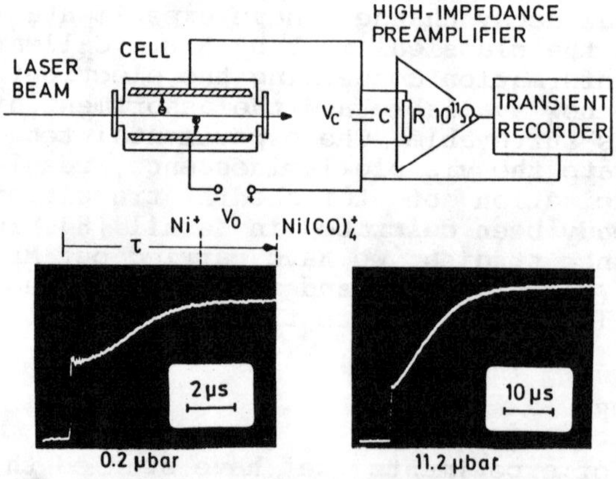

FIGURE 4. Schematic diagram of the cell experiment and two characteristic signal shapes at different pressures. At low pressures a TOF analysis is possible.

are produced, they will begin to move under the influence of the constant electric field applied across the electrodes, thus inducing an electric current in the external circuit. If the time constant of the external circuit RC is sufficiently large ($RC >> t^f$, time-of-flight of the charged particles), the time dependence of the voltage on the effective circuit-capacitance, C, is given by the expression:

$$(1) \quad U_c(t) = \frac{1}{C}\int[j_e(t) + j_i(t)]dt = \frac{e_o}{Cd}\left[\int V_e N_e\, dt + \int V_i N_i\, dt\right]$$

In eq. (1) $j_e(t)$ and $j_i(t)$ are the electron and ion components of the electric current, $N_{e,i}$ is the number of the charged particles, $v_{e,i}(t)$ their momentum velocity, and e_o is the electronic charge. Depending on the gas pressure in the cell, two characteristic regimes can be specified:

a) Free flight regime at low gas pressures. In this case the mean free path, l_p, of the particles is much larger than the distance between the electrodes d ($l_p >> d$) and their momentum velocity, given by $v(t) = [Ue_o/dm]*t$, will increase continously (U is the applied voltage, m is the mass of the particle). When $N_e = N_i = N$, the temporal shape of U_c will be given by (c.f. the 0.2 µbar signal on the figure):

$$(2) \quad U_c(t) = \frac{N_e e_o}{2C} + \frac{N_i e_o^2 U}{2Cd^2 m} t^2 \quad \text{for } t_e^f \le t \le t_i^f << RC$$

$$U_c(t) = \frac{N_e e_o}{2C} + \frac{N_i e_o}{2C} = \frac{N e_o}{C} = \text{const} \quad \text{for } t_e^f << t_i^f \le t << RC$$

Since the time-of-flight of the electrons $t^f_e=[m_e d^2/e_o U_o]^{1/2}$ is much shorter than that of the ions, the electron and ion components of the overall signal (see fig. 4) are very well distinguished and, using eq. (2), one can determine the absolute number of the electrons and the ions. It is also possible to identify different ionic species in the cell since t^f_i is mass dependent. However, as indicated in the figure, only Ni^+ ions are observed.

b) Drift regime at relatively high pressures. In this regime, l_p is much less than d, and, due to the collisional interaction of the charged particles with the gas molecules, they move with a constant drift velocity, w, towards the electrodes. The temporal shape of the voltage signal is then given by (c.f. the 11.2 µbar signal on the figure):

$$(3) \quad U_c(t) = \frac{N_e e_o}{2C} + \frac{N_i e_o W_i}{Cd} t \quad \text{for } t_e^d \le t \le t_i^d << RC$$

$$U_c(t) = \frac{N_e e_o}{2C} + \frac{N_i e_o}{2C} = \frac{N e_o}{C} = \text{const} \quad \text{for } t_e^d << t_i^d \le t << RC$$

Since the drift time of the electrons t^d_e is again much shorter than that of the ions t^d_i, it is also possible in this regime to determine the absolute number of the electrons and of the ions as well as their drift velocities. It is interesting to note that the deviation from the 1:1 ratio between electrons and ions may be interpreted as an electron attachment reaction $e^- + Ni(CO)_4 \rightarrow Ni(CO)_3^- + CO$.

4.2. MULTIPHOTONIONIZATION OF Ni(CO)$_4$

Let us start the discussion of our MPI results with the absorption spectrum of $Ni(CO)_4$ shown in fig. 5. We have

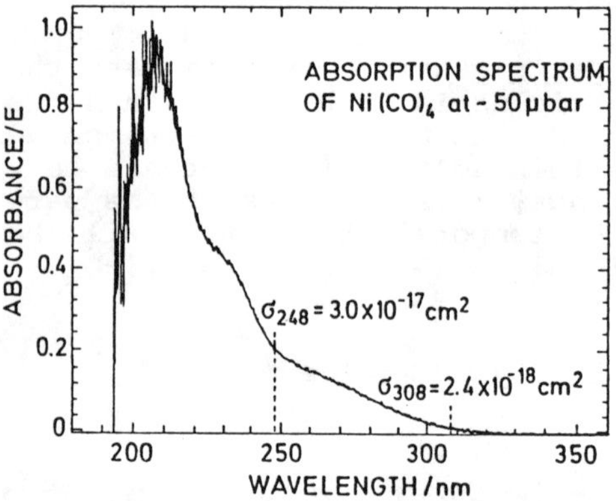

FIGURE 5. Absorption spectrum of $Ni(CO)_4$ at room temperature, cell length 10 cm. The noise in the high energy part of the spectrum is artifactial.For a possible assignment see ref. [52,53]

already discussed that this complex decays into the electronically excited fragment $Ni(CO)_3^*$ upon one-photon excitation with a XeCl laser [7]. An estimate of the quantum yield for this process revealed a value close to unity. This was evident, by comparing the fluorescence of the laser dye LC 6400 with the luminescence of $Ni(CO)_3^*$ excited at identical optical density, geometry and input energy. The quantum yield for KrF laser excitation is lower, indicating further decomposition in competition to the formation of $Ni(CO)_3^*$.

The oscillator strength of the first electronic transition, $f \approx 0.1$, deduced from the integrated absorption spectrum ($\sigma_{max} \approx 2.0 \times 10^{-17}$ cm^2, $\approx 10^4$ cm^{-1}) according to $f = 5.7 \times 10^{11}\ \sigma_{max}$, corresponds to a hypothetical fluorescence decay rate, $k_{fl}(s^{-1}) = 0.67\ ^2$●$f$, of approximately 10^8 s^{-1}. However, resonance fluorescence has never been observed. This means that the dissociation rate of $Ni(CO)_4^*$ into $Ni(CO)_3^*$ + CO, at nearly 100 % efficiency, should be at least 10^{10} s^{-1}. From these experiments we may conclude that the excess energy for one-photon excitation of $Ni(CO)_4$ is primarily retained as electronic excitation of $Ni(CO)_3^*$. This means, the absorbed energy is not consumed to break as many Ni-(CO) bonds as would be energetically possible.

Multiphoton excitation of $Ni(CO)_4$ results in the formation of ionic species. This is shown in fig. 6. As mentioned

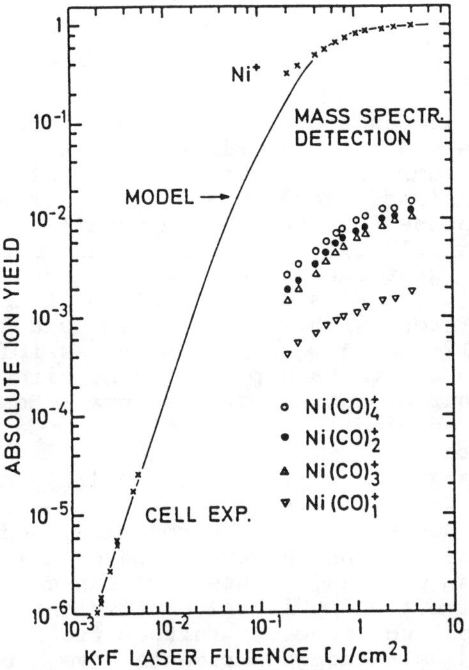

FIGURE 6. Fluence dependence of $Ni(CO)_4$ multiphoton ionization. The solid line is obtained from model calculations.

in the experimental section, we have performed cell experiments, yielding the absolute ion density, and beam experiments, yielding a relative ion density and the ionic fragmentation pattern. It was not possible to use only one technique because increasing the laser fluence in the cell could only be accomplished by focusing the laser beam, thus loosing specificity. On the other hand, the mass spectrometer was not sensitive enough to allow measurements at lower fluences. Our data points demonstrate that the initial fluence (or intensity) dependence of the ion signal is exactly of third order.

To adapt both data sets (from the cell and from the beam experiment) to one another, we took the saturation of the Ni^+ yield as an indication that all excited $Ni(CO)_4$ molecules have been transformed to Ni^+. This procedure would be conflicting for a polarized light source [60]. An ideally polarized light beam interacting with an ensemble of molecular transition dipole moments, randomly oriented in space allows for a maximum excitation probability p_{ex} = 1/3 per molecule after spatial averaging. For an n-step multiphoton process, the corresponding maximum excitation probability will be $p^{c}_{ex} = (2n + 1)^{-1}$ for a completely coherent process and $p^{ic}_{ex} = (1/3)^n$ for a completely incoherent process.

With this restriction we were able to determine the absolute ion yield as well as the fragmentation pattern for the whole range of laser fluences. From the TOF analysis of the experimental results in fig. 4, it is evident that even at very low fluences the Ni^+ signal is predominant.

4.3. RATE EQUATION MODEL

We shall now try to derive an appropriate rate equation model to account for the observed results. First we note that Ni^+ is generated in a third order process even, at so-called offresonance excitation conditions. This means that if we assume a second order process for the decomposition of $Ni(CO)_4$ (H^o= 140 kcal/mol = 6.1 eV) according to $Ni(CO)_4$ --->---> 4 CO + Ni, we have just one photon left to ionize the Ni atom (I.P. = 7.63 eV). This is only possible if the nascent Ni atoms are generated with an electronic excess energy above 2.63 eV (21212 cm^{-1}). There are accessible states in this energy region and we have presently no direct experimental evidence in favour or against this scheme. However, the model must also be adapted to the concerted appearance of the $Ni(CO)_n^+$ fragments.

It has already been suggested [4] that collisional charge transfer, like Ni^{+*} + $Ni(CO)_4$ (I.P. = 8.03 eV) ---> Ni + $Ni(CO)_4^{+*}$ and subsequent decay of the parent ion, can account for the various molecular ions. However, conservation of energy would rule out this process, since Ni^{+*} can carry at most 8.9 eV (15 eV - 6.1 eV) which is only sufficient to ionize the parent molecule but not to cause further fragmentation. One must admit that, although experiments have been performed under molecular beam conditions, a 1% collision probability is conceivable, assuming a density of 10^{12} cm^{-3}, an interaction length of 2 cm and a collisional cross-section of 50 $Å^2$.

A collision-induced mechanism would readily explain that the fragmentation pattern or the ratio between Ni^+ and $Ni(CO)_n^+$ is independent of the applied laser fluences (within a factor of two due to the rather large experimental error). If it appears, in the course of future experiments, that collisions are not a substantial reaction path, we must conclude that all ions are generated from the same precursor state. Furthermore, since $Ni(CO)_4^+$ is also observed, we infer that the three excitation steps proceed within the parent molecule.

Let us discuss this idea in more detail. It is interesting to note that a two photon ionization process which would also yield $Ni(CO)_4^+$ is obviously missing, thus indicating that dissociation at this level is faster than ionization. This is in accordance with similar observations in the case of $Fe(CO)_5$ and $Cr(CO)_6$. Since the adiabatic ionization potential of $Ni(CO)_4$ is 8.03 eV we have to assume an electronically excited $Ni(CO)_4^{+*}$ species to be the virtual precursor which decays into smaller ionic fragments, mainly into Ni^+ + 4 CO.

Another point where we can only speculate at the present time is the excitation process by which $Ni(CO)_4^{+*}$ is created.

Is it a stepwise or a coherent three photon process? For a coherent process to be operative, Jortner et al. [60] assumed that power broadening (μ_{Debye} x E) at each level exceeds the intrinsic lifetime broadening ($\pi k_i/c$) due to the dissociative decay rate k_i. This means, inserting the appropriate conversion factors, the following inequality should hold: $3.2 \times 10^6\ \mu_{Debye}\ [I_o(W/cm^2)]^{1/2} > k_i$. Since k_i should be larger than $10^{10}\ s^{-1}$, we notice that a coherent process is deterious at the lower end of the investigated fluence scale (10^5 --> 10^9 W/cm^2).

It is certainly illustrative to compare the coherent and the stepwise processes by means of the corresponding rate equation models (a rate equation model would be conflicting in the case of atomic excitation as the effective absorption cross section would decrease due to power broadening [26]).

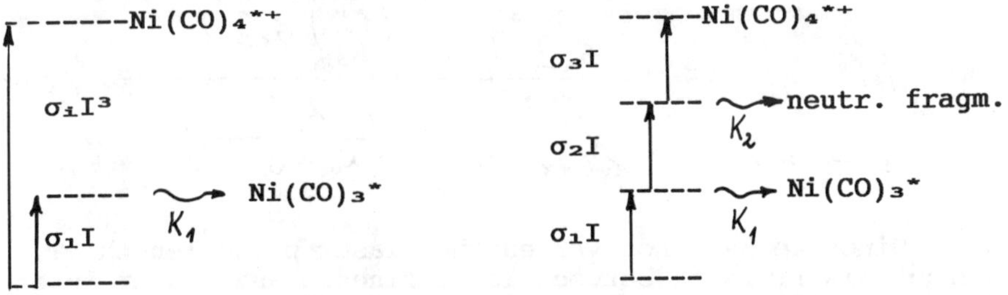

coherent excitation   stepwise excitation

Assuming a rectangular laser pulse shape (T = 15 ns), and including one photon excitation ($Ni(CO)_3^*$) in the coherent ladder, the rate equations yield the following $Ni(CO)_4^{+*}$ concentrations after the end of the laser pulse:

<u>coherent</u>

$$Ni(CO)_4^{*+} = \frac{\sigma_i I^3}{\sigma_1 I + \sigma_i I^3}\left\{1 - e^{-(\sigma_1 I + \sigma_i I^3)T}\right\} Ni(CO)_4$$

<u>stepwise</u>

$$Ni(CO)_4^{*+} = \frac{\sigma_3\sigma_2\sigma_1 I^3}{K_1+\sigma_2 I-\sigma_1 I}\left\{\frac{1}{K_3+\sigma_3 I-\sigma_1 I}\left[\frac{1}{\sigma_1 I}\left(1-e^{-\sigma_1 I T}\right)-\frac{1}{K_2+\sigma_3 I}\right] - \frac{1}{(K_1+\sigma_2 I)(K_2+\sigma_3 I)}\right\} Ni(CO)_4$$

These relations have been simplified assuming $k_i T \gg 1$. Due to the fact that we cannot specify in our step model the excitation cross sections $\sigma_{2,3}$ and the decay rates k_i we may assume, without changing the structure of the final solution, $\sigma_{2,3} = \sigma_o$ and $k_i = k_o$ and thus obtain:

$$Ni(CO)_4^{*+} = \left(\frac{\sigma_o I}{K_o}\right)^2 \left\{ \frac{K_o^2}{(K_o + \sigma_o I)^2} - e^{-\sigma_1 I T} \right\} Ni(CO)_4$$

Evaluating these expressions at low and high laser intensities we find:

$$\begin{array}{llcl} & & \text{coherent} & \text{stepwise} \\ I \to 0 & & \sigma_i I^3 T & \left(\frac{\sigma_o}{K_o}\right)^2 \sigma_1 I^3 T \\ & Ni(CO)_4^{*+} = & & \qquad\qquad Ni(CO)_4 \\ I \to \infty & & \frac{\sigma_i I^3}{\sigma_1 I + \sigma_i I^3} & \left(\frac{\sigma_o I}{K_o + \sigma_o I}\right)^2 \end{array}$$

Since we have not varied the laser pulse length T, we cannot discriminate between laser fluence and laser intensity. Therefore, we can discuss the final result in terms of laser fluences Φ_{KrF} (1 mJ/cm^2 = 1.25×10^{15} phot/cm^2). From the experiments we can evaluate the following data: $\sigma_1 IT = 3.75 \times 10^{-2} \Phi$ and $\sigma_i I^3 T = 1.93 \times 10^{-7} \Phi^3$ where Φ is expressed in mJ/cm^2. Since $\sigma_i I^3 T$ must be identical to $(\sigma_o/k_o)^2 \sigma_1 I^3 T$, we find for the ratio between excitation crosssection and decay rate $\sigma_o/k_o = 3.1 \times 10^{-26}$ cm^{2}s. This implies that the average absorption cross-section for the second and the third step should exceed 3.1×10^{-16}cm^2 since k_o is certainly larger than 10^{10}s^{-1}. However, this enormous value is very unlikely for the anticipated rather broad absorption bands of the transient fragments. Therefore, we prefer the coherent excitation model where the three photon process is in competition to the one photon process ($Ni(CO)_3^*$), both interacting solely with the ground state of the parent molecule. With this simplification there is no free parameter left because the efficiencies of both channels are known. The solid line in fig. 6 corresponds to this model. The agreement between model and experiment is quite satisfactory, but of course, it does not exclude the validity of other models.

4. SUMMARY

In this manuscript, we have outlined our present knowledge concerning the primary photochemical decay processes of metal carbonyls. This is a rapidly growing and fascinating field of research which is related to a variety of practical applications like photocatalysis and laser induced metal depositon. Moreover, if electronically excited fragment states (and states of the bare atoms) are the key to the photodynamics, these states must be generated with an inversion of excited state population. This means a new class of photolysis lasers might be within reach [9b]. Nevertheless, a better experimental characterization of the various neutral fragments on the photolysis path to the ionized atom is needed.

REFERENCES

[1] see for example: W.E. Hollingworth and V. Vaida; J. Phys. Chem. 90, 1235 (1986)

[2] M. Wrighton; Chemical Reviews 74, 401, (1974)

[3] D.M. Manuta and A.J. Lees; Inorg. Chem. 25, 1354 (1986)

[4] R.L. Whetten, Ke-Jian Fu and E.R. Grant; J. Chem. Phys. 79 4899 (1983)

[5] G.L. Geoffrey and M.S. Wrighton; Organometallic Photochemistry (Academic Press, New York, 1979)

[6] W.C. Torgler, S.R. Desjardins and E.I. Solomon; Inorg. Chem. 18, 2131 (1979)

[7] N. Rösch, M. Kotzian, H. Jörg, H. Schröder, B. Rager, S. Metev; J. Am. Chem. Soc. 108, 4238 (1986)

[8] K.L. Kompa, H. Lamprecht, H. Schröder, A.A. Puretzky, V.V. Tyakht; J. Chem. Phys. 84, 2020 (1986)

[9]a H. Schröder, H. Lamprecht, K.L.Kompa; Appl. Phys. B 28, 180 (1982)
b H. Lamprecht; Doktorarbeit, LMU München, 1985

[10] H. Schröder, I. Gianinoni, D. Masci, K.L. Kompa; Laser Processing and Diagnostics, ed. D. Bäuerle, (Springer Ser. Chem. Phys. 39, 1984), p. 257

[11] K.E. Lewis, D.M. Golden, G.P. Smith; J. Am. Chem. Soc. 106, 3905 (1984)

[12] A.E. Stevens, C.S. Feigerle, W.C. Lineberger; J. Am. Chem. Soc. 104, 5026 (1982)

[13] J.A. Connor; Curr. Top. Chem. 71, 71 (1977)

[14]a J.P. Day, F. Basolo, R.G. Pearson; J. Am. Chem. Soc. 90, 6927 (1968)
b J.P. Day, R.G. Pearson, F. Basolo; ibid 90, 693, (1968)

[15]a R.E. Winters, R.W. Kiser; J. Chem. Phys. 44, 1964 (1966)
b R.N. Compton, J.A.D. Stockdale; Int. J. of Mass Spectr. and Ion Physics 22, 47 (1976)
c P.M. George, J.L. Beauchamp; J. Chem. Phys. 76, 2959 (1982)

[16] J.L. Beauchamp, P.M. George; US patent, 4324854 (1982)

[17] H. Schröder, K.L. Kompa, D. Masci, I. Gianinoni; Appl. Phys. A38, 227 (1985)

[18] M.A. Duncan, T.G. Dietz, R.E. Smalley; Chem. Phys. 44, 415 (1979)

[19] P.C. Engelking; Chem. Phys. Let. 74, 207 (1980)

[20] G. Nathanson, B. Gitlin, A.M. Rosan, J.T. Yardley; J. Chem. Phys. 74, 361 (1981)

[21] Y.T. Yardley, B. Gitlin, G. Nathanson, A.M. Rosan; J. Chem. Phys. 74, 370 (1981)

[22] M.A. De Paoli, S.M. de Oliveira; J. Chem. Phys. 80, 730 (1984)

[23] Y. Nagano, Y. Achiba, K. Kimura; J. Phys. Chem. 90, 615 (1986)

[24] Y. Nagano, Y. Achiba, K. Kimura; J. Phys. Chem. 90, 1288 (1986)

[25] D.P. Gerrity, L.J. Rothberg, V. Vaida; Chem. Phys. Let. 74, 1, (1980)

[26] G.J. Fisanick, A. Gedanken, T.S. Eichelberger IV, N.A. Kuebler, M.B. Robin; J. Chem. Phys. 75, 5215 (1981)

[27] T.R. Fletcher, R.N. Rosenfeld; J. Am. Chem. Soc. 107, 2203 (1985)

[28] T.A. Seder, S.P. Church, A.J. Ouderkirk, E. Weitz; J. Am. Chem. Soc. 107, 1432 (1985)

[29] S.K. Huang, M.L. Gross; J. Phys. Chem. 89, 4422 (1985)

[30] W.H. Breckenridge, G.M. Stewart; J. Am. Chem. Soc. 108, 364 (1986)

[31] see for example: M. Poliakoff, A. Ceulemans; J. Am. Chem. Soc. 106, 50 (1984)

[32] J.S. Griffith, The Theory of Transition Metal Ions, (Cambridge University Press, Cambridge, 1961).

[33] C.J. Ballhausen, H.B. Gray, Molecular Orbital Theory, (Benjamin, New York, 1965)

[34] R. Hoffmann, Angew. Chem. 94, 725 (1982); Angew. Chem. Int. Ed. 21, 711 (1982)

[35] T.A. Albright, J.K. Burdett, M. Whangbo, Orbital Interactions in Chemistry (Wiley, New York, 1985)

[36] A. Veillard, Ed., Quantum Chemistry: The Challenge of Transition Metals and Coordination Chemistry (Reidel, Dordrecht, 1986)

[37] F.A. Cotton, Chemical Applications of Group Theory, 2nd ed. (Wiley, New York, 1971)

[38] A. Szabo, N.S. Ostlund, Modern Quantum Chemistry: Introduction to Advanced Electronic Structure Theory (Macmillan, New York, 1982)

[39] K. Faegri, Jr., J. Almlöf; Chem. Phys. Lett. 107, 121 (1984)

[40] C.W. Bauschlicher, Jr., P.S. Bagus, J. Chem. Phys. 81, 5889 (1984)

[41] H.P. Lüthi, P.E.M. Siegbahn, J. Almlöf, K. Faegri, Jr., A. Heiberg; Chem. Phys. Lett. 111, 1 (1984)

[42] L. Hedberg, T. Iijima, K. Hedberg; J. Chem. Phys. 70, 3224 (1979)

[43] W.E. Dasent; Inorganic Energetics (Cambridge University Press, Cambridge, 1982)

[44] H. Jörg, N. Rösch; Chem. Phys. Lett. 129, 359 (1985)

[45] S. Lundqvist, N.H. March, eds. Theory of the Inhomogeneous Electron Gas (Plenum Press, New York, 1983)

[46] B.I. Dunlap, J.W.D. Connolly, J.R. Sabin; J. Chem. Phys. 71, 3396 (1979); 4993 (1979)

[47] N. Rösch, H. Jörg, B.I. Dunlap; in ref. 36, p. 179.

[48] N. Rösch, H. Jörg; J. Chem. Phys. 84, 5967 (1986)

[49] R.O. Jones, O. Gunnarsson; Phys. Rev. B 31, 7588 (1985)

[50]a C. Daniel, M. Benard, A. Dedieu, R. Wiest, A. Veillard, J. Phys. Chem. 88, 4805 (1984) b C. Daniel, A. Veillard; in ref. 36, p.363

[51] P. Madhavan, J.l. Whitten; Chem. Phys. Lett. 127, 354 (1986)

[52] M.C. Zerner, G.H. Loew, R.F. Kirchner, U.T. Mueller-Westerhoff; J. Am. Chem. Soc. 102, 589 (1980)

[53] B. Dick, H.-J. Freund, G. Hohlneicher; Mol. Phys. 45, 472 (1982)

[54] N. Rösch, M. Kotzian, M.C. Zerner; unpublished results

[55] N. Rösch, M. Kotzian, H. Jörg; J. Chem. Phys., submitted

[56] M. Elian, R. Hoffmann; Inorg. Chem. 14,1058 (1975)

[57] A.B. Callear; Proc. Roy. Soc. A265, 71 (1961)

[58] B. Rager, W. Robers, H. Schröder, K.L. Kompa; E-MRS Proceedings, Straßburg 1984

[59] L.G. Christophorou ed. Electron-Molecule Interactions and Their Applications, (Academic Press, New York, 1984). Vol 2, p. 105

[60] S. Leutwyler, U. Even, J. Jortner; Chem. Phys. 58, 409 (1981)

KINETICS OF LASER-INDUCED PYROLYTIC CHEMICAL PROCESSES AND THE PROBLEM OF TEMPERATURE MEASUREMENTS

J. DOPPELBAUER and D. BÄUERLE, Angewandte Physik,
Johannes-Kepler-Universität Linz, A-4040 Linz, Austria

ABSTRACT

A detailed analysis of the fundamentals of pyrometric temperature determination in laser processing is presented. In particular, we discuss the effects of an inhomogeneous temperature distribution and of photogenerated excess carriers on the apparent emissivity. The implications for experimental temperature determination are outlined. The technique has been applied to kinetic studies on laser-induced deposition of carbon from various hydrocarbons, and of silicon from silane.

1. INTRODUCTION

Laser-induced chemical processing such as, for example, laser-induced chemical vapor deposition (LCVD) and laser-induced reactive etching, is an emerging new technique with many practical applications /1,2/. Therefore, kinetic studies are not only of academic interest but also help to optimize processing variables. A unique tool for studying the reaction kinetics in <u>pyrolytic</u> LCVD is steady growth of rods /3/. Here, apparent chemical activation energies and reaction orders can be derived from the temperature and pressure dependence of the axial growth rate. The technique also allows for the <u>in situ</u> observation of localized high temperature reactions in otherwise cold environments. Such investigations yield valuable information on fundamental processes that are also relevant to technologies which are widely used at present such as, for example, standard chemical vapor deposition (CVD).

One fundamental problem in investigating laser-induced photothermal processing in general is related to the difficulties in measuring laser-induced temperature distributions. For this reason, a great variety of experimental techniques have been investigated to solve this problem /3-14/. Today, the most promising approach appears to be photoelectric pyrometry. In this technique, the local temperature is derived from the intensity or the spectrum of the optical radiation emitted from the laser heated surface. The advantages of this method are (i) a high temperature sensitivity, (ii) its suitability for in situ measurements, and (iii) its relative insensitivity (as compared with other techniques) to surface properties. The technique of measuring the thermal radiation that is emitted from laser-heated solids has already been reported in the literature /3,11-14/. However, a detailed analysis of the fundamentals of the method is still lacking.

In Chapter 2, we discuss the differences between pyrometry on laser-heated solid surfaces and conventional pyrometry. In

Sect. 2.1, we give a phenomenological treatment which relates the emission of radiation from solid surfaces to the local values of the energy level populations. We use radiative transfer theory /16/ together with detailed balance arguments. In Sect. 2.2 it will be shown that the apparent emissivity of laser-heated solids may differ considerably from that of the same material in thermal equilibrium. Some experimental aspects of pyrometric temperature measurements on laser-heated surfaces are discussed in Sect. 2.3. The applicability of the technique to laser-induced steady growth of rods is treated in Sect 2.4. In Chapter 3, the technique of local temperature measurements is applied to kinetic studies. After a few comments on the experimental setup, we present in Sects. 3.2 and 3.3 detailed kinetic measurements for the examples of Ar^+ laser-induced deposition of carbon from various hydrocarbons, and of silicon from silane.

2. THE PROBLEM OF TEMPERATURE MEASUREMENTS

Various optical techniques have already been applied to measure laser-induced temperature distributions. These include (i) the measurement of the optical constants /4-6/, (ii) Raman spectroscopy /7,8/, (iii) thermal dilatation measurements /9,10/, and, (iv) measurements of the intensity or the spectrum of the radiation emitted from the laser heated surface /3,11-14/. In most cases, the determination of the maximum temperature is sufficient to characterize the process. This is due to the exponential temperature dependence of reaction rates in pyrolytic processing.

Kinetic studies of laser-induced pyrolytic chemical processes require highly precise temperature measurements. For example, it has been shown that for LCVD of Si from SiH_4 the axial growth rate increases by about an order of magnitude when increasing the local temperature from 1200 to 1300K /11/. The precision obtainable with methods (i) and (ii), however, is only about ±100K /5,6,8/. This is mainly due to the rather weak temperature dependence of the parameters which enter the evaluation of the temperature.

Pyrometric measurements, on the other hand, are highly temperature sensitive. However, the emission of optical radiation from laser-heated solids may differ considerably from that of the same material in thermal equilibrium. This difference is mainly due to the localization of the laser-induced temperature distribution and direct, non-thermal excitations induced by the laser light.

2.1. Phenomenological description of thermal radiation

The usual way to treat the emission of thermal radiation from solids is based on equilibrium arguments /17/. Obviously, such a treatment becomes inadequate in the presence of temperature inhomogeneities and of deviations from the thermal equilibrium population of energy levels. Instead, the emitted radiation has to be directly related to the local values of the energy level populations. In the following, this will be accomplished by using radiative transfer theory and the principle of detailed balance.

The propagation of electromagnetic radiation and its interaction with matter can be described by the phenomenological theory of radiative transfer (Chandrasekhar /16/). The fundamental quantity is the spectral radiance L_ω (the subscript ω denotes the differential, i.e. $L_\omega = \partial L/\partial\omega$); $L_\omega d\omega \cos\theta d\Omega dA$ is the flux of radiation in the frequency range between ω and $\omega + d\omega$ through an area element dA, into a solid angle $d\Omega$ around the direction θ, θ being the angle with respect to the normal on dA. The propagation of radiation is described by the so-called equation of transfer

$$\cos\beta \; \partial L_\omega/\partial z = -\alpha(\omega)\, L_\omega + \varepsilon_\omega \qquad (1)$$

where β is the angle between the direction of light propagation and the z-axis (see Fig. 1), $\alpha(\omega)$ the absorption coefficient, and ε_ω the spectral emission coefficient. $\varepsilon_\omega d\omega d^3x d\Omega$ is the flux of radiation emitted from the volume element d^3x into the solid angle $d\Omega$. Scattering of radiation is ignored. The radiance emitted from the surface of the material into the direction β', $L_{\omega e}$, is obtained from the general solution of (1), giving us

$$L_{\omega e}(z_o,\beta') = (1 - R)\, n^{-2}\, L_\omega(z_o,\beta) \qquad (2)$$

$$\text{with } L_\omega(z_o,\beta) = L_\omega(0,\beta).\exp(-\tau(z,0)) + \int_0^z dz'\varepsilon_\omega(z',\beta)\exp(-\tau(z,z'))$$

$$\text{and } \tau(z,z') = \int_{z'}^{z} dz''\, \alpha(z'')/\cos\beta.$$

R and n are the reflectivity and the index of refraction, respectively. β and β' are related through the law of refraction; the angular and polarization dependences of R are given by the Fresnel formulae. Here, the direction of light propagation is from the interior to the surface ($z = z_o$). $L_\omega(0,\beta)$ is the spectral radiance at z = 0 inside the material. The factor n^{-2} is based on the law of refraction: on crossing the surface the divergence is increased by a factor n, reducing the radiance by n^2. Since radiance is a concept within geometrical optics, the validity of (2) is restricted to material volumes with dimensions larger than the wavelength of the radiation. For a semiinfinite medium in thermal equilibrium at temperature T, Eq. (2) becomes

$$L_{\omega e} = (1 - R)\, n^{-2}\, L_{\omega b} = (1 - R)\, L_{\omega bo} \qquad (3)$$

Here, $L_{\omega b} = (n^2\hbar\omega^3/4\pi^3c^2)\,(\exp(\hbar\omega/kT) - 1)^{-1}$ is the spectral radiance of the blackbody radiation as given by Planck's law; $L_{\omega bo}$ is that in vacuum (n = 1). Eq. (3) is equivalent to Kirchhoff's law. Here, the emissivity e, defined as the ratio of the emitted radiance to that of the blackbody, is simply given by (1 - R). For media that are not in thermal equilibrium, such as laser-heated solids, appropriate local values of α and ε_ω have to be used in (2).

Microscopically, the quantities α and ε_ω are related to stimulated and spontaneous transitions between the energy levels of the material. There are two definitions of the term emission: one counts spontaneous and stimulated emission as emission, the second counts stimulated emission as negative absorption. Consistency of the radiation laws is obtained most simply if stimulated emission is treated as negative absorption (Baltes /18/). Using detailed balance arguments, a relation can be derived between the absorption coefficient and the emission coefficient, which for an electronic system is given by /19,20/

$$\varepsilon_{\omega}^{i}(\omega) = \alpha^{i}(\omega)\,(n^2\hbar\omega^3/4\pi^3c^2)(f_u(1 - f_l)/(f_l - f_u)) \qquad (4)$$

Here, ε_{ω}^{i} and α^{i} are the contribution of the ith pair of levels to emission and absorption of radiation, and f_u and f_l are the occupation probabilities of the upper and lower levels respectively. The energy difference between levels u and l is $\hbar\omega$. Eq. (4) is valid even if the system is out of thermal equilibrium. In local thermal equilibrium, $f_u(1 - f_l)/(f_l - f_u)$ reduces to $(\exp(\hbar\omega/kT) - 1)^{-1}$.

2.2. Apparent emissivities

Laser radiation in the near infrared, visible and ultraviolet spectral range, which is most commonly used in laser chemical processing, interacts primarily with the electrons of the solid. The energy transfer mechanisms in highly photoexcited solids have been discussed in detail in connection with pulsed laser annealing (see for example /4,21/). The electron-phonon energy relaxation time is of the order of 10^{-12}s. Therefore, in most laser processing situations, carriers and phonons can be described by the same local temperature T.

The absorbed laser power is deposited in a spatially localized region near the surface and thus leads to high temperature gradients. As a consequence, the emissivity from this region will differ from that of the same material in thermal equilibrium, and it will henceforth be described by an apparent quantity. In semiconductors this apparent emissivity may be additionally affected by a high concentration of photogenerated excess carriers. Apart from the paper of Babelot et al. /15/, the implications of these facts have not yet been considered in the literature.

The ratio of the apparent emissivity e_{app} and the emissivity e of a solid with homogeneous temperature can be obtained from (2) and (3)

$$e_{app}/e = L_{\omega}(z_o) / L_{\omega b} \qquad (5)$$

2.2.1. The consequences of localization. First, we discuss the effect of an inhomogeneous temperature distribution on the emissivity. In many cases it is sufficient to consider the temperature profile perpendicular to the surface (i.e. in z-direction). If the refractive index of the material is high, the light rays propagating within the material in a small solid angle around the surface normal are refracted away from the normal outside and fill a solid angle which corresponds to typical observation geometries. Temperature variations along the surface are not taken into consideration.

To demonstrate the order of magnitude and the temperature and wavelength dependences of the effect, we have calculated the ratio (5) for cw-laser processing and the wavelength range 500nm to 3.5µm of the emitted radiation. The temperature distribution induced by a Gaussian cw-laser beam at normal incidence in a semiinfinite solid has been calculated by Lax /22/ and by Moody and Hendel /23/. The laser-induced temperature profile along the z-axis (center of the laser beam) is given by

$$T(z) = (T_s - 300)\pi \exp(2((z_o-z)/w_o)^2) \times$$
$$\times (1 - \mathrm{erf}(2^{1/2}(z_o-z)/w_o))/2 \;\; + 300 \qquad (6)$$

where T_s is the temperature at the surface ($z = z_o$), and w_o the Gaussian beam radius ($1/e^2$ point in irradiance). A schematic is shown in Fig. 1. If we introduce a characteristic depth w of the temperature distribution, defined as the distance from the material surface to the point within the solid at which the temperature rise is 90% of the peak value, we obtain from (6) $w = 0.07w_o$. Thus w is directly related to the laser beam radius. The more the beam is focused, the steeper is the temperature gradient. The situation is somewhat different for multilayer, combined or quasi-onedimensional structures that are relevant in laser chemical processing /1/. In such cases the temperature gradients may be much less pronounced than in homogeneous bulk material.

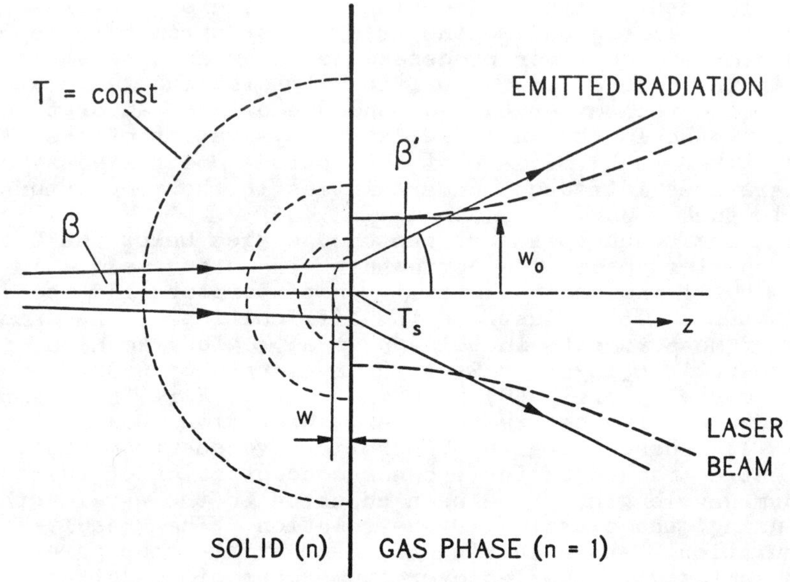

FIGURE 1. Schematic showing the emission of thermal radiation from cw-laser-heated bulk material. T_s is the laser-induced temperature in the center of the beam, w_o the Gaussian beam radius at the $1/e^2$ point in irradiance, w the characteristic depth of the temperature distribution, and β and β' are the angles between the light ray drawn and the z-axis. Some isotherms are indicated by the dashed lines.

We now further assume that the absorption coefficient α is temperature-independent. The apparent emissivity can then be described by the ratio of w and the optical absorption length α^{-1}. Since from (2) effective contributions to the emitted radiation are expected mainly from distances up to α^{-1} from the surface and because for most materials α increases with increasing temperature, one should take the value of α at the surface temperature. If $\alpha^{-1} << w$, then e_{app}/e will approach unity. Detailed knowledge of the temperature distribution is required only for $\alpha^{-1} > w$. The dependence of e_{app}/e calculated from (5) for the temperature profile (6) is shown in Fig. 2 for various values of w/α^{-1}. Here, the integrals in (2) were numerically evaluated. The figure shows that e_{app}/e increases with increasing temperature and wavelength. With $w/\alpha^{-1} = 10$, the deviation from unity is less than about 10%.

The consequences of the results shown in Fig. 2 will now be discussed for different cases.

For highly absorbing materials, with α higher than about $10^5 cm^{-1}$ (metals, pyrolytic carbon, semiconductors at wavelengths of the emitted radiation far above the band gap), the localization of the temperature distribution does not much affect the emissivity if the characteristic depth of the temperature distribution is larger than a few microns. For example, for silicon, at photon energies of the emitted radiation above the band gap, the absorption coefficient α is higher than $10^4 cm^{-1}$, and, with increasing temperature, may reach up to $10^6 cm^{-1}$ (Jellison and Modine /24/). For pyrometric temperature measurements on laser processed zones of an even smaller depth and in the case of low absorption, theoretical temperature profiles and the temperature dependence of the absorption coefficient should be included in the evaluation of the temperature. The lack of data of absorption coefficients at high temperatures introduces uncertainties in the temperatures derived in such a way.

With semiconductors, at photon energies below the band gap, free carrier absorption dominates. In this case, α is relatively small and depends heavily on temperature and the mode of excitation. To illustrate the difficulties in the interpretation of measurements in this spectral region, we have calculated the ratio e_{app}/e for Si at a wavelength of 3.0µm, using the temperature profile (6) with $w_o = 14$µm. For the temperature dependence of the free carrier absorption coefficient, the power laws describing the theoretical values from Bhattacharyya and Streetman /25/ for a donor concentration of $10^{18} cm^{-3}$ and 10.6µm wavelength, have been adjusted to the wavelength of 3µm by using the classical Drude relation. The results of our calculation, shown in Fig. 3, demonstrate that here the apparent emissivity varies over two orders of magnitude between 500K and the melting point. In reality, the increased carrier density due to laser excitation causes an increased free carrier absorption and thereby a higher apparent emissivity. However, in general no quantitative information is available about the actual excess carrier density during laser processing.

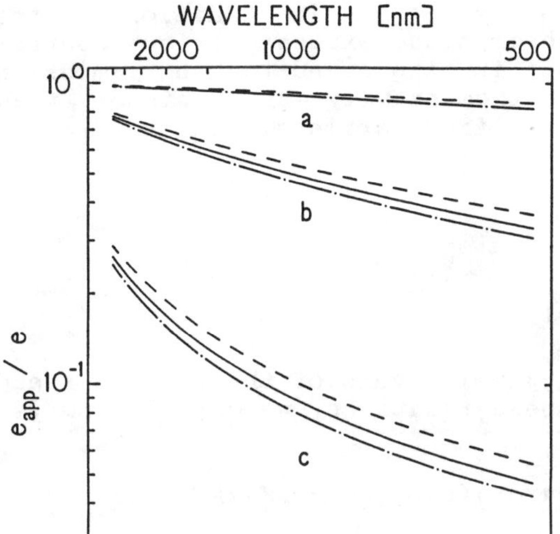

FIGURE 2. Ratio of the apparent emissivity due to localization e_{app} and the emissivity e of the same material at homogeneous temperature equal to the surface temperature as a function of the photon energy. The temperature variation perpendicular to the surface is given by Eq. (6), the surface temperatures are 1000K (dot-dashed), 1200K (solid), and 1500K (dashed). The optical absorption coefficient is assumed to be constant, the parameters (w/α^{-1}) of the curves are (a) 10, (b) 1, and (c) 0.1.

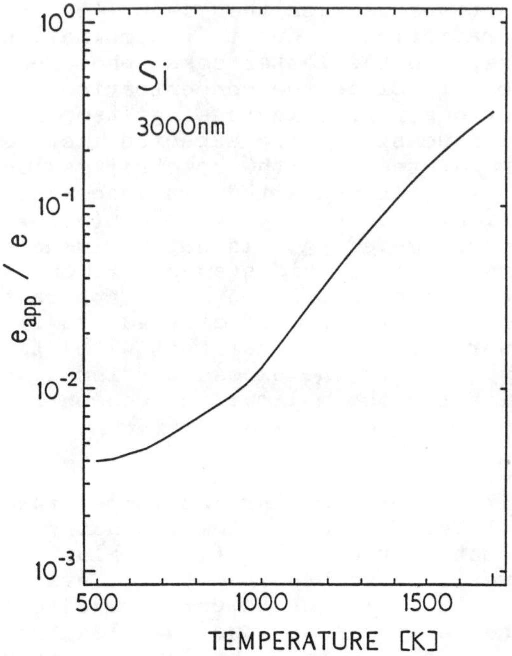

FIGURE 3. Calculated ratio of the apparent emissivity at 3000nm of laser heated Si and the emissivity at homogeneous temperature. The laser beam radius used in the calculation of the temperature profile is w_o = 14µm, the theoretical values for the free carrier absorption coefficient of Si (N_D = $10^{18}cm^{-3}$) are taken from Bhattacharyya and Streetman /25/

2.2.2. The influence of excess carriers. In addition to the enhancement in free carrier absorption, excess carriers contribute directly to an increase in the band-to-band emission coefficient, and the apparent emissivity may be higher than that in equilibrium. For thermalized carriers, f_u and f_l are given by

$$f_u = (\exp((E_u - F_u)/kT) + 1)^{-1}$$

and (7)

$$f_l = (\exp((E_l - F_l)/kT) + 1)^{-1}$$

where F_u and F_l are the quasi-Fermi levels of the electrons and holes, respectively. Using these relations, we obtain from (4) for band-to-band transitions

$$\varepsilon_\omega^{bb} (\omega,T) = \alpha^{bb} (\omega,T)(n^2\hbar\omega^3/4\pi^3c^2)(\exp((\hbar\omega - \Delta F)/kT) - 1)^{-1} \quad (8)$$

where $\Delta F = F_u - F_l$. For $(\hbar\omega - \Delta F) >> kT$, we may use Wien's approximation. Here, the emission coefficient equals that for thermal equilibrium multiplied by the factor $\exp(\Delta F/kT)$. This factor, in turn, equals the ratio of the product of the non-equilibrium carrier densities NP (N and P are the concentrations of electrons and holes respectively) to the product N_oP_o in thermal equilibrium. Note that the microscopic origin of thermal radiation at frequencies above the band gap and of photoluminescence from thermalized delocalized carriers is the same. Both result from transitions between the same levels. While in the former case the radiation is due to recombination of thermally generated carriers, in the latter case photogenerated excess carriers recombine. Since the concentration of thermal carriers increases strongly with increasing temperature, the photoexcited carrier densities are expected to be less important at higher temperatures. In the spectral region of high absorption, with $\alpha^{-1} < W$, the emitted radiance is, according to Wien's approximation, proportional to $\omega^3\exp(-(\hbar\omega - \Delta F)/kT_o)$. That means, the spectral dependence is not influenced by thermalized excess carriers. This provides an experimental test for the influence of excess carriers on the apparent emissivity. If the absolute intensities of the emitted radiation for different laser powers are also consistent with the temperature derived from the spectrum, the excess carrier density is negligible compared with the thermal carrier concentration.

2.3. Experimental aspects

There are several possibilities for deriving the temperature from the emitted radiation: (i) evaluation of the intensity at a certain wavelength (monochromatic pyrometry), (ii) evaluation of the intensity integrated over a broad band, and (iii) evaluation of the spectral dependence, including temperature determination from the ratio of the radiances at two wavelengths. For monochromatic pyrometry, we obtain from the law of error propagation

$$\Delta T = (\lambda T^2/C_2)\ ((\Delta I/I)^2 + (\Delta e/e)^2)^{1/2} \qquad (9)$$

where ΔT is the error in temperature, and λ the wavelength at which the measurements are performed. $C_2 = 1.4388.10^7$nmK is the second radiation constant and I the measured intensity. Uncertainties in the emissivity or noise in the measured intensity are less important for shorter wavelengths λ. By defining an effective wavelength which, however, will be temperature dependent, the same expression applies to broad band pyrometry.

For the temperature evaluation from the ratio of the radiances at two wavelengths, we obtain

$$\Delta T = T^2\ /\ (C_2/(1/\lambda_2 - 1/\lambda_1)).\ \Delta R_i/R_i \qquad (10)$$

with $R_i = (I_1/e_1)/(I_2/e_2)$ where I_1 and I_2 are the intensities and e_1 and e_2 the emissivities at wavelengths λ_1 and λ_2 respectively. In conventional pyrometry, the error in temperature is mainly determined by uncertainties in the emissivity, so that twin-wavelength (ratio) pyrometry gives the best results (Battuello and Ricolfi /26/). However, the situation is different for small areas. In this case, the signals are very weak so that the effect of noise can be larger than the uncertainties in the emissivities. If inaccuracies in the intensity measurements dominate, we find from (9) and (10) that regardless of the difference between the two wavelengths, the evaluation of the temperature from the radiance at the smaller of the two wavelengths will yield a smaller error than the evaluation of the ratio of the radiances. In addition, chromatic errors in the imaging system become more serious for small spot sizes. Monochromatic pyrometry is therefore the superior technique. Additional advantages are the simple experimental setup employed and the ease of calibration (by using a radiance standard such as a tungsten band lamp). However, sufficient accuracy can only be obtained if the influence of the inhomogeneous temperature distribution and of photogenerated excess carriers on the apparent emissivity is known. These requirements are fulfilled for pyrometric temperature measurements during laser-induced steady growth of rods. This will be outlined in Sect. 2.4.

Because of the strong localization of laser-induced temperature distributions, the light fluxes to be measured are rather small. Visual micropyrometry, using disappearing hot wire-type instruments, is a simple means of measuring the temperature during laser annealing /12/ and LCVD /3,11/. However, such measurements are rather inconvenient, their applicability is restricted to temperatures higher than about 1200K, and the procedure tends to fail if the hot spot to be measured becomes smaller than the wire diameter. The required sensitivity of the detector for photoelectric pyrometry can be estimated from the spectral radiance of the blackbody in the temperature range in question. For example, at a wavelength of 1000nm, the blackbody radiance (in watts per $1\mu m^2$ area, 1sr solid angle, and 1nm spectral bandwidth) is 1.41×10^{-16} at 700K, 6.7×10^{-14} at 1000K, and 8.1×10^{-12} at 1500K. Therefore, above 1000K, high quality silicon photodiodes (noise equivalent power about 10^{-14} to 10^{-13}W/Hz$^{1/2}$ /27/) will be sufficient. Lower temperatures, down to about 700K, can be measured using cooled Ge diodes and an increased spectral bandwidth. Photomultipliers, even those with GaAs photocatodes, would not reduce this low temperature limit appreciably (at 900nm and 700K the blackbody radiance is $2x10^{-17} W/\mu m^2 nmsr$). An extension of the range of measurements to even lower temperatures by using infrared detectors (e.g. InAs) as in conventional pyrometry, cannot be recommended for temperature measurements on laser-heated semiconductors because of the above-mentioned difficulties (Sect. 2.2.1) with the interpretation of measured signals in the spectral region below the band gap. On semiconductors, temperatures lower than about 700K can be measured only at the expense of spatial resolution.

2.4. Validity of pyrometry

To derive accurate values of the local surface temperature from the emitted radiation, information about the emissivity is required. High temperature values of the emissivity are available from the literature only for a limited number of materials, e.g. some metals and graphites. Only a few references on the spectral and temperature dependence of the emissivity of semiconductors are available /28,29/. The error in the temperature derived due to uncertainties in the emissivity is given by (9) and (10). From Sect. 2.2 we know that for pyrometric determination of laser-induced temperatures additional information is required. This includes (i) the characteristic depth of the temperature distribution, (ii) a lower boundary on the optical absorption coefficient, and, for measurements on semiconductors, (iii) the ratio of the laser-induced excess carrier concentration and the thermal carrier concentration. In most cases, the information has to be taken from model calculations. Direct measurements of the quantities (i) and (iii) are possible only in a few cases. One example is found in the laser-grown rods. Here, the depth of the temperature distribution along the axis can be estimated from visual observation of the glowing part of the rod. The influence of excess carriers can be experimentally tested by the procedure described above (Sect.2.2.2).

In order to confirm the validity of pyrometric temperature measurements during LCVD we performed spectrally resolved measurements of the radiation emitted from laser heated silicon rods. The TEM_{oo} beam of a cw Ar^+ laser (488nm) was focused onto the sample to a Gaussian spot diameter of $2w_o = 20\mu m$. The samples were freshly prepared Si rods (diameter 100 - 300µm, length about 5mm), laser deposited in situ from SiH_4 as previously described /11/. The light emitted from the laser heated region at the tip of the rod was chopped, analyzed by a 0.25m monochromator (bandpass 2nm) and detected by a liquid nitrogen cooled Ge pin diode. The signal from the detector was measured by means of a lock-in amplifier, the output of which was monitored by a microcomputer which also controlled the wavelength setting of the monochromator. The measured spectra were automatically corrected for the spectral response of the entire optical detection system as determined by using a calibrated tungsten band lamp.

Typical spectra are shown in Fig. 4a for various laser powers. From the visually observed length of the glowing tip of

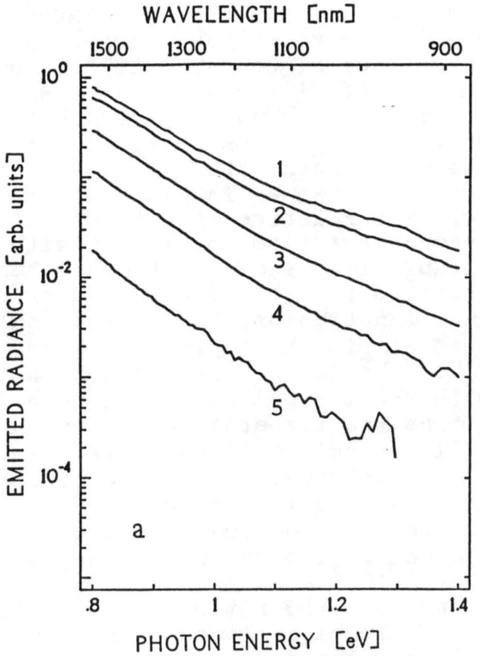

FIGURE 4. a) Spectra of the radiation emitted from the laser heated tip of a Si rod (diameter 100μm, length 5mm) grown by LCVD from SiH_4. (1) P = 150mW (1360K), (2) 140mW (1300K), (3) 120mW (1150K), (4) 100mW (1080K), and (5) 80mW (900K). The temperature values given in parentheses are derived from fits to Planck's law.

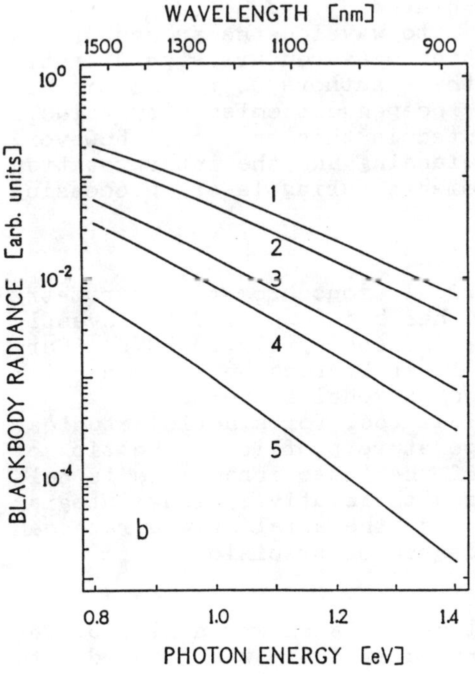

FGIURE 4. b) Blackbody radiance for the temperature values given in a), the curves are labelled accordingly.

the rod, the characteristic depth of the temperature distribution along the axis of the rods is estimated to be larger than about 100μm (this holds even in the worst case, when the reactor is filled with a gas of high thermal conductivity like H_2). Since the energy of the band-gap of Si strongly decreases with increasing temperature (the extrapolated value at 1000K is about 0.8eV /30/), the condition $w \gg \alpha^{-1}$ seems to be satisfied over the whole spectral range of measurements.

The spectral dependence of the radiance shown in Fig. 4a can be well fitted with Planck's law. The relative magnitudes of the spectra for different laser powers are also consistent with Planck's law (Fig. 4b; the slight differences between the measured and the calculated spectra are probably due to uncertainties in the spectral response function of the detection system). This result indicates that within the tip of the rods, for temperatures higher than 900K, the excess carrier concentration is negligible compared with the thermal carrier concentration. Therefore, we conclude that the temperatures measured on the rods by visual pyrometry /11/ do not contain systematic errors due to excess carriers. Monochromatic photoelectric pyrometry during LCVD of silicon rods will therefore yield accurate temperature values. Investigations on the kinetics of pyrolytic LCVD, using this approach to temperature measurements, are presented in Sect. 3.

The finding that for laser-grown rods the ratio e_{app}/e is close to unity is, of course, not always applicable in laser processing. One example where this condition does not hold are the measurements on Kr-ion laser-heated Si and Ge ($2w_o$ = 3μm), which use an InAs diode sensitive to wavelengths around 3μm as detector /13/. Here, the intensities measured are significantly below the ones calculated by these authors on the basis of Planck's law using a temperature-independent emissivity value.

The theoretical analysis presented in this section, however, provides the basis for both the planning and the interpretation of pyrometric temperature measurements during laser processing in general.

3. KINETIC STUDIES

The technique of photoelectrical monochromatic pyrometry discussed in Sects. 2.3 and 2.4 has been employed to investigate the growth kinetics in laser-induced pyrolytic LCVD. Carbon, deposited from some of the lower hydrocarbons, and silicon, deposited from silane, served as model systems.

Steady growth of rods is a unique tool for kinetic studies. In this mode of growth, the temperature profile in the tip of the rod is independent of time if the laser irradiance is held constant. Therefore, measurements with relatively long integration times are possible. In addition, the axial growth rate can be measured <u>in situ</u> with a high degree of precision.

3.1 <u>Experimental setup</u>

A schematic of the experimental setup is shown in Fig. 5. The TEM_{oo} beam of a cw-Ar^+ laser operated at 488nm was focused onto the tip of the growing rod. In most experiments a focus of $2w_o$ = 20μm was employed.

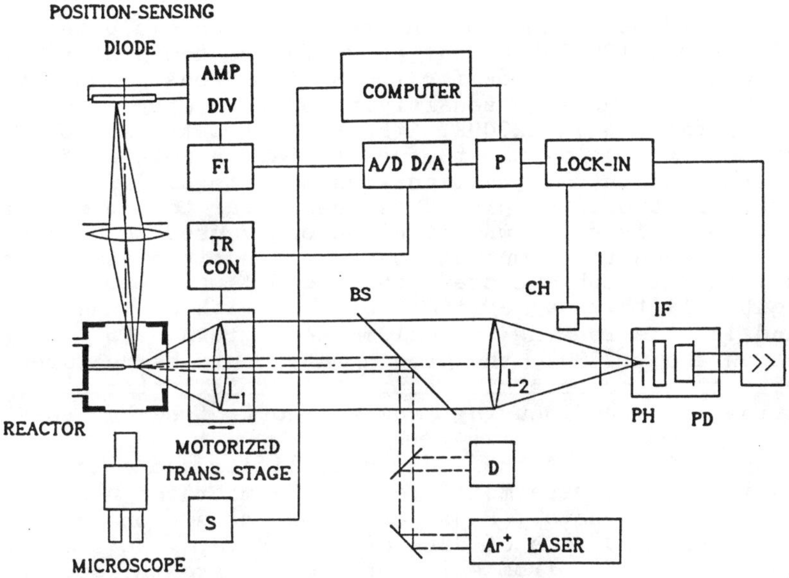

FIGURE 5. Schematic diagram of the experimental setup. AMP DIV: amplifier and analog divider, BS: beamsplitter, CH: light chopper, D: power meter, FI: electronic filter, P: peak detector, PD: Si-photodiode, PH: pinhole, S: switch, TR CON: translation control

The thermal radiation emitted from the hot tip of the rod was collected by the focusing lens, transmitted through the beamsplitter (BS) and focused onto a pinhole (PH) of 36µm diameter. The magnification of the approximately infinite conjugate lens system was 4x, the spatial resolution of the measurement was then about 9µm. The light transmitted by the pinhole was filtered by an interference filter (IF). For the measurements on pyrolytic carbon deposition (temperature range about 2000 - 3000K) a center wavelength of 700nm was used. With silicon rods (1100-1700K) the center wavelength was 1000nm. In both cases a bandpass of 10nm was employed. The radiation was detected by a high quality silicon photodiode, whose output was measured with a lock-in amplifier. A tungsten band lamp was employed to calibrate the response of the detection system. From the measured spectral radiance of the thermal radiation, the local temperature was evaluated by using Planck's law. The angular dependence of the emissivity was neglected. For the measurements on the carbon rods we used a constant value of e = 0.85 for the emissivity. For silicon the linear relation

$$e(T) = 0.946 - 2.76 \cdot 10^{-4}\, T \quad \text{for } 1000\text{K} < T < 1688\text{K} \qquad (11)$$

based on the data of Allen /28/, Sato /29/ and Lampert et al. /31/ was used. The accuracy of this relation is estimated to be within about 10%. This results in uncertainties in the evaluated temperatures of about 7K at 1000K and of about 20K at

1700K.

The standard deviations of the temperatures evaluated were typically about 10K for C-rods grown in the temperature range 2000 - 3000K, and about 5K for the silicon rods. These values again illustrate the high sensitivity of pyrometry.

For temperatures T > 2300K, the axial growth rate of C rods was measured by imaging the thermal radiation emitted from the tip of the rod onto a position-sensing lateral effect photodiode (UDT PIN LSC 30 D) placed perpendicular to the rod axis. The output of this diode was simultaneously used to control the distance between the focusing lens and the tip of the growing rod in order to achieve stationary conditions throughout the experiment. In this way cylindrical rods with constant diameters and lengths ranging up to 10mm were grown. For temperatures below about 2300K, the growth rate could be measured more precisely by means of a microscope. In this latter case, the translation of the focusing lens was controlled by the computer.

The output signals of the lock-in amplifier and of the position-sensing diode were monitored by the microcomputer for the processing of the data. Rods were grown at a constant laser power, and about 10 sets of data of growth rate and temperature were taken per rod, from which an average growth rate and an average temperature as well as their standard deviations were calculated.

3.2 <u>Deposition of pyrolytic carbon</u>

The kinetics of C deposition was investigated for the hydrocarbons CH_4, C_2H_2, C_2H_4, and C_2H_6. Experiments were performed at pressures ranging from 50 to 1000mbar. With C_2H_2, admixtures of H_2, He and Ar were also investigated. The temperature range covered extended from about 2000 to 3200K. Below 2000K non-stationary growth is observed and prevents precise data being taken. In this regime the diameter of the rods becomes smaller than the diameter of the laser beam waist. Above 3200K, carbon evaporation which leads to deposits on the windows becomes the limiting factor. With C_2H_2, at high temperatures homogeneous nucleation of carbon particles is observed in the gas phase near the tip of the rod. The threshold temperature for particle formation decreases with increasing pressure from about 3150K at 300mbar to about 2900K at 1000mbar.

The growth rate decreases in the order $C_2H_2 > C_2H_4 > C_2H_6 > CH_4$. For all of these gases, a linear region in the Arrhenius plot of the deposition rate was found. This regime can be described by

$$W = k_o(p).\exp(-\Delta E/R_gT) \qquad (12)$$

where W is the axial growth rate, p the gas pressure, and R_g the gas constant. The pre-exponential factor k_o and the apparent activation energy ΔE were determined from least-squares-fits to the data. We used the method of Williamson /32/, allowing for uncertainties in both variables ($x = 1/T$, $y = \log W$). To eliminate any arbitrariness in the choice of the temperature range used in the calculation, a reduced chi-square-test was performed /33/. Apparent activation energies determined in this way are summarized in Table I together with the apparent reaction orders derived from the pressure dependence of k_o.

TABLE 1. Activation energies and reaction orders for Ar^+ laser-induced chemical vapor deposition of carbon and silicon

Species	Temperature range+ (K)	Pressure range (mbar)	Activation energy (kcal/mol)	Reaction order
CH_4	2850-3100	500-1000	119 ± 2	1.25
	2400-2750	500-1000	43.5 ± 1.4	2.2
C_2H_2	1900-2450	50-1000	47.3 ± 0.6	0.8
C_2H_4	2000-2250	300-1000	58.3 ± 1.3	0.8
C_2H_6	2200-2650	300-1000	78.9 ± 4	2
SiH_4	1150-1350	30-300	43.5 ± 1	0.6

+) The linear range in the Arrhenius plot is pressure-dependent

The data for C_2H_2 are in excellent agreement with previous measurements performed by Leyendecker et al. /3/. The precision of the measurements, however, has been considerably improved.

An Arrhenius plot for the laser-induced deposition of carbon from gas mixtures of 300mbar C_2H_2 with 300mbar of H_2, He or Ar is shown in Fig. 6. In Fig. 7, we have plotted the pre-exponential factor k_o as a function of the partial pressure of the admixed gas. As can be seen from the figures, admixtures of both H_2 and He reduce the growth rate, while Ar increases it. The shape of the curves, however, remains unchanged. From these results we conclude that addition of hydrogen does not change the chemistry of the process. The observed influence of the gaseous admixtures on the growth rate can be explained by thermal diffusion of species. When we ignore terms of higher order in the gradients, the diffusion flux of species A in the binary mixture A-B is given by /34/

$$j_A = - c\, D_{AB}\, (\nabla x_A + k_T \nabla T / T) \qquad (13)$$

where c is the molar concentration, D_{AB} the binary diffusion coefficient, x_A the molar fraction of A, and k_T the thermal diffusion ratio. Because of the high temperature gradients in the gas phase near the hot tip of the rod, the second term which describes thermal diffusion cannot be neglected. We now identify A with the reactant and B with the gaseous product of the deposition reaction A -> D + b B, where b is the stoichiometric coefficient. In steady growth, the number of molecules of the reactant A decomposed per unit area at the surface

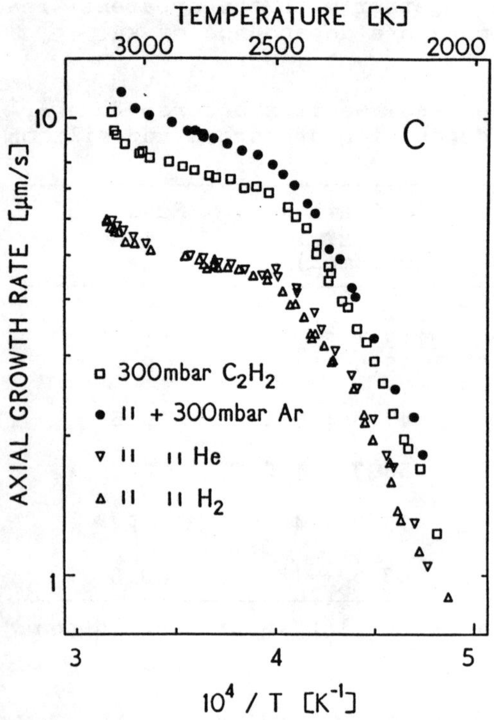

FIGURE 6. Arrhenius plot for the laser-induced deposition of carbon from gas mixtures of 300mbar C_2H_2 with 300mbar H_2, He or Ar

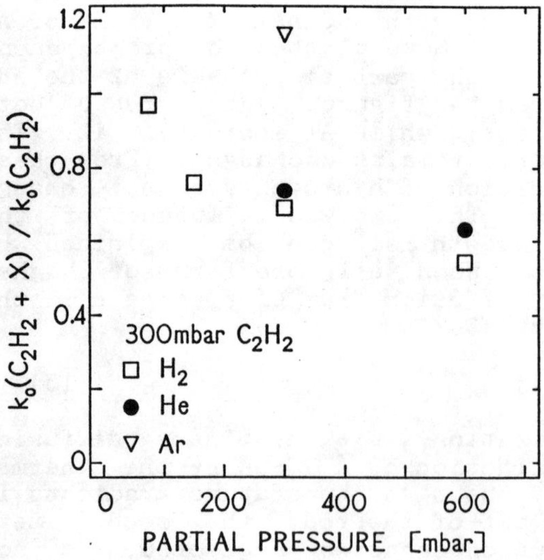

FIGURE 7. Pre-exponential factor k_o for different mixtures of C_2H_2 with H_2, He and Ar. k_o is normalized to the value for pure C_2H_2.

equals the diffusion flux of these molecules. Modelling the gas-phase transport during steady growth of rods according to the reaction given above with b = 1 /36/, and using (13) we obtain for the axial growth rate

$$W = \gamma_A \frac{M_D}{\rho} \frac{c\, D_{AB}(x_{Ao} - k_T \ln(T/T_o))}{r_o + r.F(T/T_o)} \tag{14}$$

Here γ_A is the number of atoms D per parent molecule (γ_A = 2 for C from C_2H_2), M_D the molecular weight of the deposit, and ρ its density. x_{Ao} and T_o are the molar fraction and the temperature far away from the surface, respectively. T is the gas temperature at the surface, and r the radius of the rod. $r_o = c\, D_{AB}/K$, where K is the first order rate constant, and $F(T/T_o)$ is a correction factor due to the temperature dependence of the gas density and the diffusion coefficient. The magnitude of $F(T/T_o)$ is between about 1 and 3. In the derivation of (14), k_T has been assumed to be temperature independent.

The magnitude of the thermal diffusion ratio is of the order of a few percent; its sign is positive for $M_A > M_B$ /35/. With the deposition reactions investigated in this work, the product B is hydrogen. In this case, k_T is always positive, and therefore the effect of thermal diffusion reduces the growth rate. With admixture of H_2 ($x_{Ao} < 1$), the growth rate decreases with increasing partial pressure of H_2. Eq.(14) yields about the right order of magnitude of the effect of thermal diffusion when using values of T/T_o, about 5 to 10. However, (14) cannot be used to describe cases where a third gas is admixed. With respect to thermal diffusion, the admixture of He will have about the same effect as H_2, since both gases have nearly equal transport properties. Since Ar has a higher molecular weight than C_2H_2, it is depleted near the hot surface, thus increasing the local partial pressure of C_2H_2, and thereby the deposition rate. However, from our experimental results, we cannot rule out an additional contribution of Ar to collision-induced reactions. Since k_T depends on pressure and concentration, thermal diffusion may also influence the apparent reaction order. One reason for the deviation of the reaction orders from integer numbers may be due to the effect of thermal diffusion on the growth rate (see Table I).

There seems to be general agreement in the literature that the first step in carbon formation is thermal decomposition of the hydrocarbon in the gas phase. Extensive shock-wave studies have been performed on the kinetics of thermal decomposition of CH_4 /37/, C_2H_2 /38/, and of C_2H_6 /39/. When we take into account the high temperature gradients in the gas phase in our experiments, the literature values for the rate of the initiation step /37 - 39/ give correct orders of magnitude for the carbon growth rate. Therefore, homogeneous decomposition of the hydrocarbon might be the rate-limiting step in pyrolytic LCVD of carbon. While the activation energy found for C_2H_2 is very close to that of the bimolecular initiation step, being about 46kcal/mol /38/, the apparent reaction order of 0.8 is in disagreement with second order kinetics. Thermal decomposition of C_2H_6 is the prototype of a unimolecular decomposition reac-

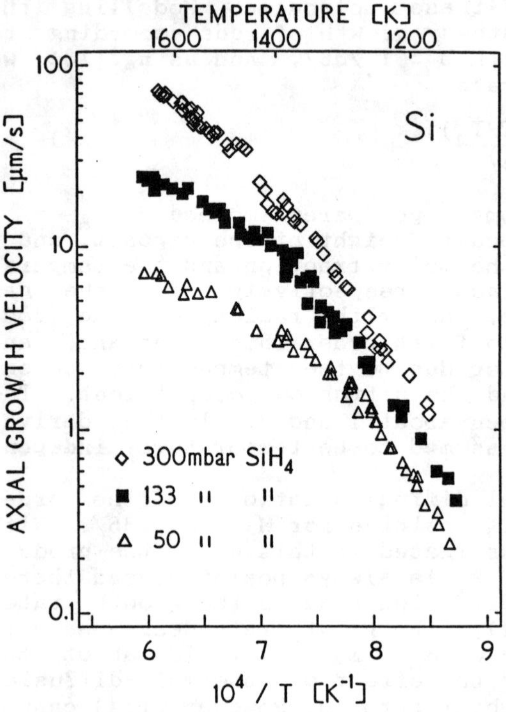

FIGURE 8. Arrhenius plot for the laser-induced deposition of silicon rods from SiH_4

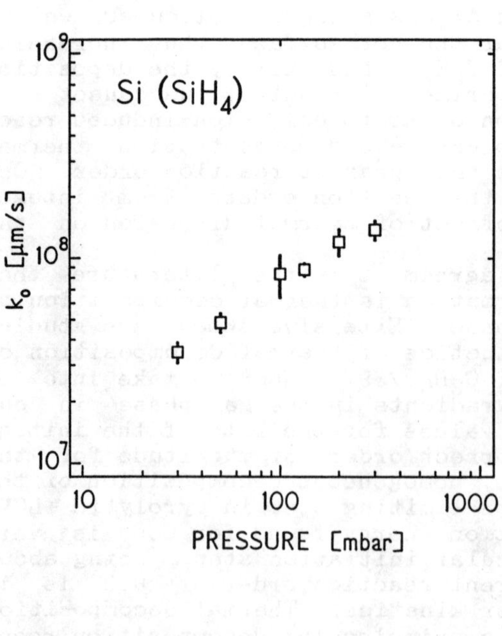

FIGURE 9. Pressure dependence of the pre-exponential factor k_o for laser-induced deposition of silicon from SiH_4

tion with an activation energy of 87kcal/mol in the high pressure limit /39/. Again, our experimental activation energy is in reasonable agreement with this value, while the reaction order disagrees. An explanation of the difference in reaction orders derived from our experiments and those given in /38/ and /39/ remains speculative.

3.3 Deposition of silicon

The first investigations on the kinetics of Ar^+ laser-induced deposition of silicon rods were reported by Bäuerle et al. /11/. In the present paper we have extended these earlier investigations and studied the pressure dependence of the deposition rate in the range 30mbar < p < 300mbar.

Figure 8 shows an Arrhenius plot of the deposition rate at three different pressures of SiH_4. The data for 133mbar are in excellent agreement with those reported in /11/. Note that in the present experiments the temperature has been derived from the emitted radiance at 1000nm, while the earlier investigations /11/ employed visual pyrometry at an effective wavelength of 650nm.

The temperature range of the present measurements was extended to about 50K below that in /11/. The low temperature limitation does not originate from the capabilities of the temperature measurement. As with the carbon rods, this limitation arises from non-stationary growth. The highest temperatures measured were 1690 $\pm$ 5K, which corresponds to the melting point of Si. Deposition near melting was accompanied by the formation of a yellowish dust, which probably consists of polysilanes. At pressures higher than 200mbar and temperatures above 1650K extremely irregular growth and formation of polysilanes was observed.

The apparent activation energies were derived by using the procedure described in Sect. 3.1. Within the error of measurements, the apparent activation energy is independent of the gas pressure in the range of pressures investigated. The average value obtained from measurements at six pressures (30 - 300mbar) is 43.5 $\pm$ 1kcal/mol. The pressure dependence of the pre-exponential factor k_o is shown in Fig. 8. From these latter data, an apparent reaction order of 0.6 is derived.

The goal of any measurement of deposition kinetics is to identify the rate controlling mechanism. For silicon deposition from SiH_4, Coltrin et al. /40/ have emphasized the importance of gas phase decomposition of silane as the primary step in silicon CVD. These authors have also summarized the previous work on this subject. One common feature of many CVD systems, namely that the apparent activation energy for deposition rates at high temperatures is much smaller than that at low temperatures, is also found in our results. Before the publication of the paper by Coltrin et al. /40/, deposition at low temperatures had been attributed to surface reactions. Using the empirical silane surface reaction coefficient of Farrow /41/, we obtain growth rates of about an order of magnitude below our measured axial growth rates of silicon rods. On the other hand, the rate constants for the initiation reaction ($SiH_4 \rightarrow SiH_2 + H_2$) cited by Coltrin et al. give correct orders of magnitude of silicon deposition rates. The literature value of activation energy of the initiation reaction (52kcal/mol) is slightly higher than that derived from our measurements.

4. CONCLUSION

The investigation of the fundamentals of pyrometry on laser-heated solids allows us to confirm the validity of such measurements. The technique can well be applied to accurate temperature measurements during steady growth of rods, which is a unique technique for kinetic studies of LCVD.

ACKNOWLEDGEMENTS

We wish to thank Prof. U. Titulaer and Dr W. Jantsch for valuable discussions, and the Fonds zur Förderung der wissenschaftlichen Forschung in Österreich for financial support.

REFERENCES

1. D. Bäuerle, Chemical Processing with Lasers, Springer Series in Materials Science 1 (Springer, Berlin, Heidelberg 1986)
2. D. Bäuerle, ed., Laser Processing and Diagnostics, Springer Series in Chemical Physics 39 (Springer, Berlin, Heidelberg, New York, Tokyo 1984)
3. G. Leyendecker, D. Bäuerle, P. Geittner, H. Lydtin, Appl. Phys. Lett. 39, 921 (1981); G. Leyendecker, H. Noll, D. Bäuerle, P. Geittner, H. Lydtin, J. Electrochem. Soc. 130, 157 (1983)
4. H. Kurz, Festkörperprobleme XXIII, 115 (1983)
5. A. Moritani, C. Hamaguchi, Appl. Phys. Lett. 46, 746 (1985)
6. K. Murakami, Y. Tohmiya, K. Takita, K. Masuda, Appl. Phys. Lett. 45, 659 (1984)
7. A. Compaan, J. Luminesc. 30, 425 (1985)
8. D. Kirillov, J.L. Merz, Mat. Res. Soc. Symp. Proc. Vol. 17, 95 (1983)
9. M.A. Olmstead, N.M. Amer, S. Kohn, D. Fournier, A.C. Boccara, Appl. Phys. A32, 141 (1985)
10. M. Bertolotti, A. Ferrari, C. Sibilia, M. Tamburini, Appl. Phys. A37, 109 (1985)
11. D. Bäuerle, P. Irsigler, G. Leyendecker, H. Noll, D. Wagner, Appl. Phys. Lett. 40, 819 (1982)
12. T.O. Sedgwick, Appl. Phys. Lett. 39, 254 (1981)
13. R.P. Salathė, H.H. Gilgen, Ref. 1, pp. 425-436; R.P. Salathė, H.H. Gilgen, Y. Rytz-Froidevaux, IEEE J.-QE 17, 1989 (1981)
14. M. Kemmler, G. Wartmann, D. von der Linde, Appl. Phys. Lett. 45, 159 (1984)
15. J.F. Babelot, J. Magill, R.W. Ohse, M. Hoch, in J.F. Schooley, ed., Temperature, its Measurement and Control in Science and Industry, Vol. 5 (American Institute of Physics, New York 1982) p. 439
16. S. Chandrasekhar, Radiative Transfer, (Clarendon Press, Oxford 1950)
17. F. Reif, Statistical and Thermal Physics (McGraw-Hill, New York 1965) p. 381

18. H.P. Baltes, Progress in Optics XIII, 1 (1976)
19. F. Stern, Elementary Theory of the Optical Properties of Solids, in F. Seitz, D. Turnbull, eds., Solid State Physics, Vol. 15 (Academic Press, New York 1963)
20. H. Barry Bebb, E.W. Williams, in R.K. Willardson, A.C. Beer, eds., Semiconductors and Semimetals, Vol. 8 (Academic Press, New York 1972) p. 181
21. H.M. van Driel, this volume
22. M. Lax, J. Appl. Phys. 48, 3919 (1977)
23. J.E. Moody, R.H. Hendel, J. Appl. Phys. 53, 4364 (1982)
24. G.E. Jellison, F.A. Modine, Appl. Phys. Lett 41, 180 (1982)
25. A. Bhattacharyya, B.G. Streetman, Solid State Comm. 36, 671 (1980)
26. M. Battuello, T. Ricolfi, High Temperatures - High Pressures 12, 247 (1980)
27. W. Budde, Physical Detectors of Optical Radiation, in F. Grun, C.J. Bartleson, eds., Optical Radiation Measurements, Vol. 4 (Academic Press, New York 1983)
28. F.G. Allen, J. Appl. Phys. 28, 1510 (1957)
29. T. Sato, Jap. J. Appl. Phys. 6, 339 (1967)
30. W. Bludau, A. Onton, W. Heinke, J. Appl. Phys. 45, 1846 (1974)
31. M.O. Lampert, J.M. Koebel, P. Siffert, J. Appl Phys. 52, 4975 (1981)
32. J.H. Williamson, Can. J. Phys. 46, 1845 (1968)
33. P.R. Bevington, Data Reduction and Error Analysis for the Physical Sciences (McGraw-Hill, New York) p. 188
34. R.B. Bird, W.E. Stewart, E.N. Lightfoot, Transport Phenomena (Wiley & Sons, New York 1960)
35. J.O. Hirschfelder, C.F. Curtiss, R.B. Bird, Molecular Theory of Gases and Liquids (Wiley & Sons, New York 1954), p. 541
36. J. Doppelbauer, D. Bäuerle, to be published
37. P. Roth, Th. Just, Ber. Bunsenges. 79, 682 (1975)
38. T. Tanzawa, W.C. Gardiner Jr., J. Phys. Chem. 84, 236 (1980)
39. J.R. Cao, M.H. Back, J. Phys. Chem. 88, 3074 (1984)
40. M.E. Coltrin, R.J. Kee, J.A. Miller, J. Electrochem. Soc. 131, 425 (1984)
41. R.C. Farrow, J. Electrochem. Soc. 121, 899 (1974)

DIFFUSION IN LIQUIDS

F.P.RICCI, D.ROCCA
Departimento di Fisica,
Università degli Studi "La Sapienza"
P. le Aldo Moro, 2
OU185 Roma - Italy.

Text already printed in the NATO ASI:
"Molecular liquids" (A.J.Barnes, W.J.Orville-Thomas, J.Yarwood, Eds.)
(Reidel, Dordrecht, 1984), pp.35-58.

THE SOLID-SOLID INTERFACE UNDER LASER-IRRADIATION.

L.D. LAUDE
Université de l'Etat,
Faculté des Sciences,
23, avenue Maistriau,
7000 MONS, BELGIUM.

Abstract

Problems encountered in analyzing the behaviour of Laser-irradiated solid-solid interfaces are reviewed and examplified. The accent is put on the limits of thermodynamic approaches to these problems and on a phenomenological description of the onset of non-equilibrium in such inhomogeneous systems both in time and space. A model for Laser-induced compound synthesis is derived pointing at the novelty and potential of the process against conventional annealing schemes.

1. Introduction

Among the interface problems encountered in Laser processing of matter, the atom migration accross a clearly defined solid-solid boundary is one of particular interest for a number of reasons. From a practical point-of-view, atom migration, or more precisely atom diffusion, is a very common technique in doping semiconductors for instance and, therefore, its fundamental description has long been established although on a macroscopic scale via the thermodynamics of irreversible processes. In first approximation, one would then welcome the idea that the concepts of diffusion theory would be applicable in laser assisted diffusion and, consequently, understanding of the latter would present no difficulty. From there the first question : is there anything different with laser-induced atom migration and is diffusion theory applicable? A second point of interest is the variety of solid-solid interfaces to be processed giving access, in principle, to the formation of various compound or alloy phases which may be drastically different from the initial phases in contact. Via conventional diffusion processes, such a phase forming is often damped or even annihilated due to contradictory effects, the reacted material contituting a barrier to reactant diffusion. Second question : could laser-processing over come that sort of difficulty?

Last on that list is the problem of forming metastable phases. Thermodynamically, the only phase to form and stabilize along a diffusion process is, among all possible structural phases made out of the same set of atoms, the one possessing the smallest free energy per atom. However, the probability to form one of the other (metastable) phases in never null and, from time to time a metastable phase may form via oven annealing. With a higher free energy, such a phase is characterized by a lesser degree of order compared to the stable phase, i.e. it presents structural anisotropy or even no long-range order like in a glass or perfect alloy. The physical properties of such metastable phases are not only different from the ones of the stable phase, they may also present distinctive advantages over the others, like for instance metallic glasses, and one would find some interest in producing them. Third question : could laser-processing lift the uncertainty and allow the formation of any metastable phase at any reliable and reasonable rate?

The three questions raised above are quite challenging as such and, although they each point at a specific aspect of the problem, they all address the same issue : what is the real potential of laser means in terms of atom mixing? So much on the waiting, how much effective? To some point, the same challenge is to be faced in the other domains treated in this course, i.e. solid-liquid or solid-gas interfaces being laser-irradiated. The prospect of this lecture is not to explore in depth the fundamentals of atom migration in a reacting medium submitted to laser irradiation but rather,
i) to underline the specific conditions of laser processing which may introduce non-equilibrium around a solid-solid interface, and ii) to extrapolate a model able to describe the variety of interface reactions which may be obtained in current

laser experiments. It must be stressed that a rationalized and exhaustive theory of these processes is, by far, out of reach due to the extreme complexity of the interaction of laser light with a strongly inhomogeneous system. Nevertheless, the phenomenological model presented here provides for an appropriate description of the reaction mechanism which develops about a solid-solid interface submitted to laser irradiation.

2. Experimental

A number of laser-initiated atom mixing processes have been experimented. In all such processes, the prospective object is an homogeneous mixing of two or more species in an either crystalline phase or amorphous ("glassy") network. The energy source to such mixing is provided by the laser beam itself being optically absorbed within matter. The initial structure of the target is a regular stack of elemental layers deposited on glass or any other thermally insulating substrate. Each layer contains one element, and the layers are assembled in a sandwich film the overall composition of which is as close as possible to a given stoichiometry. The latter may correspond, for instance, to the chemical composition of a compound phase or of an alloy eutectic. In practical terms however, the initial composition may lie within a few per cent from the aimed stoichiometry when using conventional e-gun evaporation sources and quartz monitoring. Layer thicknesses are typically a few hundreds Å and the overall thickness may exceed one micron. The number of layer couples (and interfaces) may run, thus, from unity to twenty or so. The films are laser-irradiated in the direction perpendicular to the film plan and it may be approximated that both sides of each interface is equally coupled to light at IR or visible laser wavelengthes. The irradiation parameters (beam energy, irradiation time, processed area) may be systematically varied and the products obtained for each set of parameters are characterized with respect to atom configurations, electronic and vibrational structures.

Following the choice of the components, drastic changes may be expected in the film properties upon atom mixing. For instance, semiconducting compound formation may occur when elements are taken in columns III and V, or II and VI of the Periodic Table to form III-V or II-VI compounds[1,2], respectively. Without going into details at this stage, the remarkable and recurring observation which best characterizes a laser-induced interface reaction is that complete reaction of the components is achieved at threshold values in either one of the processing parameters, the others remaining fixed. In other words, the central parameter which governs the reaction is the energy per unit time and per unit volume, i.e. power density. Within a few per cent from such a threshold power density, full compound formation is or is not achieved. This behaviour definitely stresses the importance of energy gradients in time and space in forming phases via laser means.

An example of this characteristic is shown in fig.1.[4,5].

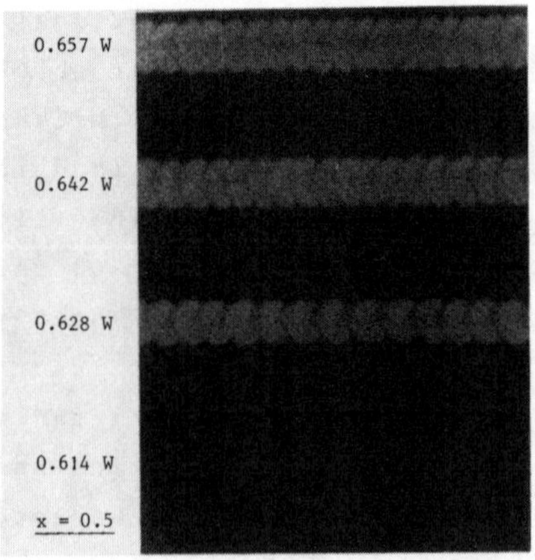

Fig.1.

A set of Ge and Se elemental layers in 1:1 stoichiometric proportion is deposited on glass with an overall thickness of 0.2 μm. The whole film is then irradiated with a CW Ar^+ beam in TEM_{00} mode. The gaussian spatial profile of the beam is about 30 μm wide at 1/e from beam maxima. It is scanned onto the film at 1 m sec^{-1} along linear and parallel pathes, distant from each other by 120 μm. From one path to the next, the beam power is increased by a very small increment (14.3 mW) without changing other characteristics. As evidenced in the optical micrograph of fig.1, an abrupt transformation into (orthorhombic) GeSe is obtained at P = 628 mW and the lowest energy required for transformation may be defined as P_t=621±7mW. Above that power, very large crystallites (≃ 10 μm) are formed. They are oriented radially and in the forward direction to the beam scanning. Upon increasing P, their dimension reduces but the overall volume of transformed matter increases proportionally to the power increase or, at constant scanning velocity, to the energy increase. A complementary information is that, about the above threshold, the energy absorbed by the film seems, from both experimental and computed figures, sufficient to "melt" the original material[(2)]. It would then be apparent that the abrupt compound formation would result from a very rapid mixing induced via the melting of one or both types of layers. At power densities lower than the threshold ones, interface atom diffusion takes place which produces partial reaction as evidenced for In-Se couples[(3)].

Another characteristic of laser-induced synthesis from solid phases is its ability to select phases. In particular, it may allow the formation of metastable phases. As an example of this, one may quote synthesis of hexagonal CdSe coumpound[(6)]. A film consisting of Cd or Se layers alternately deposited on NaCl crystal is detached from substrate and irradiated with a pulsed dye-laser beam (pulse duration $\tau = 10^{-6}$ sec). Above some 20 mJ pulse energy, the (metastable) hexagonal phase is formed. Note that this figure is the direct measure of the energy required for destabilizing and forming a given volume of hexagonal CdSe. An identical film deposited on glass transforms into cubic CdSe when irradiated with a chopped CW Ar^+ beam (irradiation time $\tau = 3.6 \times 10^{-2}$ Sec). Like in metallic glass formation, the metastable phase is here obtained upon very fast cooling.

A final aspect of laser-synthesis is the extremely short reaction time. Preliminary results(7) indicate a reaction time of the order of 10^{-8} sec in 0.1 μm thick Al-Sb films. Would diffusion theory be advocated to account for that time, this would correspond to an atom velocity of 10 $msec^{-1}$, i.e. orders of magnitude beyond usual diffusion figures. This sheds additional light on the, now obivous, non-equilibrium conditions which may be readily attained in matter upon laser-irradiation at and above power density threshold.

That threshold remains to be understood and described in energetic terms which will be the object of the two following sections.

3. Thermodynamic of irreversible processes

Let us consider a set of two solid layers, each made of a given type of atom A or B. The layer thicknesses are adjusted in such a way that the numbers of atoms A and B, present in each layer respectively, are equal. These layers are then brought to contact and form a sandwich with a perfect, continuous interface. The prospect is now to form an homogeneous (i.e. stable) mixture of A and B atoms. For this mixture to be stable requires its Gibbs free energy to be lesser than the one of either atom A or B system. For instance, this would be achieved via covalent chemical bonding, owing to the choice of the species involved and their particular electronic configurations.

Mixing the A and B species may be achieved thermally, e.g. in an oven. The "reaction" mechanism is irreversible and, therefore, its phenomenology could be approached by using concepts of the thermodynamic of irreversible processes (TIP). In there, the independent variables (temperature, pressure, atom concentrations) may vary in the system along gradients, this resulting into fluxes of matter and energy (i.e. heat). This is described by saying that entropy is produced in the system, and that part of the entropy production which is characteristic of the irreversibility is the source of entropy σ. TIP allows to evaluate σ under certain conditions.

The basic hypothesis is that of local equilibrium. The system is subdivided into atom cells and, at anytime, all variables have known and constant values within each cell. In addition, "acceptable" boundary conditions are set between adjacent cells allowing these values to slightly vary from one cell to the next and approximately trace the imposed gradients, e.g. temperature may vary by ΔT_{ij} between cells i and j, then ΔT_{jk} between cells j and k, etc... This means that within cell j, temperature is allowed to vary reversibly in the range $[T_j - \frac{1}{2}\Delta T_{ij}, T_j + \frac{1}{2}\Delta T_{jk}]$ without perturbing the thermodynamic equilibrium of cell j against cells i and k. Although it is difficult to quantify those boundary conditions, clearly one must face the problem : the steeper the gradients, the smaller the cells, and the question : how small can be a cell to maintain local equilibrium and reversibility?

Although the Gibbs free energy per atom, G_j, is homogeneous within cell j (all atoms in j are equivalent), its value decreases when increasing the number of atoms present in j, i.e. the coupling energy reduced to every atom increases in that case. In a Lennard-Jones medium, G_j is only a few percent from G_∞ (infinite cell) when up to 12th nearest neighbours may be coupled. Taking this criterion to qualify the lower limit of the cell dimension in TIP approximation, would give a minimum cell size of some 50Å diameter, containing 10% of the atoms on the surface of the cell. At lower dimensions, the surface-to-bulk atom ratio would evidently be larger, thus increasing G_j. In that case, the energy necessary to disperse the atoms of the cell would be too small to maintain its cohesion. Therefore, local equilibrium may break down upon increasing the energy gradients in the atom system.

That situation is never effective in an oven annealing

for three reasons. First, temperature gradients are extremely weak, by choice:in order to avoid strain between substrate and film or by incapacity:to surpass the thermal inertia of the substrate. In that case, the cell dimension may be extremely large and TIP may apply. Second, such gradients are monotonous, i.e. boundary conditions are homogeneous within the system. Gradients are too weak everywhere in system. Third, cells are large enough in time to allow atom diffusion to develop deeply across the A-B interface, thus inducing AB reaction within every time cell. The atom system progressively evolves from a two-species system into a three-species one (A-AB-B). Upon increasing T, compound AB forms around the original A-B interface, thus limiting further atom diffusion and reaction which may only be overrun by further increasing the processing time. Here, the AB formation proceeds in solid state phase only by classical atom diffusion.

It is under these conditions that the source of entropy may be written as a sum of scalar products between fluxes $\underline{J}$ and forces $\underline{X}$:

$$\sigma = \sum_k \underline{J}_k \cdot \underline{X}_k + J_q \cdot \underline{X}_q \qquad (1)$$

where $\underline{J}_k$'s are fluxes of the diffusing atoms k ($\sum_{k=1}^{n} \underline{J}_k = 0$); $\underline{X}_k$ forces acting on such atoms, $\underline{J}_q$ a heat flux and $\underline{X}_q$ the thermodynamic force associated with the temperature gradient induced in the system.
Explicitely, we write :

$$\underline{X}_k = \frac{\underline{F}_k}{T} - \underline{\text{grad}}\ (\frac{\mu_k}{T}), \qquad (2)$$

where $\underline{F}_k$ is a force external to the system which acts on atom k, and μ_k the chemical potential of these species k :

$$\mu_k = \frac{\delta G}{\delta n_k}\bigg|_{T,p}$$

(n_k : concentration in k, and G is the free energy),

$$\underline{X}_q = - \frac{1}{T^2} \underline{\text{grad}}\ T. \qquad (3)$$

One now examines the atom system within a given time cell, at temperature T. Here, σ reduces to :

$$\sigma = \sum_{k=1}^{n} \underline{J}_k \cdot \underline{X}_k = \sum_{k=1}^{n-1} \underline{J}_k \cdot (\underline{X}_k - \underline{X}_n) \qquad (4)$$

Since, the origin of the fluxes are the forces $\underline{X}_k$ (no force, no flux), there must exist a relation between forces and fluxes. Fluxes are therefore developed in series of forces $\underline{X}_k$'s. When such forces are weak, i.e. $\underline{F}_k$ is negligible and n_k is small, the series development of $\underline{J}_k$'s may be limited to first order terms :

$$\underline{J}_i = \sum_{k=1}^{n} L_{ik}\ \underline{X}_k \qquad (i \text{ or } k = 1,2,\ldots,n)$$
$$\underline{J}_q = \sum_{k=1}^{n} L_{qk}\ \underline{X}_k , \qquad (5)$$

L_{ik} and L_{qk} are Onsager phenomenological coefficients which are independent of $\underline{X}_k$.

In a binary system : $\underline{J}_A + \underline{J}_B = 0$ and $\sigma = \underline{J}_A \cdot (\underline{X}_A - \underline{X}_B)$

$$\text{or } \sigma = \underline{J}_B \cdot (\underline{X}_B - \underline{X}_A); \qquad (6)$$

$$\underline{J}_A = L_{AA}(\underline{X}_A - \underline{X}_B)$$
$$\text{and} \quad \underline{J}_B = L_{BB}(\underline{X}_B - \underline{X}_A) = -\underline{J}_A . \qquad (7)$$

Since $\underline{X}_A = - \underline{\text{grad}} \left(\frac{\mu_A}{T}\right)$ and $\underline{X}_B = - \underline{\text{grad}} \left(\frac{\mu_B}{T}\right)$, we may write in a one dimension system like here (z is in the direction perpendicular to the planar A-B interface) :

$$\underline{X}_A - \underline{X}_B = - \frac{1}{T} \frac{\partial}{\partial z} (\mu_A - \mu_B) = - \frac{1}{T} \frac{\partial}{\partial n_A} (\mu_A - \mu_B) \cdot \frac{\partial n_A}{\partial z}$$

$$\text{and} \quad \underline{X}_B - \underline{X}_A = - \frac{1}{T} \frac{\partial}{\partial n_B} (\mu_B - \mu_A) \cdot \frac{\partial n_B}{\partial z} ; \qquad (8)$$

from which, $$\underline{J}_A = -\frac{L_{AA}}{T}\frac{\partial(\mu_A-\mu_B)}{\partial n_A}\frac{\partial n_A}{\partial z}.$$

Since $$\sum_{k=1}^{n} n_k \underline{grad}(\mu_k)\Big|_T = 0$$

and $n = n_A + n_B$, (atom concentrations),

one finds : $$\underline{J}_A = -\frac{L_{AA}}{T}\frac{n}{n_B}\frac{\partial\mu_A}{\partial n_A}\frac{\partial n_A}{\partial z} = -D_A\frac{\partial n_A}{\partial z}, \qquad (9)$$

where $D_A = \frac{L_{AA}}{T}\frac{n}{n_B}\left(\frac{\partial\mu_A}{\partial n_A}\right)_T$ is the diffusion coefficient of atoms A at temperature T, and a similar expression for D_B. This constitutes the first Fick's law.
The time evolution of the concentrations is described by the 2nd Fick's law :

$$\frac{\partial n_A}{\partial t} = D_A\frac{\partial^2 n_A}{\partial z^2} \quad\text{and}\quad \frac{\partial n_B}{\partial t} = D_B\frac{\partial^2 n_B}{\partial z^2}, \qquad (10)$$

assuming D_A or D_B to be independent of z, i.e. at z large.

A more precise evaluation of n_A (or n_B) can be performed using statistical mechanics. After defining a path probability f(z,t) as :

$$f(z,t) = (2\pi\,\overline{z^2})^{-\frac{1}{2}}\exp\left(-\frac{z^2}{2\overline{z^2}}\right),$$

where $\overline{z^2}$ is the mean quadratic path, the diffusion coefficient D_A is now defined as :

$$D_A = \frac{\overline{z^2}}{2t}$$ where t is the path duration.

The atom concentration, which varies with time, may be developed around the interface (at $z=z_o$) as :

$$n_A(z) = n_A(z_o) + \frac{n_A}{z}\Big|_{z_o}(z-z_o)+\frac{1}{2}\frac{\partial^2 n_A}{\partial z^2}\Big|_{z_o}(z-z_o)^2+\ldots \qquad (11)$$

The flux J_A is now :

$$J_A = \frac{1}{t}\left[n_A(z_o)\bar{z} - \frac{\partial n_A}{\partial z}\Big|_{z_o}\frac{\overline{z^2}}{2!} + \frac{\partial^2 n_A}{\partial z^2}\Big|_{z_o}\frac{\overline{z^3}}{3!}\ \ldots\right] \qquad (12)$$

Reducing to the first non-zero term in the series (i.e. small variations in n_A), one finds :

$$J_A \simeq - D_A \left.\frac{\partial n_A}{\partial z}\right|_{z_o} \quad \text{where} \quad D_A = \frac{\overline{z^2}}{2t} \quad \text{as above,}$$

with a similar expression for J_B.
In the presence of a gradient of forces in the z direction (oriented from side B to A for instance) f(z,t) is no longer symetric about z_o.

After developing $f(z,t,n_A(z))$ about $z=z_o$, the flux J_A of atoms A transforms into :

$$J_A = -D_A \frac{\delta n_A}{\partial z} + n_A(z_o) \ \left[\frac{\bar{z}}{t} - \frac{\partial}{\partial z} D\right] \ ,$$

when the series development is limited to second order terms. The added part on the right side describes the oriented transport of atoms A into the layer B, where mixing occurs.
With no thermal gradient (along z), as could be approximated in the present film irradiations, it is evident, via equations (11) and (12), that non-linearity would be the rule in the presence of large forces, in particular with large chemical gradients (resulting, for instance, from strong covalent AB coupling). Additionally, the elemental layer thicknesses in the film are too small to consider the intermixing, even under near steady state conditions, to be of a semi-infinite dilution type.

The above review on diffusion theory shows the limits of the statistical approach to atom migration. It applies only in cases where : i) the diffusing species (A into layer B, B into layer A) are largely minority in the host layer, and ii) the center of mass of the species is away from the interface. In fact, the situation is far more complicated. One should take into account the vacancy behaviour (migration, production) and the solid state reaction A+B → AB, which interrupts atom migration (migration enthalpy changes abruptly from some 0.3 eV to 1.5 eV).

As a concluding remark at this point, it may be said

that, in systems of high confinement, either in space or time, the concepts of diffusion coefficient and of source of entropy (as defined above) are inapplicable which, in turn, renders problematic a correct evaluation of the system temperature. This is particularly the case in the situations described in Section 2. In such systems, laser energy injection may attain intensities at which the atom system undergoes non-linear behaviour and move into non-equilibrium upon further increasing power density. There, the source of entropy σ is in practice impossible to trace following the approximations inherent to the TIP theory. This means that the energy balance of the system should, in such cases, be approached in a different manner. The difficulty lies in the partition of the Gibbs free energy G(T,p,n) of the system, between the entropic term TS(T,p,n) and enthalpy H(T,p,n). As we have seen, for reasons of dimensionality in space-time, this partitioning is difficult to qualify and one may even raise arguments against its physical meaning in restricted systems. More relevant seems to be a projection of G(T,p,n) simultaneously, in time and space and, within this representation, to define one or more criterions of structural stability for the atom system.

Mixing the A and B species may be achieved thermally by laser-irradiating the sandwich structures, Section 2. Both species being metallic or so, photons energy is adsorbed within the layers and transforms into heat via electron-phonon coupling thus providing the excess energy required to activate atom migration across the A-B interface. This energy transfer between light and the atom networks occurs in times sufficiently short ($\sim 10^{-12}$ sec) to consider it to be instantaneous. Therefore, one may subdivide the film into identical spatial unit cells Δx, each containing the same A-B interface. This cell structure is represented in fig.2 along the x axis, parallel to the film plan. The film is now irradiated with a laser beam which hits the film at right angles. Contrary to oven annealing, the laser energy which is delivered to the system does not increase at constant rate, so that i) local equilibrium conditions vary along time and space in the material;

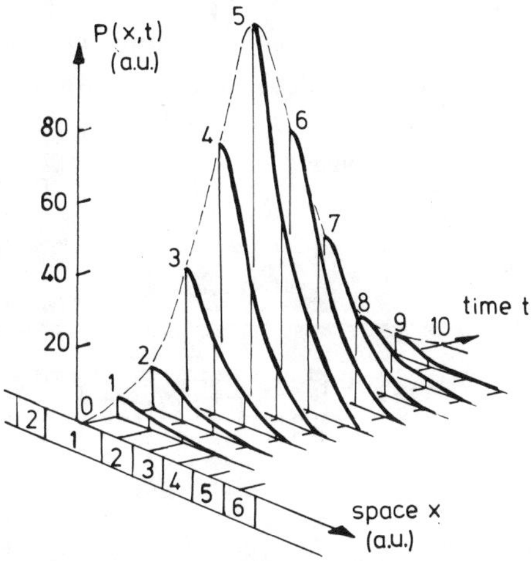

Fig.2.

ii) boundary conditions between unit cells vary also. This, together with the ability to change either energy, irradiation time or beam concentration, provides an infinite set of energy balances in space and time within the same system, using the same laser source under different space-time configurations. The beam has gaussian energy distributions in time and space, fig.2. The power density P(x,t) of that beam, deposited at time t within the cell located at coordinate x can be expressed in the form :

$$P(x,t) = (1-R)A \exp[-a(x-x_o)^2] \cdot \exp[-b(t_o-t)^2] \quad (10)$$

where a and b are laser parameters, x_o the coordinate of the cell at the center of the laser impact on the film; t_o is the time at which the beam intensity is maximum at $x=x_o$; A is a normalizing constant and R is the reflection coefficient of the outer surface of the film at the laser photon energy.

One may now trace the history of each spatial unit cell with time, allowing constant rate energy diffusion to develop between cells, along the x axis. For convenience, the time-space profile of P(x,t) can be approximated by a histogram in which constant energy is deposited within a time-space cell $(\Delta x, \Delta t)$, fig.2. At t=o, $P(x_o, o)$ is taken to be zero.

The procedure adopted in constructing the power density evolution in the film plan is described in fig.3. from t=o to t=2. The final result of this evaluation is shown in fig.4.

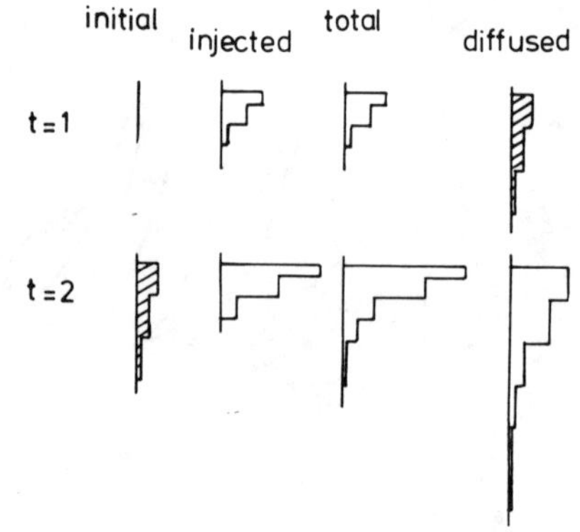

Fig.3.

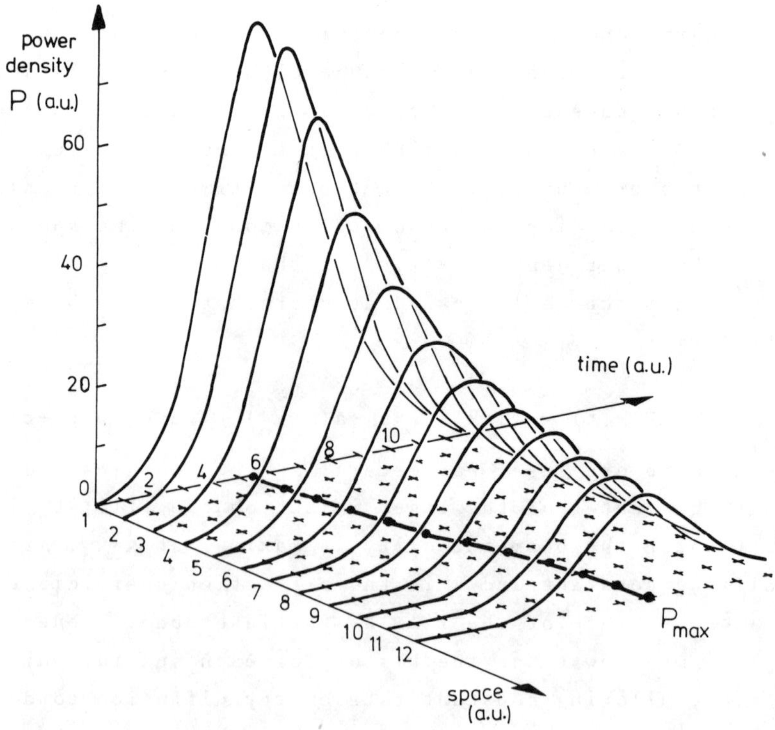

Fig.4.

The maximum power density P_{max} is obtained at $x=x_o$ at a time when the beam itself has passed its peak. Along the x direction, P_{max} is further delayed and reduced in strength due to the heat diffusion process away from the laser impact around $x = x_o$.

As mentioned in the previous section, the local equilibrium approximation of TIP may only apply in a space-time domain exceeding a minimum size, the thermodynamic equilibrium cell (TEC). Upon changing the energy injection into TEC by ΔG, equilibrium within the cell must be attained. This requires a minimum time Δt_m and a minimum size in space Δx_m. Whilst Δx_m is a characteristic of the internal equilibrium of the cell, Δt_m relates more to the equilibrium of the cell with the external medium, i.e. it would depend on the ability of the cell to equilibrate its energy exchange with the out-medium. In particular, Δt_m would depend on the heat conductance of substrate vs. the one of the cell itself. The higher the substrate conductance, the fastest the heat flow from film to substrate. Would the substrate heat conductance be larger than the inner cell one, the cell would never have time to consume the absorbed energy and change structure eventually. The time cell Δt_m would also depend on the size of the atom system in which energy is primarily injected : the larger the system, the smaller Δt_m. It is then difficult to evaluate Δt_m. To set a number, Δt_m could be around 10^{-6} sec for a 0.1 μm thick metallic film on a 1mm thick glass substrate.

At site x_o, $P(x_o,t)$ develops along t following approximately a gaussian profile, fig. 5a. The problem is to define what would be the time allocated to the system to absorb a fixed increment ΔP, at any moment along t. This is simply obtained by calculating $\left|\frac{\partial P}{\partial t}\right|^{-1}$, which is proportional to Δt. Then one determines these portions of the t axis for which $\Delta t > \Delta t_m$ to correspond to thermodynamic equilibrium. The rest of the t axis would be then associated with a non-equilibrium situation developing with time within the atom system. This is done at x= 4 in fig.5a in which non-equilibrium is attained in the time ranges (3.9,4.9) and (7.1,8) unit times, i.e. during the heating and cooling procedures, respectively.

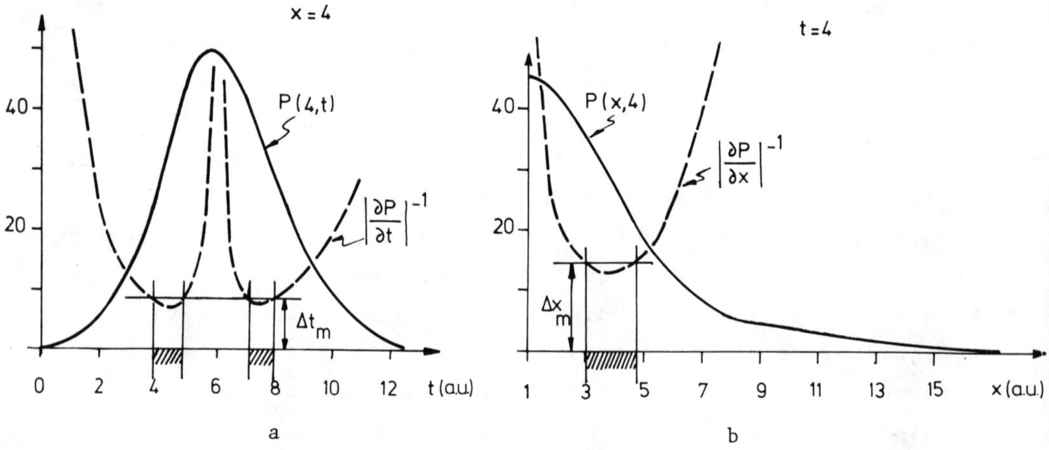

Fig.5.

A similar analysis is performed at each time interval which allows to map the time instability regions of the P(x,t) distribution, fig.4. In the same way, the space required to absorb homogeneously ΔP, Δx, is determined from the power density distributions calculated at each time interval via $\left|\frac{\partial P}{\partial x}\right|^{-1} \sim \Delta x$. Again, portions of the x axis at which $\Delta x > \Delta x_m$ correspond to equilibrium and the rest of the axis to spatial non-equilibrium, fig.5b.

Finally, the boundary lines delimiting the equilibrium and non-equilibrium regions of the map, fig.4, are projected onto the (x,t) plan, fig.6 , to constitute the spatio-temporal "phase diagram" of the system submitted to the laser beam profile defined by Eq. 10 and represented in fig.2. Other laser configurations have been tested upon varying either laser energy or laser power. The dimensions of the equilibrium cell $(\Delta x_m, \Delta t_m)$ can also be varied.

One remarkable feature is that non-equilibrium (or "melting") may be achieved by either time or space confinement of any beam, either by pulsing or scanning, i.e. depends on the

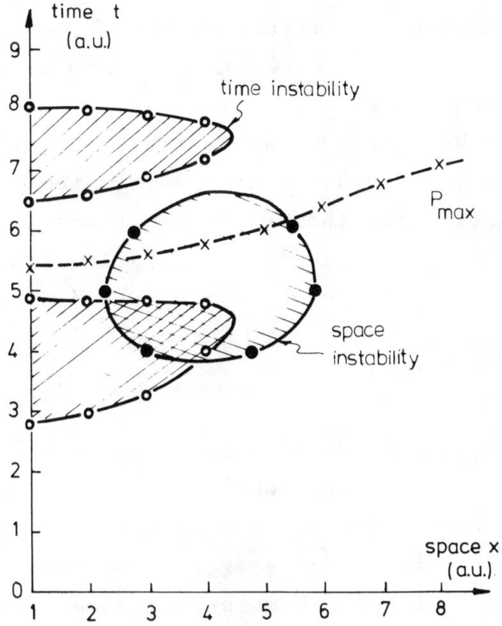

Fig.6.

gradients of energy present in the system. Another interesting point is that non-equilibrium is attainable essentially in the irradiated part of the target. Part of the energy which diffuses out of that zone never produces sufficiently strong gradients to induce non-equilibrium. In such regions, atom migration procedes in the solid phase across interfaces as described by classical diffusion theory, section 3. The product of this delayed "heating" is a non-homogeneous, partly reacted material(3). In contrast, the very broad spatio-temporal destabilization induced within the irradiated part results in a completely different product(2). In there, the onset of destabilization occurs <u>during</u> the raise of the power density over times of the order of as little as a few nsec(7). That onset is most characteristic of laser-processing solid-solid interfaces. Occuring very shortly after the onset of irradiation, the system has probably no time to progressively stabilize the atom distribution, so that early (solid state) diffusion may not induce sufficient interface reaction (e.g. compound formation) to block any further mixing.

What happens after ignition of the destabilization (within the cross-hashed domains of fig.6) is certainly not at hand using energetic arguments as done above, but rather via an atomistic approach as will be discussed in the next section. Finally, it may be noticed that destabilization re-appears eventually upon "cooling". Whatever the atom configuration during destabilization, the existing network then may be quenched depending on Δt_m, i.e. on the coupling (thermal, mechanical,...) of the atom system with the outside medium (substrate).

5. Model

Laser processing of solid-solid interfaces offers a variety of physical mechanisms which may occur competitively and simultaneously within the same experiment[3]. From figs.4 and 6, it may be recognized that energy deposition within the material is far from being homogeneous at and around the irradiation site. So will be the products of the irradiation. Focusing the attention on the laser impact region of the film, three steps in the synthesis process may be delineated.

a. destabilization of the original atom system;
b. forming a transient fluid phase;
c. solidification.

a. Destabilization. We have seen that local equilibrium breaks down via a reduction of the equilibrium cell at dimensions below Δt_m. This may be for instance, attained by a proliferation of defects[8] (dislocations, broken bonds, vacancies etc...) upon increasing the temperature of the film. This could also be associated with the changes in impurity content near interfaces. The free energy of the system would then increase, reducing the energy gap towards destabilization (or melting) in these parts of the target only. As seen in fig.6., destabilization never occurs away from the central part of the target. In these, reaction proceeds differently. After having induced a heat flux, the corresponding heat increase is delayed against the injection time.

Further, reaction may start at any interface via solid state diffusion and when eventually the energy would have sufficiently increased there to melt one or both layers the material would have been already partly reacted. In the case of compound synthesis, the occurence of covalent AB bonding would add on the complexity of the system, i) adding additional interface [A-AB] and [AB-B], and ii) freezing some of the A and B atoms into such bonds and rendering the whole systems even more difficult to destabilize. In that case, compound synthesis proceeds from these interfacial nuclei and develop by classical crystallization until completion of the reaction over extended periods of time after laser energy injection. It is probable that a few layers of transformed material may be present at the A-B interface, prior irradiation. This does not seem to be determinant in the central part of the irradiated film, probably because being discontinuous or simply having a higher free energy in relation with their low dimensionality. A rough energy diagram of this inhomogeneous network is shown in fig.7, as a function of a reaction coordinate. If the film does melt upon irradiation, the evolution of the system follows track 2. If it develops into a solid state diffusion process, it would follow track 1.

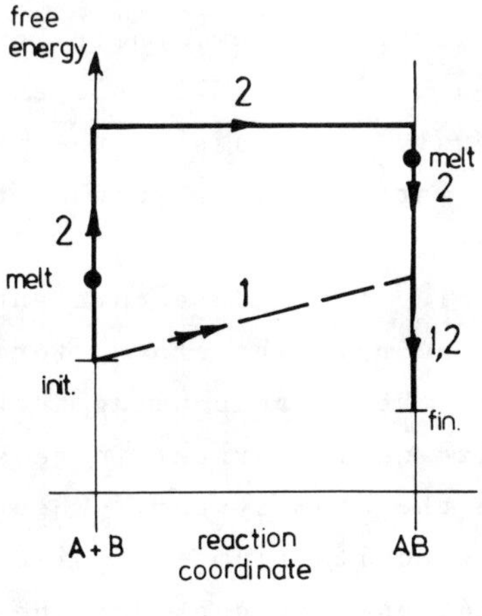

Fig.7

b. Transient fluid phase. Upon increasing temperature, the velocity of the migrating atoms (A or B) increases and the system becomes "softer". One may evaluate the capture cross-section of A on B (or vice-versa) as the average dwelltime θ an atom A would spend within 5Å from any atom B in the fluid phase. In solid state diffusion, this time θ is not less than 5×10^{-2} sec which in practice would limit the atom path to the immediate interface vicinity. Once reaching the instability regime, this solid-solid interface transforms abruptly into a solid-fluid or fluid-fluid interface. The average diffusion length may then increase by orders of magnitude. The dwell time θ would then reduce tremendously. Taking for instance the case of AlSb synthesis(7), θ could go as low as 5×10^{-10}sec.

Obviously then the AB reaction can not take place at all <u>at</u> the interface. In turn, this would produce a very deep and homogeneous mixing of both species, <u>prior</u> any reaction takes place : at any site x and any time t, the probability to find either an A or a B atom would be the same. This means also that the probability to find A and B first neighbours would be very high everywhere in the medium.

The system being homogenized, reaction may start by forming AB entities about any site. Then, AB+AB coupling could proceed to form cluster or polymers of various sizes. When clusters associate, or grow, their inner energy increases but to the expense of their kinetic energy in the fluid medium. Simultaneously, the mass of the moving clusters increases resulting into reducing further the cluster mobility with cluster size.

The clusters being formed are covalent in the nature of their interatomic bonding. Therefore, forming such structures would generate the corresponding heat of formation. Increasing their size would further increase that "inner" source of energy in the atom system. When the average cluster size reaches the equilibrium cell dimension ($\Delta x_m, \Delta t_m$) defined in Section 4, one may consider the medium to become "solid" and the heat liberation process to vanish. One may

notice here that, since the original fluid phase is homogeneous in terms of atom distribution, so would be the cluster size distribution. This has such consequences that the free energy per atom (which depends on the cluster size) would be the same for all clusters and that the energy balance between clusters be zero. In energetic terms, this would make no difference to consider the whole medium as one spatial cell in the phenomenology described in Section 4. In other words, and in contrast with a classical solid state nucleation-crystallization process, synthesis would here proceed "coherently" in the system, a process which could be termed "volume nucleation" or "coherent nucleation". This by large, explains the very large crystalline formations obtained experimentally[2,3].
Finally, optical coupling between light and that fluid phase may drop significantly precisely because of this clustering process. In effect, such clusters have inherently semiconducting properties and therefore their coupling to light would depend on the photon energy. Being at a relatively high temperature, their optical absorption edge would be reduced compared to the room temperature figure. However and whatever a precise comparative evaluation of R(T), [the reflection coefficient at temperature T], and $\Delta H_f(\Delta x,\Delta t)$, [the liberated heat of formation], competition between the two opposed effects could possibly be in favor of (1-R(T)) resulting in switching off the light heating injection soon after ignition. This would alter the phenomenology of section 4, which was essentially adapted to a metallic alloy formation. In fig.8, the power density profile of fig. 5a is represented together with estimated variations of [1-R(T)] and $\frac{1}{t}\Delta H_f$, the latter being extrapolated from a classical rate equation associating cluster size and cluster concentration. The net results would be a more pronounced onset of destabilization and a further delayed cooling compared to results in section 4. The corresponding perturbation of the phase diagram is also shown, for completeness, although such figures may be altered, one way

or another, function of the system parameters (optical absorption, clustering properties, etc...).

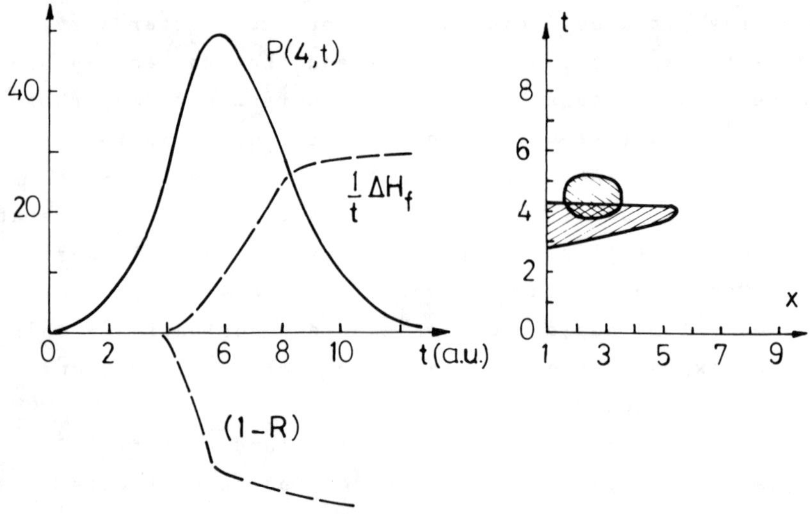

Fig.8

c. Solidification. Upon clustering, the system gains in viscosity to eventually solidify. Independently, the cluster to cluster coupling may offer a wide variety of possible configurations, as seen in fig. 9 in the case of Ge-Se clustering. Starting from the a-periodic (homogeneous) network in the fluid to the most symetric possible crystalline structure at the end of the cooling path, the system may form metastable phases which are otherwise difficult to produce by plain conventional thermal processing.

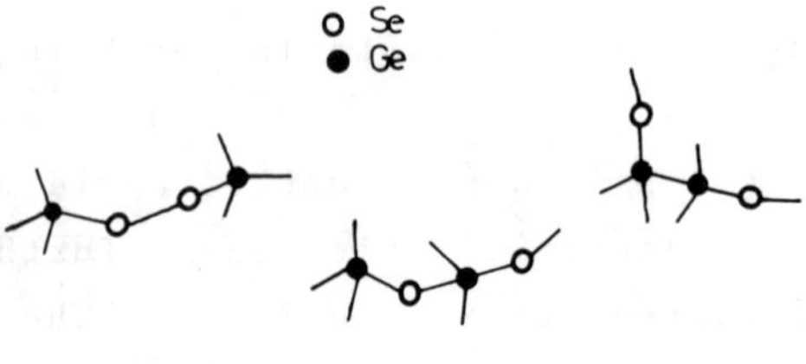

Fig.9

How can we select one of such metastable phases? The idea would be to accelerate the cooling procedure at high rate, in order to quench one intermediate phase, as has been done for instance in CdSe films[6]. This appears in figs,5 and 7 where instability domains exist in the cooling stage of the spatio-temporal energy profile. This may provide for important energy gradients and isolate glassy networks (as for metallic glasses) or less common, less symetric structures as for instance monoclinic $GeSe_2$ and orthorhombic GeSe[9]. The synthesis process which comes with the solidification regime but initiates in the transient fluid phase (clustering) may apply to multiphase systems in which several phases may co-exist. Each of these has a specific synthesis kinetics. A proper energy balance between all competing phases is almost impossible to evaluate. However such a selection is effective as, for instance, in the Cu-In-Se system (12 phases possible), peaking up the best adapted phase to the actual laser fluence. One impressive case is the forming of $CuInSe_2$[10], totally devoid of Cu-Se and In-Se binaries or oxydes. One should finally note that such laser-synthesized compounds do possess a high degree of purity[2,6] as seen from their electronic properties (intrinsic conductivity down to 200°K, for instance) despite the fact that non-stoichiometry is some 5-6%. Such extra atoms are apparently drained out of the irradiated zone, or more precisely out of the instability zone, via a very fast zone melting process which is also laser-triggered, simultaneously to the synthesis process.

6. Conclusions

It has been shown in this paper how laser sources may provide means to induce non-equilibrium situations during which drastic and irreversible changes in the atomic configurations of interfaced layers may be obtained. Not only the physical and chemical aspects of these specific laser-matter interactions are extremely appealing to the fundamentalists, but also they open brand new routes to devise new compounds or alloys and, subsequently, new material processing facilities. However, much more work is neededto fully understand the various aspects of that new brand of research activities in Materials Science.

Acknowledgements

The author would like to thank the many members of the Mons group who, over the past eight years, contributed to the development of the experimental investigations reviewed in this paper. This work has been conducted under the auspices of the Ministry for Science Policy, Brussels.

References

1. R. Andrew, M. Ledezma, M. Lovato, M. Wautelet, L.D. Laude, Appl. Phys. Lett. 35, 418 (1979);
2. R. Andrew, L. Baufay, A. Pigeolet, L.D. Laude : J. Appl. Phys. 53, 4862 (1982);
3. T. Szörényi, K. Bali, G. Radnoczi, L.D. Laude : in Laser Processing and Diagnostics (II), ed. by D. Bäuerle, K.L. Kompa, L. Laude (Les Editions de Physique, les Ulis 1986) p. 157;
4. C. Antoniadis, L.D. Laude, P. Pierrard : In Energy Beam-Solid Interactions and Transient Thermal Processing, ed. by V.T. Nguyen, A.C. Cullis (Les Editions de Physique, Les Ulis 1985) p. 387;
5. L.D. Laude : In Laser Processing and Diagnostics, ed. by D. Baüerle, Springer Ser. Chem. Phys. 39 (Springer, Berlin, Heidelberg 1984) p. 355;
6. L. Baufay, D. Dispa, A. Pigeolet, L.D. Laude : J. Cryst. Growth 59, 143 (1982);
7. P.S. Peercy : private communication;
8. M. Wautelet, C. Antoniadis, L.D. Laude : Phys. Stat. Sol. (a) 92, K43 (1985);
9. C. Antoniadis, M.C. Joliet : Thin Solid Films 115, 75 (1984);
10. M.C. Joliet, C. Antoniadis, R. Andrew, L.D. Laude, Appl. Phys. Lett. 46, 266 (1985).

PHOTOELECTROCHEMISTRY WITH PARTICULATE SEMICONDUCTORS AND ELECTRODES

M. GRÄTZEL
Institut de Chimie Physique
Ecole Polytechnique Fédérale
CH-1015 Lausanne, Switzerland

1. INTRODUCTION

Over the last ten years, the scientific community has witnessed an enormous activity in the field of photoelectrochemistry with semiconductor particles and electrodes. In the electrode domain alone one counts more than 1500 papers published during this period [1]. These investigations have been spurned by several factors of which the most important ones are light energy conversion and the initiation of selective chemistry occurring at the semiconductor solution interface by photo-excitation. Numerous reviews have appeared in recent years dealing with the progress in this field [2]. In the present lecture we emphasize the dynamics of charge carrier related reactions occurring at the semiconductor surface. Direct kinetic information concerning the rate and mechanism of interfacial redox reactions are difficult to derive from experiments where electrodes are excited by light. By contrast, this information is readily obtained from studies with dispersions of semiconductor particles in particular colloids. The transparent nature of the colloidal dispersions allows for the excitation of the semiconductor by a short laser flash producing electron-hole pairs within the particles. The dynamics of the charge carrier induced reactions at the semiconductor/solution interface can subsequently be investigated by applying fast kinetic spectroscopy. Studies of this type have been initiated only recently [3], but much important information has already been obtained. The photoelectrochemistry with particulate semiconductors is therefore the main topic of the present lecture. Before addressing this issue in more detail, we shall discuss the basic properties of semiconductor particles and give an overview of the experimental methods that are presently employed in the investigation of these systems.

2. PROPERTIES OF SEMICONDUCTOR PARTICLES

The properties of semiconductor particles being of interest here are listed in Fig. 1. They are divided in three categories comprising structural, optical and electronic aspects. These will now be discussed in more detail.

2.1. Structural properties

Particulate systems are commonly distinguished by their size, and in Fig. 2 we have applied this classification to semiconductors. the two major classes to be discerned are colloids and microparticles. The latter have a size exceeding approximately 10^3 Å and form turbid suspensions. Colloids are smaller particles and give clear solutions. Amongst the colloids we distinguish semiconductors with normal optical and electronic behavior from those that display quantum size effects ("Q" particles). Quantum size effects occur when the Bohr radius of the first excitation becomes commensurate with, or larger than, that of the particle. They

PROPERTIES OF SEMICONDUCTOR PARTICLES

STRUCTURAL	OPTICAL	ELECTRONIC
SIZE	ABSORPTION	CONDUCTION
SHAPE	LUMINESCENCE	INTRINSIC
LATTICE-TYPE	LIGHT-SCATTERING (RALEIGH,MIE)	*n* -TYPE *p* -TYPE
LATTICE-DEFECTS		FERMI LEVEL
		CHARGE CARRIER MOBILITY
		BAND POSITIONS
		CHARGE CARRIER RECOMBINATION
		DOPING

FIGURE 1

manifest themselves in a blue shift of the fundamental absorption edge and of the luminescence maximum with decreasing particle size [4]. Since the Bohr radius of the charge carriers is related to their respective effective mass (m_{eff}):

$$r = \frac{h^2 \cdot \varepsilon_0 \cdot \varepsilon}{c^2 \pi m_{eff}} \quad (1)$$

it depends on the semiconductor material. In eq. (1) ε_0 is the permeativity of the free space and ε the dielectric constant of the semiconductor. For example, in the case of CdS $*m_{e^-} = 0.2\ m_{e^-}$, $\varepsilon = 8.9$ and $r_{e^-} = 24$ Å, while for TiO_2 $*m_{e^-} = 30\ m_{e^-}$ and $\varepsilon = 170$, yielding $r_{e^-} = 3$ Å. Thus, quantum size effects are expected for particles with a radius below 25 Å and 3 Å for CdS and TiO_2, respectively.

As for the shape and crystalline structure of semiconductor particles, these are strongly influenced by the preparation methods. If so desired, materials with uniform size and high crystallinity can be prepared. Chemical precipitation or controlled hydrolysis and subsequent polymerization are the most frequently applied methods. Depending on preparation conditions, one obtains in this way semiconductor particles of spherical shape with amorphous or crystalline structure. As an example, we report in Fig. 3 electron microscopy and quasi-elastic light scattering results obtained from a hydrosol of TiO_2. The latter was prepared by hydrolysis of $TiCl_4$ at 0^oC [5]. These data reveal that there is a broad distribution of particle sizes, the average diameter being 12 nm.

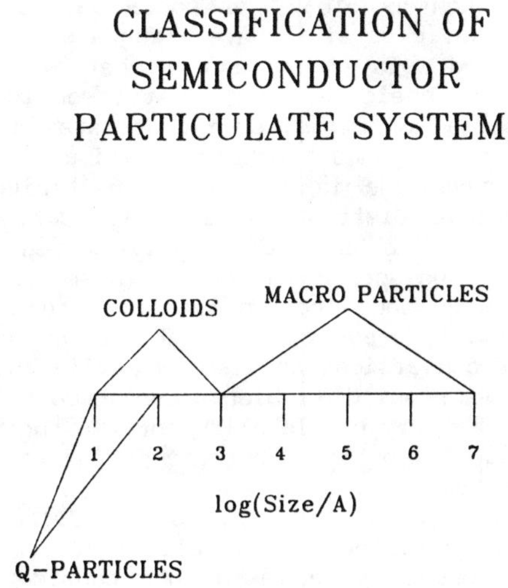

FIGURE 2

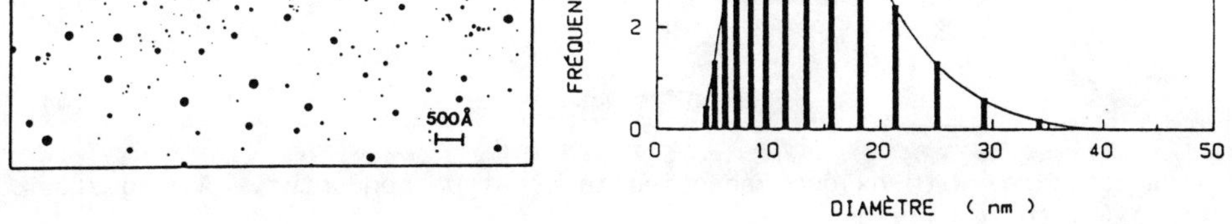

(a) (b)

FIGURE 3. Electron micrograph (a) and size distribution (b) of colloidal TiO_2 particles prepared according to ref. [5].

A particularly interesting development during recent years is the synthesis of monodisperse inorganic particles. Earlier work in this field concerned insulators, such as SiO_2 [6] or noble metals such as Pt or Pd [7]. More recently, Matijevic et al. [8] have elaborated methods for the preparation of monodispersed sols for a large number of compounds including semiconducting oxides and chalcogenides. Apart from their application in the fabrication of special ceramics, such aggregates are attractive also from the viewpoint of heterogeneous catalysis and electron transfer reactions. An important advantage is that these particles can be produced with extremely narrow size distribution and well defined geometry. For example, it is possible to obtain spherically shaped aggregates with a smooth surface. The monodisperse nature of these sols facilitates the kinetic analysis of energy and electron transfer processes whose rate is particle size-dependent. Moreover, applications in heterogeneous catalysis and light energy conversion can also be envisaged. These particles could constitute suitable building blocks or supports for more complex units. Their precise geometry should allow for the engineering of systems with suitable functionality, optimizing their performance in catalysis and heterogeneous electron transfer.

While it is true that even very small semiconductor particles can be produced with a very high degree of crystallinity, occurrence of lattice defects is common and needs to be taken into consideration. In particular, there is a link between the nature of the defect and the character of the electronic conduction in the particles, which we shall briefly elaborate in the following. We shall restrict ourselves to the discussion of the most important point defects in oxides and chalcogenides, i.e. anion and cation vacancies as well as interstitial cations. The formation of interstitial anions is energetically unfavorable and will not be considered. In order to establish a link between defect structure or deviation from stoichiometry and electronic properties, we examine the case of an oxide semiconductor of the general formular MO, Fig. 4. The formation of an oxygen vacancy can be written in the Kroger notation as:

$$O_o \rightleftharpoons V_o + 1/2\ O_2 \tag{2}$$

where O_o stands for an oxygen atom on its lattice site. The filled vacancy acts as an electron donor and ionizes in two subsequent steps producing conduction band electrons.

$$V_o \rightleftharpoons V_o^{\bullet} + e^-_{cb} \tag{3}$$

$$V_o \rightleftharpoons V_o^{\bullet\bullet} + e^-_{cb} \tag{4}$$

Thus, an oxide with oxygen vacancies is a n-type conductor. Adding equations (2) to (4) yields:

$$O_o \rightleftharpoons V_o + 2e^-_{cb} + 1/2\ O_2 \tag{5}$$

for which the mass actions law gives:

$$K = [V_o][e^-_{cb}]^2\ P_{o_2}^{1/2} \tag{6}$$

OXYGEN VACANCIES AS ELECTRON DONORS

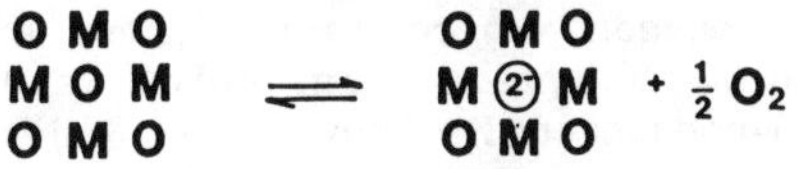

$$O_O \rightleftharpoons V_O + \tfrac{1}{2}O_2$$

$$V_O \rightleftharpoons V_O^{\bullet} + e^-_{cb} \qquad \Delta H_1^{\bullet}$$

$$V_O^{\bullet} \rightleftharpoons V_O^{\bullet\bullet} + e^-_{cb} \qquad \Delta H_2^{\bullet}$$

FIGURE 4

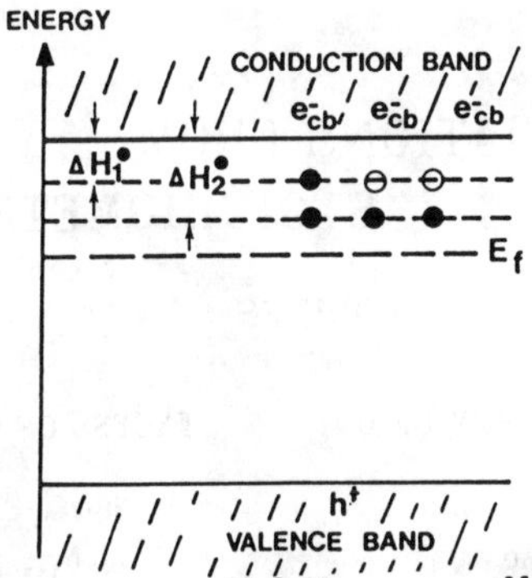

$$M_M \rightleftharpoons V_M + M_{gas}$$

$$M_{gas} + \tfrac{1}{2}O_2 \rightarrow M_M + O_O$$

$$\tfrac{1}{2}O_2 \rightleftharpoons V_M + O_O$$

$$V_M \rightleftharpoons V_M' + h^+ \qquad \Delta H_1^o$$

FIGURE 5

$$V_M' \rightleftharpoons V_M'' + h^+ \qquad \Delta H_2^o$$

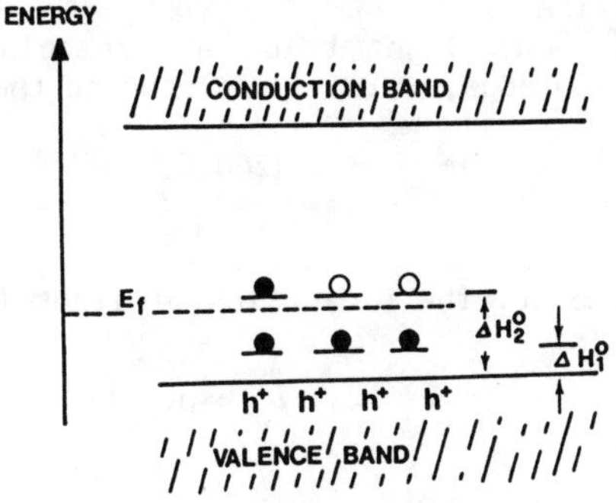

At sufficiently high temperature the first ionization is complete. Under these conditions $2[V_o] = [e^-_{cb}]$ and therefore $[e^-_{cb}] = 2K^{1/2} p_{o_2}^{-1/2}$. From this equation it is seen that the electron concentration and hence the conductance of the powder should decrease with the 1/6 power of the partial pressure of oxygen, which is confirmed experimentally. Similar to the oxygen vacancy, an interstitial metal is a donor center rendering the semiconductor n-conducting. Conversely, cation vacancies are acceptors rendering the oxide p-conducting. This is illustrated in Fig. 5. Applying analogous considerations as above for the oxygen vacancies, one finds that the concentration of holes and hence the conductance of the semiconductor should increase with the 1/6 power of the oxygen partial pressure. The relation between deviation from stoichiometry, defect structure and electronic conduction are summarized in Fig. 6.

DEVIATIONS FROM STOICHIOMETRY

OXIDES

DEFICIENCY OF O_2	EXCESS OF O_2
TiO_{2-x}	NiO_{1+x}
ZnO_{1-x}	Ag_2O_{1+x}
Fe_2O_{3-x}	
POINT DEFECTS: OXYGEN VACANCIES (V_o) INTERSTITIAL METAL (M_i)	POINT DEFECTS: METAL VACANCIES (V_M) ACCEPTOR CENTERS
DONOR CENTERS	
n -type CONDUCTION	*p* -type CONDUCTION

FIGURE 6

2.2. Optical properties

Semiconductors absorb light below a threshold wavelength (λ_g), the fundamental absorption edge, which is related to the band gap energy via:

$$\lambda_g \text{ (nm)} = 1240 \; E_g \text{ (eV)} \qquad (7)$$

Within the semiconductor, the extinction of light follows the exponential law:

$$I = I_o \exp(-\alpha l) \qquad (8)$$

where I is the distance and α the reciprocal absorption length. For example, for TiO_2, α has the value 2.6 10^4 cm^{-1} at 320 nm, which implies that light of wavelength 320 nm is extinguished to 90% after traversing a distance of 3900 Å.

Near threshold, the value of α increases with increasing photon energy. Frequently, a function of the type:

$$\alpha x h\nu \quad = \quad const\ (h\nu - E_g)^n \tag{9}$$

gives a satisfactory description of the absorption behavior in this wavelength domain. Here, the exponent has the value 1/2 for a direct transition and 2 for an indirect one. A direct transition is characterized by the fact that in the electronic energy vs. wavevector diagram, the minimum in the conduction band states occurs vertically above the maximum of the valence band. By contrast, for an indirect transition, the two extremes are displaced from each other. As a consequence, the threshold excitation requires a contribution of lattice phonons to the momentum in order to compensate for the change in the wavevector during the transition, Fig. 7. This reduces the absorption cross section and hence the value of α.

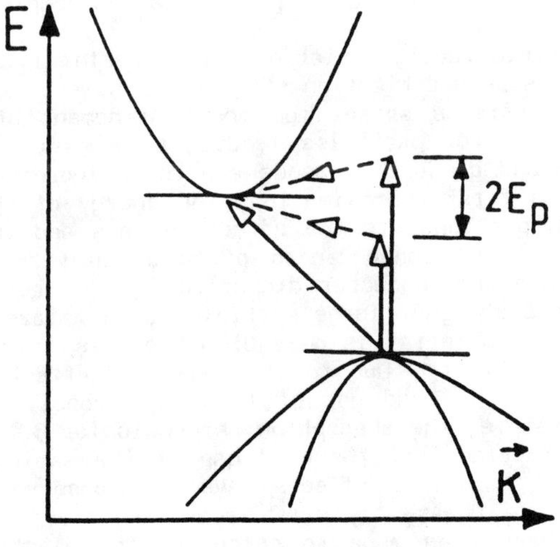

FIGURE 7. Energy vs. wavevector diagram illustrating the band structure for a semiconductor with an indirect band gap.

In some cases, deviations from eq. (9) have been observed. For example, Dutton found that for CdS, the reciprocal absorption length near band edge showed an exponential dependence on photon energy:

$$\ln \alpha \quad = \quad \beta\, h\nu \;/\; kT \tag{10}$$

The coefficient β had a value of 2.1 for single crystals with hexagonal structure. In our laboratory [9], colloidal particles of CdS have been produced by adding H_2S to an aqueous solution of $Cd(NO_3)_2$ (method 1) or by rapid mixing of NaS and $Cd(NO_3)_2$ solutions (method 2)[2]. It was discovered that hexametaphosphate was suited to stabilize the ultrafine particles preventing their aggregation by electrostatic protection. The two methods yield particles with an average size of 50 and 30 nm, respectively. Each particle is a single crystal with cubic structure. Interestingly, the reciprocal absorption length near band edge followed also the Dutton law. However, the coefficient β was found to have a significantly smaller value than that observed for single crystals. This was attributed to the large fraction of CdS atoms exposed at the surface in the colloidal solutions. Support for this interpretation comes from the observation that the β value decreases with increasing the proportion of CdS that is present at the particle surface [10].

The whole extinction spectrum of the dispersion of semiconductor particles can be calculated from Mie theory [11] which takes into account contributions from both light absorption as well as scattering. Within the dipolar limit (particle size wavelength of light) one obtains:

$$\alpha \quad = \quad 18\pi\, X\, n_o^3\, \varepsilon_2 \,/\, \left(\lambda\, (\varepsilon_1 + 2n_o^2) + i{\varepsilon_2}^2\right) \qquad (11)$$

where X is the volume fraction of the particles in solution, n_o the refractive index of te solvent and $\varepsilon(\lambda) = \varepsilon_1 + i\varepsilon_2$ is the complex dielectric constant of he particle.

At very small particle radii, deviations from the Mie equation occur, and this is due to the size quantization effects which were alluded to already earlier. These effects arise from the confinement of the charge carriers in the semiconductor particles leading to a series of discrete states in the valence and conduction bands and to an increase of the effective band gap. To a first approximation, the energy of the quantized levels is inversely proportional to the effective mass and the square of the particle diameter. The consequences of size quantization for the absorption features of a semiconductor dispersion can be quite dramatic. Thus, HgSe colloids consisting of large particles (500 Å) are black since the band gap of the bulk material is only 0.3 eV corresponding to an absorption threshold at 4130 nm. Due to the very low effective electron mass (m_{eff} = 0.05 m_{e^-}), the band gap of HgSe is strongly dependent on particle size. For example, the absorption threshold for 30 Å sized particles is at 380 nm implying that the band gap of the semiconductor has increased to 3.2 eV! [12] Similar effects have been observed with other materials such as PbSe [12], Cd_3As_2 [13], ZnO [14] and Zn_3P_2 [15].

Several attempts have been made to calculate the electronic energy levels in these "Q" particles. Brus [16] used an exciton model in order to derive the electronic energy states of an electron-hole pair, confined to a small spherical semiconductor crystallite. A quantum mechanical calculation gave the lowest eigenstate of the Wannier exciton in such clusters, which corresponds to the apparent band gap, Fig. 3b. The exciton wave function was approximated by one or a few configurations of particle-in-a -box orbitals. The energy of the first excitonic state of a semiconductor cluster with radius R is given by the approximate expression:

$$E_g(R) \simeq E_g(R{=}\infty) + \frac{h^2\pi^2}{2R^2}\left[\frac{1}{m_{e^-}} + \frac{1}{m_{h^+}}\right] - \frac{1.8\, e^2}{\varepsilon R} \qquad (12)$$

The first term in this equation is the band gap of the bulk semiconductor, the second term corresponds to the sum of the confinement energies for the electron and the hole, and the last one is their Coulomb interaction energy.

The Coulomb term shifts $E_g(R)$ to smaller energy as R^{-1}, while the quantum localization terms shift $E_g(R)$ to higher energy R^{-2}. As a result, the apparent band gap will always increase for small enough R. This is shown in Fig. 8 for various semiconductors.

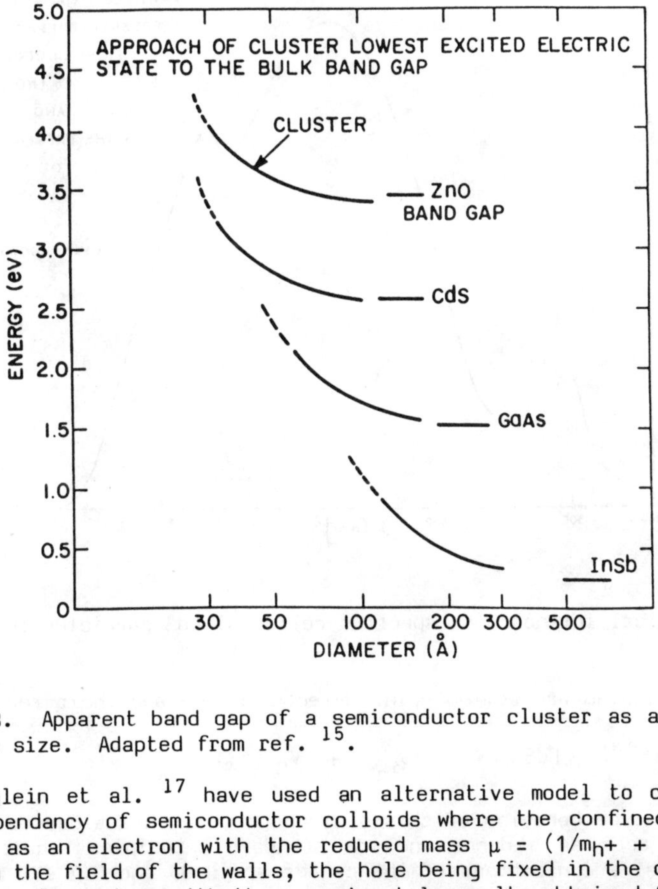

FIGURE 8. Apparent band gap of a semiconductor cluster as a function of particle size. Adapted from ref. [15].

Henglein et al. [17] have used an alternative model to calculate the size dependancy of semiconductor colloids where the confined exciton is treated as an electron with the reduced mass $\mu = (1/m_{h^+} + 1/m_{e^-})$ which moves in the field of the walls, the hole being fixed in the center of the particle. Comparison with the experimental results obtained for CdS gives better agreement than eq. (12), which predicts too large a shift in the band gap with decreasing particle radius. However as has been pointed out by Henglein et al [18], the agreement may be fortuitious since the errors introduced by the approximations (infinitely high potential energy wall at the particle surface, underestimation of the kinetic energy of the confined electron) cancel each other. Calculations with refined quantum mechanical models are presently being carried out [18].

The luminescence of semiconductor colloids has been investigated in great detail over the last few years following our initial work on CdS colloids [19]. While there is by now a wealth of experimental results available [20], only a few authors [9,21,22] have attempted to assign the observed emissions. Our analysis [9] showed that the luminescence spectrum of the CdS particles contains frequently two bands: a green emission resulting from the recombination of free carriers and a red one which arises from sulfur vacancies. Typical results are presented in Fig. 9.

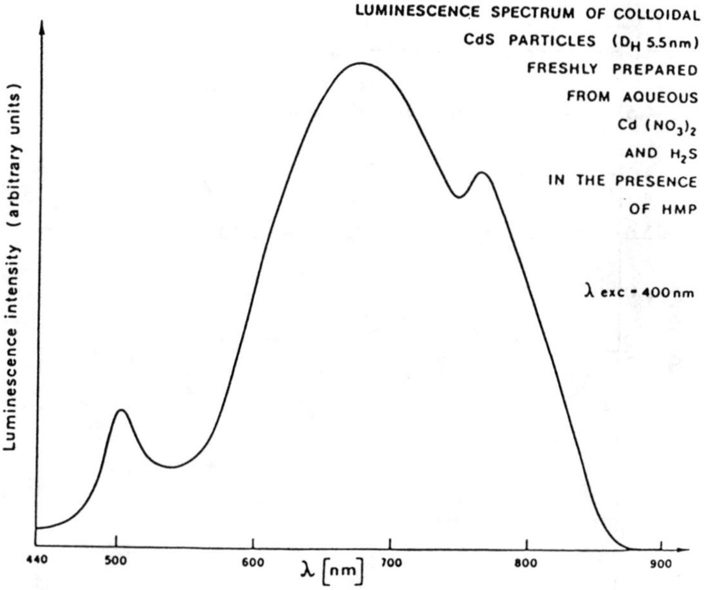

FIGURE 9. Typical luminescence spectrum colloidal CdS particles in water. (ref. [9]).

Sulfur vacancies are common point defects in CdS and their ready formation

$$CdS \rightleftarrows CdS_{1-x} + V_S \ (x \ll 1) \tag{13}$$

explains the n-type behavior of this semiconductor. In analogy to oxygen vacancies, V_S acts as a donor and the standard enthalpies required for the two successive ionizations of the center are 0.03 and 0.7 eV, respectively [23]. Red luminescence arises from the reaction of photogenerated holes with singly ionized sulfur vacancies.

$$h^+ + V^+ \rightarrow V^{2+} + h\nu \quad (\sim 700\ \text{nm}) \tag{14}$$

and hence is explained by the Lambe-Klick model, Fig. 10. Both the green as well as the red emission are quite weak. However, the red emission can be enhanced drastically, e.g. by the preparation of the colloid in a mixture of water and acetonitrile [10].

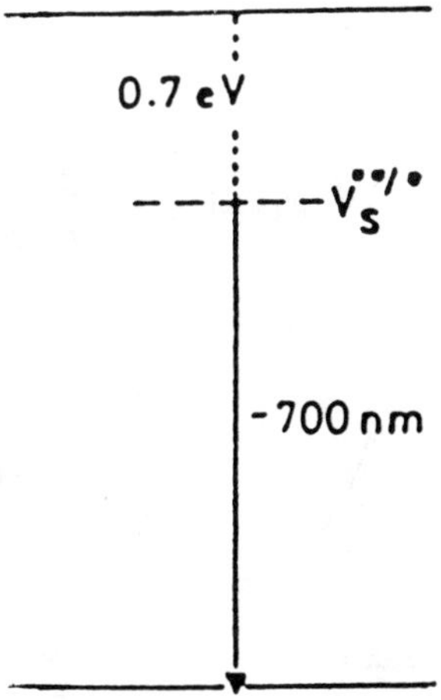

FIGURE 10. Self activated luminescence of CdS colloids; Lambe-Klick model for the operation of sulfur vacancies as luminescence centers.

A noteworthy feature of the red luminescence of CdS and other II-VI or III-V semmiconductors is that it is extremely sensitive to the presence of electron acceptors such as MV^{2+} or benzoquinone [20]. In water, a concentration of 10^{-8} M of MV^{2+} suffices to quench 50% of the emission [9]. A quantitative analysis of the results taking into account the statistics of quencher distribution over the particles showed that only one MV^{2+} molecule per aggregate is required to quench the emission[9,10]. The role of the MV^{2+} is to scavenge conduction band electrons intercepting in this way their radiative and nonradiative recombination with holes. Recent laser photolysis experiments with picosecond time resolution have shown that this reaction is very fast requiring less than a nanosecond [24,25].

Due to the quantum effects discussed above, the luminescence of colloidal semiconductors, similar to the band edge absorption, is strongly influenced by the particle size. Typically, a broad emission is observed which blue-shifts with decreasing radius of the aggregate [26]. The luminescence of these clusters is readily detectable at 298° K and can be quite strong at cryogenic temperature.

Luminescence analysis provides an important tool to study the dynamics of charge carrier recombination in such colloidal semiconductors. For CdS clusters in alcohol glasses, the emission decay time is wavelength and temperature dependent, but it becomes temperature independent below 30° K [27]. In this temperature range, the reciprocal luminescence lifetime, $1/\tau$, for 22 Å-sized clusters is 10^5 s^{-1} and 4.5×10^4 s^{-1} at 580 and 400 nm, respectively. This observation can be rationalized in terms of radiative and radiationless recombination of trapped carriers, Fig. 11. At low temperature, these two processes occur exclusively via a tunneling mechanism.

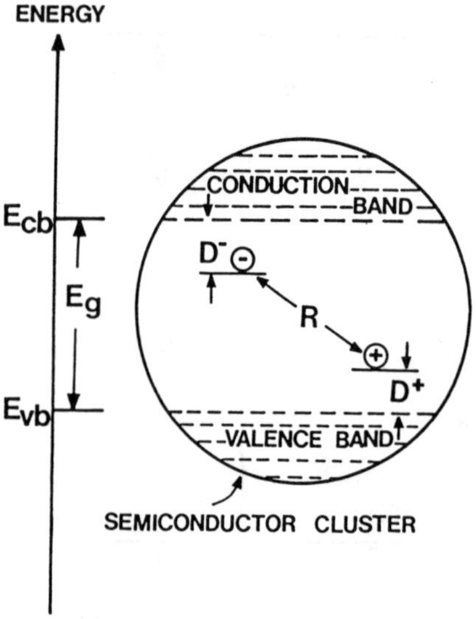

FIGURE 11. Radiative and radiationless tunneling recombination of trapped electorn-hole pairs. D^- and D^+ is the trap depth for the electron and hole, respectively.

The contribution of the radiative channel to the overall decay can be estimated by using a model derived for bulk semiconductors [28-33]. The rate constant depends on the distance R separating te electron-hole pair via:

$$k(R) \simeq 2\pi|H_{e^--h^+}(R)|^2/h \qquad (15a)$$

$$k(R) = 5\times10^8 \exp(-2R/a_o)\ s^{-1} \qquad (15b)$$

where $H_{e^--h^+}$ is the exchange integral which measures the interaction energy between the donor and acceptor, i.e. the overlap between the tails of the wave function of the trapped electron and hole, and a_o is the radius of the hydrogenic radius of the wave function for the charge carrier residing in the trap with the lower binding energy.

The dominant route for charge carrier recombination in semiconductor clusters is nonradiative. This is due to the strong coupling of the wave functions for the trapped electrons and holes to lattice vibrations. The presence of strong vibronic interactions manifests itself in the broad nature of the emission spectrum of semiconductor clusters. In the absence of phonon interaction a single emission line would be expected.

The rate constant for nonradiative exothermic tunneling is given by the general expression for nonadiabatic electron transfer. Using

Jortner's model for radiationless transitions in condensed media to derive the Frank-Condon overlap integral, one obtains [34,35]:

$$k_{nr} = \frac{2\pi}{\hbar} \mid H_{e^- - h^+} \mid^2 \exp[-S(2f+1)]\ I\ (2S[f(f+1)]^{1/2}) \times [(f+1)/f]^{p/2} \quad (16)$$

where S is the Franck-Condon displacement factor, I a modified Bessel function, f the Bose-Einstein factor giving the average occupancy of phonon energy levels and p the exothermicity of the reaction expressed as an integral number of vibrational quanta. Eq. (16) provides a satisfactory description of the temperature effect on the luminescence lifetime of CdS clusters [27] supporting the notion that trapped electron-hole pairs recombine via a radiationless multi-phonon process which is brought about by strong coupling of the wave function of the trapped charge carriers to lattice vibrations.

As for the wavelength dependence of the emission lifetime, this arises from the contribution of the coulombic energy of interaction between the electron-hole pair to the total energy of the emitted photon:

$$E(h\nu) = E_g - (D_+ + D_-) + e^2/\varepsilon R \quad (17)$$

It follows from eq. (17) that close pairs emit at higher energy than those which are further apart. Furthermore, for the close pairs the emission is faster since both k_{nr} and k_r increase with decreasing distance, eq. (15). Therefore, the emission lifetime becomes wavelength dependent and decreases with increasing energy of the photons.

2.3. Electronic Properties

We shall first assess the important problem of the nature of the potential distribution within the semiconductor. An electrostatic field is generated in a semiconductor particle or electrode as a consequence of net transfer of charge between the solid and its surroundings, in our case the electrolyte. Interfacial movement of charges is brought about either by applying a potential (bias) to the semiconductor or by redox reactions with species in the electrolyte. This, in turn, produces a space charge layer in the semiconductor. Here, four different situations may be distinguished which are outlined schematically in Fig. 12. In the absence of any space charge layer, we have the so-called "flat-band situation". If we accumulate charge on the semiconductor of the same sign as the majority carriers, we obtain an "accumulation layer". The opposite charging leads to a depletion of the mobile charge carrier and the excess charge is given only by the immobile ionized donor or acceptor states. This is the so-called "depletion layer". The depletion can go so far that the concentration of the majority carriers decreases below the intrinsic concentration n_i. If electronic equilibrium is maintained in the semiconductor up to the surface, the other charge carriers now will have the larger concentration according to equation (18):

$$c_{e^-} \cdot c_{n^+} = {}^1n_{cb} \cdot {}^1n_{vb} \exp(-E_g(kT) \quad (18)$$

where ${}^1n_{cb}$ and ${}^1n_{vb}$ are the concentration of conduction and valence band states in the solid, respectively. The position of the Fermi

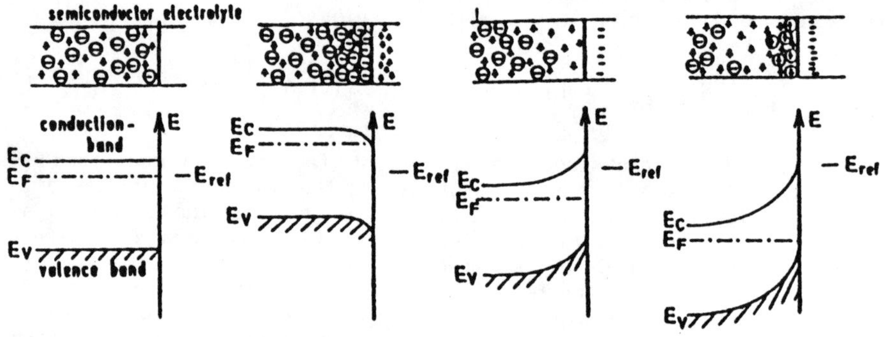

a) b) c) d)

FIGURE 12. Charge distributions at the semiconductor electrolyte contact and their consequences for the energy position of the band edges. a) neutral state = flat band position; b) accumulation layer; c) depletion layer; d) inversion layer.

level is given by the equation:

$$E_f = (E_{cb} + E_{vb})/2 + \frac{RT}{2} \ln (c_{e^-}/c_{h^+}) + \frac{3}{4} RT \ln \left(\frac{{}^*m_{h^+}}{{}^*m_{e^-}}\right) \quad (19)$$

The position of the Fermi level in Fig. 12d indicates that the semiconductor is p-type at the surface and n-type in the bulk. This is called an "inversion layer".

The inversion layer will, however, only be created if minority carriers are available at the surface either by thermal generation of electron hole pairs in the solid or by injection from the adjacent electrolyte. Otherwise electronic equilibrium cannot be obtained at the interface and the depletion layer extends beyond the thermodynamic limit.

The illustration in Fig. 12 refers to a n-type semiconductor. For p-type material, analogous pictures can be drawn with holes as mobile charge carriers and negatively charged acceptor states in the depletion layer. Then, of course, the potential profile which represents band bending is inverted in such pictures.

Using the Poisson equation, we can derive a relation between the excess space charge g_{sc} and the potential difference between the bulk ϕ_b and the surface of the semiconductor:

$$g_{sc}^2 = 2 \, \varepsilon \, \varepsilon_0 \int_{\phi_s}^{\phi_b} \rho(\phi) \, d\phi \quad (20)$$

where ε is the relative dielectric constant, ε_0 the dielectric constant of vacuum and ρ the density of electric charge. A simplified model of the depletion layer assumes that the density of ionized donors or acceptors is constant and equal to ${}^1N$. In such a case, the integration of eq. (20) gives:

$$\phi_s - \phi_b = (kT/2e)\,(W/L_D)^2 \tag{21}$$

where

$$L_D = \left(\varepsilon kT/8\pi e^2\,(^1N_d)\right)^{1/2} \tag{22}$$

is de Debye length. The same equation is valid also for large semiconductor particles (radius $r_0 >> W$). In the case of very small particles, the condition $W = r_0$ holds, and the potential difference between the surface and the center is given by [36]:

$$\Delta\phi_0 = (kT/6e)\,(r_0/L_D)^2 \tag{23}$$

For application in photocatalysis, it is useful to know the band edge positions of semiconductor particles and these are given in Fig. 13 for various semiconductors in contact with water.

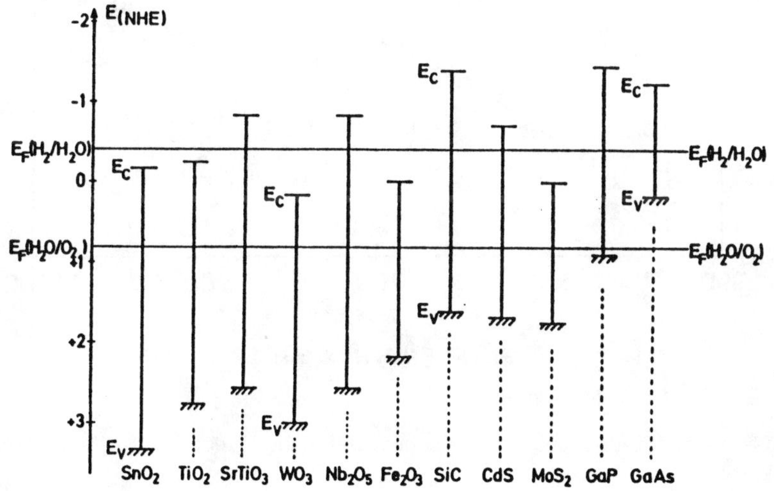

FIGURE 13. Band edge positions of several semiconductors in contact with aqueous electrolyte at pH 7.

3. DYNAMICS OF CHARGE CARRIER REACTIONS IN COLLOIDAL SEMICONDUCTORS

We shall illustrate the principle of these investigations by reporting studies with colloidal titanium dioxide (anatase) particles having a diameter of 120 Å. Irradiation of such colloidal solutions in the presence of a hole scavenger such as polyvinyl alcohol or formate ions results in the accumulation of electrons in the particles. As a result, the solution assumes a beautiful blue color under illumination. It was found that up to 300 electrons can be stored in one TiO_2 particle. (A TiO_2 particle of 120 Å size has about 3600 conduction band states. Therefore, at most 10% of the available states are occupied by electrons.) The absorption spectrum of these stored electrons is shown in Fig. 14:

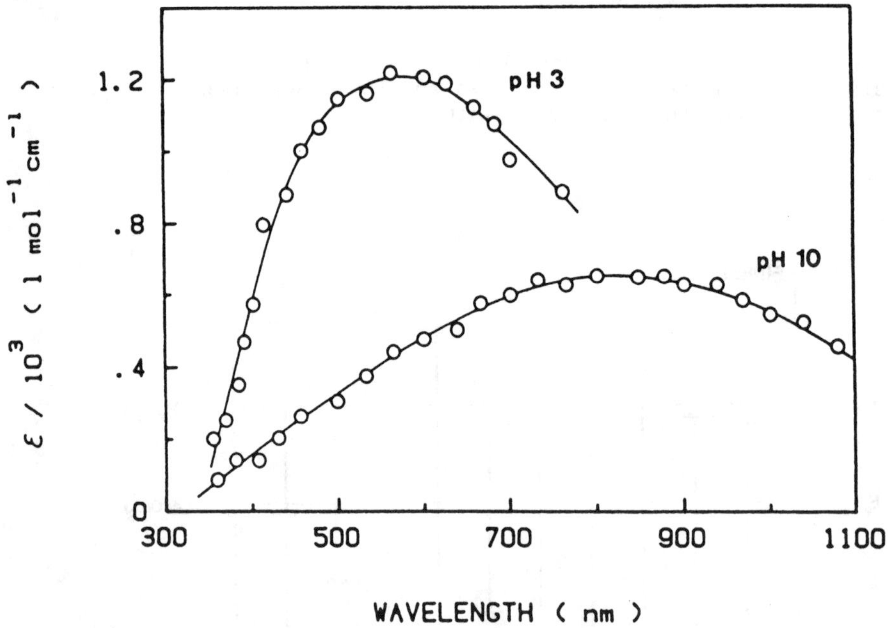

FIGURE 14. Absorption spectrum of cnduction band electrons in colloidal TiO_2 particles at pH 3 and pH 10.

The electron spectrum was found to be sensitive to the pH of the solution. Under alkaline conditions the electron absorption is very broad and has a maximum around 800 nm. Lowering the pH to 3 produced a pronounced blue shift in the spectrum which under these conditions shows a peak at 620 nm. The sensitivity of the electron absorption to the solution pH would indicate that they are located in the surface region of the particles. This has been confirmed by recent ESR experiments which show that under acidic conditions the electrons are trapped at the TiO_2 surface in the form of Ti^{3+} ions [37]. Using redox titration, we have recently been able to determine the extinction coefficient of the trapped electrons [38]. For the colloidal solutions of pH 3 the extinction coefficient at 600 nm is 1200 M^{-1} cm^{-1}.

Taking advantage of the characteristic optical absorption of trapped electrons in the colloidal TiO_2 particles, we have recorded their recombination with free and trapped holes in the picosecond to microsecond domain [39]. Fig. 15 shows the temporal evolution of the transient spectrum after excitation of TiO_2 with a frequency tripled (353 nm) Nd laser pulse of ca. 40 ps duration.

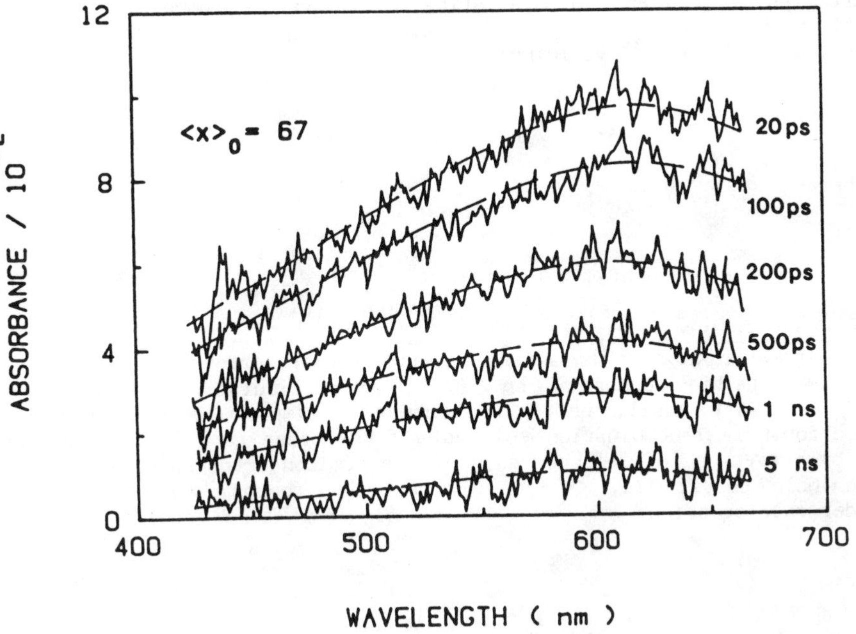

FIGURE 15. Transient spectrum observed at various time intervals after picosecond excitation of colloidal TiO_2. Conditions: $[TiO_2]$ = 17 g/l, pH 2.7, Ar saturated solution, optical pathlength 0.2 cm. Average number of electron-hole pairs present initially in one TiO_2 particle is 67.

In Fig. 15 the spectrum of the trapped electron develops within the leading edge of the laser pulse indicating that the trapping time of the electron is less than 40 ps. Subsequently, the electron absorption decays due to recombination with valence band holes.

$$TiO_2\ (e^-_{tr} + h^+) \xrightarrow{k_r} TiO_2 \qquad (24)$$

We have conceived a stochastic model to analyze the kinetics of this reaction. Since the recombination takes place between a restricted number of charge carriers restricted to the minute reaction space of a 120 Å sized colloidal TiO_2 particle, it cannot be treated by conventional homogeneous solution kinetics. The time differential of the probability that a particle contains x electron-hole pairs at time t is given by:

$$dP_x(t)/dt = k(x+1)^2 P_{x+1}(t) - kx^2 P_x(t) \tag{25}$$

where x = 0,1,2, ...

This system of differential equations is to be solved subject to the condition that the initial distribution of electron-hole pairs over the particles follows Poisson statistics. The average number of pairs present at time t, $\langle x \rangle (t)$, can be calculated by means of the generating function technique [39] yielding:

$$\langle x \rangle (t) = \sum_{n=1}^{\infty} c_n \exp(-n^2 kt) \tag{26}$$

where

$$c_n = 2 \exp(-\langle x \rangle_o) (-1)^n n \sum_{i=n}^{\infty} \frac{\langle x \rangle_o^{\,i}}{(n+i)!} \prod_{j=1}^{n} (n-i-j) \tag{27}$$

The parameter $\langle x \rangle_o$ is the average number of pairs present at t = 0.

Two limiting cases of eq. (26) are particularly relevant: When $\langle x \rangle_o$ is very small, eq. (26) becomes a simple exponential and the electron-hole recombination follows a first order rate law. Conversely, at high average initial occupancy of the semiconductor particles by electron-hole pairs, i.e. $\langle x_o \rangle > 30$, eq. (26) approximates to a second order rate equation:

$$\langle x \rangle (t) = \frac{\langle x \rangle_o}{1+\langle x \rangle_o kt} \tag{28}$$

In Fig. 15 the initial concentration of electron-hole pairs was sufficiently high to allow for evaluation of the recombination process by the second order rate equation, eq. (28). This analysis gives for the recombination rate coefficient the value 3.2×10^{-11} cm^3 s^{-1} corresponding to a lifetime of 30 ns for an electron-hole pair in a colloidal TiO_2 particle with a size of 120 Å. Experiments were also carried out at low laser fluence where on the average less than one electron-hole pair was generated initially by the laser pulse. Under these conditions, hole trapping, presumably by surface hydroxyl groups, competes with recombination leading to a product whose reaction with trapped electrons is relatively slow. The trapping rate constant for the valence band hole was derived as 4×10^6 s^{-1}.

With regards to the use of colloidal TiO_2 as a photocatalyst for the light induced cleavage of water, the fact that recombination of electrons with free holes is about ten times faster than hole trapping is disadvantageous. It explains the experimental finding (vide infra) that deposition of highly active redox catalysts, such as Pt or RuO_2, on the surface of the colloidal TiO_2 particles is required to obtain good yields in the water cleavage process. The role of these catalytic deposits is to intercept electron-hole recombination by trapping the charge carrier and to accelerate their reaction with water leading to the formation of hydrogen and oxygen.

The characteristic optical absorption of electrons in TiO_2 particles can be used to monitor directly interfacial electron transfer reactions. This opens up the way to determine the heterogeneous rate constants for fast charge transfer from the conduction band of the semiconductor to acceptors present in solution. Thus, laser excitation of colloidal semiconductors combined with fast kinetic spectroscopy offers a very useful complement to the application of conventional electrochemical techniques which achieve only a relatively low time resolution to probe electron transfer events at the semiconductor/solution interface. In the following, we use the reduction of cobalticenium dicarboxylate ($Co(CpCOO)_2^-$) as an example to illustrate this procedure:

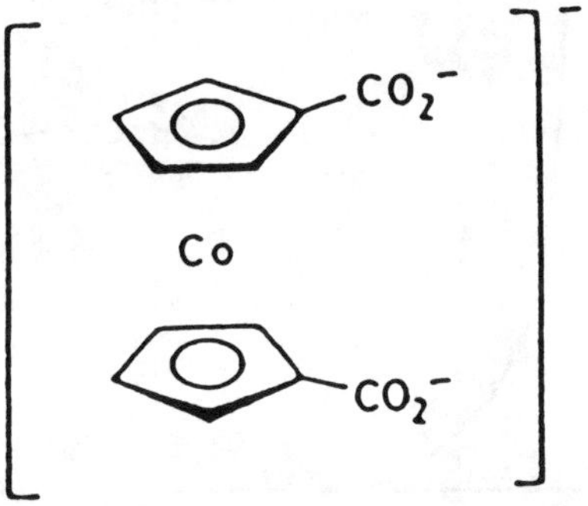

The kinetics of cobalticinium reduction were recorded by following the decay of the electron absorption at 780 nm and the growth of cobaltocene at 484 nm.

$$\underbrace{e^-_{cb}(TiO_2)}_{\lambda_{max}\ 780\ nm} + Co(CpCOO)_2^- \rightarrow \underbrace{Co(CpCOO)_2^{2-}}_{\lambda_{max}\ 484\ nm} \quad (29)$$

Fig. 16 shows transient absorption spectra obtained from the laser photolysis of colloidal TiO_2 in the presence of 5×10^{-4} M $Co(CpCOO)^-$. The solution contained also 0.1% polyvinylalcohol which acts as a hole scavenger. Immediately after the laser flash one obtains the spectrum of the conduction band electron with a maximum at 800 nm. Concomitantly with the decay of the electron absorption one observes the formation of the spectrum of the cobaltocenedicarboxylate. From the kinetic evaluation a second order rate constant of $k_6 = 4\times10^4\ m^{-1}\ s^{-1}$ is obtained. This is much smaller than the rate of a diffusion controlled reaction which for 120 Å sized TiO_2 particles is expected to be about $5\times10^{10}\ M^{-1}\ s^{-1}$. From this one infers that the interfacial charge transfer is the rate determining step. In such as case, the heterogeneous rate constant is related to k_6 via:

$$k_{het} = k_6\ [Co(CpCOO)_2^-]\ /\ (4\pi R^2) \quad (30)$$

where R is the semiconductor particle radius.

The value derived is $k = 2\times10^{-5}\ cm^{-1}\ s^{-1}$ indicting a relatively slow rate for the interfacial redox reaction at pH 10.

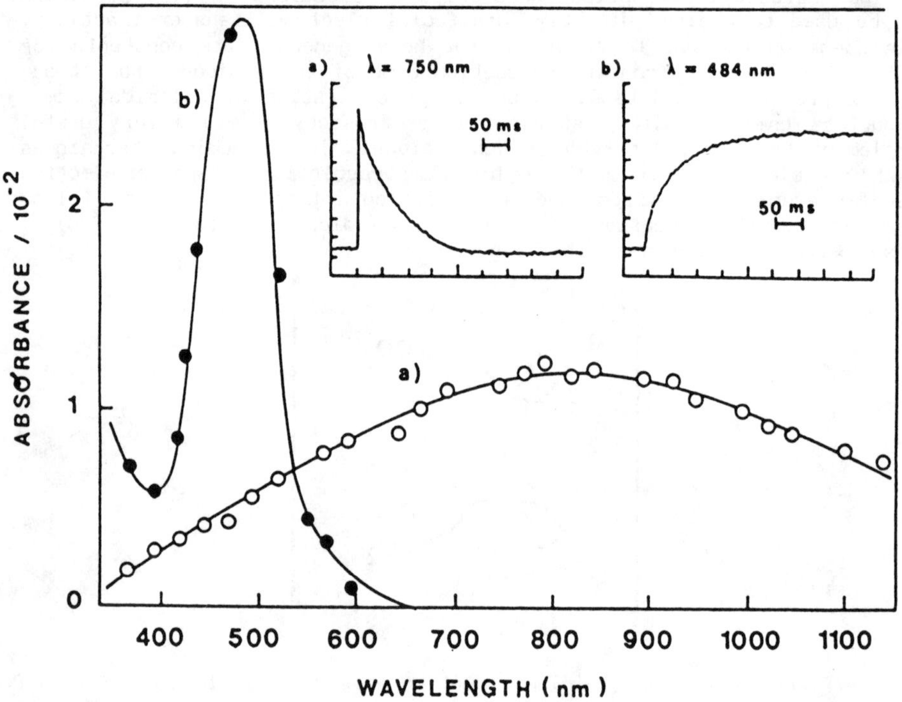

FIGURE 16. Transient spectra obtained from the laser photolysis of alkaline (pH 10) solutions of colloidal TiO_2 (0.5 g/l) in the presence of $5x10^{-4}$ M $Co(CpCOO)_2^-$ (concentration of PVA is 0.5 g/l): o, spectrum ~ 10 μs after the laser pulse; o, spectrum 400 ms after the laser pulse. Insert shows temporal evolution of te absorbance at 750 and 484n nm.

ACKNOWLEDGMENT

This work was supported by a grant from the Gas Research Institute, Chicago, USA (subcontract by the Solar Energy Research Institute, Golden, Colorado, USA) and by the Swiss National Science Foundation.

REFERENCES

1 Kalyanasundaram K: Solar Cells 15, 93, 1985.
2 Grätzel M(ed): Energy Resources Through Photochemistry and Catalysis. New York: Academic Press, USA, 1983.
3 Duonghong D, Borgarello E, Grätzel M: J.Am.Chem.Soc. 103, 6547, 1981.
4 Hayashi S, Nakamori N, Kanamori H, Yodogawa Y, Yamamoto K: Surface Sci. 86, 665, 1979.
5 Moser J, Grätzel M: Helv.Chim.Acta 65, 1436, 1982.
6 Stäber W, Fink A, Bohn E: J.Colloid Interface Sci. 26, 62, 1968.
7 Turkevich J, Aika K, Ban LL, Okura I, Namba S: J.Res.Inst.Catal. Hokkaido Univ. 24, 54, 1976.
8 Matijevic E: Langmuir 2, 12, 1986.
9 Ramsden JJ, Grätzel M: J.Chem.Soc., Faraday Trans. 1 80, 919, 1984.
10 Ramsden JJ, Webber SE, Grätzel M: J.Phys.Chem. 89, 2740, 1985.
11 Kreibich U, Fragstein C: Z.Phys. 224, 307, 1969.
12 Nadeljkovic JM, Nenadovic MT, Micic OI, Nozik, AJ: J.Phys.Chem. 90, 12, 1986.
13 Foitik A, Weller H, Henglein A: Chem.Phys.Lett. 120, 552, 1985.
14 Koch U, Foitik A, Weller H, Henglein A: Chem.Phys.Lett. 122, 507, 1985.
15 Weller H, Foitik A, Henglein A: Chem.Phys.Lett. 117, 485, 1985.
16 Brus LE: J.Chem.Phys. 80, 4403, 1984; J.Chem.Phys. 79, 5566, 1983.
17 Foitik A, Weller H, Koch U, Henglein A: Ber.Bunsenges. Phys.Chem. 88, 969, 1984.
18 Weller H, Schmidt HM, Koch U, Foitik A, Baral S, Henglein A, Kunath W, Weiss K, Diekmann E: Chem.Phys.Lett., in press.
19 Duonghong D, Ramsden JJ, Grätzel M: J.Am.Chem.Soc. 104, 2977, 1982.
20 Kalyanasundaram K, Grätzel M, Pelizzetti E: Coord.Chem.Rev. 69, 57, 1986.
21 Becker WG, Bard AJ: J.Phys.Chem. 87, 4888, 1983.
22 Rossetti R, Brus LE: J.Phys.Chem. 86, 4470, 1986.
23 Vuysteke AA, Sikronen YT: Phys.Rev.C 113, 400, 1959.
24 Serpone N, Sharma DK, Jamieson MA, Grätzel M, Ramsden JJ: Chem.Phys. Lett. 115, 473, 1985.
25 Rossetti R, Brus LE: J.Phys.Chem. 90, 558, 1986.
26 Papavassiliou GC: J.Solid State Chem. 40, 330, 1981.
27 Chestnoy N, Harris TD, Hull R, Brus LE: J.Phys.Chem., in press.
28 Thomas DG, Hopfield JJ, Augustyniak WM: Phys.Rev.A 140, 202, 1965.
29 Henry CH, Faulkner RA, Nassau K: Phys.Rev. 183, 798, 1969.
30 Era K, Shionoya S, Washizawa Y, Ohmatsu H: J.Phys.Chem.Solids 29, 1843 1968.
31 Taguchi T, Yokogawa T, Yamashita H: Solid State Comm. 49, 551, 1984.
32 Moroz M, Brada Y, Honig A: Solid State Comm. 47, 115, 1983.
33 Broser I, Gutowski J, Reidel R: Solid State Comm. 49, 445, 1984.
34 Jortner J: J.Chem.Phys. 64, 4860, 1976.
35 Jortner J: Biochem. Biophysica Acta 594, 193, 1980.
36 Albery WJ, Bartlett PN: J.Electrochem.Soc. 131, 315, 1984.
37 Howe RF, Grätzel M: J.Phys.Chem. 89, 4495, 1985.
38 Kölle U, Moser J, Grätzel M: Inorg.Chem. 24, 2253, 1985.
39 Rothenberger G, Moser J, Grätzel M: J.Am.Chem.Soc. 107, 8054, 1985.

LASER ENHANCED ELECTROPLATING

M.H. Gelchinski*, R.J. von Gutfeld, L.T. Romankiw
IBM T.J. Watson Research Laboratory,
P.O. Box 218, Yorktown Heights, New-York 10598

1. INTRODUCTION

Over the past several years we have investigated methods of high speed maskless deposition using a focused laser beam to produce both localization and substantial deposition enhancement compared to non-irradiated background (1-3). The patterns, produced from solution can be divided into two main categories; laser enhanced electroplating and laser enhanced electrodless plating. The latter category includes depositions produced without an external electomotive force and has been described in several other references (4). The enhancement mechanisms for laser enhanced electroplating have been investigated in considerable detail for the system Cu/Cu^{++} in conjunction with a focused argon laser using microscope slide glass substrates with a thin predeposited metal layer (3).

It is well known that the electronic industry uses large quantities of precious metals in the fabrication of contacts and connectors. As the price of precious metals soared during the late seventies and the number of plated parts increased, a search for high speed, selective deposition techniques, with the objective of providing precious metal only in the contact region, was initiated. The intention was to realize 90% to 95% saving of the precious metal. Our approach has been to study laser enhanced gold plating using commercially available gold solutions deposited onto nickel and nickel plated Be-Cu substrates. These substrates are typical of those used in the microelectronics industry and the use of laser plating provides a mean for localizing the gold to the essential contact areas. The significant difference between these substrates and the metallized glass substrates used earlier is ~3 orders of magnitude higher thermal conductivity of the present metallic samples. In additions, the surface roughness of the metal substrates is considerably higher, affecting the surface morphology of the deposits. We have obtained in our experiments plating rates of 10-12 $\mu m sec^{-1}$ for patterns consisting of 0.8mm diameter spots. Detailed examination of the depositions indicate they are crack free, non-porous and have excellent adhesion to the substrate.

* Permanent address: Israel Aircraft Industries Ltd., Lod, ISRAEL.

2. LASER ENHANCED PLATING

The apparatus used in our experiments is shown in Fig. 1. Here a quartz plating cell is designed to transmit the laser beam (argon, multiline) through the electrolyte on the substrate with minimal absorption prior to the beam's incidence on the substrate. Patterning was obtained by: 1) scanning the beam using a variable speed oscillating mirror and 2) movement of the sample relative to the beam by way of a computerized precision table. Plating was carried out in a potentiostatic mode using a pure gold wire as a reference electrode. For pulse plating experiments a chopper was used to provide laser pulses in synchronism with the applied negative voltage by way of a beam splitter and photodetector as shown in Fig.1.

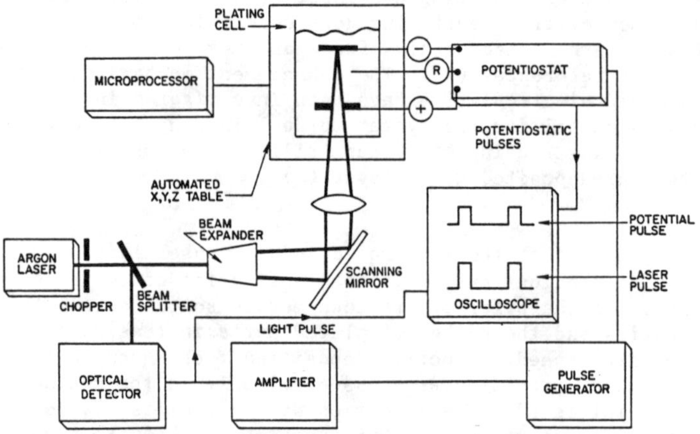

Fig. 1 Schematic of laser plating set up with a computer controlled table for sample motion.

2.1 Copper plating

To study the effect of the laser on the polarizaation (I/V) curve a special cathode was designed consisting of a small circular conducting area on a metallized slide otherwise electrically insulated by photoresist. . The system Cu/Cu^{++} was investigated and a polarization curve was obtained both with the laser off and periodically applied through the opening in the photoresist (Fig. 2). It is clear from Fig. 2 that the laser produces a large plating current enhancement, up to nearly 3 orders of magnitude larger than the current with the laser off. Furthermore, the laser is effectively enhancing the plating rate over the entire polarization curve. From this investigation it was determined that the laser increases the kinetics of plating at the lower values of the overpotential curve while increasing the diffusion or mass transport at the higher overpotentials. These observations have the following explanation: Increased temperature due to laser absorption

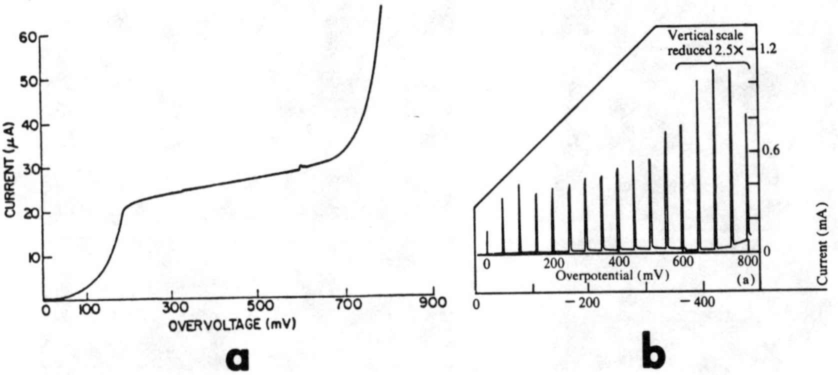

Fig. 2 (a) Polarization curve for Cu/Cu^{++} for a 500 μm diameter electrode.

(b) Polarization curve as in (a) with the periodic radiation of the cathode by a laser. The laser is applied every 50mv as shown by the large periodic current spikes.

by the cathode increases Arrhenius type processes as described by the Butter-Volmer equations relating the plating curved and overpotential in the kinetics or charge transfer region (4). Localized heating also increases mass or ionic transport at the higher overpotential values due to the large local thermal gradients in the region of the laser absorbing cathode. These gradients produce hydrodynamic stirring which serves to replenish the otherwise ion depleted boundary layer. This effectively shifts the mass transport limited region to much higher overpotential values. A third enhancement mechanism has also been identified for certain electrolytes and consists in a shift in the rest potential to more positive values. This permits plating to occur with the referenced potential set at zero. At this setting no background plating occurs while local plating occurs in the region absorbing laser radiation.

We have used the laser enhanced plating technique to demonstrate the feasibility of copper circuit repair on a micron size scale (3). Two parallel 12 μm wide, 5 μm high copper lines were connected by a laser enhanced copper bridge made by relative movement of the circuit board with respect to the beam. Electrical conductivity of the laser formed connection was experimentally determined to be as high as that of the conventionally plated copper lines of the circuit board.

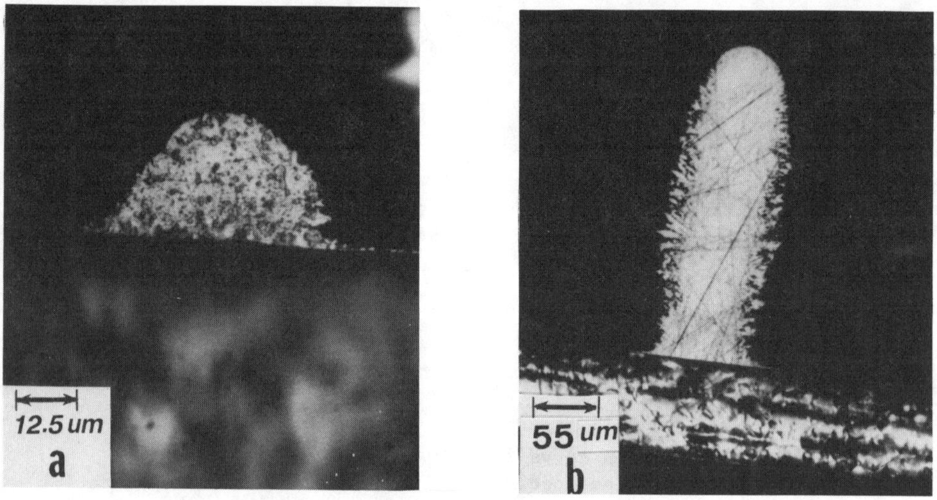

Fig. 3 Metallurgical cross sections of laser plated gold deposits. On time - 28 msec, 25% duty cycle.

(a) 30 seconds of plating (b) 300 seconds

2.2 Gold plating

Pure gold and gold alloys (Ni-Au and Co-Au) were deposited to produce soft and hard films. The following parameters and their effects on plating speed and deposition morphology were investigated: gold concentration of the electrolyte, laser power, laser spot diameter incident on the substrate, applied overpotential, scanning mirror speed. For pulse plating, the magnitude of the anodic pulse (periodic reverse plating) as well as the duty cycle of the laser and plating voltage were additional variables. Fig. 3 are optical microscope photographs of cross-sectioned pure gold deposits obtained with the commercial plating solution, Temperex HD diluted 1:2(5). Patterning was obtained using pulse plating (with synchronous laser during the cathodic pulse) to yield a 25% duty cycle. A 28ms on time with an overvoltage of -650mv was alternated with a 80ms laser off-time and an anodic pulse of 150mv overvoltage. It was found that for small laser-plated gold spots, on the order of 50 μm in diameter, two problems arise, revealed by the optical micrographs of the deposits: 1) dendritic growth at the outer edges, Fig. 3b and 2) cratering with increased on-time while maintaining the same duty cycle. This longer on-time approaches a CW plating condition and exhibits the pronounced cratering morphology. Fig. 4. Both the cratering and the appearance of dendritic growth are due in part to a depletion of available gold ions, with the depletion becoming more severe with increasing plating times. In the case of Fig. 3b, dendritic growth occurs as the hydrodynamic stirring of the solution by the laser beam is limited to the top of the very thick needle like deposit with no stirring therefore occuring at the sides near the substrate. Thus, an ionic depletion at the sides of the deposit causes plating to occur above the limiting current density and the observed texture. Cratering occurs in part due to overheating of the substrat to produce intense boiling which creates an agitation pattern favoring the sides rather than the center of the deposit. Similar cratered effects have been observed with laser chemical vapor deposition (LCVD) for somewhat analogous reasons using a gas atmosphere instead of liquid (6).

Fig. 4 Metallurgical cross section of a laser plated gold deposit with 160 msec pulse and 25% duty cycle. Note cratering at center.

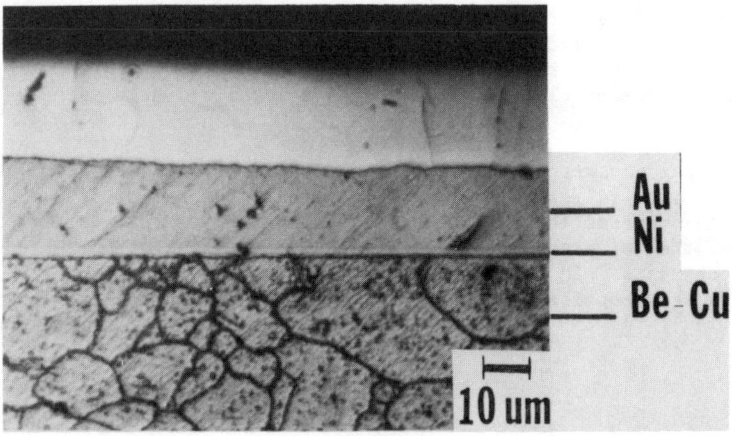

Fig. 5 Metallurgical cross section of a laser plated gold deposit under D.C. condition with a defocused beam.

To overcome the aforementioned problems while simultaneously generating large spot diameters, better suited for microelectronic contact applications (on the order of 0.5-1.0mm diameter) experiments in which overvoltage, scanning frequency, laser power and power density were varied yielded the following conclusions: 1) Laser power densities should be maintained at 10^4 Wcm^{-1} or less to prevent burning and or extensive cratering of the deposit) 2) high density deposits can be obtained by scanning the laser over the area to be plated. This overcomes, to a large extent, the limiting current of a stationary beam since the beam's movement causes it to continuously sample an ionically resupplied portion of the electrolyte, 3) highest plating speeds are obtained by using a high gold concentration electrolyte. We have demonstrated plating rates of up to 1.0 µm sec^{-1} using Autronex 55GV without Ni or Co additives with a plating current density of $5Acm^{-2}$. This solution contains 4 Troy oz. of gold per gallon. Optical micrographs of a cross section of a laser-plated spot using this solution in conjunction with 20W of laser power are shown in Fig.5. The surface morphology of the deposit appears fine grained and smooth.

Several attempts to increase still further the laser plating rates have been made by increasing the mass transport to the region of plating by: (1) high speed mechanical stirring of the electrolyte near the laser cathode impingement region, and (2) the use of a submerged recirculating jet with flow directed tangentially across the laser-cathode impingement area. The submerged jet produced an increase in the laser plating rate (20-40%), while no noticeable increase was observed from the high speed stirring.

3. LASER ENHANCED JET PLATING

To improve the mass transport and consequently the plating rate, we developed a new technique, laser enhanced jet plating (7,8), which combines the earlier laser enhanced technique with a freestanding jet of electrolyte to achieve localized high speed gold plating.

The principal advantage of the freestanding jet is its ability to provide a rapid resupply of fresh ions onto the region of plating to overcome the mass transport limitations. In addition, the jet stream limits the region of plating without the use of masks. In this present application, the jet stream also acts as an optical wave guide or light pipe for the laser beam.

The experimental set-up is shown in Fig. 6. A CW argon laser beam with maximum output power of 25 watts was passed through a beam expander prior to entering the cell, and focused approximately at the center of the jet orifice. The flow system is comprised of a solution container, a pump which recirculates the electrolyte after impingement on the cathode and the jet cell, which is a lucite hollow cylinder, consisting of an inlet chamber, maintained under pressure with the jet nozzle. The nozzle is made from a section of capillary tube glassed onto a flat pyrex plate. The end plates of the cell are quartz windows. A platinum wire serves as an anode and is inserted into the fluid chamber. A nozzle orifice of 0.5mm diameter is used, giving rise to gold spots of similar diameter. To maintain a constant current we used a PAR 173 potentiostat set to operate galvanostatically. The cathode is attached to a microprocessor controlled XYZ table via an extension arm. The microprocessor also controls the on-off gating of the argon laser and can be made to operate the on-off controls of the potentiostat. Plating takes place only when the current and the laser beam are on. For all electroplating runs, we used a high speed cyanide acid solution (Sel-Rex Autronex 55GV without addition of the cobalt complex) containing 4 Troy oz. per gallon of metallic gold. The electrolyte was preheated to 60 °C. The linear flow velocity was 1.0×10^{3} cm sec^{-1}, which corresponds to a Reynolds number of 5.5×10^{3} at the nozzle orifice. The nozzle-cathode spacing was typically 0.5cm.

Spots were made with current densities as high as $16 Acm^{-2}$. With a 0.05cm diameter nozzle orifice and 25W of laser power, plating rates of 10 µm sec^{-1} were observed. This should be contrasted to conventional gold plating where maximum current densities are $0.25 ACm^{-2}$ and plating rates are more than two order of magnitude slower (9). We also found that with a smaller nozzle, 0.035cm diameter, rates as high as 30 µm sec^{-2} were obtained. This increase in plating rate over that found for the 0.05cm nozzle indicates that plating over the larger area was limited by the available power density at the cathode in turn limited by the total available laser power.

An example of laser-jet plated spot is shown in cross section in Fig. 7. Examination of their morphology indicates a dense deposit consisting of small grains, free of cracks and pores. Upon electroetching, the samples were found to possess a columnar structure, typical of pure soft gold deposits. In addition excellent adhesion between the gold and 2 µm thick nickel was found based on the "scotch tape" pull test.

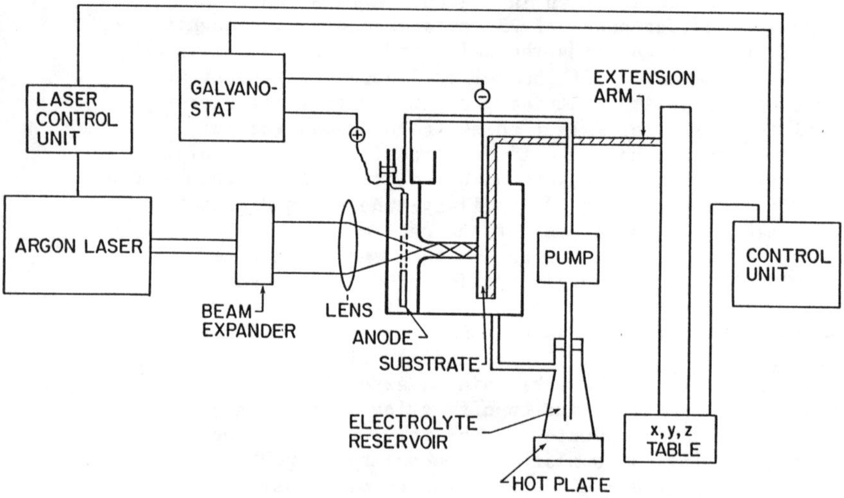

Fig. 6 Schematic of laser-jet plating system.

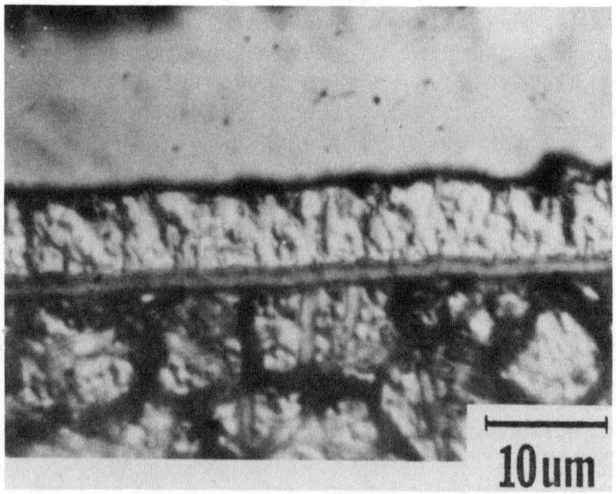

Fig. 7 Laser jet electroplated spot on nickel-plated Be-Cu shown in cross section. Plating time 1 sec. Laser power 25W current density 11ACm^{-2}. Note the columnar structure.

In contrast, cross section of deposition using the same jet and the same high current density, as in Fig.7 but without the laser during plating is shown in Fig.8. Here voids are present throughout the deposit irrespective of film thickness, apparently initiated at the nickel-gold interface. This type of columnar deposit with columns physically separated from each other is typical of metal depositions near the limiting current density for a given agitated solution, particularly of deposited on substrates with incomplete oxide removal. Scotch tape pull-testing of these deposits showed poorer adhesion.

The three significant experimentally observed results from these laser jet gold plating experiments are: 1) an increase in plating speed due to laser heating of the cathode 2) improved adhesion, probably caused by a cleaner cathodic surface resulting from local heating providing a gold to nickel metal bond rather than gold to nickel oxide bond and 3) crack free, dense deposits, at much higher current density than is possible for the jet alone. This may be due to potential shift caused by the laser irradiation. Plating at more positive potentials mean less hydrogen evolution, i.e., less current consumed for side reactions and plating below the limiting current density.

The new laser-jet technique has been demonstrated to have the highest plating rate for gold reported to date. The technique is also applicable to numerous other plating systems such as copper for which plating rate of 40 μm sec^{-1} was achieved. We envision this technique to have valuable applications for circuit repair as well as circuit design.

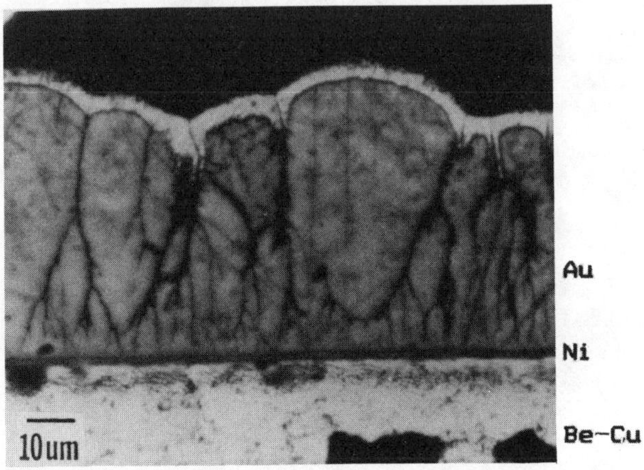

Fig. 8 Jet plating with same substrate and current density as in Fig.7, but without the laser. Plating time 20 sec. Thickness 50 μm. Note the voids and pores in the gold deposit.

R E F E R E N C E S

1. R.J. von Gutfeld, E.E. Tynan, R.L. Melcher and S.E. Blum, Appl. Phys. Lett., 35, 651 (1979).

2. R.J. von Gutfeld, E.E. Tynan and L.T. Romankiw, Abstr. No. 472, P. 1185 ECS Extended Abstracts, Los Angeles, Ca. (Oct. 14-18, 1979).

3. J.Cl. Puippe, R.E. Acosta and R.J. von Gutfeld, J. of Electrochem. Soc. 128, 2539 (1981).

4. R.J. von Gutfeld, R.E. Acosta and L.T. Romankiw, IBM Journal 26, 136 (1982)

5. L.T. Romankiw, M.H. Gelchinski, R.E. Acosta and R.J. von Gutfeld, Proceedings of the Symposium on electroplating Engineering and Waste Recycle - New Development and Trends, Electrodeposition Division, Proc. Vol. 83-12, Electrochem. Soc. Pg 66.

6. S. Allen, JAP, 52, 6501-6505 (1981).

7. R.J. von Gutfeld, M.H. Gelchinski, L.T. Romankiw and D.R. Vigliotti, Appl. Phys. Lett., 43, 876 (1983).

8. M.H. Gelchinski, R.J. von Gutfeld, L.T. Romankiw and D.R. Vigliotti, J. Electrochem. Soc., 132, 2575 (Nov. 1985)

9. D.R. Turner, Thin solid Film, 95, 143 (1982).

UV LASER ABLATION OF POLYMERS

R. Srinivasan
IBM Thomas J. Watson Research Center
P.O. Box 218
Yorktown Heights, New York 10598

1. INTRODUCTION

A simple and convenient source of laser radiation in the ultraviolet region became available with the invention of the excimer laser in the 1970's. Studies on the interaction of ultraviolet laser pulses with solid organic matter such as synthetic polymers and biological tissue led to the discovery in 1982 (1-4) of the phenomenon of "ablative photodecomposition" which results in the break-up of the structure of the organic solid by the photons and the expulsion of the fragments at supersonic velocities. The result is an etch pattern in the solid with a geometry that is defined by the light beam. The principal advantages in using ultraviolet laser radiation rather than visible or infrared laser radiation for this purpose lie in the precision (± 2000 Å) with which the depth of the cut can be controlled and the lack of thermal damage to the substrate to a microscopic level.

2. MECHANISM OF ABSORPTION OF LIGHT

Synthetic organic polymers are made up of molecules which consist of 10^3 to 10^5 atoms, principally carbon, hydrogen, oxygen, and nitrogen. A small molecular unit (monomer) of 6 to 40 atoms is repeated over and over along a chain to form a polymer. In Fig.1, two typical polymers called poly(methyl methacrylate) (=PMMA) and polyimide are shown along with the monomer units in each case. A polymer molecule

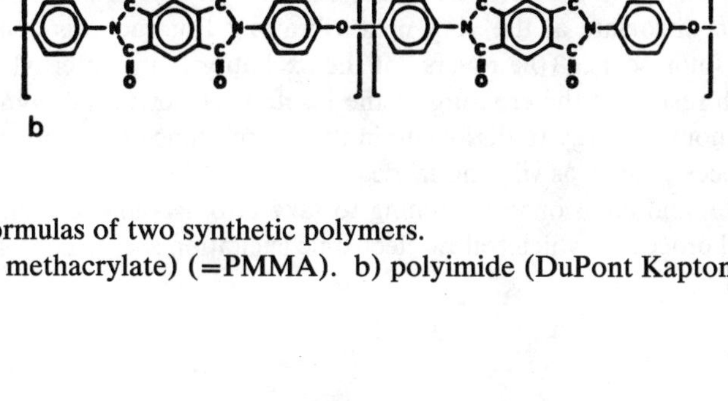

FIGURE 1. Formulas of two synthetic polymers.
a) poly(methyl methacrylate) (=PMMA). b) polyimide (DuPont Kapton™).

may consist of one hundred to 1000 monomer units. The bonding between atoms within a polymer is covalent (shared electrons) in nature and strong (60-150 Kcal/mole) but the forces between the molecules are weak (<10 Kcal/mole). Absorption of photons of ultraviolet wavelength excites the bonding electrons in polymers and specific absorptions correlate with specific groups of atoms (chromophores) in the molecules. An energy diagram (Fig.2) can describe the absorption of a UV photon by one bond A-B in the molecule. The lower curve represents the lowest (ground) energy state for the bond with the horizontal lines marking vibrational levels within a specific electronic

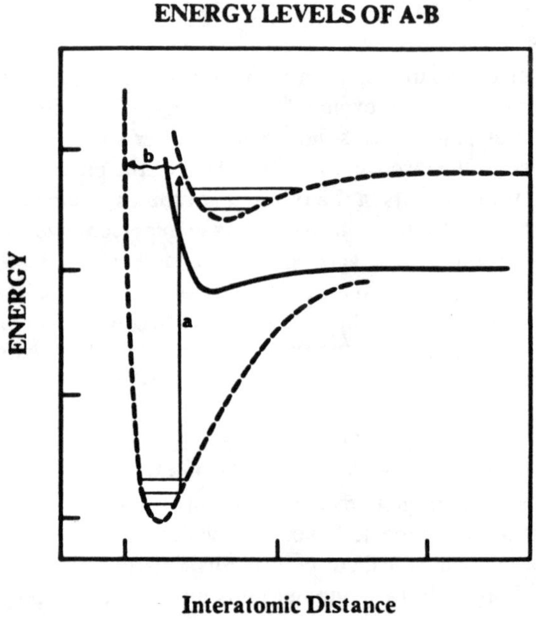

FIGURE 2. Energy level diagram for hypothetical bond A-B.
The lower broken line represents the ground electronic state; the upper broken line and the solid line represent excited electronic states.

level. Absorption of a UV photon (vertical arrow 'a') is such a rapid process that it proceeds with no nuclear motion and leads to an upper electronic level. This is still a bound state as shown here but metastable with respect to the initial state. In the the upper state the energy in the bond obviously exceeds the bond energy and therefore the two atoms can dissociate at the very next vibration. But such dissociation processes compete with fluorescence (the reverse of the excitation), and internal conversion (arrow 'b') which results in the crossing of the excited species to its ground state energy surface with enough energy to dissociate in that state. Since there are numerous other competing processes such as vibrational deactivation, intersystem crossing to a different electronic state, and collisional quenching to take into account, the photophysics and photochemical processes which follow electronic excitation can be complex and difficult to sort out (5).

The details of the interaction of the photon with the electrons in A-B were not specified in the discussion above. Such an interaction can lead to a promotion of a valence electron from a bonding to an anti-bonding orbital which is the most common process in the wavelength region >200 nm. With decreasing wavelength, the number of valence excitations increase rapidly and it may not be possible to assign absorptions to specific transitions in a polyatomic molecule. A further complexity becomes evident below 200 nm at which point, Rydberg transitions become increasingly important. Rydberg transitions result when an electron from a bonding orbital, or an electron from a non-bonding or weakly bonding orbital is promoted to a Rydberg orbital. The convergence limit to the Rydberg transitions is, as in atomic spectroscopy, a (molecular) ion and an electron (6). In polyatomic organic molecules, some or all of these transitions can occur simultaneously.

3. LASER ABLATION CHARACTERISTICS

Laser pulses from an excimer laser are typically of ~20nsec half-width. Fig.3 is a schematic representation of the impact of such a pulse of light on a polymer surface.

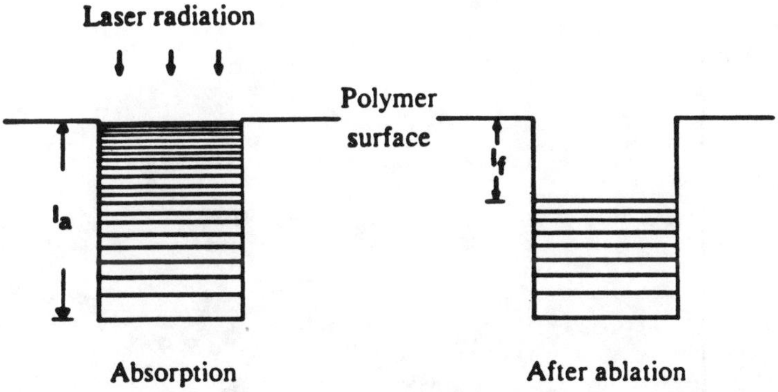

FIGURE 3. Schematic representation of impact of laser pulse on polymer surface.

The penetration of the radiation through the solid follows a simple relationship which is known as Beer's Law:

$$I_t = I_0 \; 10^{-\alpha l} \tag{1}$$

where I_0 and I_t are the intensities of the beam of light before and after transmission through a slice of material of thickness l, and α, called the absorptivity is a characteristic property of the material. In weak absorbers such as PMMA, a pulse of laser radiation of 193 nm wavelength will penetrate to 6.5 μm depth before 95% of it would have been absorbed whereas in a strong absorber such as polyimide, the penetration depth, l_a, will be only few thousand Ångstroms. If the fluence, F of the laser beam exceeds a certain threshold value, F_0, then a depth, l_f of the material is ablated by the pulse. At the same time an audible report is heard and when the irradiation is carried out in air, a spot of visible light is observed at the point where the laser beam impacts the surface. If the fluence is such that $l_f < l_a$, which is usually the case, then a depth $l_a - l_f$ which had been exposed to the light will be left behind. The next pulse will go through this partly irradiated material as well as through virgin material underlying it. The first pulse is therefore unique. But after the first few pulses, there is a linear relationship between the number of pulses and the depth that is etched. In practice, the depths etched by varying numbers of pulses are averaged and noted as the etch depth per pulse for that polymer at that wavelength and fluence. This value is reproducible within the uncertainty (± 8%) in the measurement of the etch depth and the fluence of the laser pulse. This is the reason that the etch depth can be reproduced to ± 2000 Å in most materials.

A typical plot of the etch depth/pulse as a function of fluence is shown in Fig.4 for polyimide. The three wavelengths shown correspond to the strong output lines of the

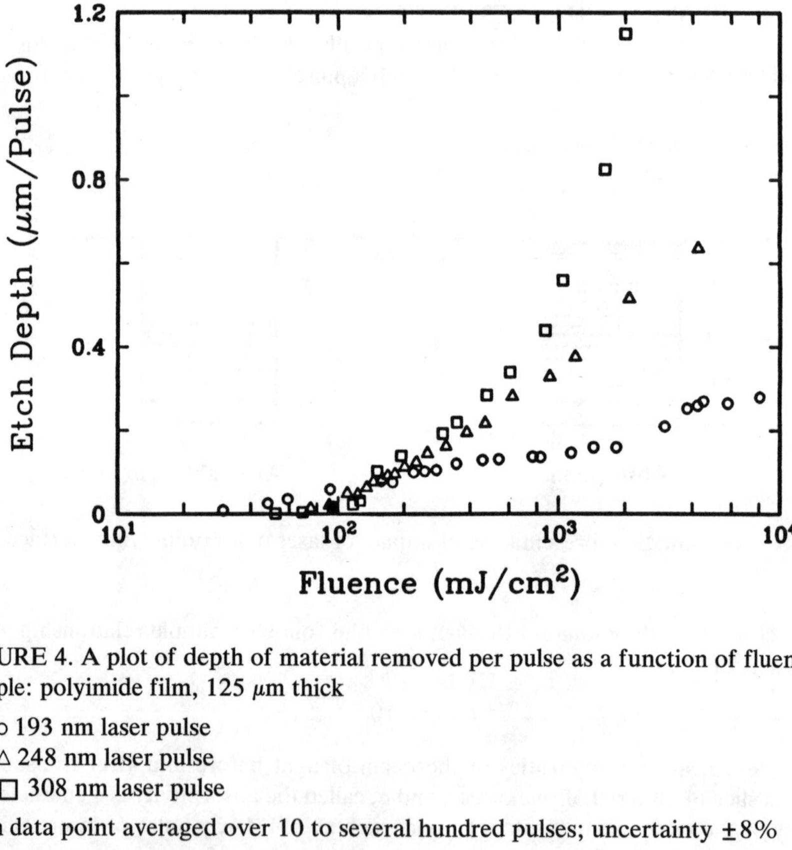

FIGURE 4. A plot of depth of material removed per pulse as a function of fluence. Sample: polyimide film, 125 μm thick

o 193 nm laser pulse
Δ 248 nm laser pulse
□ 308 nm laser pulse

Each data point averaged over 10 to several hundred pulses; uncertainty ±8%

excimer laser depending upon the gas fill. The etch depth is independent of the atmosphere in which the experiment is performed which is one of the attractive features of this process. The threshold fluence for the onset of etching decreases with decreasing wavelength. But, in general, the susceptibility to etching of a material increases with increasing wavelength and fluence. Fluence thresholds for etching and etch depth plots for a variety of synthetic polymers have been reported (2,3,7-10).

In one study (11), the thresholds and etch rates in PMMA at 90°K and 273°K were compared. At a laser wavelength of 193 nm, there was no difference in the etching characteristics at the two temperatures. But, at 248 nm, the etch rate at a given fluence at 90°K was only half of the value at 273°K. The thresholds remained unaffected.

The products of UV laser ablation have been analyzed in a few polymers (1,8, 12-14). In every instance there appears to be a diversity of products ranging from atoms and diatomics to small polyatomic molecules and small fragments of the polymer. The product composition is also wavelength dependent. From PMMA of initial molecular weight ($\overline{M}_n$) = 10^6, at 193 nm, the products are oxides of carbon, MMA (the monomer) and low molecular weight ($\overline{M}_n < 1500$) fragments of the polymer. At 248 nm, the principal product is a low molecular weight fraction ($\overline{M}_n = 2500$) of the polymer. From polyimide, the products are oxides of carbon, benzene, HCN, and elemental carbon. In addition to these stable products, transient intermediates such as C_2, CN, CH, and C have also been detected in the same systems using fast methods of analysis which are described below. Reference has already been made to the depth $l_a - l_f$ under the etched surface which is transformed by the laser photons but is left behind in the substrate. This freshly created surface has been analyzed by X-ray photoelectron spectroscopy (XPS) and by wet chemical methods (15). In oxygen containing polymers such as polyimide, this surface is found to be depleted in oxygen and the change is attributed to the loss of oxides of carbon. The penetration depth for XPS analysis is ~100 Å. Since the etched surface of a polymer can be irregular to the extent of several hundred Ångstroms, it has not been established to what depth this chemical change actually extends. Wet reactions show that there may even be a very thin (<50 Å) layer that has undergone oxidation by air on top of the altered layer that is seen by XPS.

The dynamics of the ejection of the material during the etching of a polymer by a laser pulse has been studied in several ways. Even at fluence levels below the threshold for etching of PMMA or polyimide, the ejection of diatomic species such as C_2 and CN at peak velocities as large as 6-7 x 10^5 cm/sec can be observed by laser-induced-fluorescence (LIF) (13, 16). The peak velocity changed by less than 20% over a 14-fold increase in fluence (0.05 to 0.70 J/cm^2; 248 nm) which caused a several hundred-fold increase in the amplitude of the signal. The corresponding translational temperature would be many tens of thousands of degrees. An analysis of the rotational spectrum of the fluorescence indicated that the rotational temperature was only 1000°±200°K. As the fluence is progressively increased, these species are observed in emission (17). Atoms (C) and ions (C^+) are also seen. CH has been observed in emission from PMMA at 193 nm. An analysis of its spectrum showed that its vibrational and rotational temperatures are both equal to 3200°±200°K (18). The translational temperature was estimated by timing its velocity to be ~11000°K. The velocity of the polyatomic product, MMA formed in the laser ablation of PMMA at 193 nm was measured by time-of-flight mass spectrometry (19). At fluences <0.12 J/cm^2 which are near the threshold for ablation, the translational temperatures ranged from 1000°-3000°K. There is thus a difference in the translational temperatures between a polyatomic product and a transient diatomic product from the same polymer which suggests that they may be formed by different reaction paths.

The temporal profile of ablation was determined by a photoacoustic measurement of the stress wave induced in a polymer film by the ejection of the ablating material (20). A thin, Piezo-electric transducer attached to the polymer film timed the stress wave with an uncertainty of ± 1 nsec. The delay due to travel of the stress wave through the film was compensated. Below the threshold for ablation, a pulse of UV radiation merely heats the polymer sample (10, 21) and gives rise to a sinusoidal stress pulse. But above the threshold, a strong positive signal is obtained which increases rapidly with increasing fluence. The temporal delay between the onset of the laser pulse and the onset of ablation is 7 nsec which is well within the halfwidth of the laser pulse. This behavior is common to PMMA and polyimide, even though these materials differ widely in their absorption characteristics and chemical composition.

The "doping" of a polymer to change its UV absorption and thereby increase its sensitivity to UV laser ablation shows interesting potential. In one investigation (22), it was found that the introduction of acridine, a small molecule (Mol. Wt.=179) in concentrations of 1-8% by weight in PMMA progressively reduced the threshold for ablation at 248 nm and also sensitized it to laser etching at 308 nm. Undoped PMMA is transparent to radiation of 308 nm wavelength.

4. MECHANISM OF UV ABLATION

The mechanism by which UV laser pulses bring about the etching of polymer surfaces with a minimum of thermal damage to the substrate has been a matter of controversy. There are two aspects to this problem which are: (i) the reaction path in which the polymer bonds are actually broken, and (ii) a quantitative model of etch behavior as a function of pulse width, wavelength, and fluence.

The principal reaction paths which have been proposed can be understood by referring to the energy diagram in Fig.2. It is generally accepted that the absorption of UV photons results in electronic excitation (path a). The excited electronic state can undergo decomposition in that state itself which would be a purely photochemical reaction. Or, if the excited molecule undergoes internal conversion (path b) to a vibrationally excited ground state, any subsequent decomposition can be considered to be the equivalent of a thermal process. This is the so-called photo-thermal mechanism in which the photons merely act as a source of thermal energy. Along either pathway the products can be the same (although they do not have to be) and any excess energy over that needed for bondbreak will remain in the products and will be dissipated in the ablated fragments. If the time for ablation is of the order of the duration of the laser pulse (~20nsec), the diffusion of the thermal energy to the substrate would be minimal - a diffusion length of a few hundred Ångstroms has been estimated (8) - and therefore the lack of thermal damage to the substrate would be expected (8, 21).

The expression

$$l_f = \frac{1}{\alpha} \log \frac{F}{F_0} \tag{2}$$

which relates the etch depth/pulse to the fluence, the fluence threshold, and the absorptivity at that wavelength has been derived by more than one group of workers (2,3,7,8,23). The derivation merely relates to the deposition of the photon energy according to Beer's Law (Eqn.1) and is independent of the assumed mechanism. The test of Eqn.2 lies in a plot of l_f vs. log F being a straight line with a slope equal to $1/\alpha$. Fig.4

shows that the linear relationship is certainly not true for polyimide but earlier work in which data were obtained over a smaller range of fluences allowed a linear relationship to be forced through the experimental points. The slope of such plots was invariably not equal to $1/\alpha$.

These analyses of the mechanism implicitly assumed that ablation followed the deposition of all of the photons of the laser pulse in the solid. But the photoacoustic experiments mentioned in the previous section typically showed (Fig.5) that when a small fraction of the pulse energy had been deposited in the film, the material started to

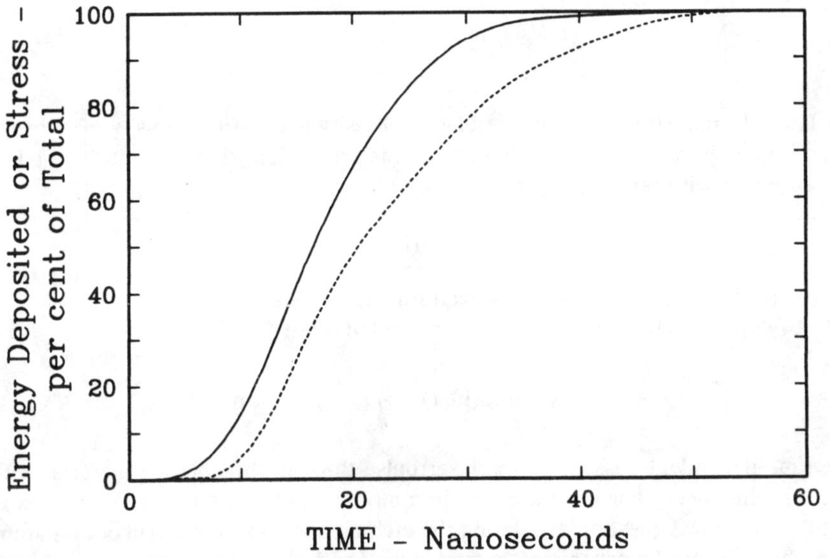

FIGURE 5. Plots of the fraction of energy in one pulse deposited in polymer film and fraction of integrated amplitude of stress pulse as a function of time.
Solid Line ______ Energy
Dashed Line - - - - Stress Pulse
Sample: Polyimide Film 7.5 μm thick; 193 nm laser pulse. Data from Photoacoustic Measurement (Ref. 20.

ablate. This temporal behavior of the system is also consistent with a simulation of UV laser ablation that has been carried out (24) in which if it was assumed that at t=0 the photons were already absorbed in the first several layers of the material and the polymer bonds were broken, then the monomer ablated layer-by-layer and the first layer was free from the surface in less than 3 psec.

The bearing of the time profile of ablation to the mechanism lies in calculations of the temperature rise in the volume of material that ablated. These calculations which form the basis for the photothermal mechanism significantly overestimated the temperature. It can be shown that the actual rate of thermal decomposition would not be adequate to lead to ablation in the time span of the laser pulse. For example, the data on polyimide (Fig. 5) show that material begins to ablate in < 10 nsec and at that point, based on the amount of energy absorbed, the temperature rise would be 1200° in the absorption depth (95%) of 0.07 μm. However, the half-life for the thermal decomposition of a polyimide at this temperature has been measured to be ~20sec (25). Such a discordance between laser ablation and pyrolysis data are observable at other wavelengths for polyimide (13). An analysis for PMMA which leads to similar conclusions has been reported (12).

A realistic model of the interaction of UV laser pulses with an organic solid has been proposed (26) which takes into account the fact that as the photons penetrated the material and bonds are broken, the photon intensity at any depth l will evolve with time:

$$I(l,t) = I_0(t) \exp\left[-k' \int_0^l dl' n(l',t) \right] \quad (3)$$

where I_0 is referred to as the unshielded laser intensity and n is the density of unbroken polymer bonds. If n is not a function of time this expression will reduce to Beer's Law. The change in n with time is given by

$$\dot{n}(l,t) = -kI(l,t)\ n(l,t) \quad (4)$$

When (3) is substituted into (4) and solved for the initial condition $n(l, t=0) = 0$ when $l<0$; $n(l, t=0) = n_\infty$ when $l>0$ then

$$\frac{n(l,t)}{n_\infty} = \left[1 + \{ \exp(kI_0 t) - 1\} \exp\{ - k' n_\infty l\}\right]^{-1} \quad (5)$$

The time-dependent passage of the laser pulse through the ablating material which is taught in this theory has been adapted in a more recent approach (27) which is tailored to the practical question of relating the etching behavior to experimental parameters. A photochemical mechanism is assumed here and the rate of decomposition of the polymer in its excited state is viewed as a process which competes with alternative processes for the degradation of the energy which do not lead to decomposition. The width of the laser pulse is a critical factor because it is only the flux of absorbed photons above a threshold value that will be available for bond-breaking. The laser pulse is considered to be made up of a continuous succession of short (~20 psec) pulses whose passage through the solid is followed by its absorption in successive layers of the material (Fig.6). The deposition of photon energy in each layer can be calculated from Beer's Law. The ablation condition is met when the number of absorbed photons in a given volume exceeds a certain value. It is necessary to determine the ablation condition from the experimental data at one wavelength which simultaneously gives the threshold level of the absorbed photon flux. The model can then be applied to UV etching at other wavelengths for the same polymer without further adjustment of any

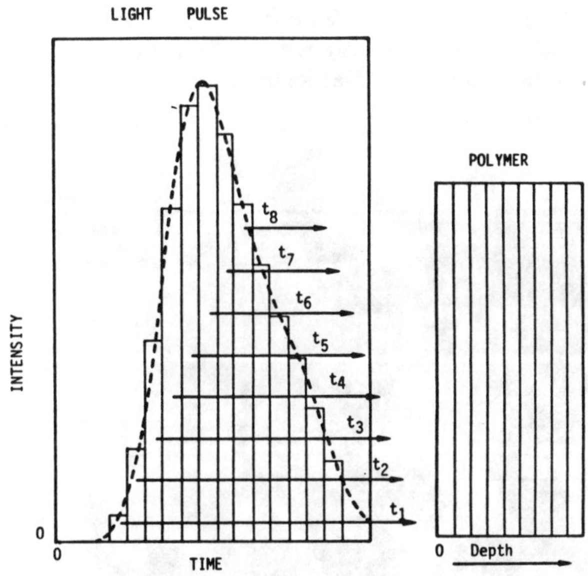

FIGURE 6. Schematic of interaction of a single laser pulse with a polymer sample. Successive slices (in time) of the light pulse are shown to be absorbed in hypothetical layers of the material according to Beer's Law. When the concentration of absorbed photons in any one layer exceeds an ablation threshold, that layer will ablate.

parameter. The predictions made from this model regarding the relationship between the pulse width and fluence thresholds (a factor of importance in practical applications) are being tested and have yielded encouraging results.

In summary, at present there is very little evidence to support a purely photothermal mechanism for the UV laser ablation of any polymer. A photochemical mechanism is more probable but it may be much more complex in nature than originally proposed (7). The process of excitation may include not only the first upper excited electronic state but, through multi-photon processes, higher states as well. Thermal activation of photochemical decomposition is another consideration in the near-UV wavelengths.

5. TECHNOLOGICAL APPLICATIONS

A reliable excimer laser which can be built into an industrial tool is a prerequisite if UV laser ablation is to make an impact as a technology. When the first commercial excimer lasers became available at the end of the 70's, they were little more than research instruments. The developments that have taken place in the intervening years have led to the offering of the first industrial model by the mid 80's which is a sure indication of the promise for the future that exists in this laser. Currently several models of excimer lasers which will deliver a constant (±10%) preset out-put of energy are available from different manufacturers. These machines are internally controlled by a microprocessor and are said to be good for several tens of millions of pulses before any serious maintenance will be needed (28).

The most obvious application of UV laser etching may seem to be in photolithography (29). It has been demonstrated (Fig.7) that photoresists can be patterned using proximity printing (30) or projection printing. Resolution extending down

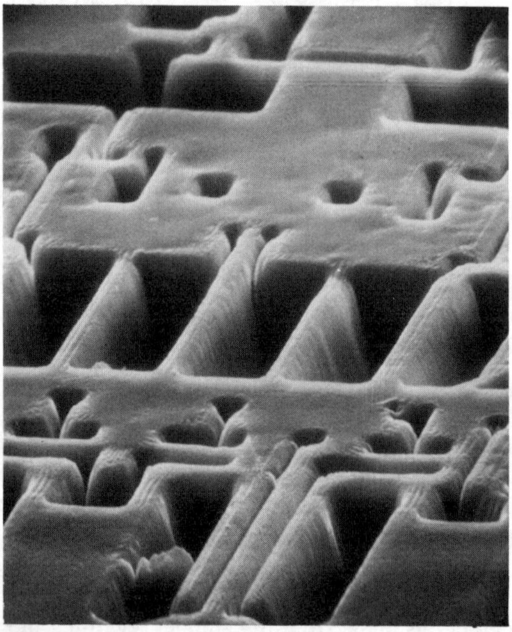

FIGURE 7. Sample of photoresist film patterned by 193 nm laser pulses.
Sample: Polyacrylate Material (DuPont Riston^{TM}); lines shown are 5 μm thick (scanning electron microphotograph by Karen Brown).

to 0.3 μm has been claimed. Another feature of this technique which can be exploited is the high aspect ratio (up to 20:1) that is attainable with wall angles overcut 1° to 30° off the vertical. However, photolithography using existing wet methods is so highly developed that it will be many years before this dry etching process will become competitive.

The applications which look attractive at present are those in which UV laser ablation offers unique possibilities. For example, the class of polymers called polyimides, one example of which is represented by Fig.1b, are important industrially because of their toughness and electrical properties. But their resistance to chemicals and heat make them difficult to pattern by any conventional method (31). The only procedure by which a polyimide film could be geometrically altered was by patterning a precursor material which could subsequently be chemically converted to the final polyimide structure. Now, UV laser etching offers a convenient solution to the problem (8, 14, 31). An important advantage that the UV laser (in contrast to an infrared laser) offers lies in the lack of carbonization at the new surfaces that are created because a surface film of carbon alters the electrical properties of the film in an undesirable way.

The economics of the polymer industry is such that an expensive technique such as UV laser ablation would benefit only a specialty polymer that would be part of a high value-added product. Such opportunities may be expected to develop gradually in optics, electronics, and aerospace industries.

6. CONCLUSION

UV laser ablation of organic matter promises to be of continuing interest as research extends into the physics and chemistry of the process in greater detail. Its importance to technology will grow with increasing availability of dependable lasers and sophisticated delivery systems.

7. ACKNOWLEDGMENT

The assistance of Bodil Braren in the preparation of this manuscript is gratefully acknowledged.

REFERENCES

1. R. Srinivasan and V. Mayne-Banton, Appl. Phys. Lett. **41**, 576 (1982); R. Srinivasan and W. J. Leigh, J. Am. Chem. Soc. **104**, 6784 (1982).
2. M. W. Geis et al, Appl. Phys. Lett. **43**, 74 (1983); T. F. Deutsch and M. W. Geis, J. Appl. Phys. **54**, 7201 (1983).
3. J. E. Andrew, P. E. Dyer, D. Forster, P. H. Key, Appl. Phys. Lett. **43**, 717 (1983).
4. M. Murahara, Y. Kawamura, K. Toyoda, S. Namba, Ohyo Butsuri **52**, 83 (1983).
5. J. G. Calvert and J. N. Pitts, Jr., Photochemistry, Wiley, New York, N.Y. 1966.
6. M. B. Robin, Higher Excited States of Polyatomic Molecules, Academic Press, New York, N.Y. 1974
7. R. Srinivasan, J. Vac. Sci. Tech.B1, **4**, 923 (1983); R. Srinivasan and B. Braren J. Polymer Sci.(Chem), **22**, 2601 (1984).
8. J. H. Brannon, J. R. Lankard, A. I. Baise, F. Burns, J. Kaufman, J. Appl. Phys. **58**, 2036 (1985).
9. H. S. Cole, Y. S. Liu, H. R. Philipp, Appl. Phys. Lett. **48**, 76 (1986).
10. P. E. Dyer and J. Sidhu, J. Appl. Phys. **57**, 1420 (1985).
11. B. Braren and D. Seeger, J. Polymer Sci. Lett. xxx (1986) (in press).
12. R. Srinivasan, B. Braren, D.E.Seeger, R.W.Dreyfus, Macromol. **19**, 916 (1986).
13. R. Srinivasan, B. Braren, R.W.Dreyfus, To be published.
14. J. T. C. Yeh, J. Vac. Sci. Tech. May-June (1986).
15. S. Lazare, P. D.Hoh, J.M.Baker, R. Srinivasan, J. Am. Chem. Soc. **106**, 4288 (1984); S. Lazare, R. Srinivasan, J. Phys. Chem. **90**, xxxx (1986) (in press).
16. R. Srinivasan, R. W. Dreyfus, Springer Ser. Opt. Sci. **49**, 396 (1985).
17. G. Koren, J.T.C.Yeh, Appl. Phys. Lett. **44**, 1112 (1984); J. Appl. Phys. **56**, 2120 (1984).
18. G. M. Davis, M. C. Gower, C. Fotakis, T. Efthimiopoulos, P. Argyrakis, Appl. Phys. A **36**, 27 (1985).

19. B. Danielzik, N. Fabricius, M. Rowekamp, D.von der Linde, Appl. Phys. Lett. **48**, 212 (1986).
20. P. E. Dyer, R. Srinivasan, ibid., **48**, 445 (1986).
21. G. Gorodetsky, T. G. Kazyaka, R. L. Melcher, R. Srinivasan, ibid., **46**, 828 (1985).
22. R. Srinivasan, B. Braren, L. Hadel, R. W. Dreyfus, D. Seeger, J. Opt. Soc. (1986) (in press).
23. H. H. G. Jellinek, R. Srinivasan, J. Phys. Chem., **88**, 3048 (1984).
24. B. J. Garrison and R. Srinivasan, Appl. Phys. Lett., **44**, 849 (1984); J. Appl. Phys. **57**, 2909 (1985).
25. H. H. G. Jellinek, S. R. Dunkle, J. Polym. Sci.(Chem), **21**, 2111 (1983). The polyimide used in this study differs from the structure shown in Fig.2b in having a methylene bridge instead of an ether bridge.
26. T. Keyes, R. H. Clarke, J. M. Isner, J. Phys. Chem. **89**, 4194 (1986).
27. E. Sutcliffe, R. Srinivasan, (to be published).
28. H. Pummer, U. Sowada, P. Oesterlin, U. Rebhan, D. Basting, Laser Optoelektr. **17**, 141 (1985).
29. S.E.Blum, K. H. Brown, R. Srinivasan, U.S. Patent 4,414,059 (1983)
30. S. Rice, K. Jain, Appl. Phys. A **33**, 195 (1984).
31. J. H. Brannon, J. R. Lankard, A. I. Baise, F. Burns, J. Kaufman, U.S. Patent 4,508,749

THERMOCHEMICAL LASER LITHOGRAPHY ON THE BASIS OF LOCAL OXIDATION OF THIN METAL FILMS

S. METEV*, S. SAVTCHENKO

Sofia University, Faculty of Physics, Sofia-1126, Bulgaria

1. INTRODUCTION

The laser lithography is a combination of methods of laser-assisted direct patterning of thopological structures. The different laser lithography methods, which include variety of physico-chemical processes /1/ are under intensive investgation in the recent years. Of especially high effectiveness and quality are the laser-induced thermo- and photochemical lithographic techniques /2/. The most developed of them is the process of laser-induced local oxidation of thin metal films.

The essence of this method consists in local laser irradiation of a thin metal film in the presence of oxidising environement (air, oxygen, etc.). Due to the thermal action of the laser radiation a thin oxide layer builds up in the zone of laser action on the metal surface, and this layer enhances the chemical stability of the irradiated film area. If the sample is placed in a suitable etchant, the unirradiated zones of the metal film dissolve, while the irradiated ones remain on the substrate (Fig. 1), forming on this way a metallic thopological structure on the substrate surface.

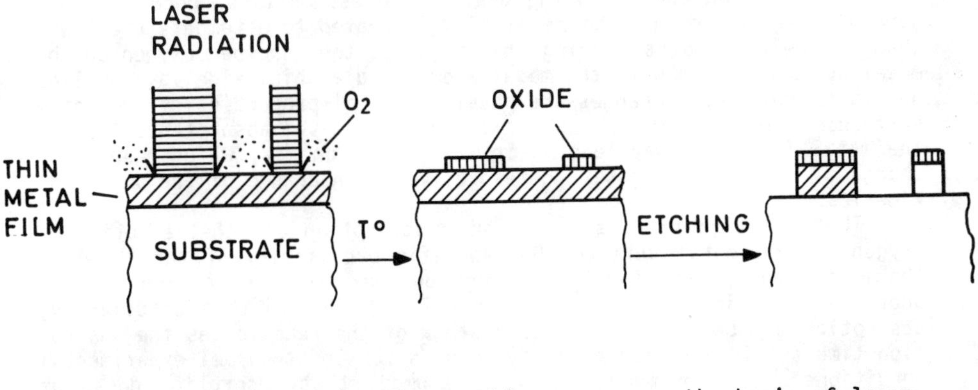

FIGURE 1. Principle of the lithographic process on the basis of laser-induced oxidation of thin metal films.

The aim of this paper is to present the results, obtained in the course of development of this method.

*present address: Max-Planck-Institut für Quantenoptik, 8046 Garching, W. Germany.

2. INVESTIGATION OF THE OXIDATION PROCESS /3/

The aim of the experimental study of the process of pulsed laser radiation activated oxidation (PLRAO) was to obtain some results, important for the use of this process in the practice.

In the experiments a Nd:glass laser has been used, operating in free generation or in Q-switch regimes at pulse duration τ of 1 ms or 50 ns, respectively.

Thin Cr films of thickness H_f from 1000 to 3000 Å, deposited by evaporation in vacuum (10^{-6} torr) on quartz substrates, have been used in the experiments. After deposition, samples have been heated for 1h in 10^{-6} torr vacuum at 350-400°C in order to avoid film recrystallization by the laser treatment. After preparation the samples have been irradiated in air by laser radiation of uniform intensity distribution in the zone of action.

In the first series of experiments the results of laser action on the metal film have been qualitatively examined. The surface of the irradiated films has been investigated by ESCA method /4/. By this method one can identify the chemical compounds on the metal surface.

The electron spectrograms have shown that after pulsed laser irradiation of Cr films in air, a thin oxide layer of Cr_2O_3 builds up in the zone of laser action. A small amount of CrO_3 (< 5%) has also been identified.

The oxide layer, which is built up after laser action, is very stable, and it enhances the chemical stability of the irradiated film area. Samples have been placed in different etchants to examine the solubility of the metal film. The results have shown that the etching time of the irradiated, even with one laser pulse part of the film, is much greater (up to 50 times) than that of the unirradiated ones, and depends on the experimental conditions (film thickness, laser pulse duration, pulse energy, type of the etchant and others).

To describe the PLRAO process a physical model has been developed which is based on the following qualitative assumptions. Under the absorbed laser radiation, the metal film is heated to a temperature lower than its melting point. During this heating, the adsorbed oxygen on the metal surface reacts with the metal atoms, and a thin oxide layer builds up. This oxide layer changes the physicochemical properties of the metal surface. In particular, it enhances the "effective" absorptance A_{eff} of the metal for laser wavelengths in the IR gerion /5/, which results in change of the heating rate and respectively, in change of the oxidation kinetics.

The oxidation process begins with adsorption and chemisorption of oxygen on the metal surface. One specific peculiarity of the PLRAO is the continuously changing temperature during the laser action and the short heating time of the metal (10^{-8}:10^{-3}s). Due to this peculiarity, adsorption may become the limiting stage of the reaction as the adsorption time t_{ad} is a finite quantity (~10^{-8}s for Cr at normal experimental conditions). In our model we have assumed, that adsorption does not limit the oxidation process in the investigated region of pulse durations τ of 5.10^{-8}:10^{-3}s.

In the theoretical model, the oxidation process has been described by the Cabrera-Matt oxidation theorey /6/. This theory describes initial stages of oxidation formation and, according to it, the oxidation reaction is limited by the diffusion of metal cations through the oxide layer to the chemisorbed oxygen. The activation barrier depends on the intensity of the electric field appearing in the oxide layer due to charge separation by oxygen chemisorption. Oxidation kinetics are described in this case by the equation:

$$\frac{\partial h_{ox}}{\partial t} = U_o \exp\left[-\frac{T_d}{T_s(t)}\right]\left\{\exp\left[\frac{h_c(T_s)}{h_{ox}}\right] - \exp\left[-\frac{h_c(T_s)}{h_{ox}}\right]\right\} \quad (1)$$

In (1), h_{ox} is oxide layer thickness, U_o is a quantity, approximately equal to sound velocity in solids, Td is diffusion activation energy, expressed in kelvin, T_s is metal surface temperature, h_c is the characteristic oxide layer thickness, separating conditionally the regions of strong ($h_{ox} < h_c$) and weak $h_{ox} > h_c$) influence of the electric field on the oxidation process. At room temperature ($T_s = T_s^o$) $h_c = h_c^o(T_s^o/T_s)$.

The temperature distribution on the metal surface may be found by solving the three-dimensional heat conductivity equation for the film and the substrate

$$\frac{\partial T_1(X,Y,Z,t)}{\partial t} - \varkappa_1 \Delta T_1 = -\int_0^t \frac{A_{eff} Q(X,Y,t')}{\rho_1 C_1} dt'$$

$$\frac{\partial T_2(X,Y,Z,t)}{\partial t} - \varkappa_2 \Delta T_2 = 0 \quad (2)$$

In (2) subscript 1 refers to the thin metal film and 2 refers to the substrate. Q is laser radiation flux density. $\varkappa_{1,2}$ are thermal diffusivities of film and substrate, respectively, ρ_1, C_1, are density and thermal capacity of the film. A_{eff} is the "effective" absorptance of the metal film, depending on the presence of oxide layer on the metal surface. As the oxide layer thickness obtained by PLRAO is much less than the laser wavelength, one can use the phenomenological theory of the optical properties of thin absorbing layers, deposited on a metal mirror /7/, which gives:

$$A_{eff} = A_o + 2\alpha_{ox} h_{ox} \quad (3)$$

where A_o is the absorbtance of pure metal and α_{ox} is the absorption coefficient of the oxide layer at the laser wavelength used, For Cr_2O_3, $\alpha_{ox} \cong 2.10^4$ cm^{-1} at 1 μm laser wavelength.

The system of differential equations (1)÷(3) describes the kinetics of PLRAO process. Because of its considerable nonlinearity this system has been solved numerically and a variety of important dependances and relations have been obtained.

The aim of the quantitative experimental study of PLRAO was to confirm the physical concepts of the developed theoretical model. Three different experimental methods have been used: resistance method, vibrating substrate method and reflectivity method.

By the resistance method oxidation - induced change of electrical resistance of thin Cr film of suitable geometrical configuration has been measured during laser radiation action. Special measures have been taken to avoid resistance change due to film heating by the laser radiation. The samples were two identical Cr thin film resistors of dimensions 0.8x2 mm, arranged on a quartz substrate at 1 mm distance between them (Fig. 2). One of the resistors has been covered additionally with transparent for the laser radiation thin film of SiO_2, which protected the resistor from oxidation. Both resistors have been included in two neighbor arms of a wheatstone bridge and then irradiated simultaneously by pulsed laser radiation of uniform intensity distribution. The temperature transient effects are avoided by this arrangement, and the electrical sygnal from the bridge is proportional to the oxidation-induced resistance change of the unprotected resistor.

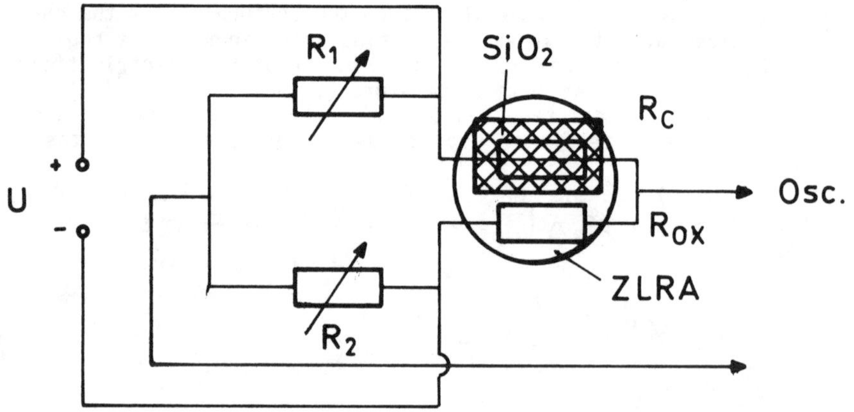

FIGURE 2: Resistance method for measuring the oxidation kinetics

FIGURE 3: Oxidation kinetics during one laser pulse of 1 ms duration (H_f = 2000 Å, Q = 6.10² Wcm^{-2}); - experiment, resistance method, ---- theory

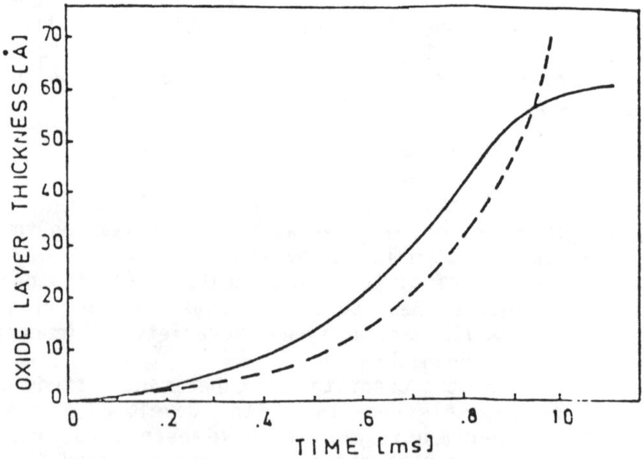

The oxidation kinetics during 1 ms laser pulse, measured by the resistance method is presented in Fig. 3. The dashed curve presents the calculated time dependence of the process. The absence of saturation region in the theoretical curve is due ot the assumption that light flux density is constant in time. In this case the film temperature is highest at the end of the laser pulse, while in the real case the heating rate (respectively, the oxidation rate) drops at the end of the laser pulse due to decrease of light flux density.

The essence of the vibrating substrate method /8/ is that the film under investigation is first deposited on the surface of a piezoelement. The frequency of the acoustic oscillations of this element depends strongly on the mass of the substance deposited on its surface. If an oxide layer builds up on the metal surface, the substrate oscilation frequency changes, and measuring this change, one can determine the mass of the oxide layer very accurately. If the dimensions of the irradiated zone are known, the mean oxide layer thickness can be calculated as well.

The oxidation kinetics and the thickness of the oxide layer have been measured also by measuring the oxidation - induced change of film reflectivity R(i.e., of "effective" absorptance A_{eff} (see (3), where A_{eff} = 1-R)).

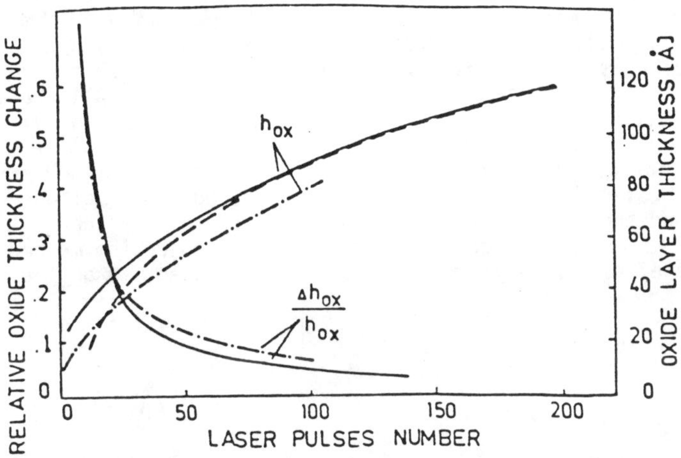

FIGURE 4: Dependence of the oxide layer thickness h_{ox} and the relative oxide thickness change $\Delta h_{ox}/h_{ox}$ on the laser pulses number (τ = 50 ns, H_f = 2500 Å, Q = 2.10^4 W.cm^{-2}): - experiment, vibrating substrate method; --- experiment, reflectivity method; -.-.- theory

Fig. 4 presents the dependence of oxide layer thickness on the number of laser pulses, obtained experimentally by simultaneous use of both the vibrating substrate and the reflectivity method. The difference in the experimental results obtained by these two methods at small oxide layer thickness may be explained suposing the oxide layer to be distributed irregularly on the metal surface. The reflectivity method gives lower values of the oxide thickness till a regular uninterrupted layer is formed. In the same figure, the relative change of the oxide thickness $\Delta h_{ox}/h_{ox}$ obtianed during one laser pulse in dependence on the laser pulses number is presented. One can see that the first several laser pulses are most effective for oxide formation, which is due probably to the fact that with increasing oxide thickness the short heating time cannot activate any considerable metal cations diffusion through the oxide layer.

The comparison of experimental and theoretical curves (Fig. 4 and Fig. 5) shows that the developed physical model describes the PLRAO process well and it can be used for theoretical optimization of the lithographic process.

3. ACCURACY AND RESOLUTION O THE THERMOCHEMICAL LASER LITHOGRAPHY ON THE BASIS OF PLRAO PROCESS

Laser-induced thermolithographic process on the basis of pulsed laser-induced oxidation of thin metal films involves four stages which determine its accuracy and resolution. The first stage is formation of the optical image to be recorded. At the second stage the optical image is converted into thermal image on the surface of the metal film. In the area of the thermal image oxidation reaction is activated, which results in recording (latent image) of the optical image at the third stage of the lithographic process. At the fourth stage of the lithographic process the recorded latent image is developed chemically using a suitable etchant, and a negative of the optical image is obtained on the substrate surface. The total accuracy and resolution of the lithographic process is determined by the accuracy and resolution of the separate stages and depends on many factors as parameters of the optical system, parameters of the laser radiation, thermophysical properties of the metal film and the substrate, type of the etchant, etc. All these factors are connected and they excercise common influence on the lithographic process.

The optical image of the pattern to be recorded may be formed by contact, projection, contour-projection or holographic methods. The projection and the contour-projection optical schemes are the most suitbale ones for laser beam image formation. In the projection method (Fig. 5) the image of a mask is formed by a projection objective on the

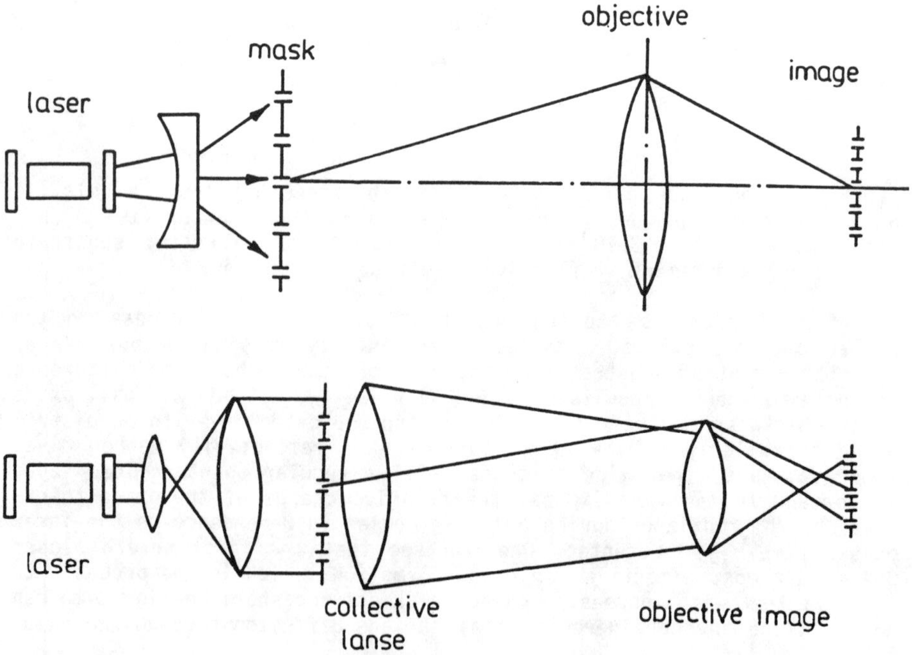

FIGURE 5: Laser projection optical systems

sample surface at suitable diminution. The projection scheme with a collective lense, placed after the mask allows up to 100 times diminution of the mask image.

One characteristic peculiarity of the laser beam image formation is the constant value of the optical contrast function for all space frequencies. This fact means that the different in size pattern features will be transmitted with the same contrast through the optical system, when coherent laser radiation is used. This fact is of great importance for the lithographic process.

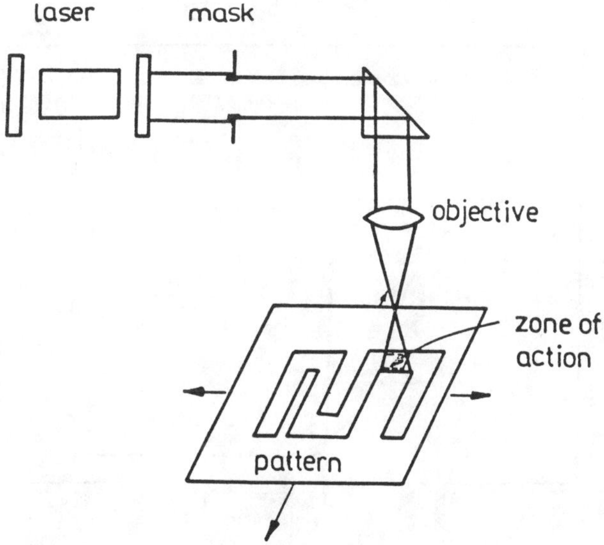

FIGURE 6: Laser contour-projection optical systems

By the contour projection method (Fig. 6) the optical image of the pattern is synthesized by computer controlled laser beam scanning along the contour of the pattern. In this case the zone of the laser radiation action has the form of a single pattern feature (square, rectangel), formed in a projection scheme. By the contour-projection method one can obtain patterns of large total dimensions and high resolution /9/.

The distortions of the optical image can be reduced to minimum using corrected optical elements in the visible region. The optical resolution enhances with reduction of the laser wavelength but from another hand the UV region requires special optics. An optimum seems to be the use of visible laser radiation in the blue-green region where high quality optical elements are developed and practically micron-scale resolution can be obtianed.

Distortion of the thermal image depends on the thermo-physical properties of the heated substrate (thermal conductivity) and on the temporal parameters of the laser radiation and are due to heat penetration outside the zone of laser radiation action (ZLRA). As far as heat penetration length is given by $\sqrt{\ t}$ (is the thermal diffusivity) a radical way to enhance the accuracy and resolution of the thermal image is to reduce the duration of the laser radiation action, t.

The accuracy and resolution of he recording process, based on the pulsed laser ixidation of thin metal films depend on the accuracy and resolution of the protective oxide layer. This oxidation process has been optimized in /10/ by a numerical computer simulation on the basis of the developed physical model of the process. The influence of different parameters as pulse duration, pulse shape, laser radiation flux density, dimensions and form of the zone of laser radiation action (ZLRA) on the accuracy and resolution of the recording process has been studied.

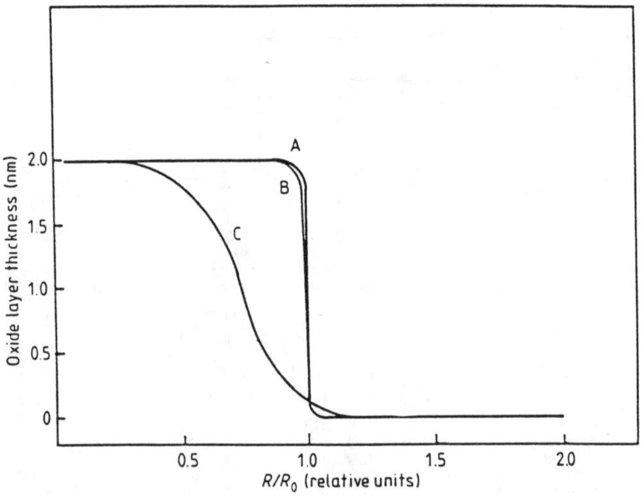

FIGURE 7: Oxide thickness distribution in the ZLRA, radius R_0 = 150 µm for three laser pulse durations. (A) 50 ns, (B) 1 µs, (C) 1 ms

Figure 7 shows the oxide layer thickness distribution in the ZLRA (of circular form) after the action of a laser pulse with gaussian temporal shape. The graphs were obtained at three different laser pulse durations and experimental conditions which cause build-up of an equally thick oxide layer in the centre of the ZLRA. The graphs show that, due to heat penetration outside the irradiated zone, long laser pulses are not suitable for accurate thermochemical image recording. One can define the accuracy of the recording process by

$$\eta = \frac{l_{rec}}{l_{opt}} \cdot 100\% \quad l_{rec} < l_{opt}$$
$$\eta = \frac{l_{opt}}{l_{rec}} \cdot 100\% \quad l_{opt} < l_{rec} \qquad (4)$$

In (3) l_{opt} is the linear dimension of the iptical image (ZLRA) to be recorded, l_{rec} is the corresponding linear dimension of the recorded image. It has been shown, /3/ that a 10 Å laser-induced oxide layer is stable enough to protect the underlying metal film from chemical etchants. Thus l_{rec} is the dimension of the zone, where the oxide thickness is greater than 10 Å. In the cases presentd in Fig. 7 the recording accuracy was 75%, 96% and 99% for 1 ms, 1 μs and 50 ns pulse durations, respectively.

A very important result of the numerical simulation are the graphs, plotted in Fig. 8. This figure presents the Cr_2O_3 layer thickness in the center of the ZLRA on the surface of Cr film in dependence on its size (radius) for three pulse durations - 50 ns, 1 μs and 1 ms at laser wave-

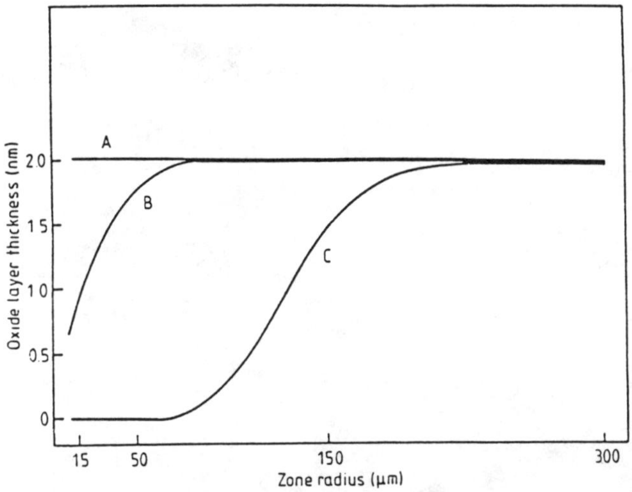

FIGURE 8: Oxide layer thickness in the ZLRA centre as a function of radius for three laser pulse durations. (A) 50 ns, (B) 1μs, (C) 1 ms

length 1.06 μm. The light flux density value has been chosen in each case such that to cause build up of equally thick oxide layer in the center of 300 μm ZLRA. Graph C in Fig. 8, which corresponds to 1 ms pulse duration, shows that different in size image features will not be recorded with the same accuracy. At this experimental conditions (graph C) features of radius less than 120 μm will not be recorded at all (according to the experimental investigations the 10 Å oxide thickness is the smallest one, which can protect reliably the irradiated zone from a subsequent etching). In the 1 μs regime (graph B) the minimum size of the recorded image feature is 15 μm. For the nanosecond regime A the thickness of the oxide layer does not depend on the ZLRA dimension. Only in this regime one can record completely complex images consisting of different in size fearures.

One of the aims of the numerical simulation was to study the influence of the image form on the recording accuracy. Three different simple feature forms have been chosen to characterize the following cases: i) a line (Fig. 9a); ii) crossing of two lines of equal width (Fig. 9b) and iii) crossing of two lines of different width (Fig. 9c).

Fig. 9 shows the oxide thickness distribution in the ZLRA after the action of 1 ms, 1 μs and 50 ns laser pulse on the surface of thin Cr film. In the same figure the 10 Å oxide thickness isolines are presented. As it has been mentioned before, the metal film will be etched in a suitable etchant up to this level. Hence these isolines represent the form of the recording image. From Fig. 9 one can see that due to the

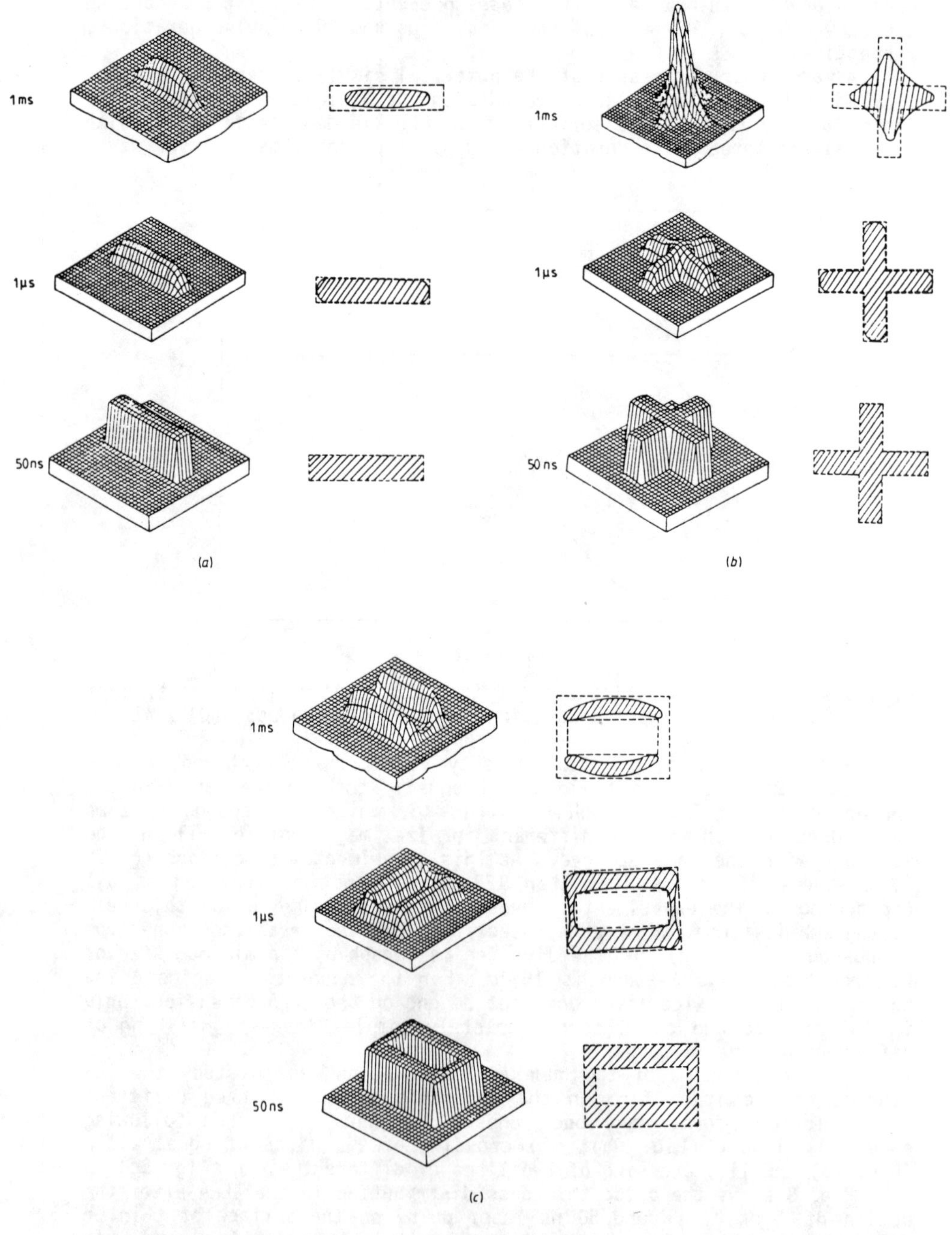

FIGURE 9: Spatial image of the oxide layer in the ZLRA for various forms. (a) Line, (b) cross, (c) rectangular hole for three laser pulse durations. -optical image to be recorded (ZLRA); -recorded image

interaction of the thermal fields from the different parts of the optical image (ZLRA) the recorded image is distored. This distortion is greater for the long laser pulses. This is due to the long time dirung which the heat penetrates outside the irradiated zone. This circumstance leads to distortion of the complex thermal field and to distortion of the recorded image, respectively. In the case of nanosecond laser pulses the heat is concentrated in the ZLRA and the recorded complex feature is distortionless.

Our investigation has shown that the nanosecond regime is the most suitable one for thermochemical image recording. In this regime the recording accuracy does not depend on the form and the size of the recorded pattern as is the case in the microsecond and especially in the millisecond regime. This regime is markedly suitable for projection laser lithography, when the recorded image consists of complex features of different form and size.

The microsecond regime is suitable for contour-projection image recording. This regime can also be used in an optical projection scheme when the image to be recorded consists of simple features. The recording accuracy depends on the size of the recorded features when it is under 30 μm.

The accuracy, the resolution and the contrast of the recorded image depends strongly on the type of the etchant. In /11/ the kinetics of the film etching has been studied in order to optimize the etching process. A study of several etchants for chromium (HCl, $Ge(SO_4)_2$ and HNO_3, $K_3F_e(CN)_6$ and NaOH, $K_3F_e(CN)_6$ and KOH) has shown that an etchant consisting of aqueous solution of $K_3F_e(CN)_6$ and NaOH has the highest selectivity coefficient K, which can be defined as the ratio of the average etching rates of unexposed and exposed chromium films. The selectivity and the etching rate depend on the relative and absolute concentrations of the etchant components.

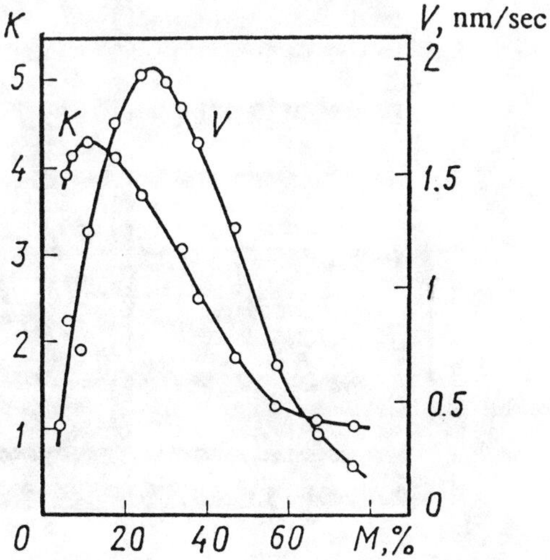

FIGURE 10: Selectivity coefficient K and etching rate V of an unirradiated Cr film versus the relative concentration of the NaOH solution in the etchant /11/

In Fig. 10 the dependence of the average etching rate V of an unirradiated Cr film and of the selectivity coefficient K on the relative concentration of the NaOH solution in etchant consisting of 25% aqueous solution of $K_3F_e(CN)_6$ and NaOH. The curves show that the selectivity of this etchant reaches its maximum value at a component ratio of 9:1, while the etching rate of pure chromium is maximum at a ratio of 3:1, which is approximately the ratio recommended in the literature /12/.

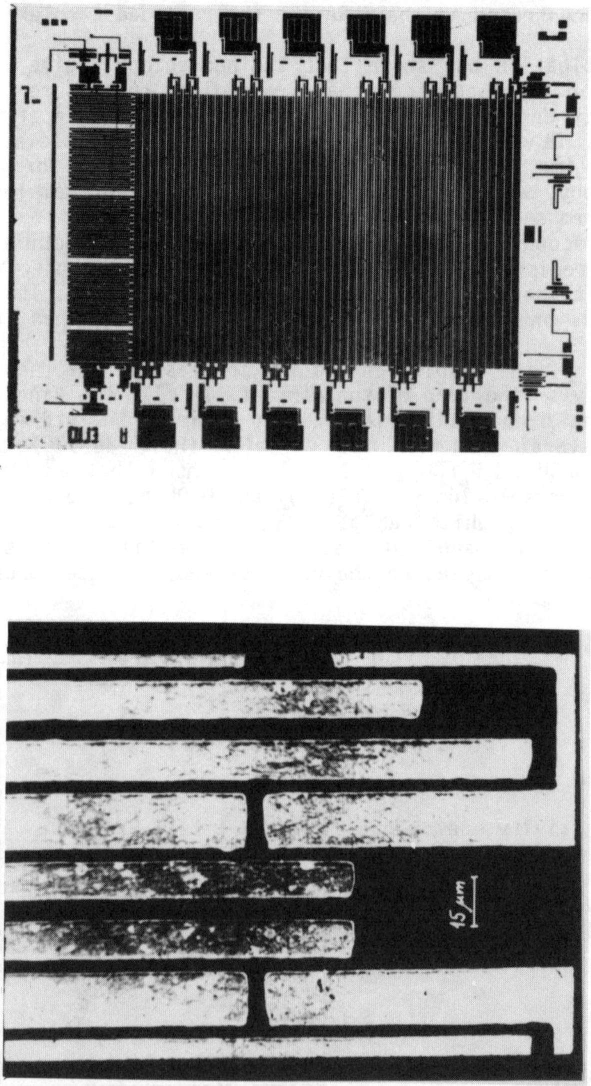

FIGURE 11: Photolithographic mask obtained by laser-induced oxidation of thin Cr film; (a) overall view of the mask; (b) a part from the same mask under greater magnification

The investigation of the influence of the various parameters on the recording process has shown that at optomum conditions complex images of high quality and resolution can be recorded.

4. APPLICATION OF THE THERMOCHEMICAL LASER LITHOGRAPHY

Thermochemical laser lithography, based on laser induced oxidation of thin metal films is a very effective method for preparation of optical elements (optical scales, gratings etc.) and photolithographic masks of high quality and resolution.

The photolithographic mask of an integral microcircuit (Fig. 11a) is obtained in a projection optical scheme in five laser pulses (τ = 50 ns, λ = 1.06 µm) by laser-induced oxidation in air of a 2000 Å Cr thin film, deposited on glass substrate [13]. This photograph demonstrates the capabilities of the method. The total dimensions of the metal pattern are 9 x 7 mm; the smallest feature is 3 µm. A part of the same pattern is shown in Fig. 11b under greater magnification. The edges of the metal lines are well formed and their roughness is smaller than 0.3 µm at a total line width of 3 µm.

One of the optical diffraction elements which is most widely used at present is the Fresnel zone plate. Previously known methods have failed to achieve success in the preparation of zone plates having a large diameter and aperture N=R/F of the order of 0.4:0.65. This is because of the necessity of depositing on the substrate a very large number of concentric zones, $(1\text{-}3).X10^4$, with a minimum period of ~ 0.7-1.2 µm and accuracy of ± 0.1 µm.

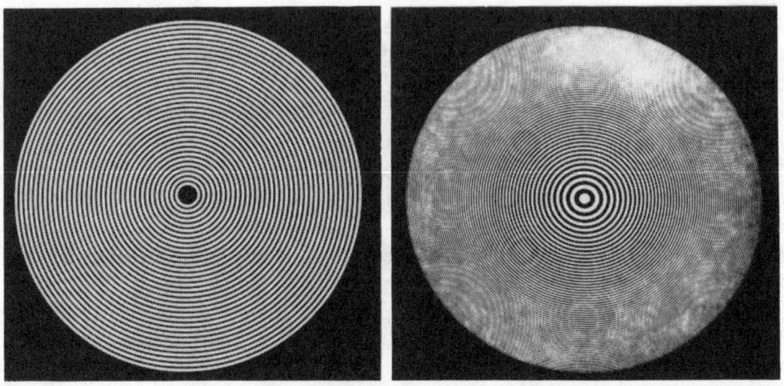

FIGURE 12: A Cr photomask of an optical diffraction element, obtained in contour-projection optical scheme, using an Ar-ion laser /14/

The thermochemical laser lithography method has been successfuly used to prepare masks of a Fresnel zone plate having a numerical aperture of N = 0.65 and a diameter of 47 mm /14/. For this purpose a CW argon ion laser has been used. The scan velocity and the power of the laser beam have been controled by a computer.

Fig. 12 presents a photograph of a mask of a diffraction Fresnel zone plate, obtained by laser-induced oxidation of Cr film on quartz substrate. The masks made in this way could then be transfered to the substrate by ion etching /14/.

This two applications of the thermochemical laser lithography method demonstrate its principle capabilities and show, that this method may become a powerful tool in the modern optics and microelectronics.

REFERENCES

1 S. Metev, "Physico-chemical basis of laser lithography" in "Physical processes in laser-materials interactions" (Plenum-New York), 1983, p. 503

2 S. Metev, V. Veiko "Thermochemical Laser lithography", Proc. Int. School on Laser-asist. mod. synth. of mater. (Gynletchitza, Bulgaria), 10-17 March 1985.

3 S. Metev, S. Savtchenko et al. "Thermochemical action of laser radiation on thin metal films", IEEE J. Quant. Electr. (1981) QE-17, 2004.

4 V. Menoshkalenko, V. Alioshin, "Electron Spectroscopy of metals" (Kiev: Naukova Dumka), 1976.

5 M. Arsuov, V. Konov, S. Metev,"Heating kinetics of metals irradiated by CW CO_2 laser in oxidizing environment", Fiz. Khim. Obrab. Mater. (1978), No.5, 19.

6 T. Benard, "Oxydation des Métaux" (G. Villars, Ed.), Paris: 1964 p. 100.

7 G. Rosenberg, "Optics of thin films", Moscow: Fizmatgiz, 1958, p. 73

8 S. Metev, V. Veiko, K. Stamenov, H. Kalev, "Investigation of laser-induced damage of thin absorbing films by quartz resonator method", Sov. J. Quant. Electron. (1977) 7, 863.

9 V. Veiko, "Contour projection method for laser treatment of materials", Kvantovaja Elektronika, (1978), 5, No. 12.

10 R. Gerassimov, S. Metev, S. Savtchenko "Optimization of the process of laser-induced thermochemical image recording", J. Phys. D Appl. Phys. (1984) 17, 1671.

11 V. Koronkevich, A. Poleshchuk, E. Churin, Yu. Yurlov, "Selective etching of laser-exposed chromium thin films", Sov. Tech. Phys. Lett. (1985), 11, 57.

12 L. Maissel, R. Glang (Eds.) "Handbook of thin film technology", McGraw-Hill, New York (1970).

13 S. Metev, S. Savtchenko, K. Stamenov, "Pattern generation by laser-induced oxidation of thin metal films", J. Phys. D: Appl. Phys. (1980), L 75.

14 V. Koronkevich, A. Poleshchuk, E. Churin, Yu. Yurlov, "Laser thermochemical technology for synthesizing optical diffraction elements utilizing chromium films", Sov. J. Quant. Electron. (1985) 15, 497.

LASER INDUCED METAL OXIDATION

R. ANDREW, M. WAUTELET
UNIVERSITE DE L'ETAT
SERVICE DE PHYSIQUE DE L'ETAT SOLIDE
MONS BELGIUM

GENERAL INTRODUCTORY REMARKS

Metal oxidation is a topic of particular interest to us since as a species we spend a great deal of time and effort trying to combat the effects of atmospheric corrosion of metals and metal structures, principally due to the formation of metal oxides. The oxide state is generally a lower energy one, though with an activation energy barrier to formation, - so that heating the metal accelerates oxide growth.

Since the advent of powerful laser sources it has also become apparent that irradiating the metal (in an oxygen containing atmosphere, of course) also accelerates oxide growth. Is this the same thing ? In other words, are lasers just inefficient and expensive heat sources, or are there other aspects of the laser/material interaction which deserve special study. The aim of this talk is to examine the overall mechanisms of laser induced oxidation. and to indicate where, why, and how laser oxidation differs from classical thermal treatments.

The starting point of any laser induced effect is the absorption of electromagnetic energy by the solid. Except possibly in the case of very powerful short duration laser pulses, where new states of matter may be involved, absorption proceeds by classical optical laws taking into account reflectivity, absorption coefficient and so on, though the nature of laser light also means that other effects such as interference might be important. Practically, the problem is that of coherent light absorption within a multilayer dielectric system which is evolving in time. This is a topic worthy of whole NATO ASI itself !

Further, at an atomic level the energy transfer between the incoming photons and the solid occurs via excitation of electrons into unoccupied states. The photon energy is such that this involves not only delocalized states but also localized ones associated with defects and impurities, often with a high concentration at the surface and other interfaces.

Since classically, oxidation proceeds via the sequence-interface reaction/diffusion via impurity of defect states/interface reaction-it is clear that optical excitation of these may drastically modify the parameters of the oxidation reaction. At the short wavelengths now available with excimer lasers we may even have photoemission across interfaces and direct bond breaking, and since charge equilibrium also plays an important role in oxidation kinetics, we might anticipate some particular effects here. The ultimate result, however, of laser irradiation is heating of the absorbing solid, and since the mean free path of excited electrons is generally much less than the dimensions of the irradiated sample, this heating may be viewed as localized in the region where absorption occurs, with subsequent behaviour controlled by classical laws of thermal diffusion, radiation and convection to the surrounding atmosphere. For high power irradiation and short times these last two can often be neglected, but nevertheless, several complications arise. Laser irradiation can offer a degree of energy coupling to the solid difficult or impossible to attain by other means, particularly whilst at the same time allowing access to atmospheric oxygen. Pulsed or focused CW beams from even moderate sized lasers can easily deliver tens of megawatts/cm^2 so that the solid is heated very rapidly and , if its absorption is high, heating is initially at least, limited to a very shallow surface region. On short time scales the total volume of the target heated is rather small and, paradoxically, the total energy delivered to the system, is also rather small, so that if oxidation proceeds the heat of formation liberated during what is an exothermic reaction may be comparable to, or can even be greater than, the energy supplied by the laser, and therefore, must be taken into account in the subsequent behaviour of the system.

In the worst case the temperature can depend on the oxide growth rate, which itself is temperature dependent, and in some circumstances the possibility of 'runaway' thermal effects cannot be discounted. The high initial heating rate means also that temperature gradients present in the solid are high and so that rapid cooling effects may arise when the laser beam is interrupted or, in a more complex situation greatly attenuated due to interference effects within the growing oxide film. The possibility exists here for 'negative feedback' or an arresting action on oxide growth. Even discounting these two effects there remains the problem of heat flow in a multilayer, time evolving system with temperature dependent thermal parameters and also, in many cases, a non-unidimensional geometry, perhaps for e.g. a gaussian profile scanning beam.

Even this relatively simple case is not easily treated (this is an understatement) and when coupled to the optical aspects of the energy coupling cited above, and the possibility of direct optical excitation of the vectors of the reaction one begins to appreciate the complexity of the subject. We can do more here than scratch at the surface of the problem.

GENERAL FORMALISM

We have indicated in the preceding section that a full treatment of laser-induced oxidation of metals involves the the solution of a set of coupled, non-linear differential equations describing the time dependent evolution of energy coupling via absorption, reaction and heat flow with the possible added complexity of optical excitaion of defects, radiation enhanced diffusion and so on. The specific features of thermochemical processes taking place under laser irradiation have been listed (Luk'Yanchuk 1982, Laude et.al. 1986.) as :

1) Non-isothermality of processes;
2) Arrhenius-type temperature dependence for the chemical reaction rate constants;
3) This eventually opens the possibility for establishing a macroscopic non-equilibrium in the system, when the temperature variations are faster that the "chemical relaxation" of a medium into a quasi-equilibrium state. Changing absorption characteristics of a

medium during the course of a chemical reaction may introduce the possibility of feedback effects between the chemical and thermal degrees of freedom of the system.

4) The establishment of new reaction regimes, due to the energy and time scales involved;
5) The possible modification of the thermodynamical factors (for example : the energy of migration of defects), by optical excitation of the vectors of the reaction;
6) Deviation from a thermodynamical description, due to the presence of a very large (spatial and temporal) gradients (at least during some transient stages).

It is worth noting that points 1,2,3 and 5 are relevant to all kinds of laser irradiation, while points 4 and 6 are more relevant to short pulsed laser irradiation. We can write :

$$\frac{\partial y}{\partial t} = F(x,y,t,T); \qquad (1a)$$

$$\frac{\partial T}{\partial t} = L(x,y,t) + R(x,y,t) - D\frac{\partial^2 T}{\partial x^2} + P(y,t,T) \qquad (1b)$$

$$\frac{\partial L}{\partial t} = G(x,y,t,T) \qquad (1c)$$

In these equations, T is temperature and t is time. L and R are specific energies provided by the laser and liberated by the (exothermal) chemical reaction, respectively; x is in the direction perpendicular to the surface; y is the fraction of the reacted matter, D is the coefficient of thermal diffusivity. P are thermal losses (convection, radiation). F depends on many factors, like the type of reaction rates (linear, parabolic, etc.), the gradients of y and T, the migration energy of the atomic vectors of the reaction. G depends on y via optical absorption and interference effects.

Due to the large number of parameters and the large imprecisions in their experimental values, numerical calculations are impossible in the general case. The formalism of equations 1(a-c) is, however, helpful in order to classify various reactions. In particular, under continous laser irradiation, R(x,y,t) plays no role, since the thermal losses to non-irradiated regions exceed by far R. At the opposite extreme, R plays the major role in explosive reactions, where L serves "only" to trigger the reaction (Fan et.al. 1980, Auvert et.al. 1981, Andrew et.al. 1982). In this case, the effect of G (interference effects) is difficult to describe, since R is much larger than L.

In the case of laser-induced oxidation of metals, the effect mostly studied is the appearance of feedback between chemical and thermal degrees of freedom of the system. This has been intensively studied first by soviet workers (Bunkin et.al. 1982, Konov 1982, Luk'Ynachuk 1982), and confirmed by us in other cases. This feedback stems form the change in the absorption properties of the system during the oxidation reaction (via interference effects), coupled to Arrhenius temperature dependence of the oxidation rate constants. When the wavelength of the incident radiation acts on one reaction component alone (or mainly), three situations may appear. If the

the radiation is absorbed by the final reaction product (here the oxide) and the reaction is accelerated by heating, then "positive feedback" occurs : as the concentration of the product increases, absorption increases, the temperature of the system increases and self-acceleration of the oxidation occurs. At the opposite, when the radiation is absorbed by the starting substrate the "negative feedback" occurs and the oxidation is stabilized. A third situation is observed under visible laser irradiation of metals : due to interference effects, the absorbed energy fluctuates when oxidation proceeds. Then the reaction self-accelerates at some times, and then diminishes when the reaction proceeds, then re-accelerates, etc. This gives rise to oscillating oxidation rates. The mathematical problem is further complicated when the laser beam is not uniform, so that the spatial profile of absorbed energy varies both in space (lateral) and time, as shown schematically in **figure 1.** In this case, the number of parameters entering equations (1) is so large that calculations are meaningless in a first stage.

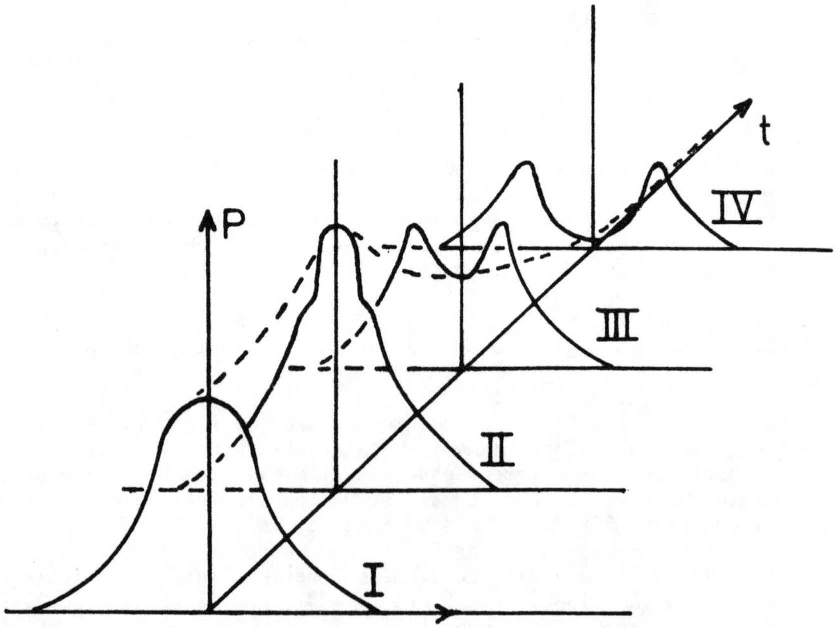

Figure 1. Schematic evolution of cross sectional power absorption due to reflectivity changes.

Another interesting case concerns situations in which more than one oxide is formed under usual conditions, like oxidation of copper. This is known to give rise to a mixture of CuO and Cu_2O, whose relative concentrations are functions, in a furnace, of temperature and pressure. Under laser irradiation, heating may occur very rapidly and the temperature fluctuate both strongly and rapidly. In this case, it has been shown theoretically and experimentally (Lun'Yanchuk 1982) that the relative concentrations of final oxides can be different from those expected from stationary diffusion. This has been found by properly taking into account the thermal history of the system under laser irradiation.

Equations (1) are non-linear. Depending on the explicit form of these equations, various types of thermochemical instabilities are expected. The reader is referred to the work of Bunkin et.al. (1982) for a more complete discussion of this subject.

In general laser-induced effects have been seen as purely "thermal" effects, albeit with some influence of the various complexities arising from the wave nature of the beam i.e. interference effects or the "quasi-adiabatic" approach to be adopted in short time scales. Clearly, thermal effects do exist and the ultimate effect of laser irradiation is to heat the target. There is, however, some evidence that so-called "non-thermal" effects also exist in laser-solid interactions of this type. See for e.g. the laser enhanced oxidations of Si (Schafer & Lyon 1982, Young & Tiller 1983, Boyd 1985), GaAs (Bertness et. al. 1985), CdS (Lichtersteiger & Webb 1985) or laser induced ablation of Silicon oxide (Fiori & Devine 1985).

Laser-enhanced (as opposed to "laser-induced") oxidation of Cr has been reported by Birjega et.al. (1986 a & b) and of Cu by Ursu et. al. (1984, 1986), using a CW CO_2 laser, though Andrew (1986) found no evidence of a non-thermal process for Ar^+ laser irradiation of Cu in shorter time scales.

Only experiment can determine whether, in a particular case, specific "laser" effects are present and the next section will outline some consirations to be taken into account when designing such experiments.

EXPERIMENTAL

The most fundamental experiment in the study of laser induced oxidaitons of a given metal is a measure of oxide thickness, or its time derivative, as a function of laser fluence and time, although a more detailed study might include influence of wavelength, oxygen pressure, background temperature and so on. Since there will in general be a heating effect of the laser it is also important to either measure or calculate the sample temperature, in order to be able to compare the results with more classical heating methods.

Oxide thickness and temperature should ideally be monitored in a time resolved, non-destructive manner, whilst a programmable temperature would also be highly desirable.

Oxide thickness

Several methods have been developed for the measurement of oxide growth after furnace annealing. Perhaps the simplest consists of a determination of the weight increase of the metal sample due to oxygen uptake or, in some cases, the weight loss after the oxide layer has been removed by scouring or chemically. Microscopy following sectioning of the sample can give more detailed information about the microstructure of the oxide layer.

In an extension of the weight increase method, the formation of the oxide layer can be followed using a specially constructed sensitive micro-balance which can be introduced into the furnace or, alternatively by gas volumetric or manometric techniques which measure the quantity of reacted oxygen.

These last two techniques can provide a thickness resolution of better than 50 Å in a time resolved way, although the response is slow. For oxide films which are thin enough to be optically transparent (typically a few thousand Å), and which develop with uniform thickness, optical interference methods offer the advantages of very fast response and the ability to probe small surface areas of the sample, and the development of laser sources makes this a particularly attractive method. The calculation of oxide thickness requires a knowledge of the optical constants of oxide and metal, although if necessary calibration can be carried out against some other method.

For laser induced oxidation studies we have developed a twin probe technique shown in **Figure 2.** Part of the primary Ar^+ laser beam is picked off using a glass slide, attenuated and focused onto the specimen surface at as near to normal incidence as possible. A second probe beam from HeNe laser is focused in a similar manner in the same place, and the two reflected beams monitored by separate detectors equipped with appropriate filters, and situated at the respective conjugate foci (with respect to the specimen) of two toroidally cut mirrors.

Figure 2

Experimental two-probe layout for thickness measurements.

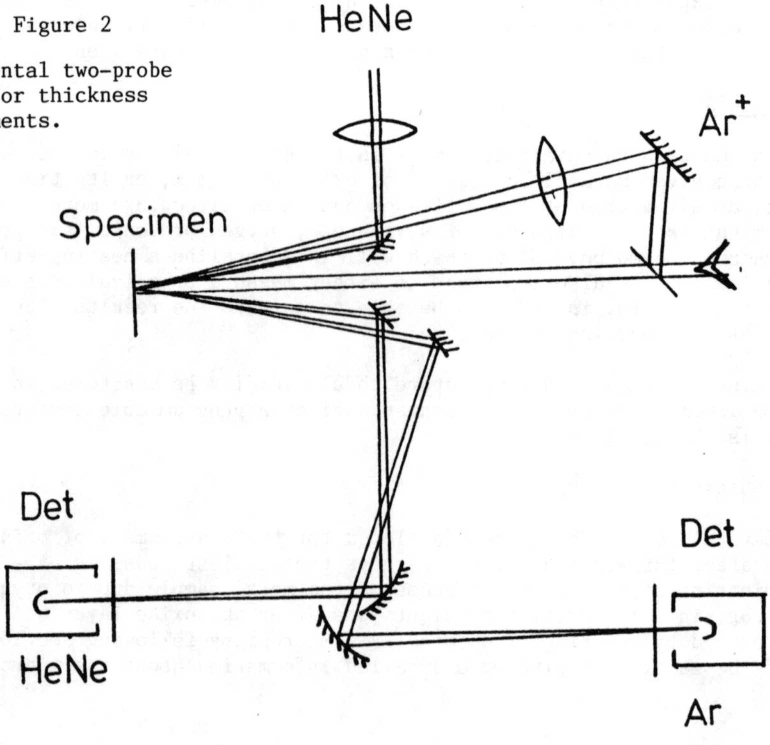

Typical probe signals during oxidation are shown in **figure 3.** The maxima and minima are due to interference effects in the growing oxide layer and may be compared to calculated spectra. A further advantage of this method is that it provides a direct measure of specimen reflectivity at the wavelength of the primary beam and thus some possibility of making corrections for the interference effects described in section I.

It is worth noting that this optical probe method works surprinsingly well even in the case where the initial metal surface is highly granular in structure, although the calculation of oxide thickness if prone to greater error.

Figure 3. Typical probe signals from the apparatus of Figure 2.

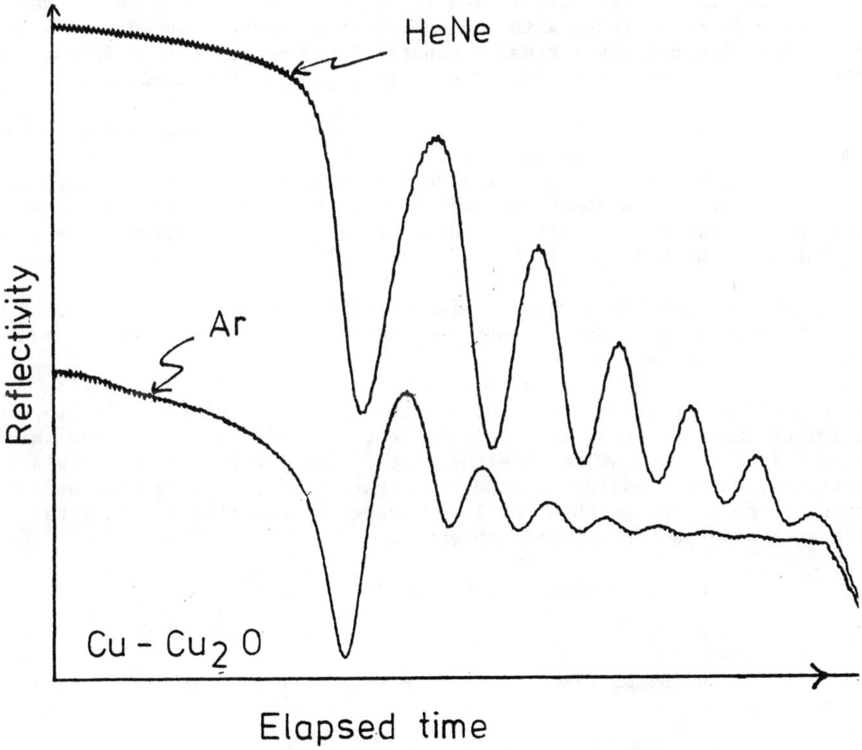

Temperature measurements and calculations

The following small equation deserves a whole line to itself :

$$x \simeq \sqrt{4Dt}$$

where x is distance, t is time, and D the thermal diffusivity of the solid since it expresses the distance that an appreciable quantity of heat once absorbed in one place will diffuse in a time t, and is a guide to useful simplifications of the temperature problem in the presence of transients. Values of D, lie approximately in the range 10^{-1}-$1 cm^2/s$ for metals and 10^{-3} - 10^{-2} cm^2/s for oxides.

Let us suppose a study consists of the growth of an oxide film up to 5 000 Å. Then, in the worst case where $D=10^{-3}cm^2/s$ we find $t \simeq 6\ 10^{-7}s$, i.e. for times significantly greater thanthe order of μs we can safely assume temperature uniformity within the thickness of the oxide layer. (For the kind of studies discussed here this is practically always the case).

Similarily, suppose that the sample is a Cu ($D=1.14cm^2/s$) target, 0.3mm thick, then on a time scale significantly exceeding the millisecond temperature uniformity can be assumed within the sample and thus a measure of rear face temperature will provide a good estimate of oxide layer temperature provided we do not ask for time resolution better than this. This is the approach adopted by Ursu et.al. (1984) where the samples are small polished metal plates mechanically supported by thermocouples on the rear face and uniformely exposed to CW CO_2 laser radiation. After a time determined by the thermocouple wires but typically of the order of seconds thermal equilibrium is established with the sample at constant temperature (as long as the absorbed power remains constant) and temperature uniformity to better than a degree established across the depth of the specimen.

An alternative "easy" grometry is that of the metallic sample in the form of a bar length L irradiated uniformly on a polished end face. For times << $L^2/4D$ the bar behaves as a semi-∞ solid with surface temperature $T(0,t) = Q/K\ 4Dt$. (Q = absorbed power flux), or alternatively if the end z=L is maintained at some ambient temperature then for times>>$\frac{L^2}{4D}$ equilibrium is established with $T(0,t) = \frac{QL}{K}$.

In either case thermocouples of negligible thermal capacity may be attached to the bar near x=0 and any necessary temperature correction made to measure the surface temperature.

A problem may arise when, for studies at a particular wavelength, insufficient laser power is available to heat the bulk metal specimen as described. In this case we may use thin metal films deposited on insulating substrates, which ideally correspond to the thin film - long time approximation. Here we assume that the laser energy is absorbed in the metal which acts as a photon/thermal converter.

For times significantly greater than $\frac{L^2}{4Ds}\left(\frac{P_f C_{pf}}{P_s C_{ps}}\right)^2$

|L = film thickness, suffixes s and f refer to film and substrate parameters.|

The thermal capacity of the film can be neglected and the thermal behaviour of the system is completely controlled by the substrate, whose front surface temperature is assumed to be also that of the film.

Thus for 2 000 Å Cu films on glass, for times much exceeding ≃ 60 nS the thin film - long time approximation is valid and T(0,t) for a bar shaped substrate is given by $\frac{Q\sqrt{4Dt}}{K}$ or $\frac{QL}{K}$ as before except that K is the thermal conductivity of the substrate and temperatures perhaps 1 hundred times higher than before are reached with the same laser power.

In the case of Gaussian profile laser beams then the one-dimensional geometry described is no longer advantageous. Heating of a semi-∞ solid with a Gaussian beam $Q(r) = Q_o \exp^{-r^2/d^2}$ gives T(r,z,t) in a non analytical form except for

$$T(0,0,t) = \frac{P_{abs}}{K\pi^{3/2}d} \tan^{-1}\sqrt{\frac{4Dt}{d^2}}$$

which formula equally well applies to a film on insulating substrate in the thin film - long time approximation.

For a substrate thickness L this formula is also valid for $t \lesssim \frac{L^2}{4D}$, or approximately 1 second for a 1mm thick glass substrate. For longer times we may calculate T(0,z,t) for semi-∞ solid and use, graphically or otherwise, a multiple folding technique using the substrate faces as thermal mirrors to obtain T(0,0,t) **figures 4 & 5.**

The above methods can give a reasonable estimate of substrate surface temperature as long as the thermal parameters are slowly varying functions of temperature. This is also an advantage of using an amorphous insulating material as a substrate.

Time dependent laser power can be catered for by application of Duhamel's theorum and for the case of a laser pulse of temporal form $t \exp^{-t^2/t_o^2}$ the maximum temperature is given by

$$T_m \simeq \frac{0.8E}{K}\sqrt{\frac{D}{t_o}}$$ where E is the total pulse energy.

Interference related changes in reflectivity can also be corrected for using Duhamel's theorum provided that the reflectivity change is uniform across the sample surface i.e. - uniform irradiation conditions.

For a Gaussian beam the drop in R first occurs only at the spot centre. Indeed at some later time R returns to its initial value whilst there now exists a low reflectivity ring around the spot centre. Evolution of the spatial profile of absorbed energy is complex and demands a precise knowledge of the radial dependence of R with time.

To further complicate tmatters this radial dependence is also a function of the time scale of the oxidation process - lower power with longer times produce larger oxidized spots. Thus the disadvantage of the Gaussian beam on thin film specimen approach is that reflectivity changes at the primary wavelength demand a precise knowledge of R(P,r,t) and some hefty calculations to obtain T(0,0,t).

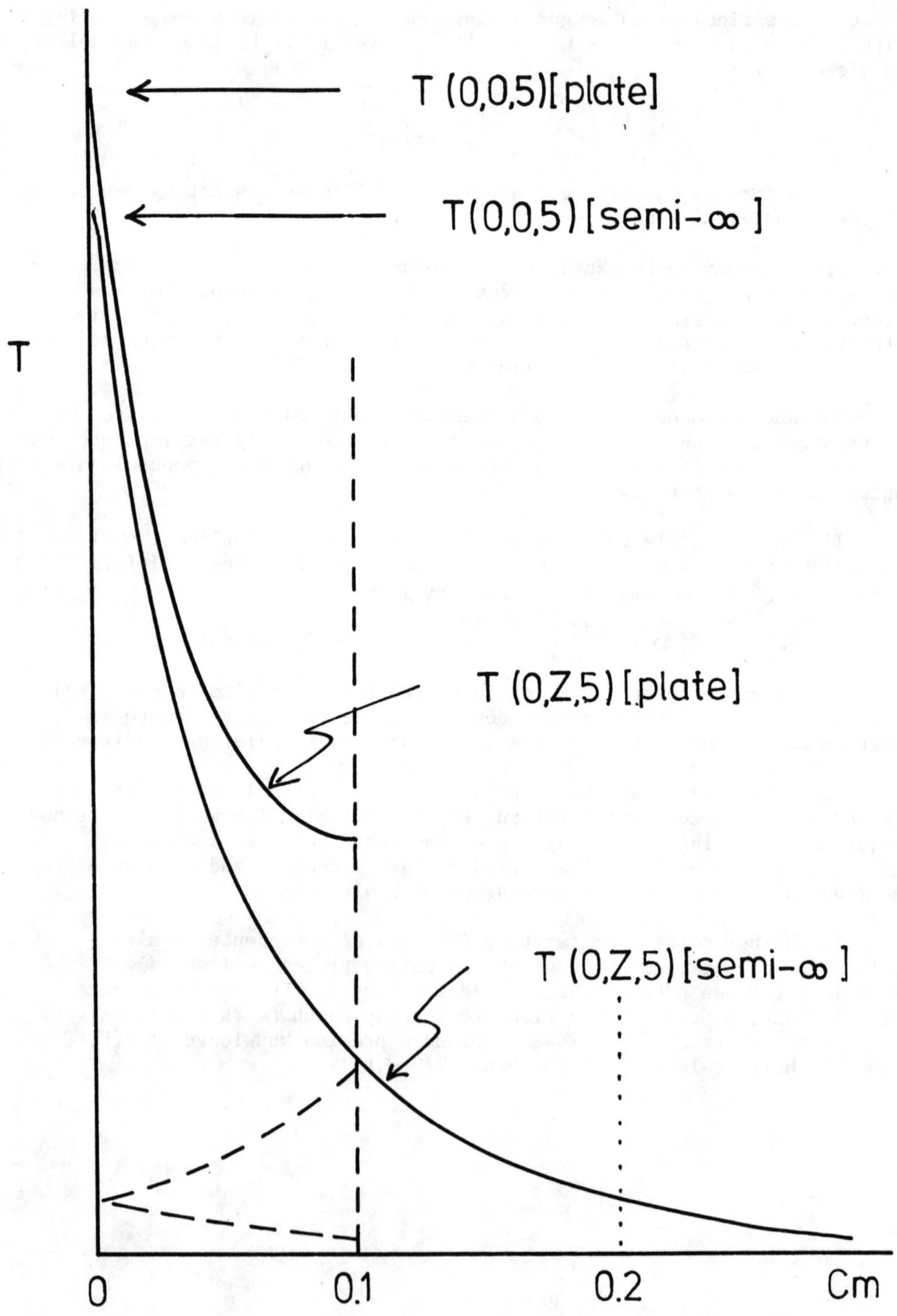

Figure 4. Multiple folding technique.

Figure 5. Evolution of T(0,0,t) for a Gaussian profile beam incident on glass substrate - various approximative geometries.

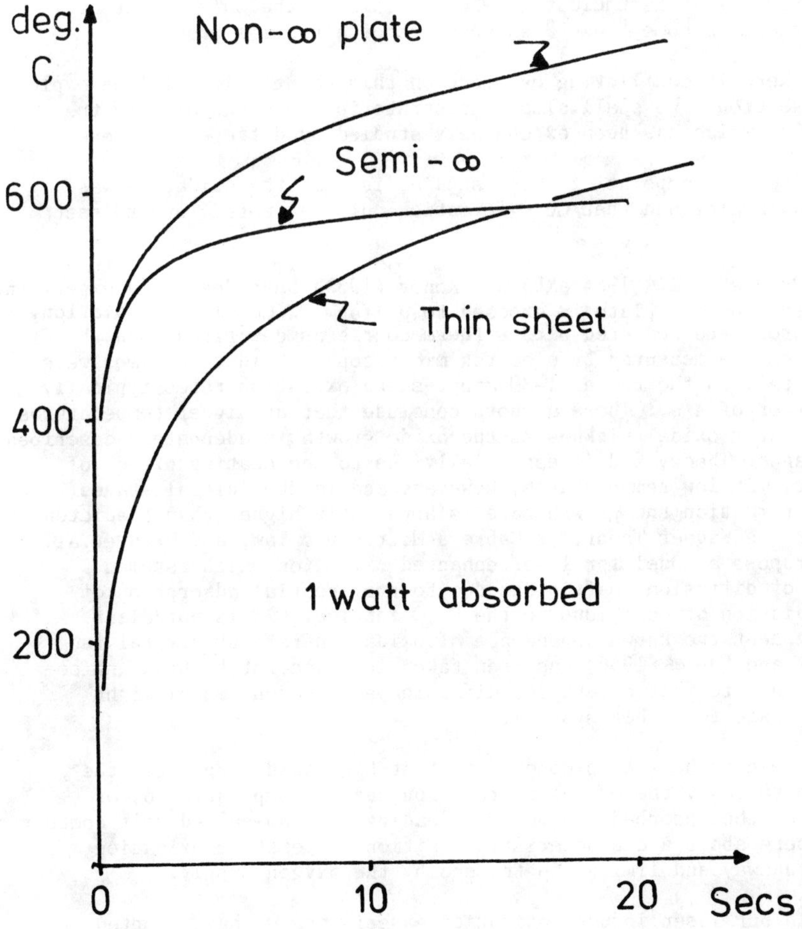

LASER INDUCED OXIDATION EXPERIMENTS

As pointed out in the preceding sections laser irradiation results in heating of the metal specimen and, at sufficiently high power densities, in oxidation.

It is generally accepted that oxidation will not occur without some heating, what is unclear is whether this is the only effect or whether there is some "laser" enhancement of a thermal process.

There is conflicting evidence on this point which will be explored in this section. We shall also concentrate in the discussion on the case of Cu, which has been extensively studied by different workers.

The common point of these studies is that the oxidation was performed in air, and that Cu_2O formation only was noted, or at least studied.

Ursu et. al. (1984 a&b) and Konov (1982) have described experiments where a polished Cu plate is exposed to uniform CW CO_2 laser radiation. Temperatures were measured with a thermocouple as described, and the oxide thickness measured by electron microscopy. Oxidation times were approximately in the range 1-30 minutes and oxidation rates typically of the order of Å/s. These authors conclude that at higher temperatures and/or greater oxide thicknesses the oxide growth is adequately described by the Wagner Theory and is essentially due to the heating effect of the laser. At low temperatures, however, and in the initial stages of oxide formation the growth rate is apparently higher than predicted by either the Wagner Theory or Cabrera-Mott cubic law, and Ursu et.al. (1986) propose a model for laser enhanced oxidation which assumes a lowering of diffusion coefficient due to preferential adsorption of 10.6μ radiation at Cu^{++} ions in the Cu_2O lattice. It is not clear to what extent the known dependence of oxidation rate or crystal face (Gwathmey and Lawless 1956) has been taken into account by Ursu and co-workers - nor to what extent the diffusion enhancement model might be appropriate for other systems.

These authors also observe that at high oxidation rates the power liberated by the oxidation reaction can be comparable to, or even exceed the absorbed CO_2 power - leading to a so-called self combustion regime where above a characteristic ignition temperature oxidation becomes runaway and limited in the end by the oxygen supply.

In our laser induced oxidation experiments we have adopted the thin film insulating substrate geometry, using the twin probe thickness monitoring system previously described and temperature calculation based on the finite plate approach. The accuracy of this calculation procedure could be checked by irradiating metal films, Cd and Sn, under inert atmospheres and correlating the jump in reflectivity occuring at the melting point with the known value of this latter.

Another important difference between our work and that of the Bucharest and Moscow groups is that we in general study oxidation of the entire thickness of the film in seconds, with oxidation rates typically 0.1 - 1μ/s.

The exposure of 2500 Å thick Cu films on glass substrates to the primary Ar^+ beam at incident power exceeding ±1W results in the formation of Cu_2O. For thick films and/or high laser powers further oxidation to CuO occurs, apparently upon completion of the Cu_2O layer, and in the form of irregular shaped islands on the film surface, often associated with buckling and partial detachment of the oxide layers from the substrate. Such formation is poorly suited to a probe experiment and here we limit the discussion to the early stages of Cu_2O formation.

The HeNe and Ar^+ probe signals from a 2500 Å thick film are shown in **figure 6** . The form of these curves remains essentially the same for incident laser powers between 1 and 14W and in this power range we calculate the oxidation rate at the start of the first decrease in HeNe reflectivity and the calculated temperature at this point ignoring for the moment the effects of reflectivity charges at 481/514 nm.

These temperatures can thus be expected to be systematically low though probably not by too much since we are dealing with the early stages of oxidation (Point A on **figure 6**)i.e. before the drop in Ar^+ reflectivity has had time to establish a large effect. Note that, particularly at high laser power, the temperatures are well above the softening point of the glass substrate but that this temperature is maintained only for a short time ≃ 20 ms at 14W and is limited to a very shallow surface region of the substrate. Thus the substrate does not deform, though it sometimes cracks.

Note also that the overall temperature fluctuations introduced by the complex changes in R(r,t,P) at 481/514 nm and subsequent thermal diffusion in the substrate cannot be too dominant since the general shape of the HeNe probe signal in **figure 6** indicates a smooth progression of the central oxidation reaction, apparently following an approximately linear growth law and exhibiting only slight asymmetry of the peaks due to these effects. This point is dramatically illustrated by **figure 7** . Here the Cu film is deposited on a 140μ thick glass substrate mounted in intimate thermal contact with a massive Cu block serving as a heat sink.

In this geometry the essentially one-dimensional nature of the heat diffusion equation and the small dimension, which allow equilibrium to be reached rapidly, mean that changes in surface reflectivity are quickly and effectively translated into fluctuations in surface temperature.

This results in large cyclic variations in the film temperature which are reproduced in the oxide growth rate. This configuration is perhaps poorly suited to kinetic studies but it already provides an important indication of the essentially thermal nature of the oxidation process. It was pointed out in the introduction that one of the problems associated with the study of laser induced reactions is that few other heating techniques, allow a comparable rate of energy coupling into the film, particularly whilst at the same time permitting access to an oxygen atmosphere, so that comparative data can be difficult to obtain.

To approach this problem a series of specimens were prepared as in **figure 8**. The 2500 Å thick Cu film was deposited via a mask

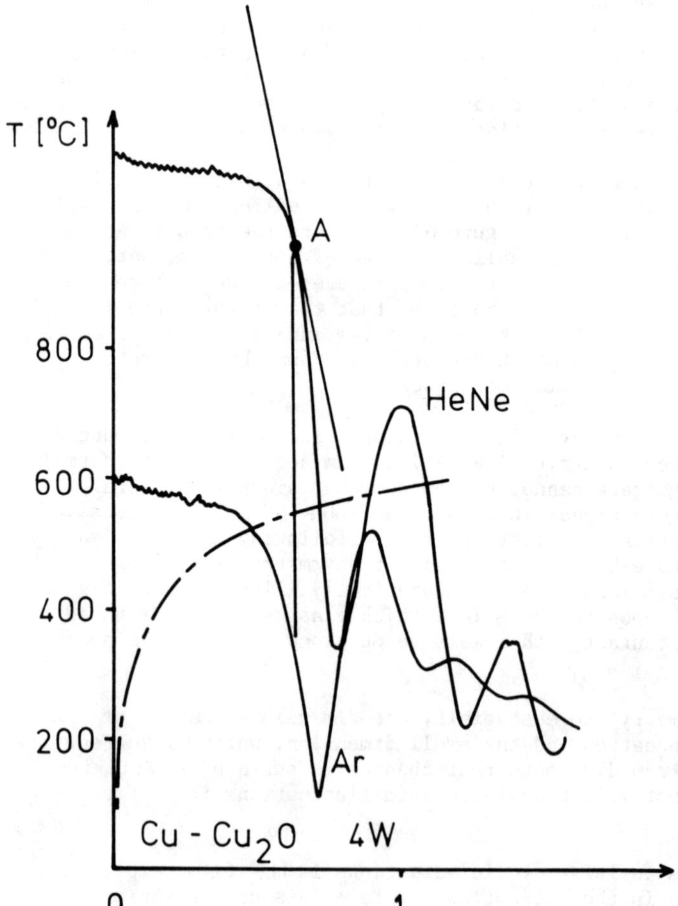

Figure 6. Two probe signals during oxidation of Cu films. Determination of the oxidation rate at A.

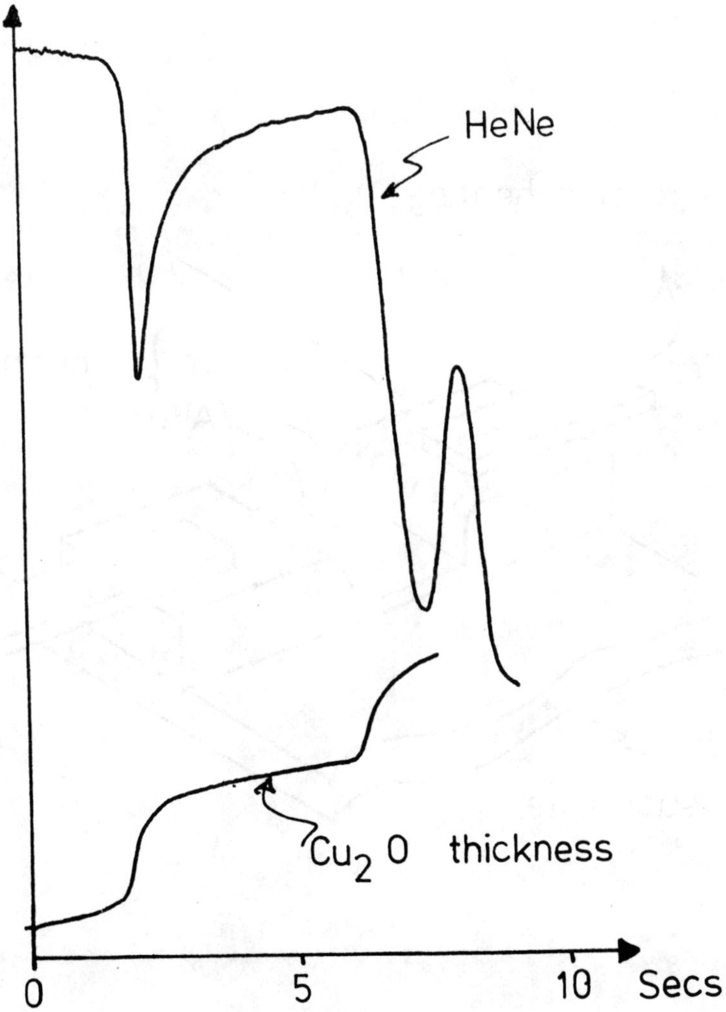

Figure 7

Oxidation of a Cu film with an efficient heat sink, showing the feedback effect.

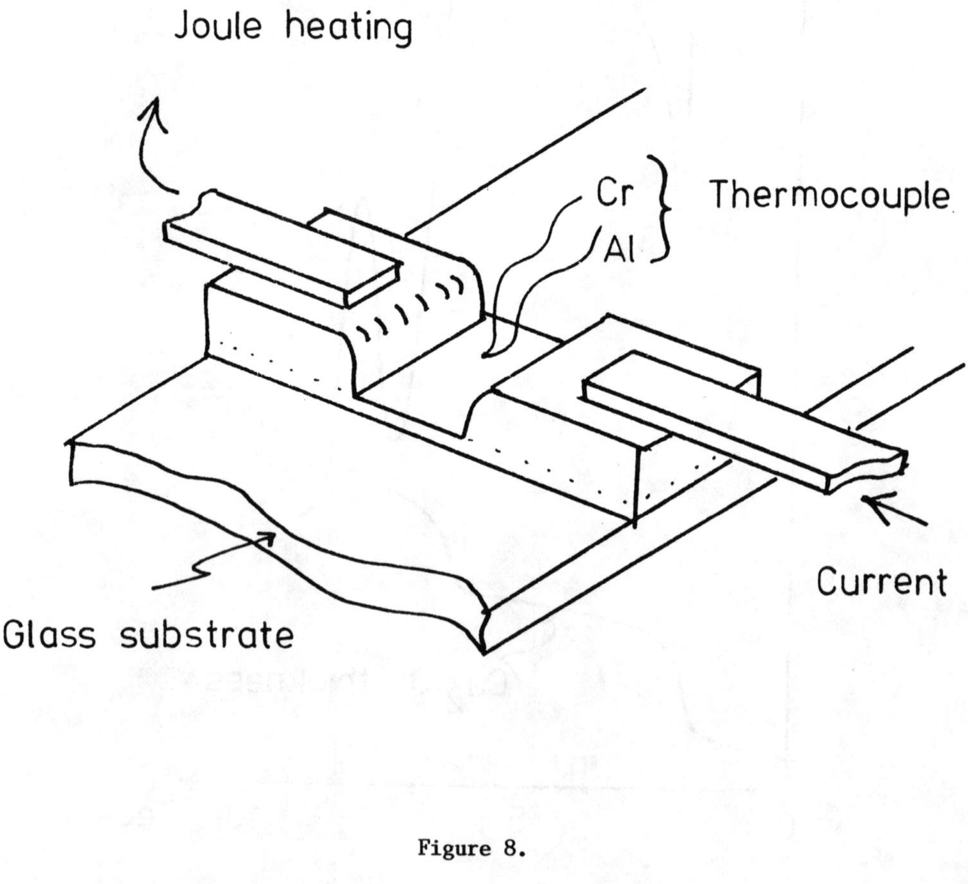

Figure 8.

Specimen for oxidation of Cu by Joule heating.

to form 5mm wide strips and ±2μ thick Cu pads deposited on either side of a 3mm x 5mm central zone. An electrical current is passed through this zone via mechanical contacts to these pads. Voltage and current were independently monitored and a system devised for holding the dissipated power constant, at least until complete oxidation of the central zone abruptly increases the overall resistance by several orders of magnitude.

The HeNe probe signal from the center of the zone is shown in **figure 9**, together with the temperature signal derived from a 25μϕ Cr/Al thermocouple in mechanical contact with the film. The measured temperature is in reasonable agreement with that calculated by assuming constant power dissipation over the area of the sample although such calculation is inexact after the start of oxidation since this occurs preferentially at the zone centre and hence presents a non-uniform resistivity of the sample. **Figure 9** shows that oxidation occurs following an approximately linear growth law and at rates and temperatures entirely compatible with Ar^+ laser induced oxidation, and thus, clearly, with a thermal model for the latter.

This view is further supported by the rather simple experiment of irradiating the sample from the rear side. Although the above mentioned interference effects in the substrate here result in non-uniform heating it is nevertheless clear that, after taking due account of the losses due to reflection by and absorption in the substrate, oxidation proceeds at a comparable rate in this configuration and therefore, since virtually no light reaches the active oxide layer, must be primarily due to heating of the thin film.

There remains only the question of whether an essentially thermal effect might be enhanced by the absorption of laser light, or indeed any high intensity light, since the precision of the temperature measurements and calculations do not allow us to exclude such a possibility. To look for possible effects, we heated a 1mm thick Cu sheet from the rear side using a hot air point stripper to ±500°C, whilst at the same time exposing the front side to the unfocussed Ar^+ beam of 10W incident power, sufficient to give additional calculated temperature rise of ±8°C.

The hot air blower was positioned asymmetrically so that after 5-10 seconds broad interference bands were produced in a wedge shaped Cu_2O layer on the copper sheet.

Close visual inspection of the specimen after this treatment revealed no trace whatsoever of the point of laser impact.

We turn finally to a closer examination of the data derived for Cu oxidation by either method, since the experiments strongly support the view that they result from the same oxidation mechanism and are therefore compatible. Using the data in tables 2 and 3 **figure 10** shows an Arrhenius plot of Cu_2O growth rate from which we derive an activation energy of ±0.65 eV. We have already noted that the growth law appears to be linear.

In **figure 9,** we see that the interval between the second and third reflectivity minima is in fact slightly less than that between

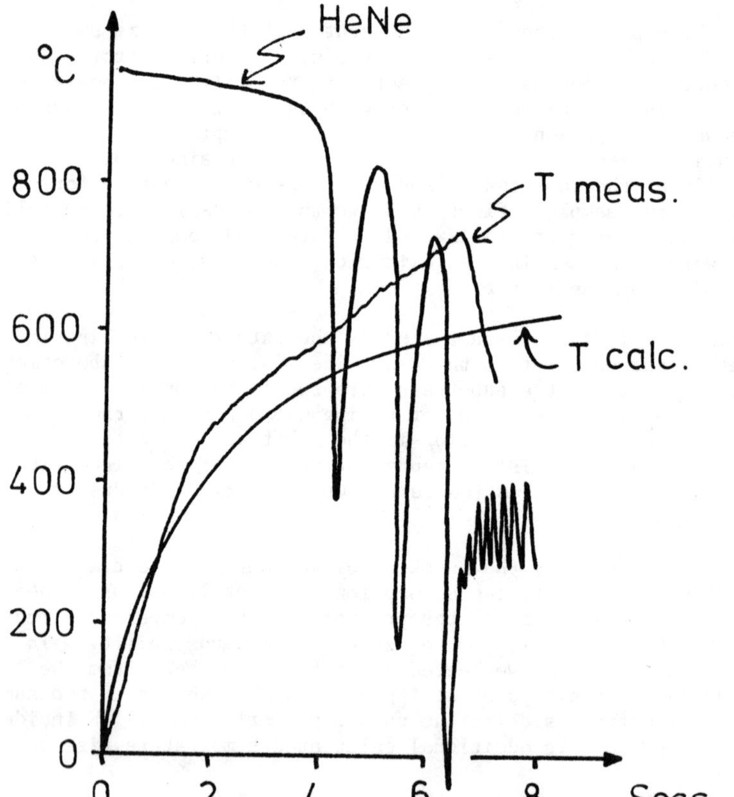

Figure 9.

Oxidation of Cu film by Joule heating with measured and calculated temperatures.

the first and second, which would imply an acceleration of growth rate at constant temperature. Using, however, the predicted temperature increase in this interval and the value of activation energy derived above we find that a linear growth law does precisely fit the data represented by fig. 9.

SUMMING UP

Our experiments described in the preceding section show that CW Ar^+ laser induced oxidation of Cu films, under our experimental conditions, is due to the heating effect of the laser. No other effect is observed. On the other hand the experiments of Ursu and co-workers using CW CO_2 irradiation of small polished Cu plates under carefully controlled conditions, show oxide growth rates, particularly in the early stages of oxidation, which are incompatible with thermal data.

Since it would indeed be curious if a photonic, rather than thermal effect was present at 10.6 μ and not at 481/514 nm we must try to reconcile these apparent discrepancies in the data.

Classical studies of Cu oxidation in oxygen atmosphere show that at temperatures and O_2 pressures in the range studied here, oxidation normally proceeds according to a parabolic growth law with the rate limiting step being the diffusion of Cu ions and electrons in the Cu_2O layer (12).

The activation energy for diffusion following the Wagner theory is given at ± 1.65 eV and the rate constant at 850°C for e.g. is ± 8.4 10^{-9} cm^2/s.

For a 400 Å thick oxide layer, i.e. about the thickness to which the data of figure 6 refers, the growth rate should thus be ≃20μ/s. The fact that we observe experimentally a rather lower growth rate and linear growth law indicates that at this combination of thickness and temperature the rate limiting step is a phase boundary reaction - probably that at the Cu_2O/O_2 phase boundary (12) occuring with an activation energy 0.65eV derived in **figure 10.**

It is worth noting that this regime of oxide growth is practically inaccessible other than by techniques such as described since it is only maintained for at most a few seconds at these temperatures.

Figure 11 depicts the situation at 850°C where the growth of the oxide in our thickness range is reaction limited. Already at 600°C the situation is different **figure 12** and for the conditions studied by Ursu, Konov and co-workers oxidation takes place entirely in the diffusion limited regime to which their recent model also applies.

Thus, the various experiments described are essentially compatible and reveal the complexity in the answer to the question "Is laser induced oxidation thermal or not"?

What is certain is that, despite complications introduced by the feedback effects both positive and negative, discussed previously, laser oxidation studies can give access to previously largely unexplored growth regimes, and offer a way to study the kinetics of interfacial reactions independently from the diffusion limited case which occupies the bulk of classical oxidation studies.

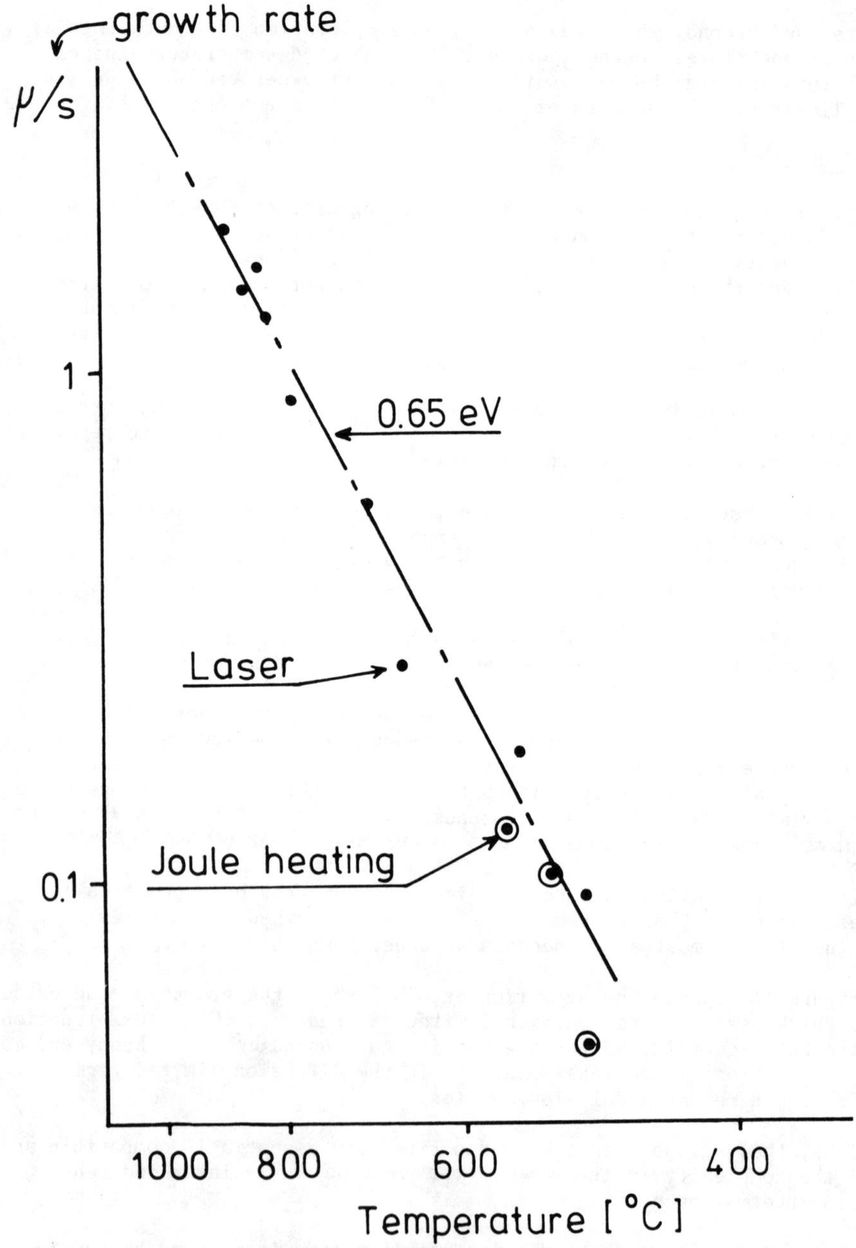

Figure 10.

Arrhenius plot of oxidation rate of Cu film vs. temperature.

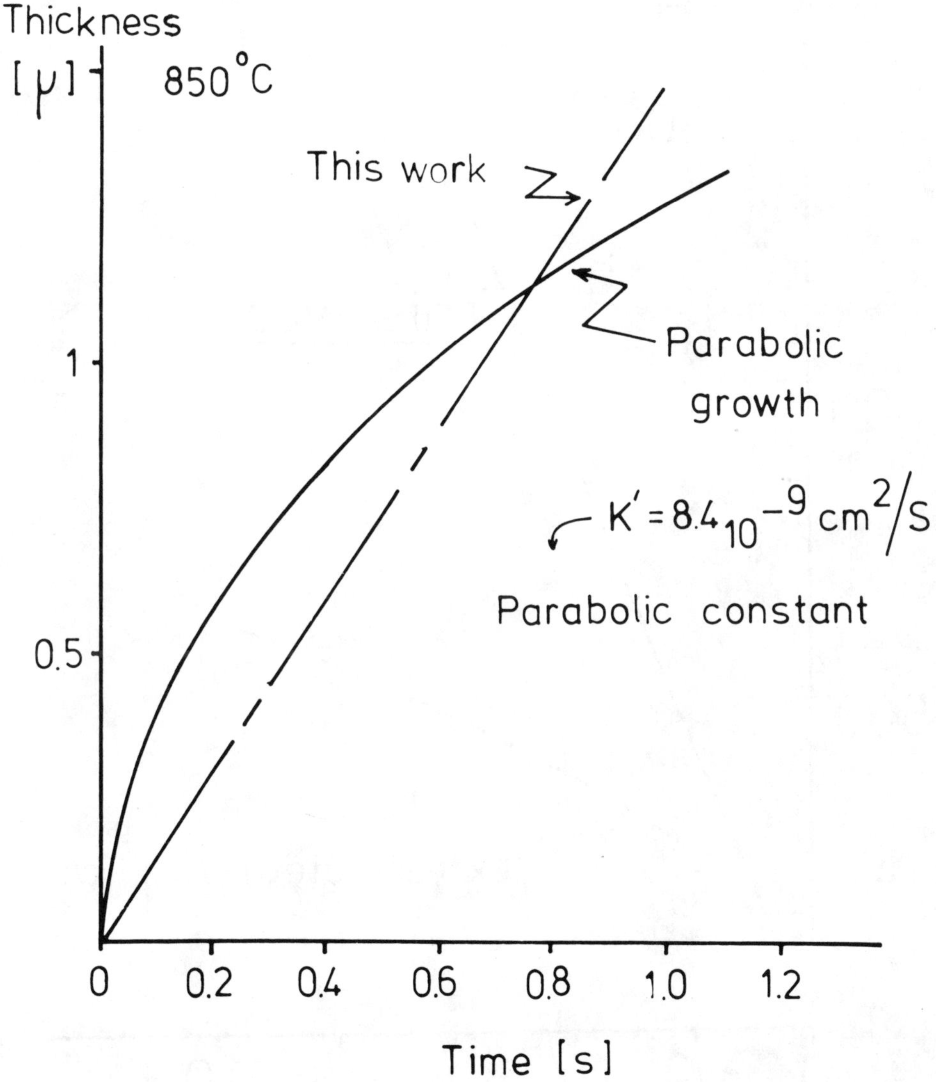

Figure 11

Comparison of the present work with diffusion limited oxide growth according to the Wagner Theory at 850°C.

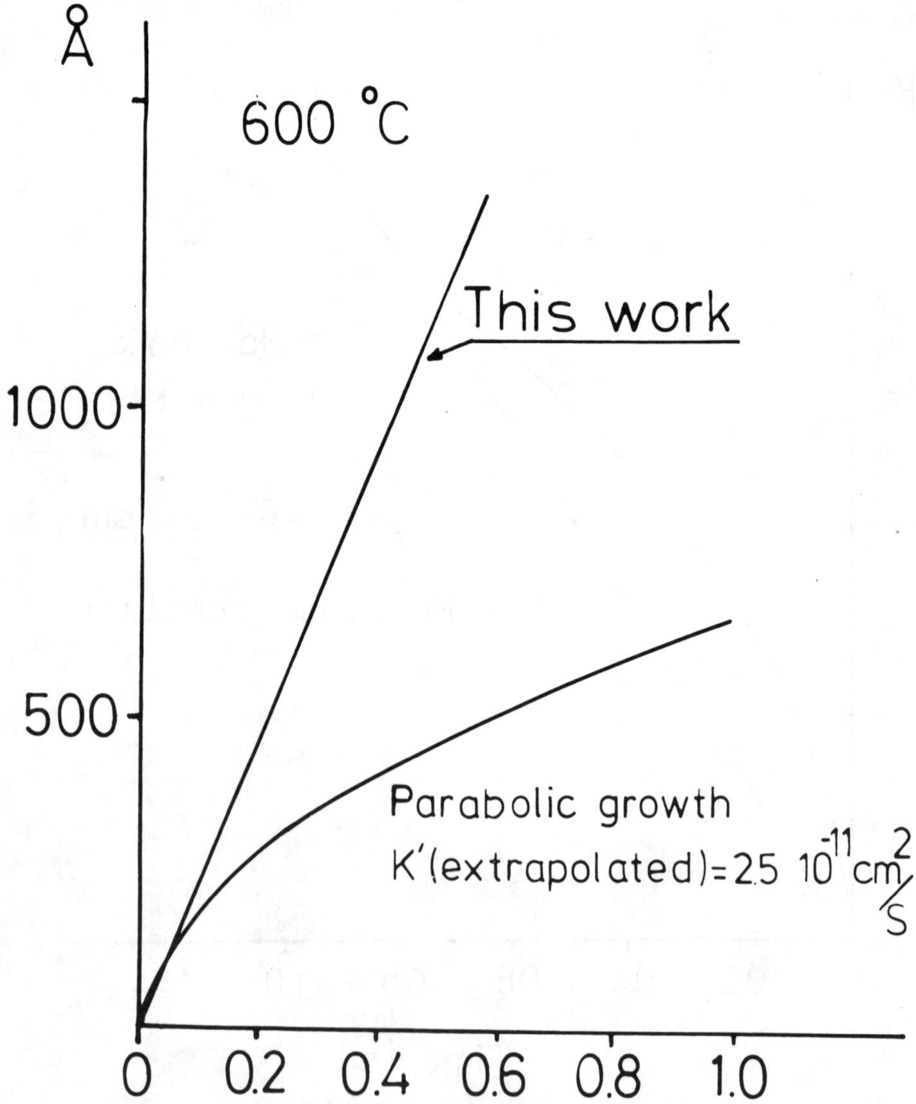

Figure 12

Comparison of the present work with diffusion limited oxide growth according to the Wagner Theory at 600°C.

REFERENCES

1) Andrew, R., 1986 Proc. of MRS Europe Strasbourg

2) Andrew, R., Baufay, L., Pigeolet, A., and Laude, L.D., 1982 J. App. Phys. 53 4862.

3) Auvert, G., Bensahel, D., Perio, A., Nguyen, V.T., and Rozgonyi, G.A., 1981 App. Phys. Lett. 39 724

4) Bertness, K.A., Petro, W.G., Silberman, J.A., Freidman,D.J., Spicer, W.E., 1985 J.Vac. Sci. Technol. A,3, 1464

5) Birjega, M.I., Nanu, L., Mihailescu, I.N., Dinescu, M., Popescu - Pogrion,N., and Sâbru, C., 1986 a Optica Acta 33

6) Bijerga, M.I., Dinescu,M., Mihailescu, I.N., Nanu, L., Constantin,C.A., Florescu, I.Th., Popescu-Pogrion, M., Sârbu,C., 1986 b Phys. Stat. Sol. (a) 95 751.

7) Boyd, I.W., 1985. Laser-assisted Modification and Synthesis of Materials ed. S. Metev. (Sofia : Sofia University Press) p. 105

8) Bunkin,F.U., Kirichenko, N.A., and Luk'Yanchuk, B.S., 1982 Sov. Phys. Usp. 25, 662.

9) Fan, J.J.C., Zeiger, J.H., Gale, R.P., and Chapman, R.L., 1980 Phys. Rev. Lett., 36 158

10) Fiori, C., and Devine, R.A.B., 1985. App. Phys. Lett. 47 361.

11) Gwathmey, A.T., and Lawless, K.R., 1956 "The Surface Chemistry of Metals and Semiconductors" ed. H.C. Gatos. John Wiley N.Y.

12) Konov, N.I., 1982 Lasers and Applications ed. I. Ursu and A.M. Prokhorov Bucharest CIP Press p. 665.

13) Laude, L.D., Wautelet, M., and Andrew, R., 1986 Appl. Phys. (in press)

14) Lichtersteiger,M., and Webb, C., 1985, Surface Sci.,154 455.

15) Luk'Yanchuk, B.S., 1982, Lasers and Applications, ed. I. Ursu and A.M. Prokhorov Bucharest CIP Press, p. 771.

16) Schafer, S.A., Lyon, S.A., 1982 J. Vac. Sci. Technol. 21, 422.

17) Ursu, I., Nanu,L., Mihailescu, I.N., 1986. App. Phys. Lett. 46.

18) Ursu, I., Nanu, L., Mihailescu, I.N., Nistor, L.C., Teodorescu,V.S., Prokhorov, A.M. , Konov, V.I., and Chapliev, N.I., 1984 b J. Phys. Lett. 45 L737.

19) Ursu, I., Nistor, L.C., Mihailescu, I.N., Teodoroscu, V.S., Nanu, L., Prokohrov, A.M., Konov, U.I., Chaplieu, N.I., 1986 Review Article.

20) Young, E.M., and Tiller, W.A., 1983, App. Phys. Lett., 42, 63.

OPTICALLY ENHANCED OXIDATION

IAN W BOYD

ELECTRONIC AND ELECTRICAL ENGINEERING
UNIVERSITY COLLEGE LONDON
TORRINGTON PLACE
LONDON WC1E 7JE

ABSTRACT

The use of intense optical radiation to stimulate the oxidation of silicon is described. By summarising all of the data published so far in the field, it is found that there must be several different mechanisms responsible for the observed effects on the oxidation process. It is emphasised that thermal effects are a dominant factor in the reactions, but that so-called photonic enhancement are evidently present when both visible and ultraviolet radiation are used. Possibile mechanisms for the enhancement modes of the reaction are proposed.

1. INTRODUCTION

The bandgap electronic states normally arising at a clean surface of crystalline silicon as a result of free radicals, or dangling bonds, can be dramatically reduced by more than five orders of magnitude under ideal conditions, by the growth of a thin oxide layer on the surface. This layer has contributed significantly to the development of the planar process and the MOS transistor as well as the high level of performance of many present day advanced microelectronic devices. The natural oxide of silicon, SiO_2, is an excellent insulator, can withstand very large electrical fields, and is extremely stable chemically.

The oxidation process itself has been the subject of many intensive investigations for several decades, and yet the simple thermal reaction of solid silicon and dry oxygen gas

$$Si + O_2 \longrightarrow SiO_2 \quad (1)$$

is not yet fully understood. Whilst many models can successfully predict oxidation rates of the silicon under certain conditions, none presently exist that can satisfactorily explain the observed behaviour of the reaction under a wider set of experimental environments.

Technologically, the reaction has traditionally been carried out in rf heated quartz tube furnaces, at temperatures between 700C and about 1200C. In recent years, the use of intense and directed optical radiation has been investigated as an alternative means of heating the silicon to the desired temperatures. There are several advantages of this. Firstly, optical radiation is more efficiently absorbed by silicon, and is now readily available and is relatively easily applied to rapid thermal processing. Heating rates can be as fast as 10^{12}K/s, using nanosecond laser radiation, even faster using pico- or femtosecond duration pulses or as modest as 500K/s using incoherent light sources. Beam technology can also give rise to localised and confined processing, presenting the opportunity of fine scale pattern production either by direct writing or projection methods. A further advantage, which will be discussed at length in this paper, is that optical technology presents a new processing condition under which to study the oxidation reaction, namely wavelength selectivity. This now means that for the first time, we can study this silicon oxidation with the possibility of initiating or altering specific reaction pathways through photon induced excitations. Consequently, with this extra processing dimension, it may now be possible to elucidate one of the most important and widely studied chemical reactions today.

2. THERMAL OXIDATION OF SILICON

Thermal oxidation of silicon is known to occur by the diffusion of the oxidising species through any pre-existing oxide layer to the Si/SiO_2 interface, where it reacts with the available silicon atoms. Several marker experiments with $^{18}O_2$ have traced the long range diffusion of this species through thick oxide layers without exchange with the already formed network, and found a 93% incorporation at the Si/SiO_2 interface, with a small amount present at the outer SiO_2/O_2 surface, which is probably due to network oxygen diffusivity from a surface layer [1]. As oxidation proceeds, more silicon is consumed as the Si/SiO_2 interface retreats further into the bulk. In fact for every unit thickness of oxide grown, 0.45 units of c-Si have been reacted.

The nature and charge state of the oxidising species has been the source of much controversy over the years. However, it presently appears from the work of Modlin and Tiller [2], that neutral species are responsible for oxidation when the oxide layer already formed is thicker than 300A. This seems to eliminate previous suggestions that charged O_2^{2-} or O_2^{-} may play a dominant role. It is not yet clear, however, whether charged species play an active role during the initial stages of the reaction, when the intervening oxide layer is thinner than 300A.

In 1965, after many independent and unrelated oxidation studies had been reported, and many different phenomenological models had been suggested, Deal and Grove (D-G) [3] presented not only results of a comprehensive study of silicon oxidation, but also a most successful model that continues to be applied today in many oxidation regimes. The model considers three consecutive physical and chemical processes relevant to the reaction, whose fluxes are equal under steady state conditions. These are, (1) the transfer of the oxidising species from the gas phase into the solid surface, (2) Fickian diffusion of the species through any oxide present to the reacting interface, and (3) reaction of the species with the silicon at the Si/SiO_2 interface. The general oxidation relationship is thus derived to be

$$x^2 + Ax = B(t + t_o) \qquad (2)$$

where x is oxide thickness, t is oxidation time, B/A is the linear rate constant, B is the parabolic rate constant, and

$$A = 2D_{eff}(1/k + 1/h) \qquad (3)$$

$$B = 2D_{eff}C^*/N \qquad (4)$$

$$t_o = (x_i^2 + Ax_i)/B \qquad (5)$$

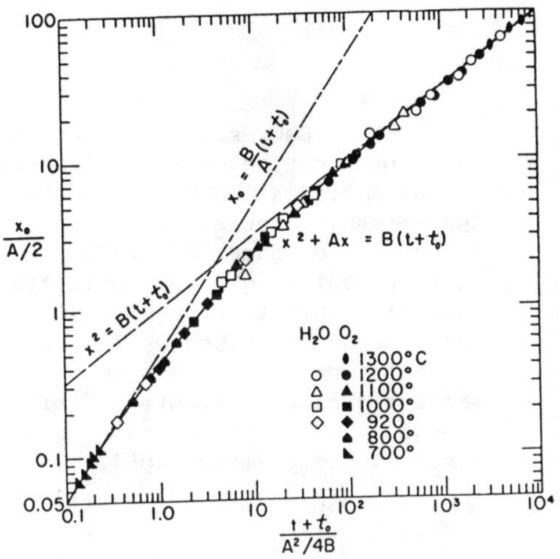

FIGURE 1. A collection of c-Si oxidation data, plotted together with the general oxidation theory of Deal and Grove [3]. This mode of plotting highlights the well known limiting forms of the D-G oxidation relationship.

and D_{eff} is the effective oxidant diffusion constant in the oxide, k and h represent rate constants at the Si/SiO_2 and O_2/SiO_2 interfaces respectively, C* is the equilibrium concentration of the oxidant in the oxide, N is the number of oxidant molecules in the oxide unit volume, and x_i is the initial oxide thickness at the onset of the reaction.

The two well known limiting forms of equation 2, are the <u>parabolic</u> oxidation regime, when $t \gg A^2/4B$, and $t \gg t_o$,

$$x^2 \approx B(t + t_o) \tag{6}$$

and the <u>linear</u> oxidation regime where $t \ll A^2/4B$,

$$x = B/A(t + t_o) \tag{7}$$

These are highlighted in the classical plot of figure 1, for oxidation of silicon in both dry oxygen and steam. Therefore at high temperatures, or the growth of thick layers, parabolic kinetics dominate, and the reaction is strongly affected by diffusion and oxidant solubility in the oxide which is known to be proportional to the ambient pressure. At lower temperatures, or for the growth of thin layers the linear rate constant B/A is important the rate is also controlled by solubility, but is also now strongly affected by the reaction rate constants at the interfaces. The activation energy for these controlling rate constants has been found to be roughly 1.23eV and 2.0eV respectively, for temperatures above about 1000C, suggesting that in this regime, a single elementary process controls the reaction. These energies correspond closely to the known diffusion of oxygen through fused silica (1.17eV), and the Si-Si bond energy of 1.83eV, although it is not clear whether the latter association in particular is merely coincidental.

3. LIMITATIONS OF THE DEAL-GROVE (D-G) MODEL

Although a significant contribution to the knowledge of silicon oxidation, the D-G model is somewhat limited for dry oxidation of c-Si. Firstly, it requires an initial oxide thickness of $x_i \approx 200A$ in the formulae above in order to successfully fit the measured reaction rates, thereby denying the application of the model in its present form to ultrathin layers. The successful use of this correction term, however,

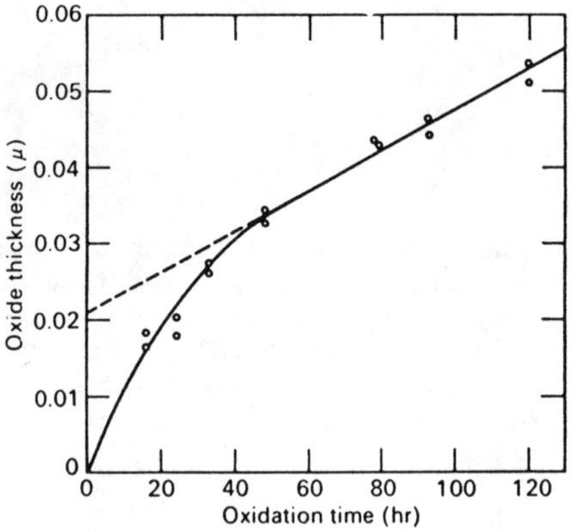

FIGURE 2. Back-extrapolation of the linear regime of the D-G model for oxidation at 700C, indicating the necessity for an initial oxide thickness of some 200A [3]. The data reveal a gradual oxidation rate for this thin regime, which can also be described by a linear-parabolic relationship with quite different rate constants. Similar trends are found at all other temperatures examined [66].

led to misleading statements referring to an "abnormally rapid oxidation regime" for the early stages of growth. Since the oxides used commercially in semiconductor devices at that time were typically 1000A thick, it was convenient to assume this essentially instantaneous oxide growth, and the effect was not extensively studied. Today however, and in the near future, it is clear that oxides of 200A and less are required to enable the drive towards ever-diminishing geometries to continue, and it is becoming more crucial to understand the optimum criteria for producing high quality thin oxide layers.

If one actually studies the growth with time of an oxide layer on silicon (see figure 2) it can be seen that the behaviour of the reaction around and below 200A can be empirically fitted by another linear parabolic curve with different constants to that used by the D-G model. Again, a very small initial oxide is present, this time of the order of 20A, but this is well known as the ever-present native oxide of the silicon. The growth of such a layer with time is shown in figure 3, below.

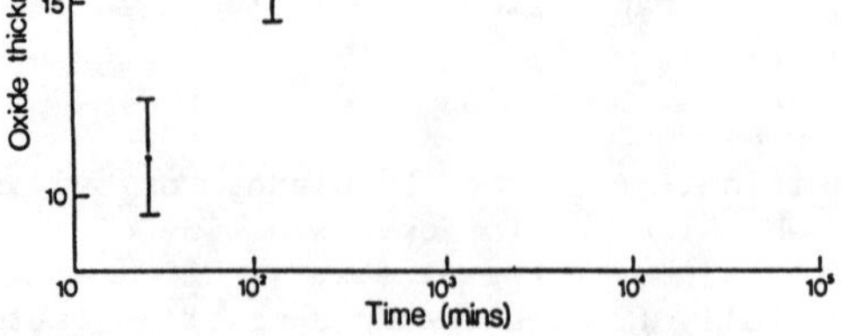

FIGURE 3. Oxide thickness on c-Si as a function of time after cleaning in 10:1 HF and a subsequent 15 min rinse in water followed by a 10 min spin dry in warm N_2 [67]. It is important to note that the growth of such layers is very dependent upon the precise surface preparation and pre-cleaning procedure followed.

Therefore, in describing the overall characteristics of the growth data, there is an initial ultrathin native oxide (20Å), a linear growth regime up to about 150Å followed by a parabolic tendency (neither of which have yet been phenomenologically explained), then the true D-G linear regime beginning around 200-250Å which leads to the D-G parabolic behaviour describing the growth of very thick films. At this time we shall not discuss the growth of the native oxide, which is a complete topic in itself, but rather concentrate upon the initial sub-200Å layer. A great many models have been proposed to attempt to explain the oxidation kinetics of this regime, and at times these have proposed significant modifications to the original D-G model, as shown in table 1. Initially, Deal and Grove proposed that space charge effects may be an important consideration during the oxidation of films up to about 150Å [3]. Other ideas based on relatively simple modifications of the original D-G model, include the incorporating changes in diffusion resulting from stress [4-6] or different diffusion mechanisms, due for example to micropores or microchannels [6,7] or the presence of additional oxidising species whose influence is affected during growth [8-13, 62-64].

TABLE 1. Summary of Silicon Oxidation Models

DESCRIPTION OF SILICON OXIDATION MODEL	REFERENCE
Diffusion of O_2 & reaction with Si + space charge effect & tunnelling/thermionic emission	Deal & Grove [3,8]
Enhanced diffusion of O_2 through micropores	Revesz & Evans [7] Irene [6]
Stress-induced diffusion effects	Doremus [5] Fargiex & Ghibaudo [4]
Creation of O_2^-/hole pair, leading to enhanced diffusion of O_2^-	Grove [8] Tiller [62]
Influence of O_2^{2-} and O^-	Lora-Tamayo et al [63]
Influence of O^-	Hu [64]
Fixed charge in oxide assists diffusion charged species O_2^-, and perhaps O^-	Lu & Cheng [10]
Fixed charge in oxide reduces available Si bonds by reducing holes with band-bending	Schafer & Lyon [11]
Both O_2 and 2-O react with silicon	Ghez & van der Meulen [12]
Equilibrium between O_2 and 2-O, only O reacts with Si for thin oxides	Blanc [13]
Two-step oxidation: Si -> α-SiO_2 + Si(I) & Si(I) -> SiO_2	Tiller [65]

The influence of charged species has been continually suggested over the years. Grove [8] suggested field effects arising from the creation of O_2^- and a hole could assist the initial reaction, since the hole could diffuse faster and then pull the slower O ion through the layer. Other charged species, such as $O_2^{=}$ [62,63] and O^- [63,64] have been considered. Alternatively, Ghez and van der Meulen [12] and Blanc [13] suggest that an equilibrium dissociation of O_2 and 2-O may exist at the reacting interface and introduce the required growth characteristics. There is evidence that the activation energy of the reaction may not be described by a single Arrhenius process after all, but may in fact be a combination of several parallel or serial mechanisms [14-16]. Recent low temperature studies indicate that although the previous activation energies appear to be valid for temperatures above 1000C, below this B appears to increase, while B/A decreases. Furthermore the pressure dependence of B/A is not linear [16,17,18] and varies with temperature and crystal orientation [19], again indicating the possibility of a more complicated multistep reaction.

It appears that at least two important factors must be included in any further analysis. The first involves the contribution of compressive stress built into the growing oxide layer at the interface with the Si during the transformation of the regular Si structure into the amorphous oxide layer. It must be remembered that each unit of SiO_2 occupies around twice the volume of the reacting silicon. Viscous flow above 970C may in fact relax this stress [20]. Upon cooling, further strain could develop because of the mismatch of the thermal coefficients of the Si and its oxide, but this should not affect any previous single stage oxidation. Any structural changes, as well as the inbuilt strain could influence not only the transport of the oxidant through the solid, but also indirectly the actual mechanism of the reaction itself.

It is also known that different types of surface pre-cleaning treatments can dramatically alter the subsequent reaction rates as well as the physical properties of the grown layers. The residual chemical species can clearly accelerate of retard the interface reaction and therefore play a dominant role during the early stages of the reaction, as well as severely affect, less directly, the later stages of the reaction. In particular, the presence of Cl or F, or a heavily doped substrate can have very significnat effect on the reaction.

It is clear that the presence of many oxidation models is a consequence of a severe lack of reliable experimental data. The required experiments are by no means straightforward to perform, and the field must wait until the measurements are performed before a new oxidation model is unanimously acclaimed. This unfortunately does not assist matters when attempting to interpret new observations resulting from laser induced oxidation of silicon. Nevertheless, the new data do present new angles to the problem, and extra parameters to aid the interpretation of the underlying basic mechanism of silicon oxidation.

3. OPTICALLY INDUCED OXIDATION OF SILICON

Intense optical beams have enabled completely unique oxidation situations to be used to study the growth of thin oxide layers in recent years. These include gas-phase deposition (photochemical and pyrolytic), evaporative growth, photo-oxidation of SiO in O_2, oxide incorporation through O implantation and annealing, or by annealing (in the solid or liquid phases) or heating of the crystal in oxygen-rich environments. These methods have been reviewed elsewhere [21, 22]. Here we will discuss the many different investigations involving the application of pulsed and continuous optical sources to induce or enhance the conventional thermal oxidation process. Several regimes will be examined, namely oxidation by (3.1) Rapid Thermal Oxidation, (3.2) Liquid Phase Oxidation, and (3.3) Photonically Enhanced Oxidation. The latter section introduces the novel concept of reaction enhancement by direct stimulation of the surface during the oxidation by impinging photons.

3.1 Rapid Thermal Oxidation

Thermal oxidation of c-Si is traditionally performed by heating the sample in an RF heated furnace usually requiring around 10 min heating-up time and 10-25 min ramping down time either side of the oxidation itself. Oxidation normally occurs at a rate of 0.3-3 A/s in production, depending upon temperature, oxygen purity, substrate orientation and doping etc. However, the intense 10 micron radiation from CO_2 lasers can be similarly absorbed by the silicon to induce direct but controlled rapid heating. This was first used in 1981 to initiate directly, the growth of thin oxide layers, in the regime of Rapid Thermal Processing (RTP).

Although 10 micron radiation is not strongly absorbed by c-Si, preheating the samples to 400C increases absorption by several orders of magnitude. The silicon can then be readily heated by the laser to very high temperatures such that it can be seen to glow red around 700C,and progressively orange and yellow at higher temperatures, and then brilliant yellow when melting is achieved. Details of this oxidation technique are published elsewhere [23, 24]. Since the reaction mechanism in this case is principally thermal, no modifications to the oxidation process are expected. Because of the Gaussian beam-shape applied, the temperature rise across the surface is radially dependent, so this technique lends itself to the study of the physical and chemical properties of oxides grown at many different temperatures simultaneously but under otherwise precisely identical conditions. The structural and electrical properties of these layers are also described elsewhere [24-26].

Rapid thermal oxidation by visible radiation is also possible, and indeed ruby and argon lasers, and intense incoherent lamp sources have recently been used to this effect [33-37]. One of the earliest studies in the field involved the use of 30 nanosecond pulses up to 1.3J/cm^2 to heat in short periods of time, and apparently not melt, c-Si in well controlled environments. In this way gradual oxide coverage on freshly cleaned c-Si surfaces could be monitored [27]. In fact, the investigators found after approximately 27 pulses, a saturated thickness of 2 monolayers (i.e.6.5A) beyond which oxidation barely progressed. It is remarkable that growth rates in this regime are some 6mm/s. Calculations showed that the number of atoms striking the surface during irradiation could not account for the total oxide growth, and since any contribution from interstitial oxygen in the bulk Si was effectively ruled out, adsorption prior to irradiation was taken to be the controlling mechanism. The growth limitation after the rapid superficial oxide coverage was thought to be a consequence of the heating times during which the oxidising species could not diffuse through the few monolayers of oxide to the reacting interface.

Whilst acknowledging the rapid oxidation induced by this technique, we wish to address the validity of this proposal for limited oxidation by considering the diffusion behaviour closely, once 6.5A of oxide has been formed. Contrary to the estimates of Cros et al, it now seems to be well established that c-Si can be melted by 30ns pulses around 0.8J/cm^2 [68,69] from the ruby laser, so the reaction will be between oxygen and liquid Si. The 1.3J/cm^2 pulses probably induce peak temperatures closer to the SiO_2

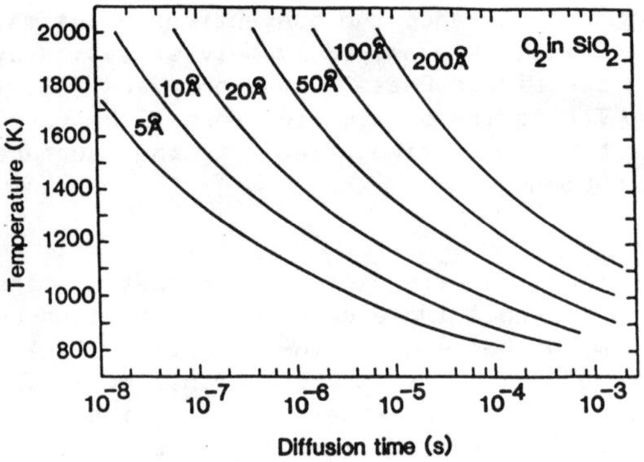

FIGURE 4. Time required by O_2 molecules to diffuse through various thicknesses of silica at a constant temperature. The curves were plotted using $t = L^2/5.4\times10^{-4}\exp(-1.16q/kT)$ where 1.16 is the activation energy and 2.7×10^{-4} the frequency factor in the usual Arrhenius relationship for diffusion [71].

melting point around 2000K, and a conservative estimate of the average temperature induced during this heating period would be closer to the c-Si melting point of 1685K. Figure 4 shows the calculated diffusion times required for molecular oxygen through silica at different temperatures for a range of oxide thicknesses. Estimating a heating time (conservatively) of 100ns, it can be seen that 10 or 100 of such pulses would significantly increase diffusion through a 6.5A thick layer. Furthermore, if the induced temperatures were higher, and the oxide actually melted, diffusion would be even faster. For the low pressures used here (5×10^{-4} Torr), it seems that surface adsorption rather than diffusion will have more control over the reaction in the early stages of the laser induced reaction. A study involving an identical number of irradiation events over different periods of time will elucidate this point. Additionally, in high vacuum, laser radiation can desorbe impurities from the c-Si surface [70]. Since there are no comparable oxidation rates for thermal oxidation in the initial few monolayers, it is not clear if the radiation itself introduces additional mechanisms other than thermal processes which may enhance the reaction in some way. These points will be discussed later in the paper.

The use of intense incoherent lamp sources has already been labelled Rapid Thermal Oxidation by the investigators. However, very recent analyses of the oxidation rates achieved by these methods, have suggested that the reaction may actually be photonically enhanced in some way, similar to that previously reported for oxidation of c-Si by visible and UV laser radiation [23, 28-32], as well as other semiconducting materials. Enhancement effects are discussed in the final section of this paper. The emergence of processing equipment incorporating intense lamp sources in the past few years, primarily for work in the annealing and alloying areas, has opened up new possibilities for rapid thermal processing in general. This is nothing more than a simple extension of the earlier work using focused laser beams to achieve rapid heating, although for large area processing it yields clear advantages. Nulman et al [33] first reported this lamp oxidation process, and others have since studied the technique. Very few groups, however, have studied the growth rates of the layers in detail, but Ponpon et al [36] have recently found a possible enhancement in their measured oxidation rate, while Chan Tung et al [37] have report a rapid initial oxidation rate of up to 11A/s during the first

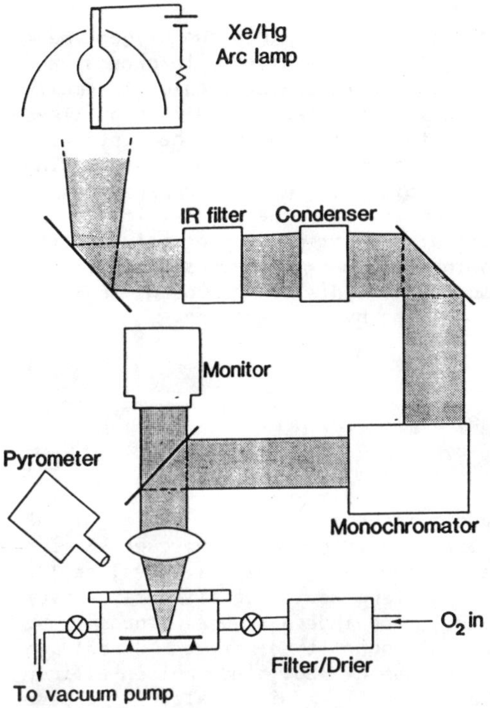

FIGURE 5. Schematic diagram of the experimental configuration used to study the growth of thin oxide layers by rapid lamp processing. The monochromator enables various wavelength dependent investigations to be performed. As with commercial systems, the sample is thermally isolated, and pyrometric methods are used to monitor the temperature

5 seconds of the reaction and just over 3A/s during the next 5-60s at 1250C. Activation energies of 0.9eV [37] and 1.2eV [33] have been extracted from the growth data for the rapid oxidation regime, while 1.4eV has been calculated for the slower post-5s oxidation [37]. These observations do not agree with extrapolations of the available data for conventional oxidation. For comparison, Massoud et al [16] find a 3.2eV activation energy for their conventional thermal oxidation at temperatures above 1000C. Clearly further work is required to establish whether this RTP procedure is purely thermal in nature. Such work is currently being performed at UCL, with regard to the wavelength dependence on the reaction. Figure 5 shows our experimental arrangement used to study the wavelength dependence of the oxidation of silicon by lamp heating [38]. The criteria for RTO are widely similar. In all cases the sample is thermally isolated such that it can be heated completely throughout during the rapid heating phase, with only a few degree differential between the front and back surfaces. It has been found that pyrometric methods of temperature monitoring are less intrusive and more direct than traditional thermocouple-based techniques. Whether or not any photonic mechanisms are present in these reactions, it is rapidly becoming clear that this "Optical Furnace" is becoming a most important piece of semiconductor processing apparatus for the future.

3.2 Liquid Phase Oxidation

As we have seen above, for oxidation to occur, gaseous oxygen must become incorporated into the oxide layer (of some arbitrary thickness) and diffuse to the oxide/Si interface to react with any available thermally excited Si atoms. Also, for thin oxide layers, where the diffusion times are negligible, the oxidation seems to be limited by the interface reaction, given that there is sufficient oxygen in the surrounding atmosphere to dissolve into the solid initially. As the temperature is increased, one would expect both the solubility and the diffusivity of the oxygen to increase, as well as the rate of production of thermally excited silicon atoms. In terms of the growth of thin oxide layers, the latter process is considered to be most important. This can be seen from by examining the linear rate constant as derived by Deal and Grove,

$$B/A = C^*/N(1/k + 1/h) \tag{8}$$

which may be reasonably approximated by

$$B/A = kC^*/N \tag{9}$$

since the gas phase transport is thought to be many orders of magnitude faster than the reaction rate at the Si/oxide interface. However, Si can be readily melted by lasers, and this introduces a completely novel regime of oxidation. The presence of a molten surface would significantly affect the usual oxidation reaction mechanisms, and indeed several groups have found this [39-46]. The rate and overall quantity of oxygen dissolving into liquid silicon would obviously increase h, but since molten silicon is much more highly reactive than c-Si, k will also increase significantly. In this case oxidation will occur extremely rapidly. However, the kinetics of the process are extremely complicated in this rapid transient situation. The silicon first of all heats up in the solid phase according to the intensity and pulsewidth of the incident radiation. It then absorbs latent heat, and melts, transforming both its optical and thermal properties. Continuing to heat up, if the radiation is still present, the temperature gradients will induce thermal convection within the liquid, and at sufficiently high temperatures the liquid will enter a further latent heating period and evaporation will occur. During this time, various species from the surrounding ambient will impinge upon the surface and dissolve or react with the atoms. Upon resolidification, the quantity of atoms dissolved, together with their precise location, and the solubility and segregation characteristics of the species as well as the velocity of the solidification front will determine the amount of oxidation. Additionally, any reactions occurring in the gas phase near to the sample surface between thermally or photo-excited species and the ambient could also be important. Therefore the precise temperature reached as well as the nature and pressure of the ambient, and the presence of additional impurities, will be very important.

Another important criterion is the presence of any surface oxide on the Si before irradiation. SiO_2 melts at 1980K, whereas c-Si melts at 1685K. If the laser beam induces temperatures between these points the silicon will melt, and the oxide will dissolve into the melt at some rate which is known to be temperature dependent. An extrapolation of the known data [59] reveals that even just below the melting point of SiO_2, the dissolution rate would be around 500A/s, and an underestimated native oxide layer of 10A would take 20ms to dissolve, which is many orders of magnitude longer than the known time duration of the molten phase. It has been reported that the presence of C in the melt (C is unavoidably incorporated into Czochralski Si at levels up to 5×10^{17}atoms/cm^3) can accelerate the dissolution of SiO_2 by a factor of 2 [60], through

$$C + SiO_2 \longrightarrow SiO + CO \tag{10}$$

In this case, the products could be lost in the gas phase. Other impurities could also have similar effects on the reaction. Neglecting these influences for the moment, it is instructive to estimate the appropriate rate constants for the reaction at a temperature just under the oxide melting temperature. From the work of Eckertova [61], the impingement rate of gas phase molecules on to a surface can be estimated by (assuming the ideal gas law)

$$M_i = 3.513 \times 10^{22} \; P/\sqrt{(MT)} \tag{11}$$

where P is the equilibrium gas pressure (in Torr) at temperature T, and M is the molecular weight of the gas. For pure oxygen at room temperature and one atmosphere this is 2.7×10^{23} s^{-1}cm^{-2}. It can be calculated that around 2.2×10^{14} oxygen atoms are present per A in SiO_2. Therefore by assuming a sticking coefficient of unity sufficient oxygen could be incorporated to grow one A of oxide every nanosecond. This of course assumes no oxygen evaporation from the melt, and a rapid diffusivity of the oxygen in the liquid silicon. We estimate the oxygen diffusivity in molten Si to be approximately 10^{-4}cm^2/s, while for diffusion in solid SiO_2 at 1975K it is calculated using [71]

$$D = 2.7 \times 10^{-4} \exp(-1.16/kT) \tag{12}$$

and found to be 3×10^{-7} cm^2/s, some 100 times faster than at 900C. Thus, oxygen can diffuse through 100A of liquid silicon in only 5ns, or 1000A in 500ns. If a 20A native oxide is present, the oxygen will take nearly 70ns to travel across into the liquid. Therefore in this regime the maximum oxidation rate will be limited by solid phase transport.

If sufficient energy is deposited by the pulsed laser, then melting of both the silicon and the oxide will occur. Since there is no longer any solid barrier to oxygen incorporation, one would expect increased dissolution and reactivity. Estimates have shown that only a small energy increase is required to achieve this for an insulating film on top of an extremely efficient bulk metallic conductor [43]. At higher powers evaporation will occur, and significant losses of material would be expected. Eventually it would be reasonable to assume that the evaporated material would shield the silicon from the tail end of the pulse. In this way, the amount of oxygen incorporated would be limited somewhat by a saturation fluence, as observed experimentally [42, 43].

Resolidification processes will also affect the mechanism of oxygen incorporation into the solid. At low levels of oxygen incorporation, the silicon will solidify either in crystalline form, or in an amorphous state, depending upon the cooling rate The oxygen can be incorporated in many forms, including interstitial sites, precipitates, or polymerised complexes sometimes involving additional impurity atoms. If the concentration of oxygen is sufficiently greater than the solubility level at a particular temperature, then we have a supersaturated condition, and precipitation will occur. The preferred location for the nucleation of these oxides, or conglomerates is not yet known. If a thin surface film is formed, this will introduce a diffusion barrier to the oxidant if a second laser pulse attempts to repeat the process. This clearly points to an important processing difference between single shot and multishot liquid phase oxidation.

Cohen et al [43] have recently argued that a large fraction (80%) of the oxygen impinging upon the Si when it is molten does not become incorporated in a grown layer. This introduces questions regarding the precise nature of incorporation and loss of oxygen from the silicon under these transient heating conditions. Westendorp et al [39] for example, have found no evidence for oxygen diffusion into silicon (with a native oxide layer) irradiated by 20ns ruby laser pulses at a fluence of 1.5J/cm^2. These observations can be explained by the diffusion arguments as outlined above, providing the 30A native oxide is known not to have melted. Hoh et al [40] have similarly found that the presence of a 905A oxide layer inhibited further oxidation that could otherwise be induced by 150ns pulses at 0.53 or 1.06 μm sufficient to melt but not vaporize the c-Si. However, Garulli et al [41] found, in contradiction to Westendorp et al that for identical laser irradiation conditions, an oxygen super-saturation could be obtained midway in depth through the laser melted region. One possible explanation is that Garulli et al underestimated their laser fluence, and induced surface temperatures far in excess of the melting point of c-Si, such that oxide melting, atomic evaporation, convective heat flow, or even boiling occurred. This is supported in part by the more recent data of Bentini et al [42] showing oxygen incorporation in silicon as a function of incident beam fluence. Only an extremely small of oxygen is incorporated for lower fluences, while a massive uptake of oxygen occurs once fluences around three times the melting threshold are applied. Thicknesses up to 500A have been grown this way. Cohen et al [43] have recently considered many possible oxygen incorporation and loss schemes for the reaction under these conditions. They have suggested that convection may be most important. Although the controlling mechanism has yet to be established, it is difficult to rule out effects due to the large density of highly excited particles (plasma, atoms, or groups of atoms) that are photogenerated by laser beam fluences three times that usually required to induce melting.

Ultraviolet laser radiation has also been used to induce liquid phase oxidation of silicon. Liu et al [44] used 15ns pulses at 266nm to induce a crystal-amorphous phase transition in c-Si by creating rapid quenching action, and when this was repeated in an air or O_2 ambient, oxide layers up to 300A were produced. Fogarassy et al [45] have also shown that oxygen incorporation by multishot melting in the UV is very much enhanced by the presence of specific impurities during the melting. In fact, by contrast to As and Sb doped Si, both In and Bi enabled appreciably higher rates of incorporation, and this was partly related to surface segregation effects.

Richter et al [46] have used an XeCl laser at 308nm to grow oxides as thick as 3000A at rates of more than 100A/s using a multishot processing technique. The rapid reaction rate was achieved with fluences up to 0.85J/cm^2, which is well beyond the single shot melting threshold. In fact, these are amongst the thickest oxides grown by laser, and it is not yet understood why these particular experimental conditions are conducive to such efficient oxide growth. Ultraviolet radiation has been known to enhance the solid-phase reaction measureably, as will be discussed in the final section of this paper, but the dramatic rates recorded here are orders of magnitude greater than the enhanced solid phase reaction, indicating that a novel processing regime has indeed been achieved.

Although the melting of the silicon provides a unique environment to study oxidation, it seems to have introduced many more questions than it has answered regarding the fundamental nature of the reaction. So far in this paper, we have only alluded to photo-enhanced effects to the reaction, and it is suggested that some of these are already present in several of the situations described above. In the last section, we will review the possibilities regarding the present understanding of photonic enhancement of the oxidation reaction.

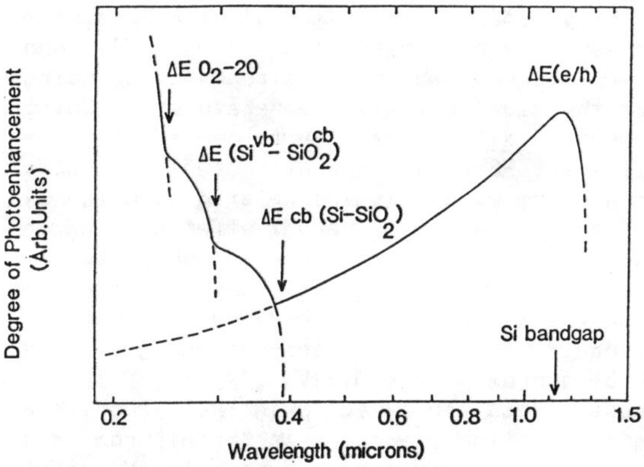

FIGURE 6. Spectral summary of the various mechanisms for the enhancement of the c-Si oxidation reaction. Assuming the validity of each process, an estimate of the relative contribution of each has been made for a constant incident photon flux.

3.3 Photonically Enhanced Oxidation

There are now more than 20 publications showing that laser radiation not only induces oxidation of various materials, but that it also enhances the reaction somewhat. This has been suggested not only for silicon, but also for GaAs [47-50], and other III-V binary, ternary and quaternary compound semiconductors [51], InP [52] and SiO [53], as well as metals [54] and the not unrelated reactions of laser enhanced nitridation [55] and etching [56] and the reaction of hydrogen with graphite [57]. Whilst some of the mechanisms may apply to one or more of the reactions, we shall concentrate here only on photonically enhanced oxidation of c-Si.

The term Photonic Enhancement (PE) used here embraces all reported effects in a light induced reaction that cannot be attributed to simple thermal mechanisms. Hence, as in most non-thermal reactions the specific characteristics of the individual quanta of the radiation are extremely important. Therefore PE embraces all forms of photo-generation involving excited species or states within a solid, liquid or gas, whose presence in that state leads directly to a nonthermal contribution to the reaction. Figure 6 is a brief summary of most of the proposed PE mechanisms in the literature, in terms of spectral response. An attempt has been made to estimate the relative contributions of the various effects for a constant incident beam power, and the reader is warned that not only is the relative importance of each open to debate, but in several cases, so are some of the enhancement effects. These are discussed below. It is important to recognise at this stage, that the photon flux directed at the c-Si surface during the enhancement will inevitably also induce some thermal effects, since none of the reactions reported to date are 100%

quantum efficient. Therefore, one must also bear in mind that in almost every experiment, there will be a background oxidation enhancement due to an increase in sample temperature.

The first report of light enhanced oxidation of silicon was by Oren and Ghandi [58], who used UV radiation during a conventional thermal oxidation reaction. They attributed the observed increase in oxidation rate to the photogeneration of electrons which are then liberated into the oxide leaving the c-Si in a highly charged state. They suggested a possible energy threshold for this to be that required by electrons to make a transition from the valence band of the silicon into the conduction band of the oxide, i.e. 4.25eV. This corresponds to a wavelength of 292nm 292nm, and one of the most prominent lines of their mercury arc lamp near 254nm. A similar proposition was made by Schafer and Lyon [31], who applied low power radiation to a Si sample which was already undergoing thermal oxidation in a furnace at the usual elevated temperatures. Owing to the level of irradiation, however, allowances have to be made for the increase in temperature, and only small PE rates were extracted, typically of the order of 20-50%. A major problem with this and related techniques, is the satisfactory isolation of thermal and non-thermal effects. As a consequence of their study, it was noted that there was a threshold to the enhancement around 3-3.5 eV, which was attributed to the photo-excitation of conduction band electrons from the silicon into the conduction band of the oxide. Since this does not require the silicon band-gap energy, it is thought to occur at a wavelength of approximately 3.2eV.

The majority of the present models are based principally on the indirect laser photogeneration of conduction electrons in the Si from the valence band, a process that occurs at photon energies near to and above 1.1eV. In chemical terms, this may be likened to antibonding. Gibbons [28] proposed photo-ionisation. The models relate the enhancement to the new equilibrium of photo-generated carriers above that established by the usual intrinsic thermal generation, and to effects this may have upon the reaction limited oxidation regime, thought to be relevant to the growth of thin films, as discussed earlier. Whilst Schafer and Lyon [30] again obtained small percentage enhancements over furnace oxidation, Gibbons [28] and Boyd et al [23,29], who used Ar laser beam intensities many times greater and without any significant background oxidation, obtained much faster reaction rates. Here, as before, it is extremely difficult to accurately determine the increase in temperature caused by the relaxation of the photo-excited carriers and therefore the precise photonic contribution [72]. Indeed, there are regimes where the thermal effects completely dominate [29]. Because no direct method of temperature monitoring can easily be used here, heating effects have to be estimated using laser heating models [73] and a small uncertainty in the beam energy or beam radius translates into a larger error in estimated growth rate [22]. Nevertheless, PE factors of up to 5 have been crudely estimated. It is important, however, to note that this is in conjunction with a background increase in the thermal contribution to the reaction of 5 orders of magnitude, and that the PE is very much a secondary process. Unfortunately, until a method of accurately monitoring the temperature becomes available, other less direct techniques must be used. Major advances in pyrometric measurement methods for rapid thermal processing, however, may soon be available.

Young and Tiller [32], have shown that the enhancement is directly proportional to the number of photons incident during the reaction rather than the power density, in the blue/green range of the Ar laser. Thus, in order to obtain the same enhancement rates for different wavelengths, the photon fluxes can be adjusted so they are equal for each wavelength used. Thus less power is required by lower frequency radiation to achieve the same enhancement as higher frequencies. This means that if one applies the 647nm line from the Kr laser, and the same photon flux from the 350nm line of the Ar laser to induce oxidation, the same enhancements should be obtained even if the power of the former was only half that of the latter. Alternatively, if one preheats the Si mildly to around 400C, where the Si oxidation is extremely slow, and repeats the experiment with Nd:YAG laser radiation (at 1064nm) and with the 350nm Ar line, almost a factor of three ratio of beam powers would show the same effect. These effects, of course only apply where the additional heat supplied by the relaxation and recombination of the photogenerated carriers has minimal effect on the reaction. Owing to the wavelength dependence of the silicon absorption coefficient, the shorter wavelengths will tend to heat up the silicon quite efficiently. Young and Tiller also report a second photonic effect, that a bigger enhancement has also been found with <100> orientation Si than with <111> material. Because of the different density of atoms in each crystal plane, this seems to be contrary to what may be expected for a purely thermal process.

The question remains as to what precise effect these excited carriers will have on the overall reaction. Since, as stated in the first section of this paper, the basic oxidation reaction is little understood in the early stages, it is presently difficult to define a precise photonic mechanism. An initial attempt at this [29] relied upon the fundamental reaction proposed by Blanc [13]. It has since been determined that the Blanc model does not contain the correct pressure dependence of the linear rate constant of the reaction [19]. However, since this model was essentially a special case of the more general model of Ghez and van der Meulen [12], the latter remains a possible starting point for analysis. Basically, the proposed mechanism involves O_2 diffusion to the Si/SiO_2 interface, where an equilibration reaction produces some O atoms from the O_2, both of which react with the silicon. Similarly, the more recent model of Hu [64], which also encompasses the D-G law, as well as the Blanc model, may also be valid. Hu assumes the intermediate step in the reaction is the chemisorption of O_2 molecules, so that the pressure dependence of the linear rate constant takes the form of the chemisorption isotherm, which is taken to be the Freundlich isotherm. For ultrathin films, Hu has suggested that the transported oxidant may no longer be molecular oxygen.

At this stage, it seems to be relatively straightforward to introduce the effects of visible radiation on the reaction to most of the models requiring the presence of electrons, or holes, or partially ionised silicon. The observations in this regime, however, do appear to confirm the importance of such reactions in the overall oxidation scheme. It is not so easy to accommodate models involving strain-affected diffusion as the limiting factor in the growth of thin layers. Under these conditions, the photogeneration of extra oxidant would not enhance the reaction, unless these species were appreciably faster than those prevalent before the photoexcitation.

The final category of enhancement modes discussed in this diagram will be the photodissociation of molecular oxygen close to the silicon substrate. The bond energy of O_2 is known to be close to 5.1eV, corresponding to a wavelength of 240nm. Any photons of energy greater than this will crack the molecule into two atoms. In this state, of course, they will be extremely reactive, and if they do not react with the silicon or the oxide layer at this stage, they may alternatively form ozone (O_3), which is also a much more powerful oxidant than its allotropic cousin. Further reactions may also be induced between ozone and the UV radiation, and of course shorter UV radiation with the O_2 and/or the substrate, but these have not been included.

This section is concluded by some recent results of UV (5eV, 248nm) induced bond reorientation within an adsorbed layer of oxygen on a silicon surface [72]. It has been found that an energetically unstable oxide is formed, which, after annealing at 949K, relaxes into a more conventional amorphous SiO_2 structure. It was shown that it is not possible to induce this transformation by purely thermal means. The experiments were repeated with 4.03eV, 308nm photons at the same intensity, but no oxygen incorporation was observed. Although these effects were induced with an absorbed power density of 0.5W/cm^2, 200A of oxide were grown when 1.5W/cm^2 was applied for 300min to c-Si in an oxygen pressure of 1kg/cm^2.

3.4 Summary and Conclusion

A brief review of silicon oxidation initiated by photonic means has been given. Regimes involving solid phase thermal and nonthermal mechanisms have been identified, and by assuming all the reported observations to be valid, an overall enhancement spectrum has been derived. The reader is of course invited to remain open-minded at these early stages of the field regarding the possible controlling mechanisms in the optically initiated reactions. It has previously been shown that thermal mechanisms often dominate in laser induced oxidation of silicon [22,72] and it is emphasised here that an accurate knowledge of the surface temperature is essential in order to satisfactorily quantify non-thermal reaction rates. This also happens to be important for other laser assisted pyrolytic reactions.

In order to best understand the unique influences of optical photons during the oxidation of c-Si, the field of conventional thermal oxidation of silicon has been briefly summarised. Here, it is seen that more than 10 different models have been published that attempt to describe the Si + O_2 reaction, and that this makes it very difficult to identify the precise photonic mechanisms in the reaction. However, one is tempted to suggest that the enhancements reported should tend to support models actively involving charged carriers and electronically excited silicon, rather than diffusion processes associated with structural properties of the oxide, since the visible and UV radiation used more obviously affects the former rather than the latter.

We have already seen in these proceedings that surface and subsurface properties can extend significantly into many materials. Since the present area is concerned with the growth of films as thin as several monolayers, it seems essential that such effects be included in any modelling of the mechanisms of reaction. It is concluded that the unique near-surface electronic and atomic structure with the associated defect population, as well as band-bending, are most likely important candidates for inclusion in future models describing the growth of ultrathin dielectric layers. Although not considered here, excitation of the oxide layer itself, by UV radiation in excess of about 5.0eV for the case of silicon dioxide [74], may also have an observable influence on the reaction, as it is appears to heav in the case of optically assisted oxidation of GaAs as reported by Siejka et al in these proceedings.

3.5 References

1. J.C. Mikkelsen, Appl. Phys. Lett., 45, 1187 (1984), and references therein.
2. D.N. Modlin, W.A. Tiller, J. Electrochem. Soc., 132, 1659 (1985).
3. B.E. Deal, A.S. Grove, J. Appl. Phys., 36, 3770 (1965).
4. A. Fargeix, G.Ghibaudo, J. Appl. Phys., 54, 7153 (1983).
5. R.H. Doremus, Thin Solid Films, 122, 191 (1984).
6. E.A. Irene, J. Appl. Phys., 54, 5416 (1983).
7. A.G. Revesz, R.J. Evans, J. Phys. Solids, 30, 551 (1969).
8. A.S. Grove, "Physics and Technology of Semiconductor Devices", Wiley, New York (1967).
9. M. Hamasaki, Solid State Electron. 25, 479 (1982).
10. Y.Z. Lu, Y.C. Cheng, J. Appl. Phys., 56, 1608 (1984).
11. S.A. Schafer, S.A. Lyon, Appl. Phys. Lett., 47, 154 (1985).
12. R. Ghez, Y.J. van der Meulen, J. Electrochem. Soc., 119, 1100 (1972).
13. J. Blanc, Appl. Phys. Lett., 33, 424 (1978).
14. E.A. Irene, D.W. Dong, J. Electrochem. Soc., 125, 1146 (1978).
15. R.R. Razouk, L.N. Lie, B.E. Deal, J.Electrochem.Soc.,128, 2214 (1981).
16. H.Z. Massoud, J.D. Plummer, E.A. Irene, J. Electrochem. Soc., 132, 1745 (1985).
17. T. Smith, A.J. Carlan, J. Appl. Phys., 43, 2455 (1972).
18. M.A. Hopper, R.A.Clarke, L.Young, J.Electrochem.Soc., 122, 1216 (1975)
19 Y.J. van der Meulen, J. Electrochem. Soc., 119, 580 (1972).
20. E. P. Eernisse, Appl. Phys. Lett., 35, 8 (1979).
21. I. W. Boyd, in "Laser Processing and Diagnostics", ed. D. Bauerle, Springer-Verlag (1984).
22. I.W. Boyd, Proc. E-MRS Symposia B&D, Strasbourg 1986, to be published.
23. I.W. Boyd, J.I.B. Wilson, J.L.West, Thin Solid Films, 83, L173 (1981).
24. I.W. Boyd, J.I.B. Wilson, Appl. Phys. Lett., 41, 162 (1982).
25. I.W. Boyd, J. Appl. Phys., 54, 3561 (1983).
26. I.W. Boyd, J.I.B. Wilson, Manuscript Under Preparation.
27. A.Cros, F. Salvan, J. Derrien, Appl. Phys., A28, 241 (1982).
28. J. F. Gibbons, Japan. J. Appl. Phys. Suppl., 19, 121 (1981).
29. I.W. Boyd, Appl. Phys. Lett., 42, 728 (1983).
30. S.A. Schafer, S.A. Lyon, J. Vac. Sci. Technol., 19, 494 (1981).
31. S.A. Schafer, S.A. Lyon, J. Vac. Sci. Technol., 21, 423 (1982).
32. E.M. Young, W.A. Tiller, Appl. Phys. Lett., 42, 63 (1983).
33. J. Nulman, J.P. Krusius, A. Gat, IEEE Electron Device Letters, EDL-6, 205 (1985).
34. M.M. Moslehi, K.C. Saraswat, S.C. Shatas, Appl. Phys. Lett., 47, 1113 (1985).
35. A.M. Hodge, C. Pickering, A.J.Pidduck, R.W.Hardeman, in "Rapid Thermal Processing", ed., T.O. Sedgewick, T.E. Seidel, B-Y Tsaur, Materials Research Society, Pittsburgh 1986, p313.
36. J.P.Ponpon, J.J.Grob, A. Grob, R.Stuck, J.Appl.Phys., 59, 3921 (1986).
37. N. Chan Tung, Y. Caratini, L. Liauzu, in Proc. E-MRS (Symposium D), Strasbourg, June 1986.
38. F. Micheli, I.W. Boyd (To be Published).
39. J.F.M. Westendorp, Z.L. Wang, F.W. Saris, in "Laser and Electron Beam Interactions with Solids", ed B.R. Appleton, G.K. Celler, North-Holland, 1982, p.255.
40. K. Hoh, H. Koyama, K. Uda, Y. Miura, Japan.J.Appl.Phys.,19,L375 (1980)
41. A. Garulli, M. Servidori, I. Vecchi, J. Phys. D, 13, L199 (1980).

42. G. G. Bentini, M. Berti, A. V. Drigo, E. Jannitti, C.Cohen, J. Siejka, in "Laser Chemical Processing of Semiconductor Devices", ed F.A. Houle, T.F. Deutsch, R.M. Osgood, Extended Abstracts of MRS Symposium B, Boston, 1984, p126.
43. C. Cohen, J. Siejka, G. G. Bentini, M. Berti, L.F. Dona Dalle Rose, A.V. Drigo, Proc. E-MRS, Symposium D, Strasbourg, June 1986, to be published.
44. Y. S. Liu, S. W. Chiang, F. Bacon, Appl. Phys. Lett., 38, 1005 (1981).
45. E. Fogarassy, C. W. White, D. H. Lowndes, J. Narayan, Proc. E-MRS, Symposium D, Strasbourg, June 1986, to be published.
46. H. Richter, T. E. Orlowski, M. Kelly, G. Margaritondo, J. Appl. Phys., 56, 2351 (1984).
47. W. G. Petro, I. Hino, S. Eglash, I. Lindau, C. Y. Su, W. E. Spicer, J. Vac. Sci. Technol., 21, 405 (1982).
48. V.M. Bermudez, J. Appl. Phys., 54, 6795 (1983).
49. J. Siejka, J. Perriere, R. Srinivasan, Appl. Phys. Lett., 48, 773 (1985).
50. P.A. Bertrand, J. Vac. Sci. Technol., 132, 973 (1985).
51. M. Fukuda, K. Takahei, J. Appl. Phys., 57, 129 (1985).
52. M. Fathipour, P.K. Boyer, G.J. Collins, C.W. Wilmsen, J. Appl. Phys., 57, 637 (1985).
53. S. E. Blum, K.H. Brown, R. Srinivasan, Appl. Phys. Lett., 43, 1026 (1983).
54. M. Wautelet, in "Laser Assisted Modification & Synthesis of Materials" ed., S.M. Metev, Sofia University "Kl. Okhridski" (1985).
55. T. Sugii, T. Ito, H. Ishikawa, in Extended Abstracts of the 16th International Conference on Solid-State Devices and Materials, Kobe, 1984, p433.
56. T. J. Chuang, I. Hussla, W. Sesselman, in "Laser Processing and Diagnostics", ed., D. Bauerle, Springer-Verlag (1984), p300, and references therein.
57. C.I.H. Ashby, Appl. Phys. Lett., 43, 609 (1983).
58. R. Oren, S.K. Ghandhi, Appl. Phys. Lett., 42, 752 (1971).
59. H. Hirata, K. Hoshikawa, Japan. J. Appl. Phys., 19, 1573 (1980).
60 B. Bathey, H.E.Bates, M. Cretella, J.Electrochem.Soc., 128, 771 (1980)
61. L. Eckertova, "Physics of Thin Films", Plenum, London, 1984.
62. W.A. Tiller, J. Electrochem. Soc., 119, 591 (1980).
63. A. Lora-Tamayo, E. Dominguez, E. Lora-Tamayo, J. Llabres, Appl. Phys., 17, 79 (1978).
64. S.M. Hu, Appl. Phys. Lett., 42, 672 (1983).
65. W.A. Tiller, J. Electrochem. Soc., 128, 689 (1981).
66. E.A. Irene, J. Electrochem. Soc., 125, 1798 (1978).
67. I.W. Boyd, J.I.B. Wilson, J. Appl. Phys., 53, 4166 (1982).
68. H.C. Webber, A.G. Cullis, N.G. Chew, Appl. Phys. Lett., 43, 669 (1983)
69. P.S. Peercy, W.R. Wampler, Appl. Phys. Lett., 40, 768 (1982).
70. D.M. Zehner, in "Pulsed Laser Processing of Semiconductors", Semiconductors and Semimetals V23, ed R.F. Wood, C.W. White, R.T. Young, Academic Press, 1984, p405, and references therein.
71. J.C.C. Tsai, in "VLSI Technology", ed. S.M. Sze, McGraw-Hill, London (1983).
72. I.W. Boyd, S.C. Moss, in "Laser Chemical Processing of Semiconductor Devices", ed., F.A. Houle, T.F. Deutsch, R.M. Osgood, Extended Abstracts of MRS Symposium B, 1984, p132.
73. C. Fiori, Phys. Rev. Lett., 52, 2077 (1984).
74. N. Itoh, private communication.

U.V. LIGHT INDUCED OXIDATION OF GaAs

J.SIEJKA, R. SRINIVASAN*, J. PERRIERE and R. BRAREN*
Groupe de Physique des Solides de l'Ecole Normale Supérieure, Université Paris VII, Tour 23, 2, Place Jussieu, 75251 PARIS CEDEX 05, FRANCE.
* I.B.M., Thomas J. Watson Research Center, Yorktown Heights, N.Y. 10598 (U.S.A.)

Abstract - The effect of far-U.V. light intensity (at constant dose : 45 J/cm^2) on the oxidation rate and composition of the oxides formed on GaAs was studied using an excimer laser or a U.V. lamp. The oxidation rates were deduced both from the measurements of quantities of oxygen (using nuclear microanalysis) and of cations (using RBS in channeling geometry) present in the oxides. It was found that the amount of cations was practically independent of light intensity, but the amount of oxygen was higher for the oxides formed by lamp irradiation. Both cations and oxygen contents decreased drastically with an increase of wavelength from 193 to 308 nm. By combining XPS and nuclear microanalysis, it was found that the oxides formed by laser were non stoichiometric, whereas those formed by lamp were stoichiometric containing large quantity of As^{+5}. The mechanism of GaAs oxidation induced by U.V. light was discussed.

INTRODUCTION

Although the light-induced enhancement in the oxidation rate of GaAs has been extensively studied (1,2,3), the results do not offer a clear picture of the process. This is in part due to the use of different wavelengths and light sources (visible or U.V., pulsed or continuous), experimental conditions (low pressure, ambient atmosphere etc.) and analytical techniques (Auger, XPS, synchrotron radiation etc.) which give selective information. In this work we focussed our attention on the U.V. induced growth of relatively thick oxide films starting from a substrate that was covered by a thin native oxide layer ($\sim$ 1.5 nm). As a result, the kinetics of the oxide growth were dominated by atomic transport through the oxide and less by the GaAs surface chemistry. The samples were analyzed by nuclear microanalysis allowing independent measurements of amount of oxygen and cations (As + Ga) fixed in an oxide layer using respectively : $^{16}O(d,p)O^{17*}$ nuclear reaction and RBS in channeling geometry (3). Some XPS measurements were also performed in order to check the chemical composition of the oxide.

IRRADIATION EXPERIMENTS

The samples used were (p) or (n) doped. Some experiments have also been performed using a semi insulating substrate. All samples were 100 oriented. After cleaning they were exposed to the U.V. radiation.

Two sources of U.V. radiation were used. (i) The laser radiation was from an excimer laser (Lambda-Physik 201 E) with wavelength of 193, 248 and 308nm (pulse width $\sim$ 14 ns). (ii) The C.W. radiation was from a U.V. lamp (BHK Model 9660-OZ). It delivered 2.5 mW/cm^2 at 185 nm and 15 mW/cm^2 at 254 nm.

The temperature of the sample was lower than ∿ 50° C during the lamp irradiation. Both U.V. laser and C.W. lamp irradiations were performed in air (dry or with relative humidity ∿ 60 %).

RESULTS AND INTERPRETATION

A. Oxidation induced by far U.V. irradiation (185-193 nm).

The laser irradiations were performed at λ = 193 nm using an energy pulse of 0.045 J/cm^2 (energy fluence : 3.2 10^6 W/cm^2). The total dose of irradiation was 45 J/cm^2. The lamp irradiations were performed at λ = 185 nm. The time of irradiation was adjusted to 5 hours so that the sâme total far-U.V. energy (45 J/cm^2) was deposited on the sample. Control experiments have shown that the simultaneous presence of 254 nm radiation during the lamp irradiation had only a small effect on the oxidation rate of GaAs. The effective time of irradiation was thus 10^9 times greater than that used with the excimer laser. The atomic composition of the oxides (mean values), as deduced from channeling and nuclear microanalysis, are presented in Table 1.

TABLE 1

GaAs samples	As + Ga Contents	^{16}O Contents	^{16}O / As + Ga	XPS Data
- as prepared	4.1	6.3	∿ 1.5	
IRRADIATED by laser :				
- λ = 193 nm	25	32 (26)*	1.3 $\pm$ 0.1	As^{+5}, As^{+3}, As^0, Ga^{+3}
- λ = 248 nm	8.4	9.5 (3.2)*		
- λ = 308 nm	4.6	8 (1.7)*		
U.V. LAMP (5 h) :				
λ = 185 nm (+ λ = 254 nm)	26.8	45.6 (39)*	1.75 $\pm$ 0.25	As^{+5}, As^{+3}, Ga^{+3}
Oxidation by atomic Oxygen and ozone (5h)		11 (4.7)*		

* Oxygen uptake

All atoms content are given in units of 10^{15} at./cm^2
Energy of the laser pulse : 0.045 J/cm^2 (intensity : 3.2 10^6 W/cm^2)
U.V. lamp intensity : 2.5 10^{-3} W/cm^2 with 185 nm
15 10^{-3} W/cm^2 with 254 nm

Fig. 1 shows, for example, the R.B.S. spectra that were recorded in channeling geometry on GaAs samples that were : (1) non irradiated : (2) irradiated by 1000 laser pulses : (3) irradiated by the U.V. lamp (5 hours). Fig. 2 shows the XPS spectra obtained from the same samples as shown on Fig. 1. By comparison, the XPS spectrum of GaAs anodic oxide (∿ 15 nm thick) is also presented.

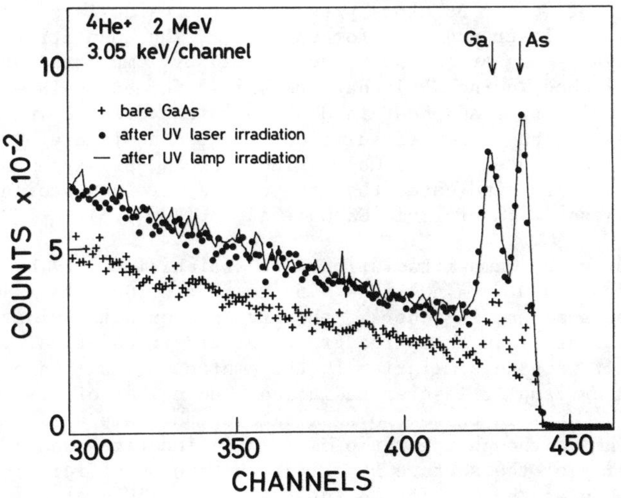

Fig. 1 - RBS aligned spectra for GaAs samples : (i) non irradiated (bare); irradiated with : (ii) laser (λ = 193 nm) ; (iii) lamp (λ = 185 nm).

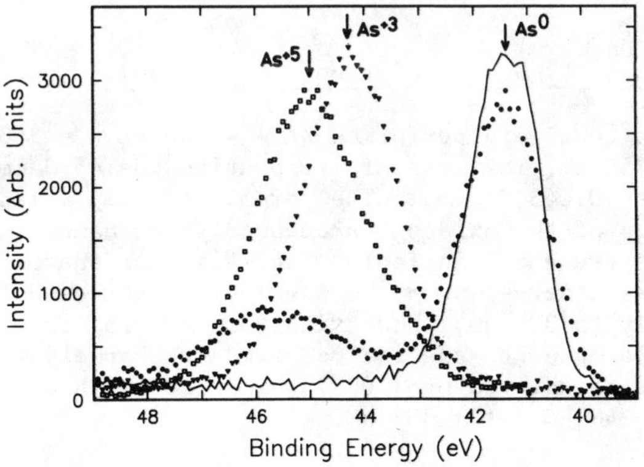

Fig. 2 - XPS peaks of As core levels for GaAs samples : ( ) non irradiated (bare) ; irradiated with:(•) laser ; (□) lamp ; (▼) GaAs wet oxide.

The quantities of Ga and As atoms present in the oxide layers were practically the same in laser and lamp formed oxides. The quantities of oxygen atoms were clearly higher in the oxides formed by lamp irradiation. It was further established (using XPS) that the oxides formed by laser irradiation were non stoichiometric and contained non oxidized arsenic in the presence of As^{+5} and As^{+3}. Thus laser irradiation produced a mixture of oxidized (As^{+5}, As^{+3}) and reduced As and Ga_2O_3. The ratio R, of oxygen atoms to cations was 1.3 which indicated that the oxides were non stoichiometric even in comparison to the formula $Ga_2O_3 + As_2O_3$ (R = 1.5).

The oxides formed by lamp irradiation were stoichiometric. Their composition varied from $Ga_2O_3 + As_2O_3$ (R = 1.5) to that of $Ga_2O_3 = As_2O_5$ (R = 2). XPS measurements shown, in general, the presence of both oxidation states of As (As^{+3} and As^{+5}) in the oxide grown by lamp irradiation. The most striking result was the similarity in the content of As + Ga atoms in the oxides formed by lamp or laser irradiation. The number of U.V. photons was essentially the same in these two experiments (the lamp at 185 nm and the laser at 193 nm) although the ratio of irradiation times was 10^9. Assuming the reflectivity of the surface 50 % (4), the photon efficiency for As + Ga cation formation at the GaAs/oxide interface was $\sim 10^{-3}$ atom./photon. This suggested that the enhancement of the oxidation rate by far U.V. radiation was determined by the photon efficiency of "cation injection" at the oxide interface and not by the oxygen transport through the oxide. "Cation injection" seemed to be limiting factor in the overall oxidation process when a lamp was used. The formation of As_2O_5 and the absence of As^0 is believed to be the best evidence of this view. It seemed that the interfacial transport (and/or diffusion) of oxygen, during the laser irradiation was not rapid enough to induce the total oxidation of the substrate atoms presented in the oxide.

B. Oxidation induced by U.V. radiation (248-308 nm).

The laser irradiations were performed at λ = 248 or λ = 308 nm and this of the lamp at λ = 254 nm. The dose of irradiation was 45 J/cm^2 delivered by 1000 laser pulses (0.045 J/cm^2, pulse) or 50' of lamp irradiation. The atomic composition of the oxides, as deduced from channeling and nuclear microanalysis was presented in Table 1 and Fig. 3. A sharp decrease in the oxidation rate was observed when the wavelength decreased from 193 nm to 248 and especially to 308 nm, when oxidation was practically absent. Fig. 4 shows, for example, the RBS spectra recorded in channeling geometry on GaAs samples that were irradiated by 1000 laser pulses with wavelength of :
(1) 193 nm ; (2) 248 nm ; (3) 308 nm.

As in the case of far U.V. light, the lamp irradiation leaded to the formation of stoichiometric oxides. The amounts of cations present were practically the same in laser and lamp formed oxides, but the amounts of oxygen were slightly higher in the lamp formed oxides. Fig. 5 shows the thicknesses of oxides formed by laser or lamp irradiation as a function of wavelength. It was assumed that the oxide thickness of 0.24 nm corresponded to 10^{15} O. atoms/cm^2.

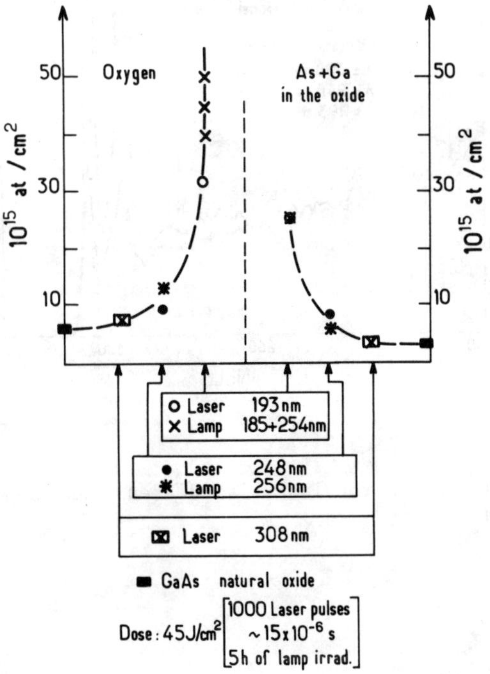

Fig. 3 - Oxygen uptakes and cation contents in the oxides formed by laser or lamp irradiation at different wavelengths.

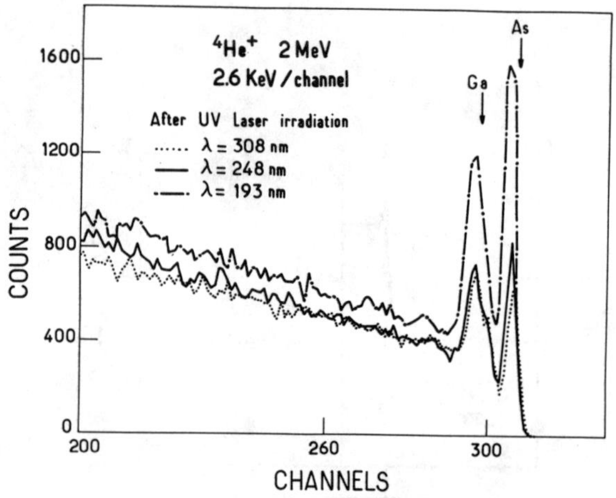

Fig. 4 - RBS aligned spectra for laser irradiated GaAs samples : effect of wavelength.

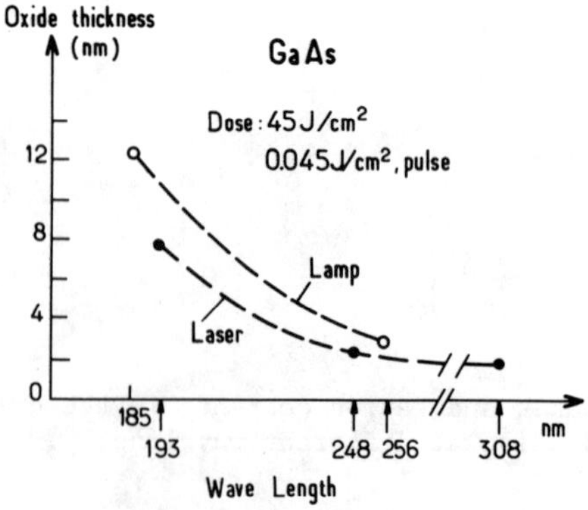

Fig. 5 - Thicknesses of GaAs oxides formed by laser or lamp irradiation : effect of wavelength.

DISCUSSION

During the irradiation by far-U.V. light, the atomic oxygen and ozone were formed in air. It seemed reasonable to assume that oxygen transport (diffusion) through the growing oxide did not control the oxidation kinetics at $\lambda = 193$ nm, since 10^3 pulses of laser irradiation (4) induced a complete oxygen exchange between a 18 nm thick ^{18}O enriched anodic oxide and surrounding ^{16}O atmosphere. Therefore the oxidation was controlled by the interface reaction between oxidized species (atomic oxygen,ozone) and Ga and As surface atoms. Since the temperature was lower than 50°C (during the lamp irradiation), the kinetics of the surface reaction was determined by the photo induced formation of "free" As + Ga atoms at the GaAs/oxide interface which were capable of undergoing oxidation. The microscopic mechanism of this process, which was believed to be the breaking of the Ga and As back bonds was independent of light intensity (ratio of intensities: 10^9). The oxidation rate involved in photo oxidation process only depended on the number of photons striking the GaAs surface (photon efficiency $\sim 10^{-3}$ atoms/photon). These results agreed with our observation that the oxidation of GaAs solely by atomic oxygen and ozone (i.e. when U.V. radiation did not reach GaAs surface) represented a small fraction of oxide formed by direct irradiation (Table 1). The laser irradiated oxides were non stoichiometric and contained a large quantity of As^0. This was probably due to the reduction of arsenic oxide by surface Ga atoms due either to the photo excitation or to the transient heating during the light absorption. Further studies are needed in order to clarify this point. These results agreed with our earlier observation (3) that during laser irradiation ($\lambda = 193$ nm) of relatively thick (280 nm) native oxide, deterioration of the GaAs interface was observed (As precipitation).

It was found that the photo oxidation rate of GaAs decreased with the wavelength and was practically absent at $\lambda = 308$ nm (for 45 J/cm^2). This effect could be explained by a dramatic decrease of the density of oxidized species (atomic oxygen, ozone) with wavelength. Therefore the kinetics of GaAs oxidation would be controlled by diffusion through the oxide and not by the interface reaction. This hypothesis agreed with our finding (to be published) that the efficiency of oxygen exchange sharply decreased when the wavelength increased and agreed with results on U.V. induced oxidation of silicon (5).

CONCLUSIONS

Photo induced oxidation of GaAs was particularly efficient using low intensity far-U.V. light (185-193 nm). The quantity of the cations present in the oxides was practically independent of light intensity. Stoichiometric oxides ($\sim$ 12 nm thick) containing large quantity of As^{+5} were formed by U.V. lamp irradiation, whereas non stoichiometric oxides were formed by low intensity laser irradiation (containing As^0). The oxidation rate sharply decreased with wavelength and oxidation was practically absent at $\lambda = 308$ nm. It was suggested that the mechanism of photo induced GaAs oxidation by far-U.V. light was controlled by the interface reaction (breaking of Ga and As back bonds) and at 248 and 308 nm wavelengths by diffusion through the oxide.

REFERENCES

1 - PETRO W.G., HINO I., EGLASH S., LINDAU I., SU C.Y. and SPICER W.E., J. Vac. Sci. Technol. 21 (1982) 405
2 - BERMUDEZ V.M., J. Appl. Phys. 54 (1983) 6795
3 - SIEJKA J., SRINIVASAN R. and PERRIERE J., in "Energy beam-solid and transient thermal processing" (V.T. Nguyen and A.G. Culis eds.) MRS Europe Symposia Proceedings, Vol. 4, p. 139, Les Editions de Physique, 91944 Les Ulis (1985)
4 - SIEJKA J., SRINIVASAN R. and PERRIERE J., Appl. Phys. Lett. 46 (1985) 773
5 - SIEJKA J., SRINIVASAN R., PERRIERE J., BRAREN R. and LAZARE S., "U.V. photon induced oxidation of GaAs and Si : effect of wavelength and light intensity". To be published in MRS-Europe Symposia Proceedings, Les Editions de Physique, 91944 Les Ulis (1986)

PARTICIPANTS

ALLAN, G.	Laboratoire d'Etude des Surfaces et Interfaces, Faculté Libre des Sciences H.E.I., Institut Supérieur d'Electronique du Nord, 3 Rue François Baes, 59046 Lille Cedex, France.
ANDREW, R.	Faculté des Sciences, Université de l'Etat, av. Maistriau, 7000 Mons, Belgium.
ANTONIADIS, C.	Faculté des Sciences, Université de l'Etat, av. Maistriau, 7000 Mons, Belgium.
ARNOLD, L.	Max Planck Inst. für Quantenoptik, 8046 Garching, Germany.
AUTERINEN, I.	Technical Research Center of Finland, Otakari 3A, 02150 Espoo, Finland.
AUVERT, G.	CNET, BP 98, 38243 Meylan, France.
BARBOSA, M.A.	Dept. Eng. Metalurigica Feup, Rua dos Bragas, 4099 Porto Cedex, Portugal.
BAUERLE, D.	Johannes Kepler Universität Linz, Angewandte Physik, 4040 Linz, Austria.
BENEDEK, G.	Universita degli Studi Milano, Dipartimento di Fisica, Via Celoria 16, 20133 Milano, Italy.
BIERNAUX, A.	Faculté des Sciences, Université de l'Etat, av. Maistriau, 7000 Mons, Belgium.
BONZEL, H.P.	Institut für Grenzflächenforschung und Vakuumphysik der Kernforschungsanlage , Jülich Gmb H, Postfach 1913, D-5170 Jülich 1, Germany.
BORCKMANS, P.	Service de Chimie Physique II, C.P. 231, Campus Plaine ULB, Boulevard du Triomphe, 1050 Bruxelles, Belgium.
BOUREE, J.E.	Lab. de Physique des Solides, 1 place Aristide Briand, 92195 Meudon Principal Cedex, France.

BOYATZIS, S	N.R.C. "Democritos", 15310 aghia Paraskevi, Attikis, Greece.
BOYD, I.W.	Dpartment of Electronic and Electrical Engineering, University College London, Torrington Place, GB-London WCIE 7JE, Great Britain.
CHAPET-ROUSSEAU, M.	IEF Bat.220, 91405 Orsay Cedex, France.
CHUANG, T.J.	(c/o Prof. Ertl), Fritz Kaber Ins.,Max-Plank Gezelschaft, Faradayweg 4-6, D-1000 Berlin 33, W-Germany.
DAGONNIER, R.	Faculté des Sciences, Université de l'Etat, av. Maistriau, 7000 Mons, Belgium.
D'ANNA, E.	Dipartimento di Fisica dell'Universita Lecce, Italia.
DE LA PORTILLA, J.M.	Escuela Universitaria Politecnica, Perez del Toro 1, 35003 Las Palmas, Spain.
DEWEL, G.	Service de Chimie Physique II, code postal 231, Campus Plaine ULB, Boulevard du Triomphe, 1050 Bruxelles, Belgium.
DUMONT, M.	Faculté des Sciences, Université de l'Etat, av. Maistriau, 7000 Mons, Belgium.
DYER	University of Hull, Cottingham road, Hull HU6 TRX, England.
FAIN, B.	Telaviv University, Ramat Aviv 69978, Tel Aviv, Israël.
FRITSCH, G.	Technischen Universität Munchen, Physics Department, 8046 Garching B., Munchen, Germany.
GELCHINSKI, M.	Dept.4982, Materials and Process Engeneering IAI, Ben Gurien International Airport, Israel.
GEORGE, T.F.	Faculty of Natural Sciences and Mathematics, the University at Buffalo, Clemens Hall, Buffalo New York 14260, USA.
GIERES, G.	Siemens A.G., Zef Tph 3, 8520 Erlnagen, Germany.
GRATZEL, M.	Ecole Polytechnique Fédérale de Lausanne, Dept. de Chimie, Institut de Chimie Physique, EPFL-Ecublens, 1015 Lausanne, Switzerland.
GUNES, M.	Uludag Universitesi, Fen Edebiyat Falkultesi, Gorukle Bursa, Turkey.
GUNGOR, A.	Uludag Universitesi, Fen Edebiyat Fakultesi, Gorukle Bursa, Turkey.

HULIN, D.	Laboratoire d'Optique Appliquée, ENSTA, Batterie de l'Yvette, 91120 Palaiseau, France.
ITOH, N.	Department of Crystalline Materials Science, Faculty of Engineering, Nagoya University, Furo Cho Chicusa Ku, Nagoya 464, Japan.
JACKSON, P.E.	Dept.of Applied Physics, University of Hull, Hull HU6 7RX, UK.
JOHNSON, T.	Dept. of Chemistry, Washington State University, Pullman, WA 99164 4630, USA.
JORDAN, T.	Inst. for History and Philosophy of Science, University of Toronto, Toronto, Ontario, Canada.
JOST, D.	Institute of Applied Physics, University of Berne, Sidlerstrasse 5, 3012 Bern, Switzerland.
KEY, P.H.	University of Hull, Cottingham road, Hull HU6 TRX, England.
KULLMER.	Johannes Kepler Universität Linz, Angewandte Physik, 4040 Linz, Austria.
LAUDE,L.D.	Faculté des Sciences, Université de l'Etat, av. Maistriau, 7000 Mons, Belgium.
LOCQUET, J.P.	Celestijnenlaan 200 D, 3030 Leuven, Belgium.
LUCHES, A.	Dipartimento di Phisica, CP 193, 73100 Lecce, Italia.
MAVROYANNIS, C.	Division of Chemistry, National Research Council of Canada, Ottawa, Ontario, Canada K1A OR6.
METEV, S.	Max Planck Institut für Quantenoptik, 8046 Garching, Germany.
MICHELI, F.	Dept. Electronic Eng., University College London, Torrington Place, London WCIE 7JE, UK.
MILNE.	Heriot Watt University, Riccarton Campus, Currie Eddinburgh, Scotland.
MONTEIRO, F.J.	Dept.Engenharia Metalurgica, FEUP, Rua dos Bragas, 4099 Porto Cedex, Portugal.
NORTON, D.P.	15881 N. King Bradford, Baton Rouge, LA 70816, USA.
PETZOLD, H.C.	Fraunhofer Institut für Mikrostructure Tech., Dillenburger Str. 53, 1000 Berlin 33, Germany.
POGGI, Y.	Labotatoire d'Electrostatique et de Materiaux Diélectriques, CNRS 166, 38042 Grenoble,France.

PRIOLO, F.	Dipartimento di fisica, Corso Italia 57, 95100 Catania, Italia.
QUENON, P.	Faculté des Sciences, Université de l'Etat, av. Maistriau, 7000 Mons, Belgium.
RICCI, F.	Dipartimento di Fisica, Universita degli Studi di Roma, "La Sapienza", Piazzale Aldo Moro 2, 00185 Roma, Italia.
ROBERS.	Max Planck Institut für Quantenoptik, 8046 Garching, Germany.
SCHRODER, H.	Max Planck Institut für Quantenoptik, 8046 Garching, Germany.
SIEJKA, J.	G.P.S. de l'ENS, Tour 23, 2 Place Jussieu, 75251 Paris, France.
SRINIVASAN, R.	IBM T.J. Watson Research Center, P.O. Box 218, Yorktown Heights, New York 10598, USA.
THUILLARD, M.	Sidlerstrasse 5, 3000 Bern, Switzerland.
TITULAER, U.M.	Inst. Theor. Physics, Joh. Kepler University, Altenbergerstrasse 65, 4040 Linz, Austria.
VAN DRIEL, H.M.	Department of Physics, University of Toronto, Toronto, Ontario, Canada, M5S 1A7.
VANGRUNDERBEEK, J.	Celestijnenlaan 200 D, 3030 Leuven, Belgium.
WALGRAEF, M.	Code Postal 231, Campus Plaine ULB, Boulevard du Triomphe, 1050 Bruxelles, Belgium.
WEHNER, M.	Fraunhofer Inst. für Lasertech., Drooselweg 87, 5100 Aachen, Germany.

INDEX